DESIGN OF MACHINE TOOLS

The principal author wishes to express his gratitude to Dr. B. Rajiv, Dr. M.D. Jaybhaye, Associate Professor, Dr. B.B. Ahuja Deputy Director & Professor, Production Engineering Department, and other colleagues in the department, who have always rendered kind help and advice in revising new edition.

S.K. Basu

College of Engineering, Pune
An Autonomous Institute of Govt. of Maharashtra
February 2014

Preface To The Sixth Edition

The present trend towards higher productivity in almost all fields of engineering manufacture has resulted in the evolution of newer and better techniques of manufacture and brought about a thorough change in the design of machine tools. Such an evaluation is characterised by the increased quality and reduced cost of production machines, as well as by the more and more use of automatic, programme – controlled or transfer machines, in the field of specialized production of complicated components. As a result the traditional method of design based on massive construction is gradually giving way to limit design.

Very few books are available to-day which can give a comprehensive method of designing machine tool elements. Author's long experience, as a teacher in Production Engineering, tells him about the need of one comprehensive textbook which can meet the requirement of a student stepping into the field of machine tool design. The book is, therefore, designed primarily to meet the requirements of a Mechanical and Production Engineering students of our universities at the undergraduate, as well as post-graduate levels.

Most of the principles discussed in this book have been explained with neat and simple sketches, which in some cases are self-explanatory. An effort has been made to give an idea to the students about probable questions that may be expected in the examinations and possible answer to many of the numerical questions. It is hoped that the book will prove to be an essential guide book of information to reach workers and designers working in the machine tool factories, laboratories and research Institutions as well.

Sixth Edition

Design of Machine Tools

S.K. BASU

Professor Emeritus, Dept. of Production Engineering
Govt. College of Engineering, Pune;
Formerly, Director, Central Mechanical Engineering Research Institute,
Durgapur; Professor, Production Engineering Department;
Jadavpur University; Head, Mechanical Engineering Department;
R.E. College, Durgapur

D.K. PAL

Formerly, Professor, Mechanical Engineering Department Regional
Engineering College, Durgapur

Oxford & IBH Publishing Co. Pvt. Ltd.
New Delhi

(A Unit of CBS Publishers & Distributors Pvt Ltd)

New Delhi • Bengaluru • Chennai • Kochi • Kolkata • Mumbai
Hyderabad • Jharkhand • Nagpur • Patna • Pune • Uttarakhand

CBS Publishers & Distributors Pvt Ltd
204 FIE, Patparganj Industrial Area, Delhi-110 092
E-mail: delhi@cbspd.com, cbspubs@airtelmail.in

Reprint 2018

ISBN 978-81-204-1777-9

Printed at Chaman Enterprises, New Delhi.

1-L4-11

Contents

Preface To The Sixth Edition v

1. Introduction **1**
1.1 The Fields of Development 2
1.2 Classification of Machine Tools 2
Establishing and Expansion of Machine Tools Industry 14
Conclusion 16

2. Determination of the Forces Acting on the Tool in Certain Machining Operations and Horse-power Requirement **17**
2.1 Determination of the Tool Forces in a Lathe Operation 18
2.2 Calculation of Forces Acting on Milling Cutter and Determination of the Power Consumption of the Machine 24
2.3 Calculation of Power Consumption in Cylindrical Grinding Operation 27
2.4 Determination of the Thrust on a Drill and the Power Consumption 28
2.5 Determination of the Forces Acting on a Broach and Power Consumption in the case of Broaching Machine[6] 31
2.6 Determination of the Forces Acting on a Shaping Tool and the Power Calculation in Shaping Machine 32

3. Kinematics of Machine Tools **36**
3.1 Classification and Choice of Driving System 36
3.2 Basic Consideration in the Design of Drives 37
3.3 Determination of Variable Speed Range in Machine Tools 38
3.4 Graphical Representation of Speed and Structure Diagram 43

3.5 Various Types of Structure Diagram 47
3.6 Selection of Optimum Ray Diagram 50
3.7 Difference Between Number of Teeth of Successive Gears in a Change Gear Block 51
3.8 Analysis of a Twelve-Speed Gear Box 53
3.9 Standardisation of speed ratios 55
3.10 Compound Ray Diagram 56
3.11 Feed Gear Boxes 57
3.12 Strength Calculation of Gears 59
3.13 Rigidity of Grinding Wheel-workpiece System in an Internal Grinder[87] 62

4. Further Studies On Kinematics **76**
4.1 Transmission in Stepped Regulation 76

5. Stepless Regulation In Machine Tools **89**
5.1 Classification 89
5.2 Mechanical Faction Drive 90
5.3 Methods of Increasing the Range of Regulation in Modern Machine Tools 95
5.3 Semi Toroidal Drive 99
5.4 Considerations in Variator Calculation 100
5.5 Friction Loss in Friction Variators due to Loss in Sliding Velocity 101
5.6 Principles of Self-locking in Variator 102
5.7 Further Analysis of Ball Variators 104

6. Machine Tool Guides **106**
6.1 Classification of Guides used in Machine Tool 106
6.2 Wearing of Guides 106
6.3 Guide Materials 110
6.4 Temperature Deformation of Guides 111
6.5 Liquid Friction in Guides 112
6.6 Kinetic Friction and Stick-slip Vibration 117
6.7 Specimen Calculation for Guides having Lubrication Wedge 119
6.8 Methods of Calculating Pressure on Guides 121
6.9 Guides having Rolling Friction 126
6.10 Behaviour of the Machine Tool Guide Friction under the Effect of Lubrication 128
6.11 Fundamental Types of Circular Guides 131

6.12 Accuracy and Wear of Machine Tool Guides 131
6.13 Hardness of the Different Guide Materials 133
6.14 Design of Guides under Hydrostatic Lubrication 134
6.15 Type of Hydrostatic Slides 138
6.16 Hydraulic Load Relief 139
6.17 Oil Pocket Shape and Dimensions 140
6.18 Gas Film Lubrication 141
6.19 Design Procedure for Aerostatic Slideways 143
6.20 Influence of Hardness of Material on Guide Wear 145
6.21 Effect of Surface Preparation 145
6.22 Effect of Micro-Structure and Chemical Composition 146
6.23 Effect of Surface Pressure 146
6.24 Seizure and Tearing of Guides 146
6.25 Error Contribution to the Job Due to Longitudinal Wear on Lathe Guides 147
6.26 Contact Deformation of Guides 147
6.27 Contact Deformations in Guides with Clearance 148
6.28 Clearance Adjustment in Guides 150
6.29 Fabricated Guides 150
6.30 Error Estimation in Guide Design 150

7. Design of Beds, Tables and Columns **153**

7.1 Various Types of Beds Used in Machine Tools-Their Construction and Design Features 153
7.2 Determination of the forces acting on the horizontal table or a vertical boring machine 157
7.3 Column Design of a Milling Machine and Maximum Deflection Error in a Milling Machine 161
7.4 Column Design of Drilling Machine 162
7.5 Stiffness and Natural Frequency of Machine Beds 164

8. Design of Power Screws of Machine Tools **169**

8.1 Types and Classification 169
8.2 Design Calculations 171
8.3 Strength of Lead-screw 173
8.4 Ball Recirculating Power Screw Assemblies 175
8.5 Calculation for the Maximum Static Load 178
8.6 Efficiency of the Ball Recirculating Power Screw 180
8.7 Compensation of 'Backlash' in Ordinary Sliding Screw Assemblies 183
8.8 Vertical Roller Feed Screw 184

8.9 Distribution of Load Between the Threads of Nut in the Power Screw with Sliding Friction 188
8.10 Load Distribution on the Threads of the Nut of Ball Recirculating Screw Assembly 192
8.11 Analysis of Axial Load and Contact Rigidity of a Recirculating Ball Screw 202
8.12 Evaluation of the Rigidity 203
8.13 Sensitivity Analysis 205
8.14 A Critical Analysis 205
8.15 Analysis of Preload 206
8.16 Standard Dimensions of Recirculating Rail Screw Assembly 211
8.17 Calculation for Dynamic Loading 213

9. Spindle Units in Machine Tools 215
9.1 Spindles and Their Supports :-Special Features, Material and Construction 215
9.2 Typical Spindle Ends 215
9.3 Spindle Supports 217
9.4 Calculation on Sleeve Bearing 220
9.5 Ball Bearings 221
9.6 Adjustments of Ball Bearings 222
9.7 Roller Bearings 223
9.8 Rigidity of Spindle Units 224
9.9 Rigidity of the Rolling Friction Supports 226
9.10 Magnitude of Deflection at the Free end of the Spindle 228

10. Lubrication and Rigidity in Machine Tools 230
10.1 Introduction 230
10.2 Steps in Selecting Proper Lubrication Oil 230
10.3 Frictional Condition of Working 231
10.4 Specification of Lubrication Oils 235
10.5 Rigidity of Machine Tool Units 237
10.6 Some Errors Affecting Rigidity 241
10.7 Overall Static Rigidity of Machine Tools 244
10.8 Dynamic Rigidity of a Machine Tool 245

11. Controlling Systems in A Machine Tool 247
11.1 Classification 248
11.2 Single-Disc Selective Speed Changing System 255

12. Electrical Equipments in Machine Tools **258**
12.1 Basic Ideas 258
12.2 Selection of Motor for any Executive Organ of a Machine Tool 259
12.3 Regulation of Speed in Electrical Control 260
12.4 Circuit Diagram for Starting the Driving Motor of a Machine Tool 264
12.5 Electrical Brakes 265
12.6 Electromagnets used in Machine Tools Control 265
12.7 Electromagnetic Clutch 267
12.8 Ferromagnetic Powder, Clutch 270
12.9 Reversing Mechanism of a Light Duty Planing Machine 271
12.10 Thermal Relay in Machine Tools 272
12.11 Electrical Automation in Horizontal Drilling Machine 273
12.12 Automatic Lifting of Tool During the Return Stroke of a Planing Machine 276
12.13 Basic Ideas into Regime of Working of Motors 277
12.14 Classification of Automatic and Semiautomatic Controls 283

13. Hydraulic Control System in Machine Tools **286**
13.1 Introduction 286
13.2 Typical Hydraulic Systems in a Machine Tool 287
13.3 Elements of Hydraulic Systems in Machine Tools 291
13.4 Resistance Encountered in Flow Through Pipe 291
13.5 Evaluation of Basic Parameters for Design 292
13.7 Energy Losses 295
13.8 Compressibility Factor 295
13.9 Efficiency of Hydraulic Pump Considering Losses 296

14. Programme Control in Machine Tools **297**
14.1 Introduction 297
14.2 Automation in Machine Tools 297
14.3 Magnetic tape Controlled Machine Tool 299
14.4 Photoelectric Tracing System 300
14.5 Principles of Numerical Control 302
14.6 Method of Disposition of Punched Holes in Cases of Binary and Decimal 306
14.7 Principle of Operation of NC Systems 307

15. Built-in-inspection Units in Machine Tools **310**
15.1 Introduction 310

15.2 Systems of automatic Inspections 311
15.3 Some Typical Built-in-inspection Equipments 311
15.4 Characteristic Features in Designing 314
15.5 Conclusion 315

16. Vibration in Machine Tools **316**
16.1 Introduction 316
16.2 Forced vibration 316
16.4 Self-excited Vibration in Machine Tools 320
16.5 Other Types of Forced and Damped Vibration 322
16.6 Stick-Slip Vibration in Machine tools 325
16.7 Minimization of Stick-Slip Vibration in Machine Tools 331
16.8 Vibration Isolated Tool Holders 339

17. Microdisplacements in Machine Tools **343**
17.1 Introduction 343
17.2 Magnetostrictive Drive 343
17.3 Thermodynamic Drive 347
17.4 Minimisation of Positional Error by the Use of Oscillating Normal Force 349
17.5 Recirculating Ball Screws 350
17.6 Surface Topography and Contact Stiffness 351
17.7 Error Due to Stick Slip Motion 352

18. New Concepts in Machine Tools Design **356**
18.1 Introduction 356
18.2 Probability Concept in Design 357
18.3 Unified Systems Approach to Machine Tools Problems 359

19. Industrial Robots and Their Applications **367**
19.1 Introduction 367
19.2 Basic functions of robotic elements 368
19.3 Mobility of Robots 370
19.4 Reliability in Operation 372
19.5 Control 373
19.6 Hierarchial Computer Control Configuration Robotic System 377
19.7 Sensing 378
19.8 Assembly and Megassembly Robots 379

20. NC-CNC-DNC Machines **381**
20.1 Introduction 381

20.2 Principles of a CNC Machine 382
20.3 Classification of CNC or NC Machines Through Control Axes 382
20.4 Open and closed loop control systems in N.C., C.N.C. Machines 386
20.5 Working of N.C. Machine Tool 387
20.6 Transducers and Monitoring of Displacement 387
20.7 N.C. Retrofitting 398
20.8 Programme and Compoter-man Interaction 401
20.9 Present status 401
20.10 Electronics Revolution and Computer Growth 402
20.11 Direct Numerical Control (DNC) 404
20.12 Part programming 406
20.13 Part Programming 407
20.14 CNC Lathe 412
20.15 Machining Centre HMT-KTM (Horizontal Machining Centre) 412
20.16 A few commonly used terminology in NC, CNC 416

21. Robot Language-state of the Art **418**
21.1 Introduction 418
21.2 Robot Language Outline 420
21.3 General Description of Programming Langnage 420
21.4 Real Time 420
21.5 Geometric Modelling 421
21.6 Tool and object sets of coordinate axes 422
21.7 Movements 423
21.8 Sensors 424
21.9 Tools 425
21.10 Example of programming 425
21.11 Some Commercial Languages 426
21.12 Conclusions 428

22. Flexible Manufacturing System **430**
22.1 Introduction 430
22.2 FMS–its Meaning, Objectives and Significance 430
22.3 Classification of FMS 433
22.4 "BUILDING BLOCK" Concept 433
22.5 CNC Machining Centre 434
22.6 CAD CAM-FMS 435
22.7 Precision Movement 437

23. Dynamic Analysis of A Few Sub-systems in Machine Tools **441**
- 23.1 System Analysis to Study Dynamic Compliance 442
- 23.2 Spindle Supports 445
- 23.3 Thermal Deformation of Supports 448
- 23.5 Tribological Considerations 450
- 23.6 Calculation of Rigidity Considering Finite Element[127] 454

24. Non-uniform Microdisplacement **460**
- 24.1 Calculations for Non-Uniform microdisplacement 460
- 24.2 Non-dimensional Parameters 461
- 24.3 Minimisation of Positional Displacement Error 463
- 24.4 Distribution of Hydrodynamic Force in Specific Lubrication Pocket 463
- 24.5 Dynamic Load Rating of Spindle Supports 465
- 24.6 Rigidity of Spindle unit and Optimum overhang of the Spindle end:(126, 127) 466
- 24.7 Accuracy of the Spindle Units 468

25. Reliability Analysis of Some Machine Tool Elements **470**
- 25.1 Introduction 470
- 25.2 Collet Chucking Analysis 471
- 25.3 Analysis of Spindle with Linear Elastic Supports and Viscous Damping 477

26. Questions **483**
- (A) Questions 483
- (B) Answers 498

Glossary **517**

References **519**

Index **529**

CHAPTER 1

Introduction

With the ever increasing complexities of manufacture and discoveries of newer and better techniques of manufacturing, there has been a remarkable need for changing the "look" and "character" of modern machine tools. By the terms "look" and "character" we mean the aesthetic quality and functional characteristics respectively. These functional characteristics are enriched by higher spindle speeds to reduce the machining time, greater rigidity, increased power output and higher efficiency of metal removal.

The accuracy of manufacturing close-to-form and close-to-tolerance jobs, uniformity in manufactured parts, minimum losses in friction in the various kinematic parts of the machine tool- require certain changes in the orthodox machine tools particularly in their drives and controlling units. The uses of different cutting tools (to mention in particular mineral ceramic tools in lathe and milling machine) require the spindle to be provided with suitable bearings capable of accepting increased power transmission and able to maintain its rigidity even under the cases of unavoidable temperature rise with the life long accuracy.

Remarkable advance has been made all over the world in the field of design and development of machine tools. This will be clear by a mere critical comparison between a machine tool built today and the same machine tool built nearly a decade ago. The directions towards which all these developments have taken place are innumerous. Gone are the days when the safety factor of four was considered essential in designing for the sections of the beds and box details of the machine tools. Designers of today are more careful in considering weight-to-rigidity ratio as an important factor of aesthetism and finer functioning rather than relying on the methods of rule of thumb evolved by mere experience and intuition. The evolution in the field of plastics and synthetic material has made possible the substitution of orthodox materials thereby reducing the time and money involved.

The study of dynamics of machine tools – a subject of comparatively recent origin enabled us to design machines which could minimize the machining errors that are liable to creep into the job under the dynamic condition, viz. deflection error, positional error, error due to vibrational instability etc.

1.1 The Fields of Development

Machine tools are nothing but instruments which have been created for the purpose of manufacturing wider range of products of all categories. Their designs therefore are variable functions dependent on time, obsolescence, technological changes, etc.

Every year more and more machines with newer and better accessories are being marketed. The production of all such machines is the outcome of specialized demand from the various engineering industries. In the manufacturing of their products, these machines are instrumental. The all - round developments in machine tool engineering can be very aptly-illustrated by Figure 1.1.

1.2 Classification of Machine Tools

Broad classification of machine tools was done by Schlesinger [2] and later on emphasized by N.S.Acherkan [1] et. al. in their book based on the functions they could perform. Such functional groupings of traditional lines was spread over distinct groups such as:

Group 1: Turning (Centre Lathe, Turret, Capstan, Semi-automatic, Automatic)
Group 2: Drilling (Horizontal, Vertical, Radial, Pillar, Multi-spindle, Jig Boring)
Group 3: Grinding & Polishing
Group 4: Gear tooth and Thread grinding
Group 5: Milling (Universal, Vertical, Surface, Horizontal, End etc.)
Group 6: Planning & shaping machines
Group 7: Sawing & cutting machines
Group 8: Miscellaneous

The broad groups of machine tools ; as given here are on the basis of functions. Other system of classification, as advocated by more and more

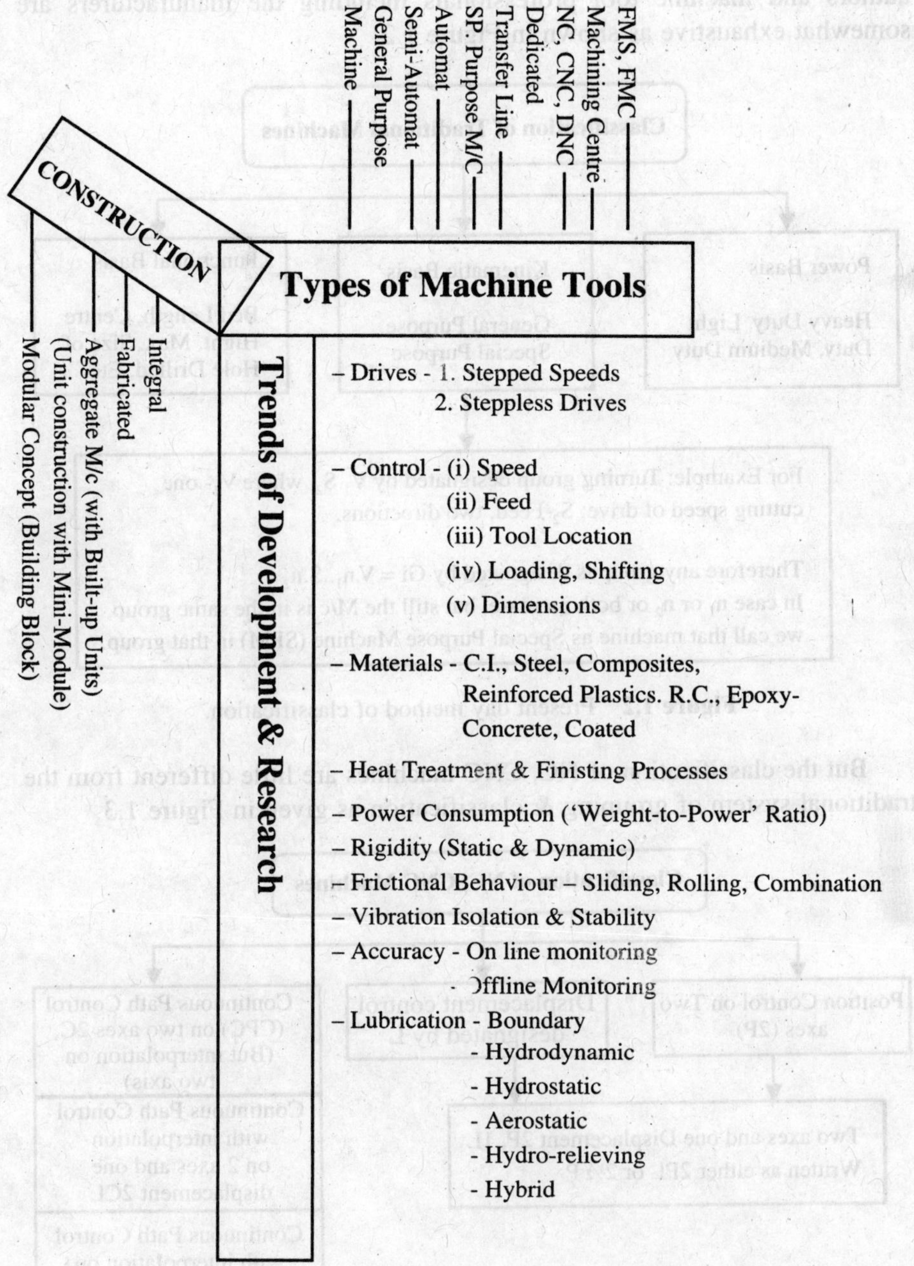

Figure 1.1 Trends in the development of Machine Tools.

authors and machine tool professionals including the manufacturers are somewhat exhaustive as shown in Figure 1.2.

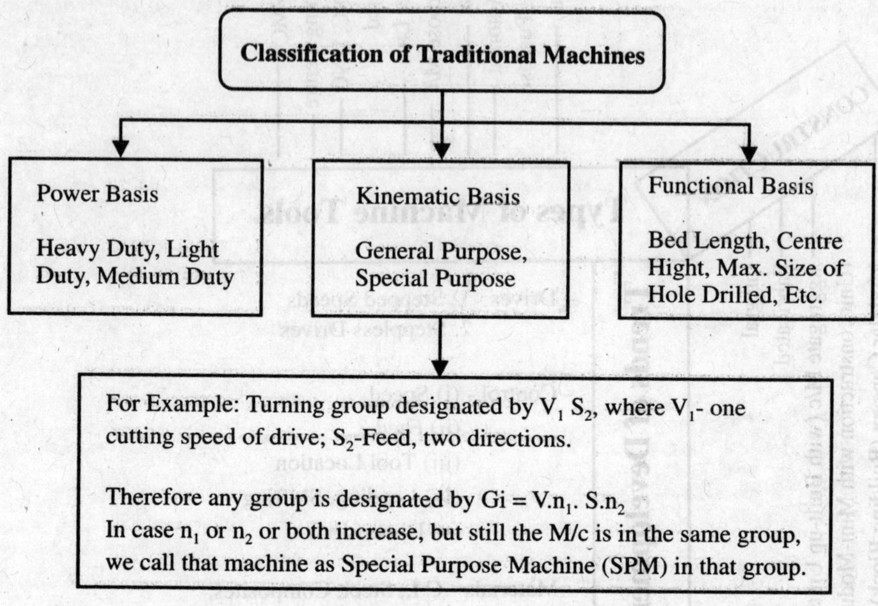

Figure 1.2 Present day method of classification.

But the classification of NC, CNC machines are little different from the traditional system of grouping & classification as given in Figure 1.3.

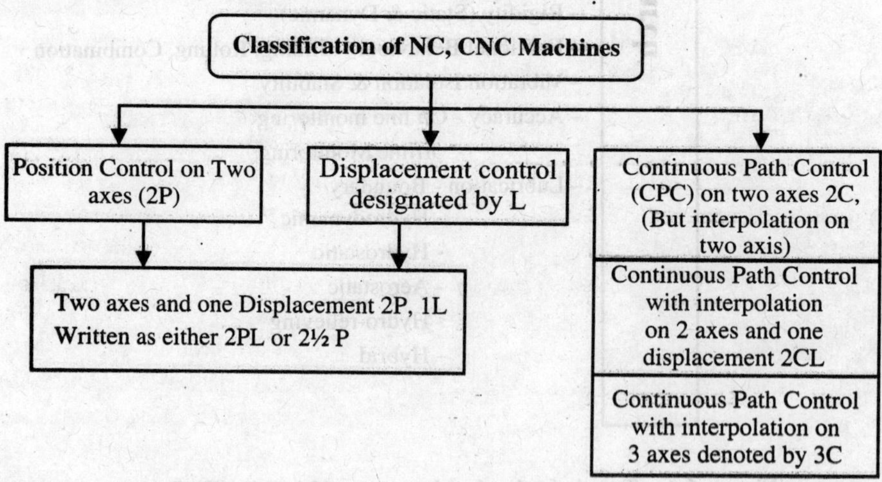

Figure 1.3 Designating the Numerical & Computer Numerical Control machines.

The NC, CNC, DNC and CNC machining centres are mainly distinguished by

(i) Positional control of various axes of operation (nP)
(ii) Displacement control (L)
(iii) Continuous path control (nC)

Where P, L, C stands for positional, displacement and continuous controls respectively and '*n*' stands for number of axes respectively as shown in Figure 1.3.

In many heavy machine tools there is a wide scope of R& D activities in the field of welded structure, plasma assisted machining, concrete filled bed, etc.

The trend of R & D in metal cutting machines is towards:

(i) Unit head construction – to some extent based on building block concept;
(ii) Development of automatic control system with or without feedback;
(iii) Design of elements including bearings, spindles, beds, etc.;
(iv) Use of adhesives for joining elements;
(v) Design and tribological consideration for life and dynamic stability;
(vi) Types and tribological characteristics of materials used , including composites, metal matrix composites;
(vii) Lubrication system including hydrostatic, hydrodynamic, aerodynamic or hybrid systems of lubrication, like aerostatic-cum-aerodynamic or hydrostatic- cum-hydrodynamic etc.

In the field of machine tool drives we are more and more drifting towards individual drive, that is to say each machine being driven by a separate motor. Every buyer of machine tools will like to have as many number of speeds as possible on his machine, even though the workers will be using only a few until and unless under strict supervision. This tendency has a natural leaning towards stepless regulation. Of course, to save machining time every machine tool should be provided with stepless regulation. The stepless regulation can be effected by friction and pressure variators, electro-mechanical regulation, hydraulic, pneumatic and electronic systems. Friction and pressure variators can give stepless regulation only to a limited range. Similarly electro-mechanical regulation gives us only a near approach to stepless regulation. Controlling of speeds can be done by three ways, namely simple, selective and pre-selective. In pre'selective control we can change a speed without going through any intermediate change or without completely stopping the machine during major manipulation. The controlling levers can be arranged best by taking into account the principle of ergonomics ; that is to say the controls should be located within the easy reach of the operator. Depending on this we can have centralized control or decentralized control.

Modernization of machine should be done from the point of view of increasing the speed of working and power consumption. There is practically no limitation of power required for a machine tool except the economic consideration. With the increase in technological possibilities the modernization of machine tools must be directed towards reducing the number of accessories and simplifying the problem of servicing.

Machine tools are nothing but instruments for production of some kind or other and therefore when we buy machines, in reality we buy production. In order to increase the level of production and thereby reduce the human labour spent on the same, we must try to automatize the technological processes. Automation is only one of the fundamental steps in all-round technological developments of our country. For the purpose of automation, it is absolutely necessary to put in energy to the executive organ of the machine tool to overcome the force due to cutting, friction, etc. and to provide certain information in some form or other depending upon the configuration of the job to be made.

Copy turning lathes manufactured by the Ordzonikidze factory in Moscow work on hydroservomechanism for the control of microfeed and microdisplacement of the tool dependent on the tracer movement. In copying lathes machining time can be shortened by using multitool for machining on the principle of single tool. The designs of these machines are so compact that in no case they occupy more space on the shop floor than an ordinary machine of the same power.

Most of such copying machines, when used for machining antifriction bearing races or universal joint crosses, are provided with special automatic loader. The copying lathe has many features which favour individual automation. Out of these the following features are worth mentioning:

(a) automatization of turning process by copying from a template;
(b) automatization of work cycle with automatic multicut recycling device;
(c) automatization of the extended work cycle by automatic spindle speed change and automatic programme control.

Progress of Machine Tool Building

Flow production using transfer mechanisms and "inline" transfer machine, rationalization of industry and standardization of- components-are but a few essentialities of the technically advanced countries like Russia ENIMS (the

Institute for Experimental and Scientific Investigation in Machine Tools) in Moscow is one of the many research institutes that Russia is proud of. Staffed with well-trained scientists and engineers, the institute is carrying out work in all possible fields of development of machine tools. The factory "Stankokonstruksia" in the name of Lenin is also associated with this research institute. Stankokonstruksia is engaged in manufacturing mostly gear cutting, thread grinding machines, etc. Side by side the factory is also engaged in manufacturing experimental prototype machine designed by ENIMS. This research institute is divided into various sections, viz. general investigation in machine tools, electrical equipments of machine tools, metal cutting section, grinding machines section, programme control section, hydraulic control section, pneumatic equipment section, prototype section, etc.

Apart from ENIMS, various other firms are also engaged in research, design and construction of new types of machine tools. Automatically controlled drum type turret lathe designed and constructed by Gorky Automatic Machine Tool Factory in Kiev is undoubtedly a new one incorporating modern accessories. Kolomna Machine Tool Works which specializes in the manufacture of heavy duty vertical lathes and boring mills is situated only a few miles from Moscow. In the research laboratory attached to this factory considerable work has been done in the field of hydrodynamic lubrication under the leadership of M.N. Tsirlin, G.A. Levit and others. In the Institute of Automatics, Kiev, considerable researches in the fields of various control systems, servo-mechanisms and micro-displacement etc. are being carried out.

More and more uses are being made now-a-days of plastic guides. Such guides have been used so far in heavy duty planing and vertical boring machines.

Plastic materials bonded to the cast iron surfaces of the table V-guides have been used in Butler No. 8A spiral-electric planing machine. In many heavy duty machines of U.S.S.R. such plastic guides have been used. But in U.S.S.R. the mostly used plastic is textolite metallurgical variety B. In ENIMS considerable research work has been carried out into the characteristic behaviours of the plastic guides by A.S. Lapiduce, G.A. Levit and U.N. Sokolov ajid others.

Concurrently *t* tremendous amount of researches have been carried at Cincinnati Milacron, in the field of adaptive control, numerical control and computer numerical machines. Various machining centres and 'in-line' transfer machines developed in various places are the ultimate fruits of such researches. In the University of Berlin under Dr. G. Spurr, a large volume of work had been done in the field Computer Aided Design (CAD) and computer aided manufacture (CAM).

In the Production Engineering Research Association, Melton Mowbray, U.K., work is being carried with great success, in automatic transfer of jobs in transfer machines and use of computer controlled robots for almost universal 'work in the field of metal processing. Progress in the application of robots to interact with sophisticated manufacturing processes has been made possible by the developments in microprocessor technology allowing large increases in the capability of robot control systems to deal with variability. This work was done under the able guidance of

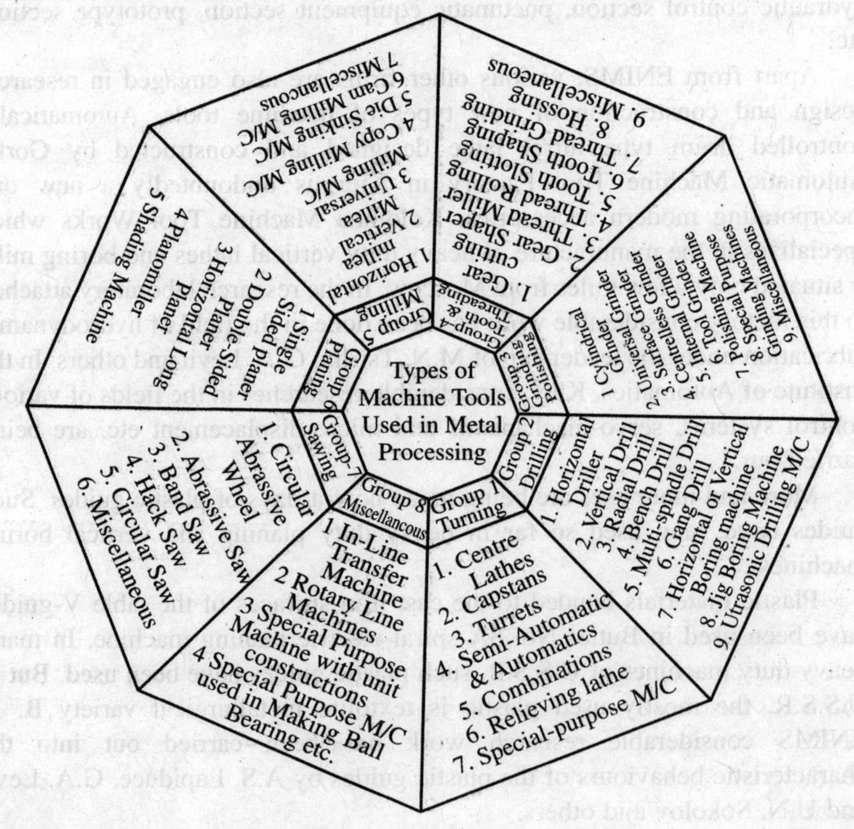

Figure 1.4 General Classification of Machine Tools.

W.B. Heginbotham, Director General of PERA. Readers may like to refer to the publication of Heginbotham, published in the proceedings of the 9th AIMTDR Conference held at IIT, Kanpur in December, 1980.

Progress of Machine Building

To overcome the shortage of skilled operators in an expanding technological sphere, the evolution of computer controlled machines is a general outcome. Manufacture of defence goods requires more and more precision and uniformity of the manufactured parts and we all believe that such demand will only increase with the passage of time. All over the world there is a trend towards manufacture and uses of computers numerical control (CNC) machine. There are firms now specialising in the development of computer controlled machining centres, and transfer machines with centrally controlled systems.

In the field of dynamic acceptance tests and stability of machine tools, the work done in the University of Birmingham under the leadership of Dr. S.A. Tobias is worth mentioning. Prof. Tobias had been doing valuable work in the field of machine tool vibrations.

Concurrently a large volume of work has been done all over the world in the field of metal cutting. The new concept of metal cutting and the basic scientific ideas connected with it have considerably influenced the design of a modern machine tool.

Hydrodynamic and hydrostatic slides are now-a-days commonly used in heavy duty machine tools. Researches are being carried out to obtain high speed air bearings (with one lakh r.p.m.) to be used in precision machine tools. Synthetic bearings and bi- or tri-metallic bearings are also finding increasing use in machine tools.

Dynamic Data System (DDS) advocated by Prof. S.M. Wu had almost revolutionised the field of condition monitoring using dynamic data for determining the malfunction or functional characteristics of system consisting of multiple subsystems. This indicates the growing use of stochastic processes in production engineering. It is a modelling technique using dynamic data in the form of a 'time series' to develop physically meaningful stochastic, difference and or differential equation. This system, which is based on time dependent data, can be used for universal application in all fields of engineering.

Machine Tools in India

Though manufacture of machine tools in India dates back to early twentieth century, planned production of the same started only after independence when the giant public sector undertaking Hindustan Machine Tools at Bangalore came into existence. Amongst the pioneers in the field, it is

Mysore Kirloskar that produced conepulley lathe in 1941. Subsequently considerable emphasis was laid on the production of the indigenous machine tools during our first five year plan and the target production of machine tools in subsequent plans. In effect there was a rapid progress in building 'mother machines' which could serve as infrastructure for manufacturing engineering and other commodities. The programme was launched with the slogan 'Build machines, Build India'. Today the Hindustan Machine Tools which is the biggest machine tools complex in India has opened up a new promise in the field of building indigenous precision machine tools in diversified field with up-to-date technological knowhow.

Though during the last 10 years considerable amount of developmental activities have taken place, the amount of export of machine tools has also gone up, as is evident from available statistics indicating the import of machine tools during the said period. This is because our R&D efforts in some specific areas of machine tools could not match with sophisticated technological know-how available in advanced countries. This, is particularly noticeable in the field of heavy machine tools wherein import of machine tools had to be done to meet the latest available technology. But this import of know-how in some cases has also helped us in developing our expertise for absorbing the associated technology, enlightening ourselves, and for effecting further imjrrovement in our developmental activities for producing indigenous machines.

Import of know-how in the form of foreign collaboration is normally characterised by a slower rate of assimilation by the recipient, thereby rendering the know-how obsolete or ineffective after a short time. The only means of combating this technological obsolescence may be 'in-house' R&D, so that assimilation of imported technology and innovative work go hand in hand.

Our Central Machine Tools Institute at Bangalore, Central Mechanical Engineering Research Institute, Durgapur, Machine Tools Laboratories at various IITs and in house R&D facilities attached to various major machine tools industries provide excellent infrastructural facilities for absorption of sophisticated technology, development of newer and better technology–appropriate to the requirement of the country, as well as development of quality products which could stand in the competition at home or in the export market.

Infrastructural facilities available in our country for research, design and development of machine tools, at present, provide an excellent opportunity to lay more emphasis on quality and ingenuity in design thereby resulting in self-reliance in R&D activities.

Under special requests from the Government of India, UNDP and UNIDO have agreed to set-up, in co-operation with CMTI, a modern NC training-cum-administration centre at Bangalore. The CMTI has already initiated its work on the developmental activities in the field of CNC machine tools during the last few years.

The R&D activities are now more focussed on the following machinery:

(1) Metal cutting machine-general purpose and special purpose including precision machine tools;
(2) Heavy machine tools;
(3) Metal forming machinery etc.; and
(4) Plastic processing machinery and foundry machinery etc.

In addition to this, development work in the other field of 'com-petence-oriented-machine-tools' is also under progress, in non-conventional machines such as EDM, ECM, NC and F.M.S., CNC machining centres, auto-tool exchanger etc. though such activities are still now limited and in some cases suffer due to technology gap.

Though India has made rapid stride in manufacturing machine tools, many of our machine tools have bold leap to take due to either design being back-dated, or obsolete, lacking in precision and not incorporating essential features of control and monitoring. Constant pressure grinding machines, machines with adaptive control are yet to be developed indigenously.

In metal forming, Hindustan Machine Tools at Hyderabad, Godrej, Scotish India Machine Tools, etc. are producing some medium and heavy machines. In the forming machines category Hindustan Machine Tools exhibited in IMTEX 79 newly developed high speed press, flow forming lathe, while Godrej and Boyce Manufacturing Company exhibited pneumatic press brake, high efficiency transfer press, etc. In many heavy machine tools there is a wide scope for R&D activities in the field of welded structure, plasma assisted machining, concrete filled bed, etc.

The trend of R&D in metal cutting machines is towards:

(i) unit head construction—to some extent based on building block concept;
(ii) development of automatic control system with or without feed back;
(iii) design of elements including bearings, spindles, beds, etc.;
(iv) use of adhesives for joining elements; and
(v) design and tribological consideration for life and dynamic stability.

Machine Tools Development Council, through various subgroups have already finalised sectoral technology report for metal forming machines, heavy machine tools, precision machine tools with modular constructions. All these reports show up-to-date demand assessment, indicate the technological gap and identify technologies requiring to be updated.

Modular concept, as stated above, is useful for building general purpose machine. A few self-contained units (unit heads) are so designed that they may be arranged in different combinations to suit the machining requirements of particular component. Thus thc requirement of a special purpose machine might be largely eliminated. This gives flexibility in design and in most cases shows techno-economic viability, with particular reference to their uses in small scale sector.

The growth in the computer technology indigenously has provided a basic support structure essential for development of good numerical control machine tools. Such technological development undoubtedly help us consolidate our path of self-reliance in the field of machine tools and effects development of indigenous know-how. Though self-reliance is essential to a great extent in machine tool R&D, 100 per cent self-reliance, which no country in the world has achieved, is not needed. In cases where requirement of some machines is negligibly small, it may be economic to import technology, rather than justify heavy expenditure on R&D.

Since machine tools industries is one of the basic industries on which depends the economic upliftment of the country, it has become necessary today to pay more attention to some of the vital points which almost challenge our existence in tne technologically competitive market.

Machine Tools Exhibition held in 1975 (IMTEX 75) which exhibited 500 machine tools and 5000 allied equipments, lend a tremendous boost to Indian machine tools industries. IMTEX now, held regularly, show further progress of the machine tools industries and explore new possibilities of export. It is really needed that we should occupy a respectable position in the export market and this requires rethinking in our export strategies to some extent.

In the field of high precision machines, the technology of manufacture calls for special plant and equipment, viz, Jig Boring, Jig Grinding and other super finishing machines, since the extent of precision in terms of dimensional tolerance varies in the range of IT-2 to IT-6.

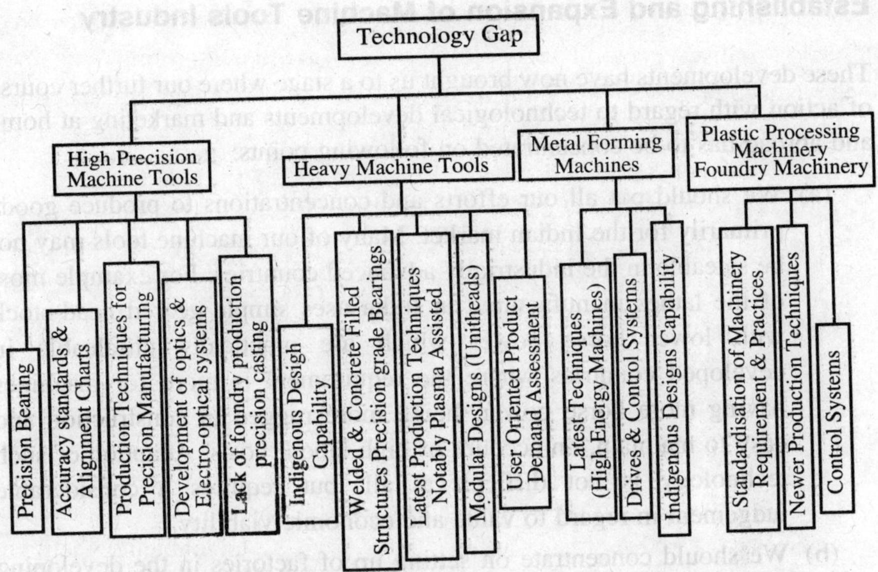

Figure 1.5 Technological gap in some areas of machine tools.

Central Manufacturing Technology Institute (CMTI) has established Acedemy of Excellence fot study & research in advanced manufacturing technology in Bangalore, where in addition to scheduled training programmes, it conducts other programmes, as well as, 'on-site' programmes at 'customer's premises, and corporate training programmes etc.

Similarly, there is a need for Research Centre for heavy machine tools, apart from promoting 'in-house' R&D in various heavy industries. This is essentially to bridge the technology gap, assimilate sophisticated technology-appropriate to our requirement, make technoeconomic viability study and to project our futuristic trend. To a limited scale foreign technical collaboration is also needed, particularly in certain areas of heavy machine tools.

For plastic processing machinery, indigenous design capability is available to a large extent. What is morereeded is researches into the production processes and their impact on newer design. Standardisation of units, and control system is also an essential need. CIPET, the Institute set up for the purpose, is already proceeding with this task with a view to achieving self-sufficiency.

It is necessary for CIPET to build up expertise in design and standardisation of toolings, moulds etc, since many users of plastic processing machinery are still not in a position to reduce the already existing technology gap, be it in small scale or medium scale sector.

Central Mechanical Engineering Research Institute has also been considerably helping manufactures to design special-purpose machines.

Establishing and Expansion of Machine Tools Industry

These developments have now brought us to a stage where our further course of action with regard to technological developments and marketing at home and abroad has to be concentrated on following points:

(a) We should put all our efforts and concentrations to produce goods primarily for the Indian market. Many of our machine tools may not be saleable in the industrially advanced countries. For example most of the lathes manufactured in India uses simple geared head-stock with lower horse power which are practically unsaleable in developed countries where the requirement is more for machines having more horse power, being more rugged in construction and easy to maintain, on account of high labour costs. Adaption of such technology is not difficult at all, but requires a considerable judgement in regard to value and economic viability.

(b) We should concentrate on setting up of factories in the developing countries in partnership with local entrepreneurs on the basis of technical know-how fee and royalty.

(c) Government of India, through its agencies like Planning Commission and Department of Science and Technology, should concentrate more in giving aids in the form of pilot projects etc. on the basis of the technological know-how deve-loped at national laboratories, to its neighbouring countries and least developed countries.

(d) We should participate more and more in world trade fairs. This is justified for boosting up our image in foreign market and exploring further directions of export possibilities. We had experienced quite an impressive boost after our participation in Hannover Industrial Trade Fair as early as in 1973.

(e) Such participation helps our engineers and designers to have an opportunity for expert comments and feedback of information related to their work from foreign users. They are also exposed to up-to-date or latest designs marketed by our competitors.

(f) For the more developed countries of Europe, America and Japan, in order to upgrade our technology we should aim at

   (i) manufacturing machines either complete (for fitment of advanced control systems, or

   (ii) manufacturing machines, partly machined and assembled (PMA) for final assembly abroad.

(g) So far as export of machinery or technical know-how, we should have stronger link with the least developing countries (LDCs). This will help us utilise the already developed technical know-how in regard to sophisticated machine and thereby gradually build up our own expertise.

(h) Value Engineering and Ergonomics study is absolutely essential. Not much work has been done in this field till now. But to survive in global competition, we must regard this as an important strategy.

(i) User's education: User should be taught to use simple machine with added functional performance and not 'anything sophisticated' machine, with all the gadgets on.

(j) Probabilistic method of determining the size and capacity range should be adopted before modernising our existing machines.

(k) Reliability and maintainability criteria should be primary consideration in developing precision machines.

(l) Design of special purpose machine/general purpose machine should be based on building block concept.

(m) Technology deal with LDCs or within our country between laboratories and industries must be completed as a package deal.

(n) In developing industries in the least developing countries, the manufacture of machine tools may profitably be carried out through a number of small scale industries, specialising in components, arranged in "a cellular structure feeding the centralised factory assembling the main product. This is needed essentially to develop captive ancillary units or small industries complex.

(o) For the positive growth of machine tool industries, information relating to these forms a vital segment of total informations required viz, technical enquiry regarding product, new information, compilation of state of art and trend reports, technoeconoinic data etc. National Information Centre for Machine Tools and Production Engineering (NICMAP) recently set up at CMTI, Bangalore, as a part of National Information System for Science and Technology (NISSAT) is a welcome step in dissemination of knowledge. It is certainly going to help transfer of technology.

There are large number of Indian agents representing Indian as well as foreign firms manufacturing machine tools and accessories. In such cases NICMAP supplements the process of exchange of information and guidance leading to transfer of technology.

Conclusion

The industrial image of a country depends to a large extent on the reputation and quality of the capital goods which it manufactures and exports.

Therefore the indirect benefit consequent on our improvement in the image of the country will greatly enhance the export of all manufacturers especially in the field of metal working engineering goods.

We should adopt a special strategy for export to the ESCAP (Economic and Social Council for Asia and Pacific) region and to other developing countries, where the technological advancement and skill are not at the same level as found in India. Now our growth has certainly reached an adequate high stage of technology, as has been seen from the research and development work done in our various National laboratories and Research Institutes and In-house application oriented researches in various industries.

That India had the potential to assist LDCs in this field is clear from UNIDO meeting held at Tbilisi-USSR as early as in October 1974, wherein it was felt that India having reached intermediate stage of technology should seriously consider assisting LDCs, in the development of machine tool technology.

Jyoti CNC Automoation Ltd., Gujarat, produces a wide range of CNC machines for any of metal cutting application. Milling machines produced by Bharat Fritz Warner Ltd. (BFW) continue to play a key role in the Jet-set age of machining across the country. During the period spread over last 25 years, in particular, in the field of CNC turning, Milling & Grinding Machines. All Micromatic groups have been making high quality machines with indigenous technology. In the field of ECM, EDM manufacture, the role played by Technofour Ltd. is remarkable. Indian Machine Tool Manufactures Association offers training programmes for machine tool design & manufacture for the benefit of the prospective young intreseted engineering. Such courses offer hands-on-session in tune with application-oriented as well as Industry endorsed curriculum. Apart from the brief informations put up here, these are a large number of small and medium scale industries catering to the needs of manufactures demands for special purpose machines as well as well as machines with NC, CNC systems.

CHAPTER 2

Determination of the Forces Acting on the Tool in Certain Machining Operations and Horse-power Requirement

In Designing a machine tool, the first thing that a machine tool designer must know is the power requirement of the machine tool. For fairly accurate determination of the energy consumption, one must know the conditions of cutting, the maximum magnitude of cutting force and torque required under the dynamic condition. On knowing these, it is possible to calculate the horse-power of the motor driving the machine tool and assign its suitable specification. Without knowing the torque it is not possible to design the various shafts in the gear box, nor is it possible to design the gears for strength. For the design of bearings on the gear box also, it is absolutely necessary to know the extent of axial thrust coming on the shafts.

Accurate determination of the various forces due to cutting in a machine tool is. not yet possible. But several empirical relationships are available. Most of these empirical relations are based on experiments and logical decisions.

In this chapter we have given the existing empirical relations for the determination of the tool forces and horse-power consumption in a few specific important cases of metal machining. We have also included a supplementary chart of efficiency in various transmitting systems.

2.1. Determination of the Tool Forces in a Lathe Operation

Determination of Forces Acting on the Cutting Tool.

In any metal cutting operation in a lathe, there acts a force R on the tool. This force R can be resolved into three components :

(1) P_y, in the horizontal plane, perpendicular to the direction of the feed;
(2) P_x, in the horizontal plane, against the direction of the feed (frictional force); and
(3) P_z, in the vertical plane, perpendicular to both P_y, and P_x.

These components of the cutting forces are shown in Figure 2.1

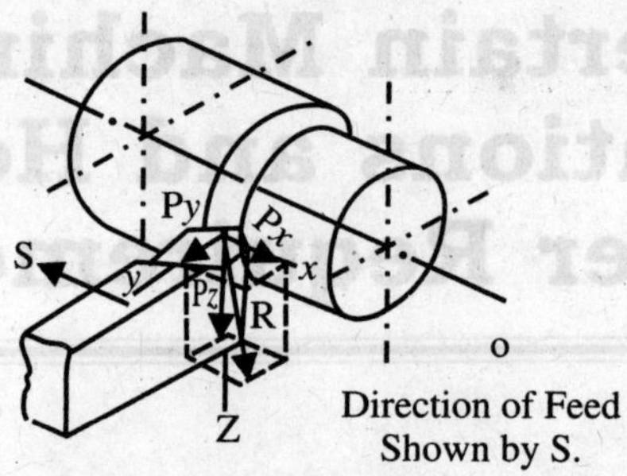

Figure 2.1 In lathe tools.

These three forces depend on five essential characteristics, *viz.* tool geometry, depth of cut, feed, work material and condition of working such as provision for coolant, etc. The influence of speed is not very appreciable.

It is difficult to calculate the exact force acting on the tool under a cutting condition, because it is difficult to take care of the different variable factors that come into play in an actual cutting operation.

A.M. Rosenberg and A.N Eremin did a considerable amount of research work in this field in U.S.S.R. A good deal of reference to their work has been done in the *Metal Cutting* by Sisoev, published from Moscow in 1960.

Empirical formula determining the force P_z can be expressed as under:
(Based on the relation given by Kronenberg)

$$P_z = C_p \times t^x \times S^y \times K, \tag{2.1}$$

where Cp – coefficient, characterized by the work material and condition of working such as tool, coolant, etc.
t – depth of cut in mm,
S – feed in mm/rev,

K – overall correlation coefficient, consisting of the actual condition of working and tool angles, which varies from 0.9 to 1.0.

K again consists of four different coefficients:

$$K = K_c \cdot K_\Phi \cdot K_\lambda . K_m$$

K_c – correction coefficient for coolant,
K_Φ – correction coefficient depending upon the entering angle,
K_λ – correction coefficient depending upon the back rake angle
K_m – correction coefficient depending upon the material.

The value of these coefficients depending upon their parameters can be interpolated from the tables below:

Table 2.1

Coolant	K_e	Φ	K_Φ	γ	K_γ
Dry	Y	Ky			
Dry	1	30°	1.05	–15°	1.40
Soda water+	1.03	45"	1.00	–10°	1.30
Emulsion	1.10	60°	0.96	+ 5°	1.23
Mineral oil	115	75°	0.94	0	1.13
				+5°	1.06
				+ 10°	1.00
Hard					
Mineral+	1.20-1.25	90°	0.92	+15°	0.94
				+20°	0.89

Table 2.2

Material	$\sigma_b \frac{Kg}{mm^2}$	K_m	*Material*	$\sigma_b \frac{Kg}{mm^2}$	K_m
	40–50	0.76		140–160	0.88
	50–60	0.82		160–180	0.94
	60–80	0.89	C.I.	180–200	1.00
Steel	70–80	1.00		200–220	1.06
	80–90	1.10		220–240	1.12
	90–100	1.18		240–280	1.17
	100–120	1.28			

σ_z in this table denotes the ultimate strength of the material.

The Values of Coefficents and Indices[46]

Table 2.3

Material	*σ_b in kg/mm²*	*Hardness Brinell H_B*	*Cp*	*x*	*y*	*Type of operation*
Steel	75	215	225	1.0	0.75	Turning and boring
			264	1.0	1.0	Facing and parting
Grey C.I.	–	190	98	1.0	0.75	Turning and boring
			135	1.0	1.0	Parting and facing

Example Calculate the magnitude of the forces acting on a H.S.S. single point cutting tool while machining m.s. The total depth cut given is 4.75 mm and the feed is 0.68 mm/rev. of the spindle.

Solution

$$K = K_m \cdot K_\lambda \cdot K_\Phi \cdot K_e$$

$$= 0.89 \,.\, 1{\cdot}00 \,.\, 1{\cdot}05 \,.\, 1{\cdot}00 = 0935.$$

From tables above:

$$\left.\begin{aligned} C_p &= 225 \\ x &= 1.00 \\ y &= 0.75 \end{aligned}\right\} = p_t = 225 \times 4.75^1 \times 0.68^{0.75} \times 0.935 = 830 \text{ K}_g$$

assuming the operation to be a turning operation.

The components P_z, P_y, and P_z are approximately connected by the following expressions

$$\frac{P_x}{P_z} \approx 0.3 \text{ to } 0.2$$

$$\frac{P_y}{P_z} \approx 0.2 \text{ to } 0.1$$

Of course the above relations depend upon the tool parameters. The data given here were obtained with tool having rake angle 10° and entering angle (main angle in the plan view of the tool) equal to 30°.

While turning or cutting plastic materials the cutting forces are reduced by 1.5 to 2 times.

The component Pt of the cutting force gives rise to the turning moment effected on the work piece. This can be found out from the formula:

$$M_{cutting} = \frac{P_z \cdot D}{2} Kgmm,$$

where D—diameter of the job in mm.

The component P_y tries to deflect the work piece in the horizontal plane and together with the component P_x gives a resultant reaction, which in turn tries to bend the job. The force P_z causes the bending of the tool also in the vertical plane. The magnitude of such bending moment is given by

$$\mathrm{M_{Bending}} = P_z \times l$$

where l is the overhang or cantilever lengh of the cutting tool. The comsumption of power in cutting can be calculated from the formula:

$$(H.P)_{cutting} = \frac{P_z \cdot v}{75 \cdot 60} \text{ Horse power } = \frac{P_z \cdot v}{60 \cdot 102} \text{ k watt}$$

where v—speed of cutting in m/min. This is to say

$$v = \frac{\pi \cdot D \cdot n}{1000}$$

where D—diameter of the job turned in mm and n—speed in r.p.m. In order to have an input of this power at the spindle, it is necessary that the power of the electrical motor is more than this, in order to cover the losses in bearings and other kinematic chains. Hence the motor horse power should be as under:

$$\text{Power of the motor} = \frac{P_z \cdot v}{60 \times 102\ \eta_{\text{overall}}} \text{ k watt"} \qquad (2.2)$$

$$\eta_{\text{overall}} = \eta_1 \cdot \eta_2 \cdot \eta_3 \cdot \eta_4 \cdot \cdot \cdot,$$

where η_{overall} — overall efficiency of the trasmission system;

η_1 — efficiency of the ball bearings used in the gear box;

η_2 — efficiency of the sleeve bearings used in the gear box pr speed units;

η_3 — efficiency of the gear trains ;

η_4 — efficiency of the belt drive from motor to the pulley of clutch;

η_5 — efficiency of clutch.

Usually $\eta_1 = 0{\cdot}98 - 0{\cdot}995$; $\eta_2 = 0{\cdot}98 - 0{\cdot}985$;
$\eta_3 = 0{\cdot}98 - 0{\cdot}985$; $\eta_4 = 0{\cdot}96$, and so on.

Therefore on an average the overall efficiency Varies between 08 to 0-9.

Supplementary chart

	Transmission system	*Efficiency*
1.	Open belt without tightening pulley	0.98
2.	Open belt with tightening pulley	0.97
3.	V-belt	0.96
4.	Spur gear drive with ground teeth	0.99
5.	Spur gear drive with machined teeth	0.98
6.	Spur gear drive with conical threads	0.97
7.	Worm-worm wheel:	
	Single start	0.68
	Double start	0.75
	Triple start	0.81
8.	Chain drive	0.96–0.97
9.	Bearing (ball or roller)	0.995

In Figure 2.2 has been shown a nomogram for finding out the r.p.m. of the spindle, if we know the relative cutting speed and the diameter of the work piece.

Deflection of the tool considering only the cantilever effect.

In a lathe, the tool is held in a tool post projecting out by a certain distance. If this distance is denoted by l then the deflection is obtained as under:

$$\delta = \frac{P_z \cdot l^3}{3EI}$$

where E— Young's modulus of the tool material,

I— area of cross-section of the tool

$\left(= \frac{bh^3}{12},\right.$ where b–width of the shank, and

h–depth of the tool shank.$\left.\right)$

P— vertical component of the cutting force.

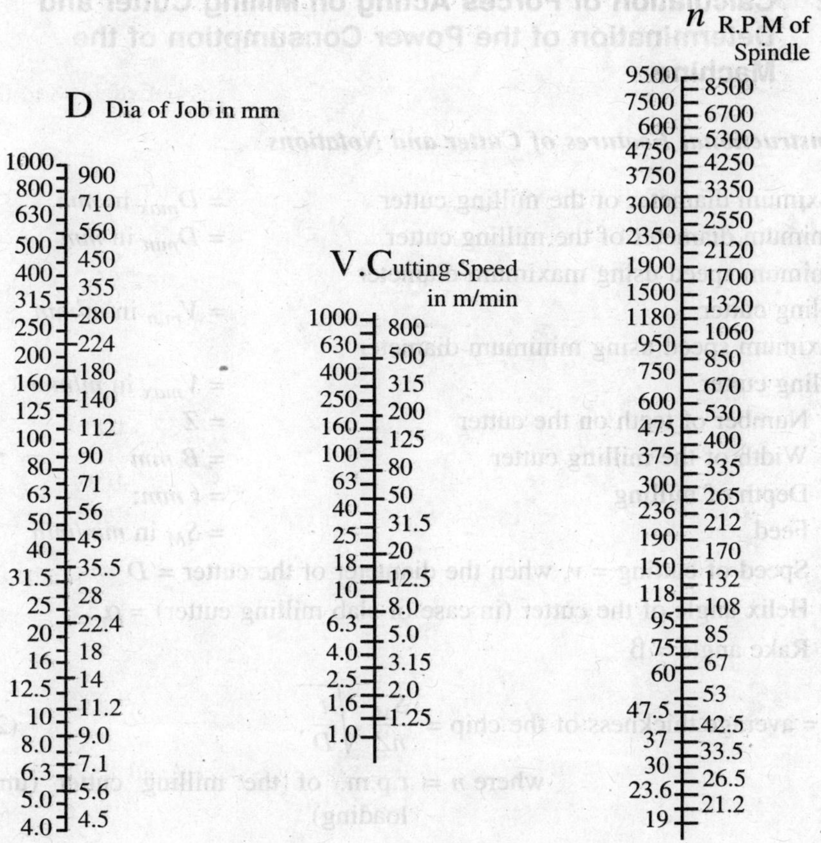

Figure 2.2 Nomogram for Speeds.

But this calculation presumes that the body of the tool post and the cross-slide (on which the tool post is fitted) are rigid and do not undergo any deformation or cross-slide does not assume any inclined position on the guides of the machine under the dynamic condition.

It should be understood that such calculations are absolutely necessary for (a) designing the tool post; (b) calculating the rigidity of the tool; and (c) determining the accuracy of the work to be done on the machine. Therefore, it should be the duty of the machine tool designer to go through this step before he can concentrate on designing the tool post and carriages or cross-slides.

From the accuracy specified for the work to be made on the machine, we can calculate very easily the optimum "overhang" of the tool and specify the same to the buyer of the machine tool.

2.2 Calculation of Forces Acting on Milling Cutter and Determination of the Power Consumption of the Machine

Constructional Features of Cutter and Notations

Maximum diameter of the milling cutter = D_{max} in *mm*
Minimum diameter of the milling cutter = D_{min} in *mm*
Minimum speed using maximum diameter milling cutter = V_{min} in *m/min*
Maximum speed using minimum diameter milling cutter = V_{max} in *m/min*
Number of teeth on the cutter = Z
Width of the milling cutter = B *mm*
Depth of milling = t *mm*;
Feed = S_M in *mm/min*
Speed of cutting = v, when the diameter of the cutter = D
Helix angle of the cutter (in case of slab milling cutter) = α
Rake angle = β

$$h_m = \text{average thickness of the chip} = \frac{S_M}{nZ}\sqrt{\frac{t}{D}}, \tag{2.3}$$

where n = r.p.m. of the milling cutter (under loading)

$$= \frac{1000\, v}{\pi \cdot D}.$$

If the intensity of cutting pressure (on the periphery) is denoted by K_M, then

$$K_M = f\,(h_m) = 400\ Kg/mm^2 \qquad [2]$$

$$\text{Effective power} = \text{N} = \frac{K_M \cdot t \cdot B \cdot S_M}{60 \cdot 120 \cdot 1000}\, K \text{ watts} \tag{2.4}$$

Average area of the chip in case of multiple teeth milling cutter

$$= \frac{S_M}{v}\, t \cdot B \cdot mm^2$$

where $v = \pi \cdot D \cdot n\ mm/min$

Knowing the parameter $\frac{S_M}{v}\, t \cdot B.$ and using the graph showing the relation of this parameter to the tangential force, we can find out the tangential force on the cutter. From the tangential force again we can find out the effective power requirement in the following manner:

$$N = \frac{P_t \cdot V}{60 \times 102} \text{ K watts} = \frac{P_t V}{60 \times 75} \text{ H.P.} \tag{2.5}$$

In this chapter we do not wish to show this graph, since it is better to calculate the tangential force from the empirical relationship shown later. (See Equation 2.6.)

Transmission of motion from the driving motor to the main spindle of the machine is carried out through belt drive, clutch, bearings and a number of gear trains, etc. From the individual efficiencies of all these elements of the machine tool transmission system we can find out the overall efficiency. Let us assume that the overall efficiency is equal to η_{overall}.

Therefore the power of the motor must be equal to:

$$\mathrm{N}_m = \frac{N}{\eta_{\text{overall}}}$$

Assuming that in a real case the power required from the motor to be equal to be 10 % more than this value, which is justifiable to safeguard against unusual heavy demand

$$\mathrm{N}_{m \text{ actual}} = \frac{N}{\eta_{\text{overall}}} \times 1.10$$

Usually, three phase electrical motor is used for milling machine drive. It is the duty of the designer to specify the speed of the motor, amount of slip and the efficiency.

Speed of the motor n_m is obtained from the relationship

$$n_m = n_{\max} \times \left(\frac{1 - S'}{S'} \right),$$

where $\eta_{\max}$ = maximum speed of the motor and S' equals slip in fraction. Usually the slip is near about 6 per cent.

In the case of milling operations the component of the cutting force in the radial direction of the cutter

$$P_r \approx (0.35 - 0.55)\ P_t,$$

where P_t is the tangential force in the milling operation.

The horizontal component of the milling force which acts on the feed mechanism of the table of the milling machine and on the fixture in which the jobs are fixed is given by

$P_h \approx (1–1.1).\ P_t$,—if the operation is done against feed;

$P_h \approx (0.8–0.9).\ P_t$,—if the operation is done in the direction of the feed;
and
$P_h \approx (0.6–0.9)\ P_t$,—in case of vertical milling operations.

The vertical component of the milling force P_v which can be calculated by knowing the angle of contact of the milling cutter with the work, is responsible for causing the bending of the fixture and the work table.

Milling force can also be calculated by empirical formula shown below.

Tangential force P_t in milling operation can be calculated from the formula:[28]

$$P_t = C_p \cdot t^x \cdot S_z^y \cdot Z \cdot B^z \cdot D^C\ Kg, \tag{2.6}$$

where, C_p – constant depending upon the characteristic of the tool material and the material of the job;
t – depth of milling in mm;
S_x – feed per tooth in mm;
Z – number of teeth in contact with work piece;
B – width of the milling cutter in mm;
D – diameter of the cutter in mm;

and x, y, z, and q are the exponents.

For steel, the values are :	For C.I. the values are :
$C_p \approx$ 40–80	$C_p \approx$ 48–70
$x \approx$ 0.86	$x \approx$ 0.83–1.14
$y \approx$ 0.74	$y \approx$ 0.65–0.70
$z \approx$ 1.00	$z \approx$ 1.00–0.90
$q \approx$ 0.86	$q \approx$ (–0.83) to (–1.14).

The effective power consumption in the case of milling is obtained from

$$N = \frac{P_t \cdot v}{60.102}\ \text{K watts}$$

where P_t is in kg and v is in m/min.

Specifications of the driving Motor

In specifying the motor we must give the following particulars: H. P., speed, power supply with phase, frequency, voltage, continuous or intermittent working, frame size, ballbearings, plane journal bearing etc.

Standard sizes for the fractional H. P. motors both A. C. andD. C. : 1/20, 1/12, 1/8, 1/6, 1/4, 1/3, 1/2, 3/4, 1. For large power, induction motors used can have standard H. P. ratings : 1, 1½, 2, 3, 5, 7½, 10, 15, 20, 25, etc.

2.3. Calculation of Power Consumption in Cylindrical Grinding Operation

Let us assume following nomenclature in respect of cylindrical

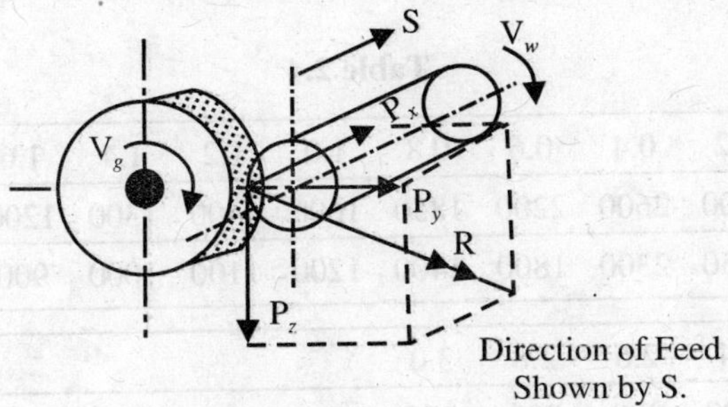

Figure 2.3 Force in Grinding.

grinding operation on a grinding machine (shown in Figure 2.3).

V_w — velocity of the work piece in m/min;

S — longitudinal feed of the wheel in mm/rev. of the work;

V_g — velocity of the grinding wheel in m/min;

t — depth of grinding per one cut in mm;

p_m — specific resistance to grinding of the work material in kg/mm^2;

P_z — vertical component of the force in cylindrical operation;

P_x — horizontal component of force against the feed;

P_y — Radial component of the force in cylindrical grinding operation, directed away from the wheel but towards the work piece. This force has the tendency of bulking the work piece and dragging the wheel on to the job. A considerable value of P_y will naturally hamper the accuracy of the grinding operation.

The tangential component P_z which constitutes the major value of the grinding force is proportional to the area of the material removed and the velocity rate $\frac{V_w}{V_g}$,

that is to say $$P_x \alpha (S.\ t),$$

and $$P_x \alpha \frac{V_w}{V_g},$$

or in other words $$P_x = p_m \cdot \frac{V_w}{V_g} \cdot s.t. \qquad (2.7)$$

The table below shows the various values of p_m dependent on the parameter ($S.\ t$) and the material of the work piece:[24]

Table 2.4

S.t in mm²		0.2	0.4	0.6	0.8	1.0	1.2	1.4	1.6	1.8	2.0
P_m	*Steel*	3300	2600	2200	1850	1600	1400	1300	1200	1100	1050
	C.I.	3050	2300	1800	1400	1200	1100	1000	900	850	800
S.t in mm¹		2.4	2.6	2.8	3.0						
P_m	*Steel*	1000	900	900	850	P_m is in kg/mm²					
	C.I.	750	750	750	725						

To evaluate the force P_z :

(1) multiply t by S;

(2) find out the value of the velocity ratio $\frac{V_w}{V_g}$;

(3) find out the value of p_m in *kg/mm²* from the table quoted above; and

(4) finally calculate the force P_z feeding the information in equation (2.7).

The power required by the grinding wheel is obtained by the equation:

$$N_g = \frac{P_x \cdot V_g}{60 \times 75} H.P. \qquad (2.8)$$

The power consumed in rotating the job:

$$N_w = \frac{P_z \cdot V_w}{60 \times 75} \qquad (2.9)$$

2.4. Determination of the Thrust on a Drill and the Power Consumption

The drill force can be approximately calculated by the following expression:[24]

Table 2.5

Work material			Keying						Spline Broaching			Circular or cylinder Broach		
	Mechanical property		$S_z \le 0.07$ mm			$S_z > 0.07$ mm			S_z as per table No. 2.6			S_z as per table No. 2.6		
Name	H_B	σ_B kg/mm^2	c_v	m	y	C_v	m	y	C_v	m	y	C_v	m	y
	upto 200	upto 70	9.8	0.87	1.4	7.7	0.87	1.4	15.5	0.60	0.75	16.8	0.62	0.62
C-Steel	200-230	70-80	8.8	0.87	1.4	7.0	0.87	1.4	14.0	0.60	0.75	15.5	0.62	0.62
	above 230	above 80	6.3	0.87	1.4	5.0	0.87	1.4	10.2	0.60	0.75	11.2	0.62	0.62
	≤ 200	—	6.2	0.6	0.95	6.2	0.6	0.95	17.5	0.5	0.6	14.0	0.50	0.60
C.I.	200	—	5.1	0.6	0.95	5.1	0.6	0.95	14.7	0.5	0.6	11.5	0.50	0.60

Force in Broaching : *In case of Key and splines broaching* : $P = C_p \cdot S_z^{\,y}\, bZ\, n_i$ kg (2.12)

In case of cylindrical broaching : $P = C_p \cdot S_z^{\,y}\, DZ$ kg (2.13)

where C_p — coefficient depending on the characteristics of the work material and the conditions of cutting:

b — width of key or spline in mm;

Z — number of teeth engaged at a time;

n_i — number of splines;

D — diameter of the broached hole in mm.

The values of the characteristic constants and exponents are shown in tables below:

Values of S_z in mm

Table 2.6

Operation	Steel							
	C–Steel			*Alloy Steel*			*C.I.*	
	$H_B \le 200$	$H_B = 200\text{-}230$	$H_B > 230$	$H_B \le 200$	$H_B = 200\text{-}230$	$H_B > 230$	$H_B \le 200$	$H_B > 200$
	$\sigma_B \le 70 \frac{kg}{mm^2}$	$\sigma_B = 70\text{-}80 \frac{kg}{mm^2}$	$\sigma_B > 80 \frac{kg}{mm^2}$	$\sigma_B \le 70 \frac{kg}{mm^2}$	$\sigma_B = 70\text{–}80$	$\sigma_B > 80 \frac{kg}{mm^2}$		
Keying	0.04-0.07	0.07-0.12	0.04-0.07	0.03-0.06	0.06–0.10	0.04-0.07	0.08-0-15	0.07-0.12
Splining	0.04-0.06	0.04-0.08	0.03-0.05	0.03-0.05	0.04-0.06	0.03-0.05	0.05-0.10	0.04-0.08
Circular broaching	0.02-0.03	0.02-0.05	0.02-0.03	0.02-0.03	0.02-004	0.02-0-03	0.04-0.08	0.03-0.06

H_B in this table denotes Brinell hardness number.

Values of C_P and Y_P.

Table 2.7

HB	σ_B in $\frac{kg}{mm^2}$	Key		Spline		Cylindrical	
		C_P	Y_P	C_P	Y_P	C_P	Y_P
Steel up to 200	upto 70	177	0.85	212	230	700	0.85
Above 230	70-80	202	0.85	0.85	0.85	762	0.85
	above 80	250	0.85	284	0.85	842	0.85
C. I. < 200	—	115	0.73	152	0.73	300	0.73
> 200	—	137	0.73	215	0.73	354	0.73

$$P = K \cdot D \cdot S^m \quad (2.10)$$

where P — drill force in kg.

K — constant, dependent on the work material, and other geometrical parameters of the drill;

S — feed in mm/rev.

D — diameter of the drill in mm (maximum size that could be drilled); and m — constant.

For C. I. material : $P = 60. D. S^{0.8}$ kg.

For steel : $P = 85. D. S^{0.7}$ kg.

The turning moment on the drill:

$$M_t = K_t\, D^{1.9}.\, S^{0.8} \text{ kg. mm} \quad (2.11)$$

where K_t — constant, equal to 33.8 for steel and 23.3 for cast iron.

Knowing the drill force and the turning moment we can find out the power requirement of the machine. But it should be always borne in mind that values given above for K and K_t are approximate values. In real character K and K_t are nothing but product of several constants depending on tool-shape or form, hardness of the drill, etc. and are susceptible to change under conditions of cutting oil or coolant. But the various constants which constitute K or K_t are not still investigated properly to give definite characterizing values.

2.5 Determination of the Forces Acting on a Broach and Power Consumption in the case of Broaching Machine[6]

The velocity of broaching is obtained from the equation:

$$v = \frac{C_v}{T.S_z^y},$$

where C_v— coefficient dependent on the conditions of metal cutting;

T— life of the broach in minutes;

S_z— feed per tooth in mm;

σ_B — referred to in the Tables 2.6 and 2.7 shows the strength of the material.

Knowing the speed of the broach and the broach force, from table 2.5 we find out the horse power required for the machine.

T can be found out from the equation

$$T = \sqrt[m]{\frac{C_v}{V.S_z^y}}$$

2.6. Determination of the Forces Acting on a Shaping Tool and the Power Calculation in Shaping Machine

The usual forces acting on a shaping tool are shown in sketch in Figure 2.4. The average velocity of ram in its middle position during its stroke

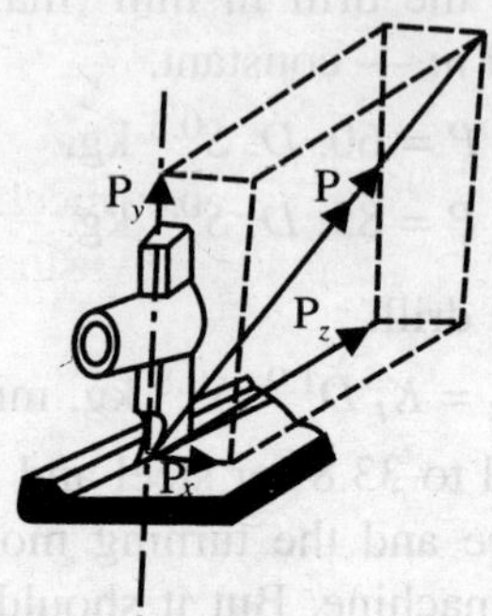

Figure 2.4 Forces in a Shaping Process.

$$V_t = \pi . n . \frac{l . h.}{l + \frac{h}{2}}$$

where n – r.p.m. (see Figures 2.5 and 2.6)

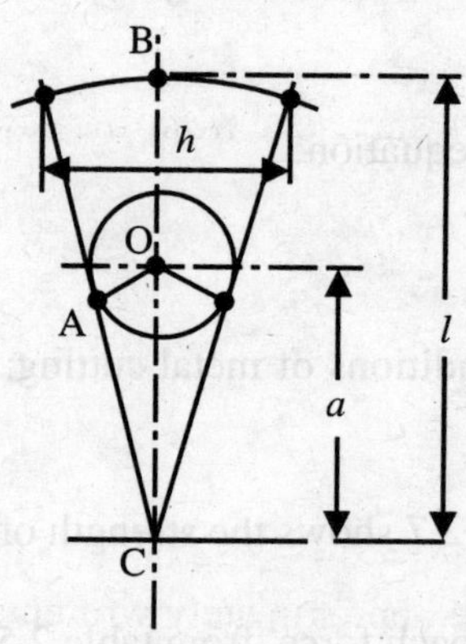

Figure 2.5

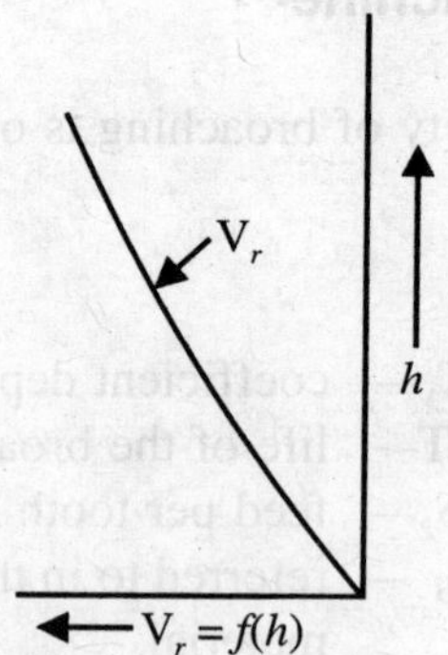

Figure 2.6

That is to say $v_r = f(h)$, a mathematical function of "h", since π and l are constants together with n

$$n_{max} = V_r \cdot \frac{1 + \frac{h_{min}}{2}}{\pi . l . h_{min}}$$

$$n_{min} = V_r \cdot \frac{l + \frac{h_{max}}{2}}{\pi . l . h_{max}}$$

for the same value of the average cutting speed V_r.

The velocity diagram for shaping machine mechanism has been shown in Figure 2.7. In this figure we have been shown the

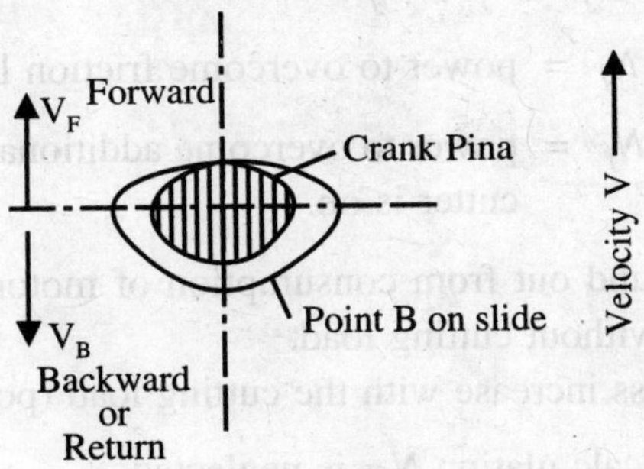

Figure 2.7

forward and backward velocity of crankpin *A* (see Figure 2.5) and of point *B* (Figure 2.5) on the slide.

Here V_F — velocity in forward or cutting stroke;

V_B — velocity in the return on backward stroke;

If the force of cutting = P_z which can be found out from the law of orthogonal cutting, then the power consumption

$$N = \frac{1}{\eta} \frac{P_z \cdot V_r}{60 \times 102} \text{ K watts;}$$

P_z can be found out empirically from the relation

$$P_z = C_P \, . \, t^x \, . \, S^y \, . \, K \text{ in } K_g, \tag{2.14}$$

where constants *x*, *y*, *K* and C_P can have the same value as in lathe tool.

For C. I. or steel the cutting force *P*, can be approximately written as

$$P_z = 190 \, . \, t \, . \, S^{0.76} \, . \, K \tag{2.15}$$

The equation 2.14 holds good for planing operation as well.

Power Capacity of Machine Tools

This is very important both to the designers and users. It is not very advisable to use lower capacity or higher capacity of motor than what is necessary.

Let N_m = power capacity of electric motor at full load;
N_e = power required for actual cutting;
N_f = power required for overcoming friction of machine elements, when in motion.

Then $$N_m = N_e + N_{f'} \tag{2.16}$$

but $$N_f = N_{f'} + N_{f''} \tag{2.17}$$

$N_{f'}$ = power to overcome friction loss without cutter.

$N_{f''}$ = power to overcome additional friction loss when cutter is on.

$N_{f'}$ — can be found out from consumption of motor when running the machine without cutting load.

$N_{f''}$ — friction loss increase with the cutting load (power).

For approximate calc ulation $N_{f''}$ is neglected

If η_{overall} = efficiency of the machine tool,

$$\eta_{\text{overall}} = \eta_b^{z1} \cdot \eta_g^{z2}$$

where η_b^{z1} = efficiency for ball bearings having Z_1 number of ball bearings

η_g^{z2} = efficiency of gear connection for Z_2 number of gears, and so on.

So, we can find the final efficiency by multiplying the efficiency in each stage. But this is only an approximate method.

$$\eta_{ov} = \frac{N_e}{N_m} = .85 \text{ to } 0.9 \text{ in low r.p.m. 40–50}$$

$$\frac{N_e}{N_m} = 0.5 \text{ when r.p.m. is high (i.e. 3000–4000 r.p.m.).}$$

In modern machine tools, it is better to find the power capacity at different speeds and from that to calculate the friction losses. N, can'also be calculated from cutting forces and speed.

To find friction losses N_f

1. This can be found out from the experimental data of similar type of machine tool. Summarizing the data we can arrive at the following formula:

$$N_{f'} = \frac{d}{10^6} (\Sigma n + 2n_z) \text{ K watts,} \tag{2.18}$$

where d = average diameter of journals of all shafts in the drive in mm;

Σn = summation of r.p.m. of all shafts including the spindle;

n_z = r.p.m. of the spindle for each case.

CHAPTER 3

Kinematics of Machine Tools

3.1. Classification and Choice of Driving System

A Machine tool usually consists of three parts, namely, (*a*) drives, (*b*) elements and (*c*) control.

The various elements of the machine tool are made and installed to make an assembled machine tool. The construction of various parts or elements can be of integral or fabricated type. Drives can be broadly classified into three categories : (1) mechanical drive (2) electrical drive and (3) hydraulic drive. The choice of a particular drive will depend upon a few important factors listed below:

(*a*) First cost (including the cost of installation), (*b*) Efficiency, (*c*) Extent of speed regulation available, (*d*) Power-to-weight ratio, (*e*) Rigidity of the system. (*f*) Maintenance and reliability, (*g*) Simplicity of control gear, (*h*) Operational environment. (*i*) Uniformity of motion. (*j*) Braking requirements, etc.

Motions required for a particular machining operation can be subdivided into categories shown below :

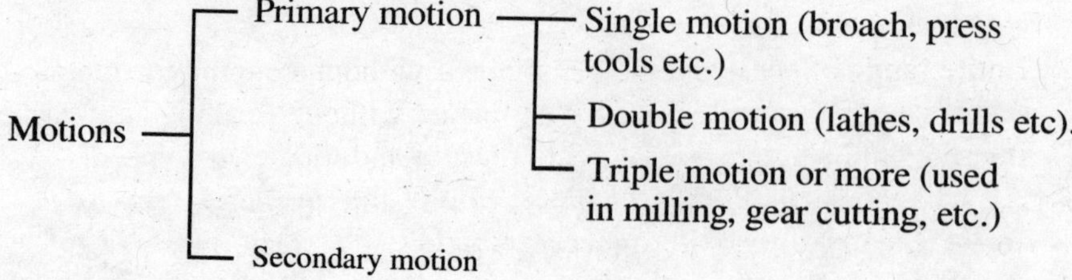

Primary motions are the essential motions required for cutting, *i.e.* to provide cutting speed and feed, etc., while the secondary motions are those basically used for other setting operations.

These motions are usually obtained from the driving units, but may be from one or more sources.

Drive to a- machine tool may be obtained from line-shaft connecting various types of machinery on the shop floor or from individual motor. In the case of line shaft, the line shaft rotates at a constant r. p. m. being run by a prime mover. In the case of drive using individual motor, the system is known as Individual drive. Each one has got its advantages and disadvantages. But in the present decade we are more and more using Individual drive. In some cases when the primary motion consists of more than one motion, each motion is given by a separate motor.

This is an absolute necessity in the case of machine tools using system of programme control.

Uniformity of motion, infinite number of speeds between maximum and minimum (stepless drive) can be easily achieved in the case of hydraulic drive or electrical drives. Mechanical drive using pressure variators can give almost stepless regulation but within a very limited range of speeds.

The use of hydraulic drive is usually limited by the straight line motion that it can execute. It is usually used in grinder, broaching machine, milling and shaping machine, etc. The mechanism includes only cylinder, pistons, pumps, filters, valves- etc

In all cases of individual motor drive, it is a practice to use A.C. motor for small or medium-sized machines. For heavy duty machines D.C. motors are used. Use of D.C. motor eases the problem of electromechanical regulation.

3.2. Basic Consideration in the Design of Drives

Before surveying any of the various arrangements of drives, it is well to consider the problem of ideal variable speed device, which should embody the following conditions.

1. There should be sufficient speed changes to divide the total range into increments between 10 and 15%.
2. Entire range of speed should be obtained without stopping the motor.
3. Any speed desired should be made without making all the intermediate changes between the present and the desired speed.
4. All changes should be obtained within the machine tool itself, without using any auxiliary countershafts.

5. Only the gears, through which the speed is actually obtained, should be engaged at one time.
6. Minimum possible number of shafts, gears and levers should be used.
7. As far as possible the control should be made centralized and to the advantage of the operators.

3.3. Determination of Variable Speed Range in Machine Tools

(i) *Mechanical regulation.* The ideal speed range is undoubtedly the one which gives an infinitely variable range between the maximum and the minimum speed selected. This approaches, as the number of speeds go on increasing, towards "stepless drive", which can be obtained by electrical, hydraulic or friction devices. But more number of speeds means more cost of the gear box, which is not likely to be commercially prospective. It is, therefore, customary to choose a set of finite speeds which enables to have the minimum speed loss, without making the number of available speeds too high. Even in this limited sphere, however, a wide choice of arrangements is available. The speeds can be arranged in A.P., G.P. or H.P. It is obviously known that when the speeds are arranged in a "geometric progression", the speed loss is minimum. Let the range of speed be denoted by R and the various speeds by $n_1, n_2, n_3 \ldots n_k, n_{k+1}, \ldots n_s$

$$R = \frac{n_{\max}}{n_{\min}}$$

where $n_{\max}$ and $n_{\min}$ denotes maximum and minimum r.p.m. respectively. The peripherical speed on a gear is given by

$$v = \frac{\pi . D . n}{1000} \, m / \min$$

where D = diameter of the gear (pitch circle diameter) in mm and n = r. p. m.

Suppose we take two diameters D_{k+1} and D_k corresponding to which the r.p.m.'s are n_k and n_{k+1} for a cutting speed v m/min.

From the relation various speed curves can be drawn as shown in Figure 3.1

Let *A, B* represent these two points. Now consider a diameter D– intermediate between D_k and D_{k+1}. Let v_k and v_{k+1} be the two cutting speeds corresponding to r.p.m's n_k and n_{k+1} for the diameter D.

$$\text{Then average loss of speed} = \frac{\delta_1 + \delta_2}{2} = \frac{v_{k+1} - v_k}{2}$$

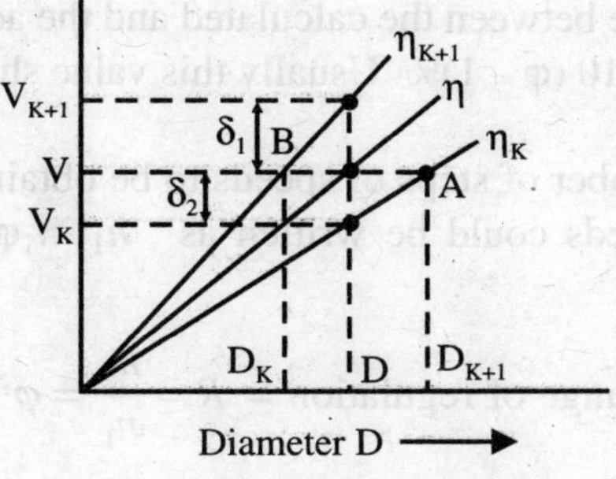

Figure 3.1.

$$\text{Proportional loss of speed} = \frac{\delta_1 + \delta_2}{2v} = \frac{v_{k+1} - v_k}{2v}$$

$$= \frac{\Delta v}{2v} = \frac{v_{k+1} - v_k}{2(v_k + \delta_2)}.$$

Now neglecting δ_2, which is small compared to v_k, we can express the proportional loss of speed in the form

$$\frac{v_{k+1} - v_k}{2v_k} = \frac{n_{k+1} - n_k}{2n_k}$$

$$\therefore \qquad \text{Proportional loss of speed} = \frac{n_{k+1}/n_k - 1}{2},$$

where $\frac{n_{k+1}}{n_k} = \varphi =$ geometrical ratio of the series.

$$\therefore \qquad \text{Proportional loss of speed} = \frac{\varphi - 1}{2}$$

To make the above loss constant and minimum, it is necessary that φ should be constant and minimum and as such the series must be in G.P. In machine tool φ lies between I and 2

$$1 < \varphi \leq 2.$$

If φ is 1, proportional speed loss = 0; if φ tends to I, then the system becomes stepless; if φ = 2, it is 50%; $\varphi = \sqrt[E]{10}$ or $\sqrt[E]{2}$ where E is a whole number.

In A.C. motor the usual speed ratio is approximately equal to 2. Standard values of φ are 1.06 – 1.12 – 1.26 – 1.41 – 1.58 – 1.78 – 2.

But in the design of machine tools only 1.26 – 1.41 – 1.58 are used. Lesser the value of φ, more complicated and expensive the gear box is bound to be. If the difference between the calculated and the actual r.p.m. is denoted by Δn, as a rule $\Delta n \leq 10\ (\varphi - 1)\%$. Usually this value should not be more than 4%.

If there are z number of steps of speeds to be obtained on a machine tool then the various speeds could be written as : n_1; $n_1\varphi^{2-1}$; $n_1\varphi^{3-1}$; $n_1\varphi^{4-1}$ ... $n_1\varphi^{z-1}$, or $n_z = n_1\varphi^{z-1}$.

$$\therefore \qquad \text{Range of regulation} = R = \frac{n_z}{n_1} = \varphi^{z-1} \qquad (3.1)$$

$$Z = 2^{E_1} . 3^{E_2} \qquad (3.2)$$

where $E_{1,2} = 0$ or any positive integer, to exclude certain speeds, viz. 5, 7, 11 etc

Normal r.p.m. of the Spindle

Table 3.1

φ = 1.06	φ = 1.12	φ = 1.26	φ = 1.41	φ = 1.58	φ = 1.78	φ = 2
1	2	3	4	5	6	7
1.00	1.00	1.00	1.00	1.00	1.00	1.00
1.06	–	–	–	–	–	–
1.12	1.12	–				
1.18						
1.25	1.25	1.25				
1 32	–	–				
1.40	1.40	–	1.40			
1.50	–	–				
1.60	1.60	1.60		1.60		
1.70	–					
1.80	1.80	–	–	–	1.80	
1.90	–					
2.00	2.00	2.00	2.00			2.00
2.12	–					
2.24	2.24					
2.36	–					
2.50	2.50	2.50	–	2.50		
2.65	–					
2.80	2.80	–	2.80			
3.00	–					
3.15	3.15	315		3.15		
3.35 '	–					
3.55	3.55					

3.75	–					
4.00	4.00	4.00	4.00	4.00		4.00
4.25	–	–	–	–	–	–
4.50	4.50					
4.75	–					
5.00	5.00	5.00				
5.30	–	–				
5.60	5/0		5.60	–	5.60	–
6.00	–	–	–	–	–	–
6.30	6.30	6.30	–	6.30		
o670	–	–				
7.10	7.10					
7.50	–					
8.00	8.00	8.00	8.00	–	–	8.00
8.50	–					
9.00	900					
9 50	–	–	–	–		

This is the basic principle of "Mechanical Regulation of speed". In many machine tools this system of mechanical regulation is combined with electrical regulation to cover a vast field of speed ranges. It gives a near approach to stepless regulation.

(ii) *Ray diagram to show the principle of electro-machanical regulation*: Say in the diagram (Figure 3.2) the thick lines show the speeds obtained by mechanical regulation and the related mechanically regulated speeds of rotation of the main spindle are denoted by n_1', n_2' ... n'_z.

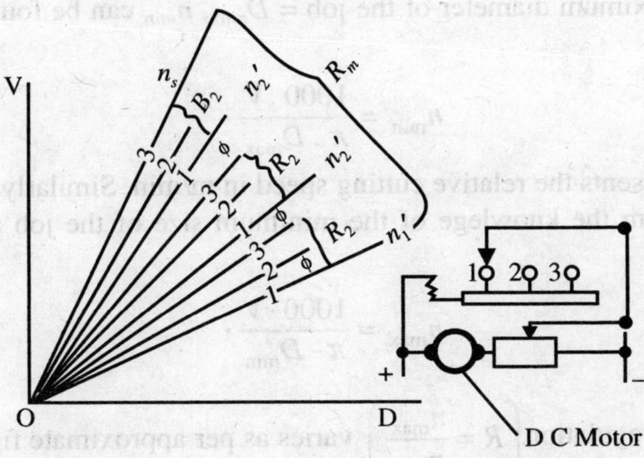

Figure 3.2 Ray Diagram Electro-Mechaical Regulation.

R_m denotes the overall regulation of the machine tool while R_e is the range of electrical regulation obtained by changing the resistances in the parallel excitation of D.C. motor.

It is evident from the above diagram that

$$n'_2 = n'_1 \,.\, \text{Re} \,.\, \Phi \;;\; n'_3 = n'_2 \,.\, \text{Re}. \,\Phi = n'_1 \,.\, R^2_e \,.\, \Phi^2$$

$$n_3 = n'_z \,.\, \text{Re} = n'_1 \,.\, \text{Re}^z. \,\Phi^{z-1} \,.$$

$$R_m = R^z_e \,.\, \Phi^{z-1} \qquad \therefore R_m. \,\Phi = (\text{Re} \,.\, \Phi)^z$$

$$Z = \frac{\log(R_m \,.\, \Phi)}{\log(Re \,.\, \Phi)} \tag{3,3}$$

Again in the ray diagram we can write $n_2 = n'_1 \,\Phi^{3-1}$ where s = total number of speeds obtained by the combined method, and

$$R_m = \Phi^{2-1} \text{ or } \log R_m = (s-1) \log \Phi,$$

and consequently,

$$S = \frac{\log R_m}{\log \Phi} + 1 \tag{3.4}$$

But

$$s = kz, \tag{3.5}$$

where k = number of contacts on the rheostat and
z = number of steps in only mechanical regulation.

Taking the case of a lathe, the minimum speed required can be assessed from the point of view of turning operation on the work having the diameter corresponding to the swing of the machine.

If the maximum diameter of the job = D_{max}, n_{min} can be found out from the formula

$$n_{\min} = \frac{1000 \,.\, V}{\pi \,.\, D_{\max}},$$

where V represents the relative cutting speed in m/min. Similarly n_{max} can be found out from the knowlege of the minimum size of the job and relative cutting speed:

$$n_{\max} = \frac{1000 \cdot V}{\pi \cdot D_{\min}},$$

The range of regulation $\left(R = \dfrac{n_{\max}}{n_{\min}}\right)$ varies as per approximate figures stated below depending upon the types of machines:

For lathe, boring and milling machine R ≈ 50-100
For drilling machine R ≈ 15-30
For shaping machines R ≈ 10
For grinding machines R ≈ 1-13
For automatic turrets R ≈ 10-30
For semi-automatics R ≈ 16-24
For planing machine R ≈ 6-10

The speed of the driving spindle should' be so chosen as to avoid too high turning moment and too drastic grear reduction.

Minimum number of teeth on a pinion should be 18-20, to avoid interference and also in order to have sufficiently uniform transmission of rotational speed. Usually in machine tools we use gears having modules 1, 1.5; 2; 2.5; 3.5; 4; 5.0; 6.0; 8.0; 10; 12; and the width of each gear in mm should vary between 6*m* to 15*m* where *m* denotes the module.

3.4. Graphical Representation of Speed and Structure Diagram

The distribution of speeds on the intermediate shafts taking into account the various considerations enumerated above can be effectively done by representing the speeds on a graph using log Φ as unit [See Figure 3.3 (*a*) and (*b*).

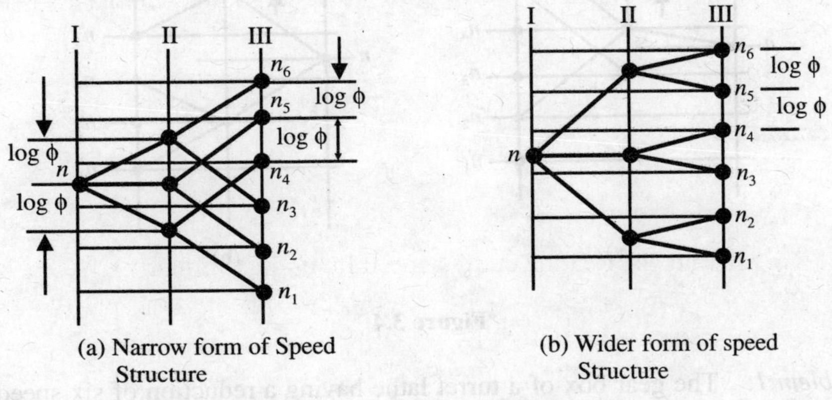

(a) Narrow form of Speed Structure

(b) Wider form of speed Structure

Figure 3.3 Graphical Representation of Speeds.

The two figures show the two different representations of the speed structure. In the first case it is narrow; while in the second case it is much more wide. Various structural representations are possible in each case of drive to a machine tool. I, II and III in the above sketches represent driver shaft, intermediate shaft and the driven shaft respectively. Though in both the cases we are having the same number of speeds and same magnitudes; in case (*a*) the intermediate shaft will have three speeds almost nearer to each other in comparison to those in the case (*b*). Similar characteristics are worth noticing in selecting the speeds that we want to obtain on each shaft of the machine tool gear box. The method is simple and decisive and therefore very much in use by the machine tool designer. Even when the number of speeds to be obtained is great; the graphical representation presents no complication.

The graphical solution is almost a necessity in order to arrive at ai definite decision regarding the speeds of the intermediate shaft.

Below are some more ray diagrams for finding the various speed rations.

The diagrams in Figure 3.4 solve the same problem as in diagrams before, only difference being the number of speeds on the intermediate shafts. The ray diagrams are very useful to a machine tool designer, since he knows how much power he may go upto and what is the lowest permissible speed on any shaft. The real diagram may take a little different shape, as is evident from the problems solved below:

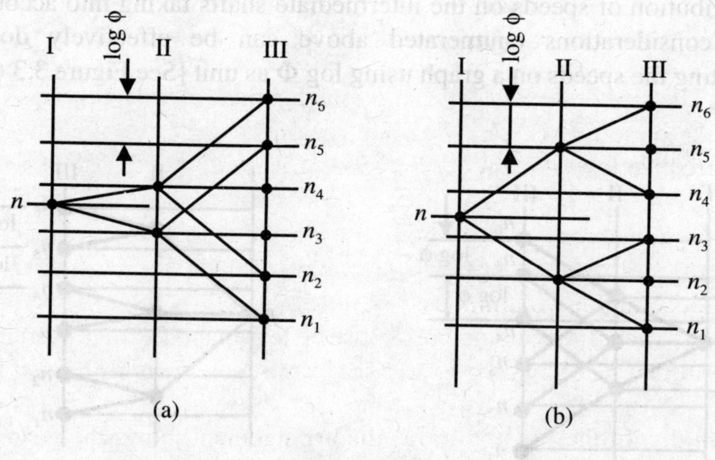

Figure 3.4

Problem 1: The gear box of a turret lathe having a reduction of six speeds from 25 to 800 r.p.m. has to be designed, the drive consisting of the A.C. motor having a speed of 1500 r.p.m. The speeds are to be arranged in geometric progression.

Two different ray dispositions are shown in the sketch by full lines and dotted lines respectively. (See Figure 3.5).

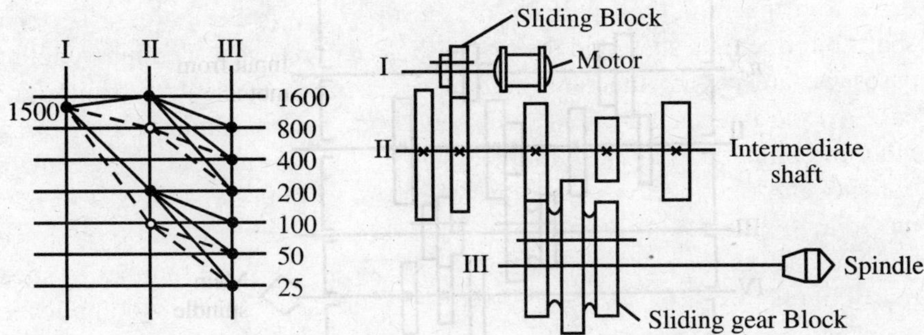

Figure 3.5

CONSTRUCTIONAL DETAILS :

(1) Put all the log values of speed on III.

(2) Put the log values of 1500 on I.

(3) Choose the block models and the type of ray diagram.

(4) Select the points on 2 and the gap in between these points.

Important points to be considered

1. The size or dimension of a particular shaft is proportional to (torque) If the torque is less, the diameter of the shaft also becomes less. To realize lower quantity of torque, it is necessary that the speed of the shaft should be high, if the power is constant.

   Therefore, it is a good proposition to have as high speed as possible on the intermediate shaft, so that the gear box does not become too voluminous.
2. The gear ratio should preferably be less than 1; though in many cases, it may be obligatory to use gear ratio more than 1 or equal to 1.

Problem 2. In the Figure 3.6 (a) the arrangement.shows the gear box of a lathe having 24 spindle speeds. Represent the speed graphically to find out the maximum and minimum reduction of speed from each shaft assuming the following relevent data:

Maximum speed to be obtained	= 2400 r.p.m.
Minimum speed	=11.5 r.p.m.
Number of speeds in the gear box	= 24
Motor has 8 K watt capacity and maximum speed of the motor	= 1455 rev/min.

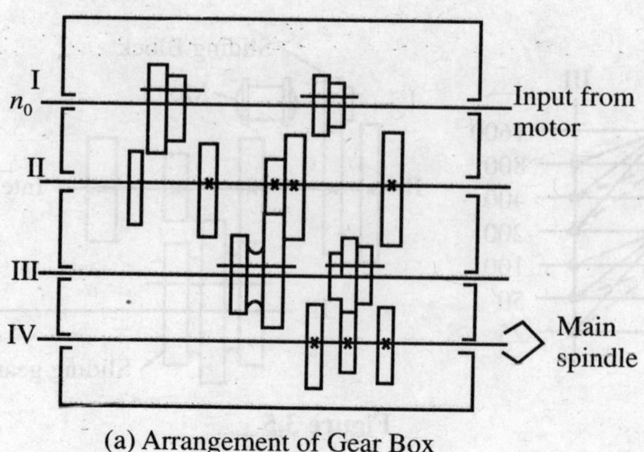

(a) Arrangement of Gear Box

Figure 3.6.

Total speeds obtained	Shaft I	–	one speed.
	Shaft II	–	4 speeds.
	Shaft III	–	4.2 – 8 speeds.
	Shaft IV	–	4.2.2 – 24 speeds.

Hints : (1) Find out the value of Φ from formula $\Phi^{23} = \dfrac{2400}{11.5}$

(2) Plot log Φ on the scale shown.

(3) Find out graphically the speeds on each shaft (see Figure 3.66*b*).

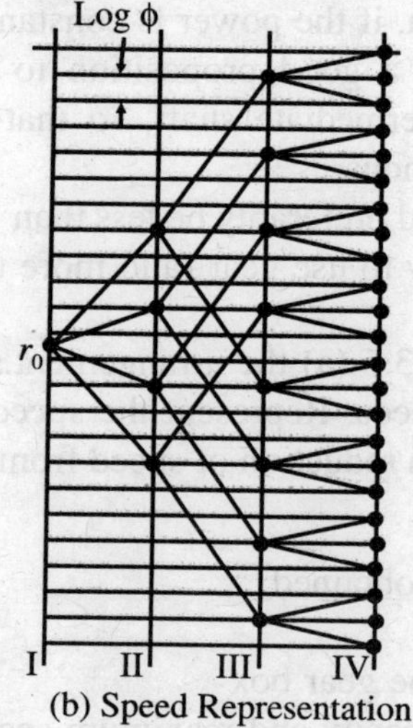

(b) Speed Representation

Figure 3.6 *b*

(4) From the arrangement of the gear box find out the gear teeth.

(5) Design the pulley size for the reduction of the motor speed to n_e.

N.B. It is not at all imperative that all the shafts will be placed in vertical planes.

The deviation diagram

As mentioned earlier the actual speeds available on the machine will be deviated from the calculated speeds by an amount Δn which can have either positive or negative value. The speeds should be so designed by choosing suitable gear trains such that

$$\sum \Delta n \approx 0$$

Such deviations while plotted give the deviation diagram as shown in Figure 3.7.

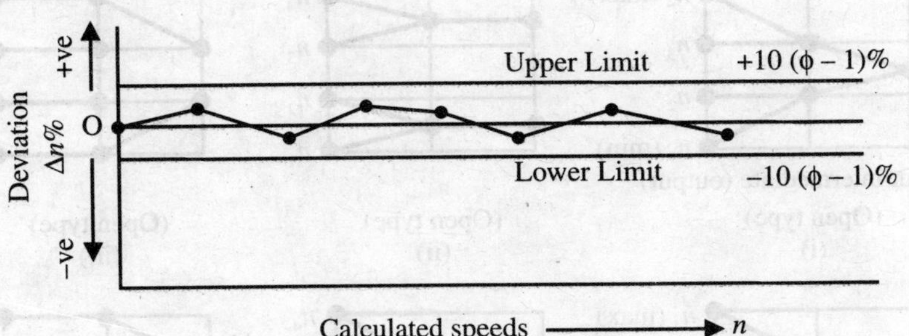

Figure 3.7 Deviation of Available Speeds from Calculated Speeds.

After all the speeds have been chosen, we can proceed with the design of gears and the layout of the gear box. But for the strength calculation of the gears we must know the tangential force coming on the gears. In the previous chapter, the author has given a number of empirical formulae for calculating turning moment due to cutting, wherefrom we can find out the tangential force on each gear.

3.5 Various Types of Structure Diagram

Structure diagram displays the pattern of connection of speeds at the input, output and intermediate stages without indicating the actual speeds of intermediate points and gear-ratios. Structure diagrams can be 'opened' or 'crossed' type. It is important that while drawing such diagram arrows must be drawn parallel when repeated for another set in the same stage to maintain same gear rafio.

Example A four-speed two-stage gear box -can have any of the following alternative structures:

$$4 = 1 \times 2 \times 2 = 2 \times 1 \times 2 = 2 \times 2 \times 1$$

Ray Diagrams

Ray diagram displays exact location of speed and then a ray dia-gram helps in calculating gear ratios. Various ray diagrams can result from a single structure diagram and these diagrams can be classified as unilateral, bilateral and skewed.

Example Let us consider 1 × 2 × 2 open type structure diagram of Figure 3.8(i).

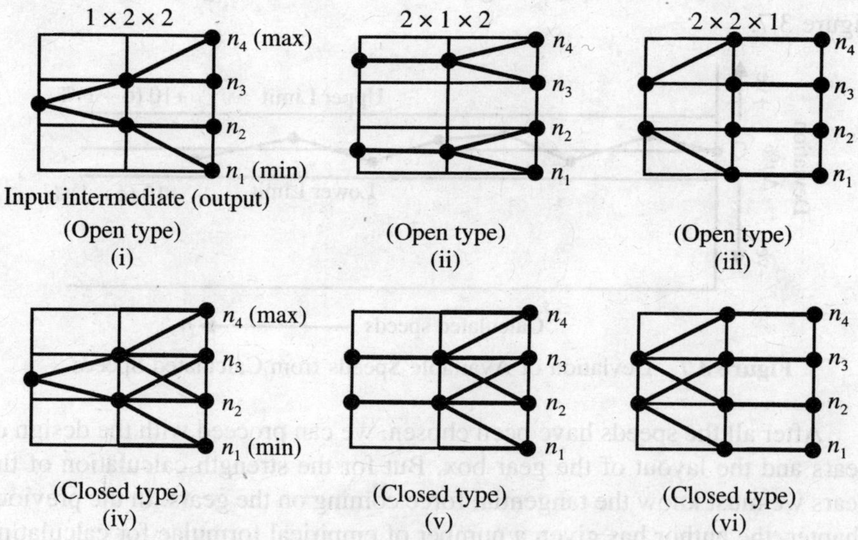

Figure 3.8 Four speeds Two stage gear box.

Numerous ray diagrams can be drawn by altering the input poir.t and intermediate points, some of these open type ray diagrams are shown below:

In this manner, many more ray-diagrams (both open and crossed type) can be drawn from a parent structure diagram. In fixing the input point it is, however, important to bear in mind that the input point should be preferably located towards higher speed to avoid large transformation ratio between the motor shaft (which runs normally at a high speed) and the input shaft. With this consideration, ray diagrams (i), (v) and (ix) are preferred ray diagrams.

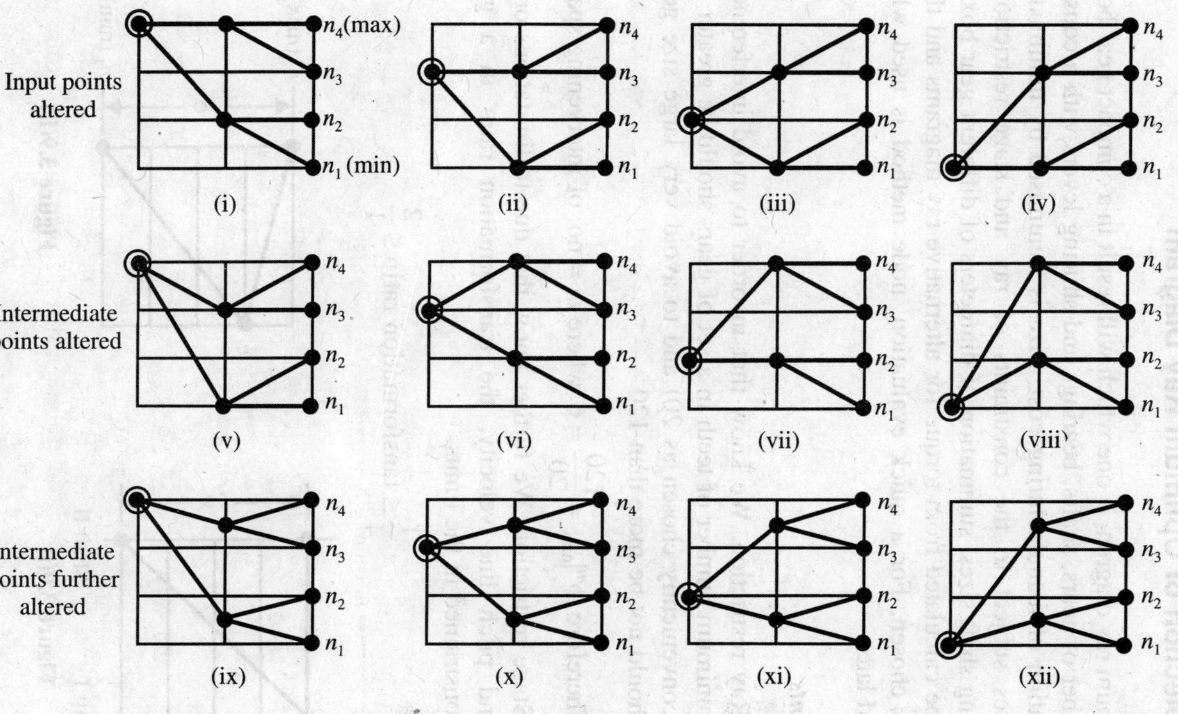

Figure 3.9 Various forms of ray diagrams altering input and intermediate points.

There are, however, other considerations to finally select the optimum ray diagram.

3.6 Selection of Optimum Ray Diagram

An optimum ray diagram is one which will result in a compact gear box with less number of gears, shafts, bearings and shifting levers with a consequent minimisation of manufacturing cost. Size is minimised by minimising the shaft sizes subject to the constraints of ray and stage restriction. For minimising shaft sizes, summation of diameters of different gear box shafts have to be calculated from torque for alternative ray diagrams and the best layout is chosen. For a quick evaluation, node method is used which is illustrated later.

Constraints

(i) Ray restriction. We know that in order to avoid interference, the minimum number of teeth in a set of gears should be greater than if (conveniently chosen as 20) and to avoid very large size gears, it should not be more than 120.

Therefore $|\phi^m|_{\max} = \dfrac{120}{20} = 6$ where m = no. of intervening spaces.

(ii) Stage restriction. We further know that due to limitations or space and pitch line velocity, the transformation ratio in a gear is constrained in the limits

$$\frac{1}{4} \leq \text{tansformation ratio} \leq \frac{2}{1}$$

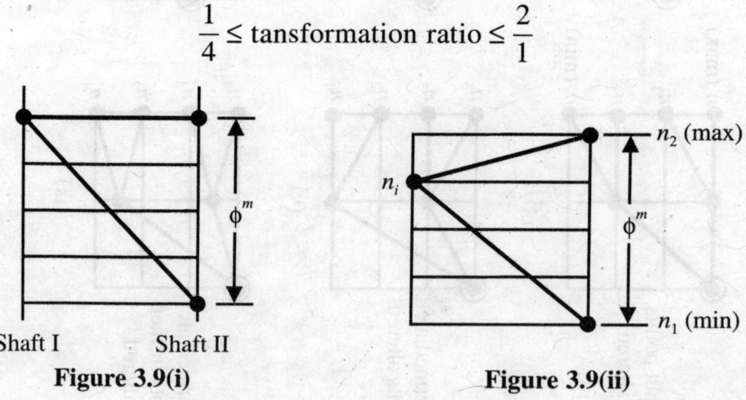

Figure 3.9(i) **Figure 3.9(ii)**

Looking into the following ray diagram.

$$\therefore \qquad \frac{n^2}{n_1} \leq 2; \text{ and } \frac{n_1}{n_1} \leq 2 \times 4 \leq 8$$

In a stage $\phi^n \leq 8$ where n is the number of intervening spaces between n_1 and n_2.

Node Method of Optimisation

Node is a point from which a ray initiates or at which a ray terminates. Nodes are numbered from maximum speed and (corresponding to minimum torque) at each shaft and comparison is carried out with nodal sum (which, in fact, represents the sum of shaft diameters) as the criterion.

Illustration: Let us compare the three ray diagrams (Figure 3.10) for a six speed gear box (1 × 2 × 3).

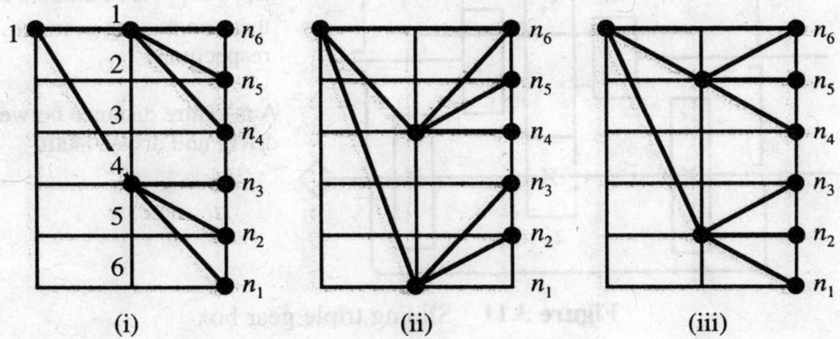

Ray Diagram	Shafts			Nodal sum	Remarks
	I	II	III	≃ Σ dia	
(i)	1	4	6	11	Best
(ii)	1	6	6	13	
(iii)	1	5	6	12	

Figure 3.10 Optimisation of Ray Diagram

Therefore, type (i) is the best layout. Depending on the value of ϕ, it has to be, however, checked whether ray restrictions and stage restrictions are obeyed or not.

3.7 Difference Between Number of Teeth of Successive Gears in a Change Gear Block

With reference to Figure 3.11, let us assume that

$$Z_1 < Z_2 < Z_3$$

m = Module of gear pairs
f_0 = Height coefficient (addendum) ratio of height of tooth to module
ξ = Coefficient of deformation of profile
ξ_e = $\xi + \xi'$ = (deformation) coefficient for the mating pairs D_e = Outside diameter of gear (suffix I or I′ denotes outside diameter of gear having number of teeth Z_1, and Z_1 respectively).

$$D_{e1} = mZ_1 + 2f_0 - 2\xi'_1 + 2A - 2A$$

$$= m[Z_1 + 2f + 2(\xi_{c1} - \xi_1) - (Z_1 + Z'_1)] + 2A$$

$$De_1 = m[Z_1 + 2(f_0 + \xi_1) - \xi_{e1} - (Z_1 + Z'_1)] + 2A$$

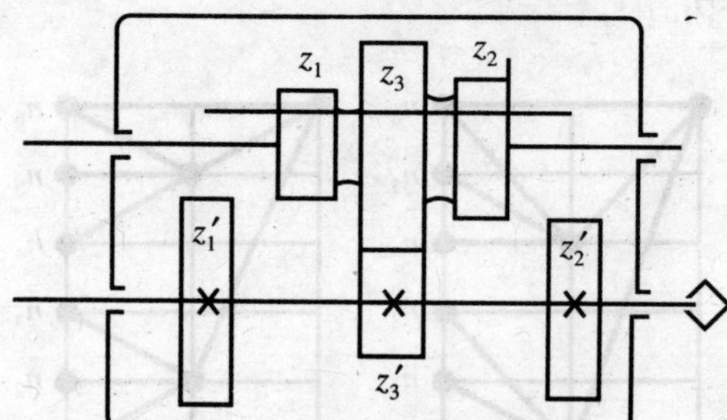

Figure 3.11 Sliding triple gear box

Consequently

$$D'_{e_1} = m[Z'_1 + 2(f_0 + \xi'_1) - 2\xi_{e1} - (Z_1 + Z_1')] + 2A \quad (3.6)$$

$$D_{e_2} = m[Z_2 + 2(f_0 + \xi_2) - 2\xi_{e2} - (Z_2 + Z_2')] + 2A$$

$$= m[2(f_0 - \xi_2') - Z_2'] + 2A \quad (3.7)$$

$$D'_{e_3} = m[Z_3' + 2(f_0 + \xi_3') - 2\xi_{e3} - (Z_3 + Z_3')] + 2A$$

$$= m[2(f_0 - \xi_3) - Z_3] + 2A \quad (3.8)$$

$$D_{e2} + D'_{e3} < 2A$$

$$m[4f_0 - 2(\xi_2' + \xi_3') - (Z_1' + Z_3)] + 4A < 2\lambda$$

$$(Z_2' + Z_3) - 4f_0 + 2(\xi_2' + \xi_3) > \frac{2A}{m} \quad (3.9)$$

Similarly

$$(Z_1' + Z_3) - 4f_0 + 2(\xi_1' + \xi_3) > \frac{2A}{m} \quad (3.10)$$

$$A = \frac{m(Z + Z')}{2} \text{ and } \xi + \xi' = 0 \quad (3.11)$$

$$Z_2' - Z_3 - 4f_0 > Z_2 + Z_2' \tag{3.12}$$

$$\therefore Z_3 - Z_2 > 4f_0 \tag{3.13}$$

Similarly

$$Z_3 - Z_2 > 4f_0 \tag{3.14}$$

f_0 is either 1 or 0.8.

$$\therefore Z_3 - Z_2 > 4 \tag{3.15}$$

$$Z_3 - Z_1 > 4 \tag{3.16}$$

Z_3 must be greater than both Z_2 and Z_1 by four number of teeth

3.8 Analysis of a Twelve-Speed Gear Box

Figure 3.12 shows the layout sketch of a gear box with usual notations for obtaining twelve speeds at the output spindle. The speed ray diagram for such a layout is given in Figure 3.12(a).

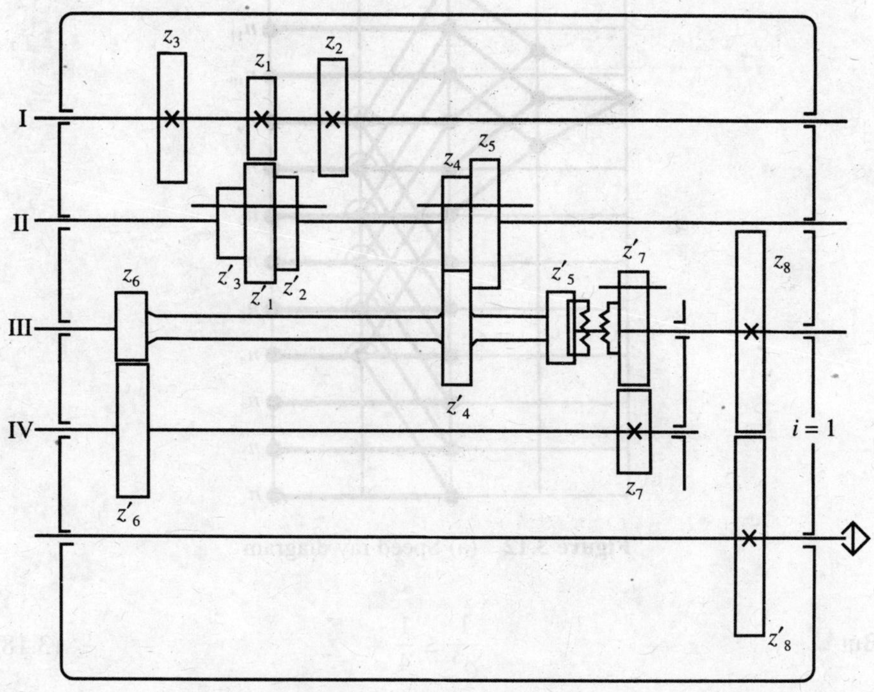

Figure 3.12 Twelve speed-gear box.

Gear speed ratios obtained, may be written as

$$i_1 = i_7 \times J$$
$$i_2 = i_8 \times J$$
$$i_3 = i_9 \times J$$
$$\vdots \qquad \vdots$$
$$\vdots \qquad \vdots$$
$$i_6 = i_{12} \times J$$

Where $$J = J = \frac{Z_6}{Z'_6} \times \frac{Z_7}{Z'_7}$$

$$\therefore \quad J = \frac{i_1}{i_7} = \frac{i_2}{i_8} = \ldots = \frac{i_6}{i_{12}} = \frac{1}{\phi^6} \tag{3.17}$$

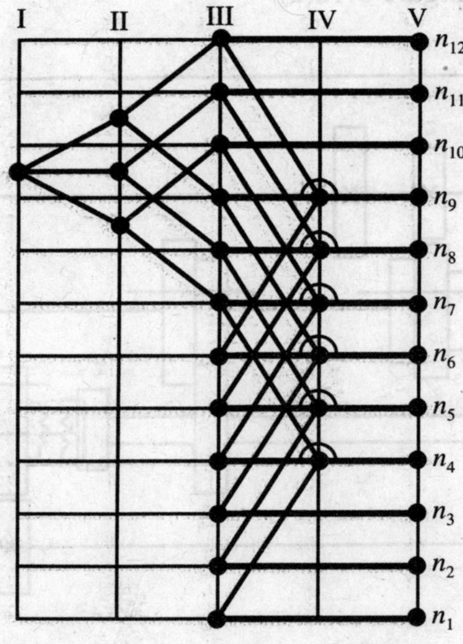

Figure 3.12 (a) Speed ray diagram

But $$\frac{1}{\phi^6} \le \frac{1}{4} \tag{3.18}$$

The Table 3.2 gives the optimum value ranges for Z_6/Z'_6 and Z_7/Z'_7 for different values of ϕ.

3.9 Standardisation of ϕ

The usual speeds obtained in a synchronous motor are: 3000 r.p.m., 1500 r.p.m., 750 r.p.m. etc. or 2000 r.p.m., 1000 r.p.m., 500 r.p.m. etc. These speeds are obtained by the use of polechanging devices.

If $\phi = 1.26$ $\quad \frac{Z_6}{Z_6'} = \frac{1}{1} \quad \frac{1}{\phi^1} \quad \frac{1}{\phi^2} \quad \frac{1}{\phi^3} \quad \frac{1}{\phi^4} \quad \frac{1}{\phi^5} \quad \frac{1}{\phi^6}$

$$\left|\frac{1}{\phi^6}\right|_{max} = \frac{1}{4} \qquad \frac{Z_7}{Z_7'} = \frac{1}{\phi^6} \quad \frac{1}{\phi^5} \quad \frac{1}{\phi^4} \quad \frac{1}{\phi^3} \quad \frac{1}{\phi^2} \quad \frac{1}{\phi^1} \quad \frac{1}{1}$$

If $\phi = 1.41$ $\quad \frac{Z_6}{Z_6'} = \frac{1}{\phi^2} \quad \frac{1}{\phi^3} \quad \frac{1}{\phi^4}$

$$\left|\frac{1}{\phi^4}\right|_{max} = \frac{1}{4} \qquad \frac{Z_7}{Z_7'} = \frac{1}{\phi^2} \quad \frac{1}{\phi^1} \quad \frac{1}{\phi^2}$$

If $\phi = 1.58$ $\quad \frac{Z_6}{Z_6'} = \frac{1}{\phi^2} \quad \frac{1}{\phi^3}$

$$\left|\frac{1}{\phi^3}\right|_{max} = \frac{1}{4} \qquad \frac{Z_7}{Z_7'} = \frac{1}{\phi^1} \quad \frac{1}{1}.$$

As given in equation (3.2), $Z = 2^{E_1} \cdot 3^{E_2}$ where $E_1, E_2 \simeq 0$ or any positive integer to exclude certain speeds, viz, 5, 7, II etc.

But $$\phi = \sqrt[e_1]{2} = \sqrt[e_2]{10} \tag{3.19}$$

From this $$e_1 = 3e' \text{ and } e_2 = 10_e' \tag{3.20}$$

Table 3.3

e_2	40	20	10	5	4.0	3.3
e'	4	2	1	0.5	0.4	0.33
e_1	12	6	3	1.5	1.2	1

Various standards in regard to value of ϕ have been compared in Table 3.4. From this table find that Renaid series 40 and Renard series 20/5 have been omitted from the ISA standard.

3.10 Compound Ray Diagram

Sometimes it is necessary to design gear box of a machine tool to cater for large number of speeds, say 14 or 16, starting from a very low value of n (minimum) to a very high value of n (maximum). To prevent the gear box becoming too bulky, in such a case, it is necessary to design a compromise gear box. To explain further it is worth mentioning that such gear box will act like two separate gear boxes. Suppose we require 14 speeds, out of which eight are in the low spectrum, while other six are in the area of high speeds. But for simplicity in construction, we want some value of ϕ to hold good for both these two regions—high speeds and low speeds.

Table 3.4

ϕ				R
Progression ratio	ISA Standard	$\sqrt[e_1]{2}$	$\sqrt[e_3]{10}$	Renard Series
1.06	–	$\sqrt[12]{2}$	$\sqrt[40]{12}$	R 40
1.12	1.12	$\sqrt[6]{2}$	$\sqrt[20]{10}$	R 20
1.26	1.25	$\sqrt[3]{2}$	$\sqrt[10]{10}$	$R\,\frac{20}{2}$
1.41	1.40	$\sqrt[2]{2}$	$\sqrt[6]{10}$	$R\,\frac{20}{3}$
1.58	1.60	$\sqrt[1.5]{2}$	$\sqrt[5]{10}$	$R\,\frac{20}{4}$
1.78	–	$\sqrt[1.2]{2}$	$\sqrt[4]{10}$	$R\,\frac{20}{5}$
2.00	2	$\sqrt[1]{2}$	$\sqrt[3.3]{10}$	$R\,\frac{20}{6}$

A ray diagram in such case is called a Compound Ray Diagram as shown in Figure 3.13. Normally we work in the low range using input shaft I, intermediate shafts II and III and final output spindle IV, when the lower ray diagram is followed. With a little modification with one additional gear pair, belting connection between shaft II and III, and a clutch connecting II and III, we can use ray diagram to obtain upper speed spectrum. To work on the high speed range, we need to declutch first and then get the drive from the shaft II transmitted to shaft III by belting.

Now-a-days probabilistic concept of designing speed range required on a machine tool has considerably simplified the design of the gear box. Selection of the same should invariably be made based on the demands of specific categories of products to be manufactured on the machine. Where the industrial work requires to have speeds varying between 30 r.p.m. and 2500 r.p.m. with as many number of steps as possible, it may be worthwhile to have high speeds and low speeds structures combined in the form of a compound ray diagram.

3.11 Feed Gear Boxes

The drive for the feed mechanism in a machine can be either a drive directly taken from a separate motor through necessary transmitting systems or mechanism or it may be based on the main spindle.

In such case its design is limited by the maximum speed and maximum torque of the spindle. There are typical devices for engaging the feed mechanism—being normally in the form of clutches, sliding gears, etc.

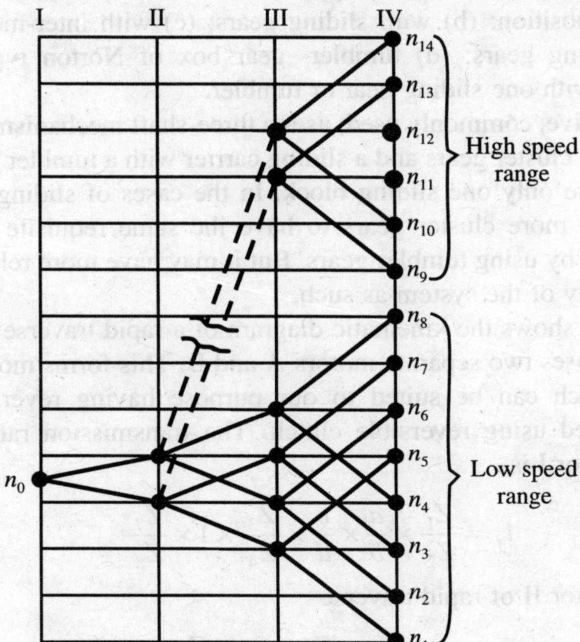

Figure 3.13 Compound speed diagram.

In case of feed drive taken from the spindle, the total transmission ratio of the feed mechanism is determined by

$$S = t\, i_f \tag{3.21}$$

where S = feed rate, mm/revolution

t = pitch of the traversing elements, mm

i_f = transmission ratio from spindle to the traversing element.

In case of separate feed drive

$$S' = n \cdot if\, t \tag{3.22}$$

where n = r.p.m. of the motor and S' = feed rate in mm/min

Normally the design of feed gear box becomes compact in case of spindle drive. Torques developed by various shafts in the feed gear box can be calculated from the equation

$$\frac{P_f \times 1}{2\pi} = \frac{M_f}{i_f} \cdot \eta f \tag{3.23}$$

where P_f – feeding force, t – pitch of traverse, M_f – torque on feed shaft, i_f – transmission ratio and n_f – efficiency of the transmission system.

Feed gear boxes are of various types, classified as under: (a) with change gear on fixed position; (b) with sliding gears; (c) with inter-mcshing gear cones and sliding gears; (d) tumbler- gear box of Norton type: and (c) Meander type with one sliding gear or tumbler.

Meander drive, commonly used, uses a three-shaft mechanism consisting identical double cluster gears and a sliding carrier with a tumbler gear, or the carrier may have only one sliding block. In the cases of sliding block, the system requires more cluster gears to have ihc same requisite number of speeds obtained by using tumbler gears. But it may have more reliability and increased rigidity of the system as such.

Figure 3.14 shows the kinematic diagram of a rapid traverse drive.

This drive uses two separate motors A and B. This forms more or less a differential which can be suited to our purpose having reverse traverse whenever needed using reversible clutch. The transmission ratio to shaft when using motor I is

$$i_f = \frac{Z_1}{Z_2} \times \frac{a}{b} \times \frac{c}{d} \times \frac{Z_3}{Z_4} \times 1 \times \frac{Z_7}{Z_8} \tag{3.24}$$

When using motor II of rapid traverse

$$i_f = \frac{Z_9}{Z_{10}} \times \frac{Z_{11}}{Z_{12}} \times 2 \times \frac{Z_7}{Z_8} \tag{3.25}$$

In this case Z_5 is made fixed and planet gears go rolling round this, the transmission ratio of the differentials being two.

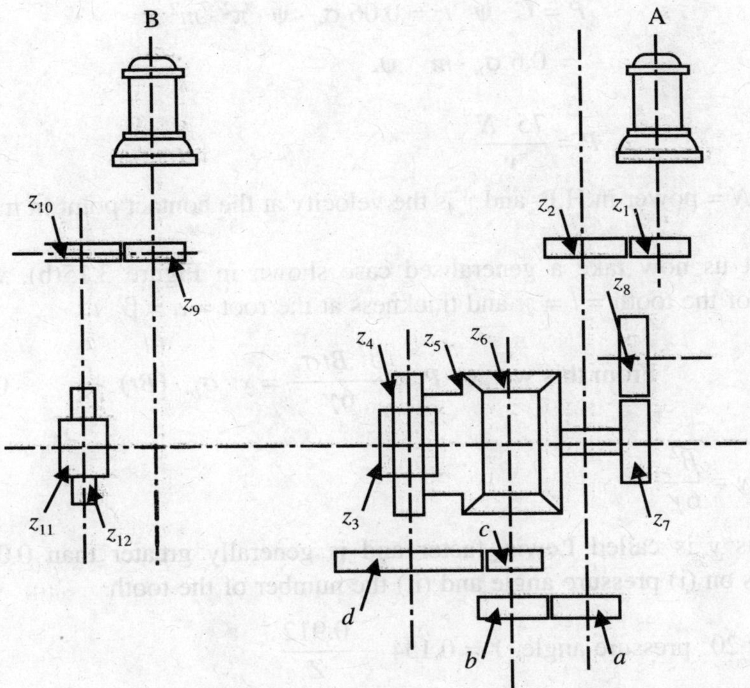

Figure 3.14 Rapid traverse drive

3.12 Strength Calculation of Gears

Calculation for the strength of the spur gear in bending is based on the consideration of the force acting on the tooth. Assuming the configuration shown in Figure 3.15(a), P is the tangential force acting at the free end of the tooth, which is considered as a cantilever, causing a bending moment at the spot.

$$\text{Bending moment} = P \times 0.7t = \frac{0.5t^2 . B}{6} \cdot \sigma_b$$

From this we get $P = \dfrac{\sigma_b \cdot 0.25}{6 \times 0.7} \cdot (B \cdot t)$

Taking permissible value of σ_b equal to ultimate strength σ_u, we got,

$$P \simeq 0.06\ \sigma_u\ B \cdot t \tag{3.26}$$

or in other words $P = CBt$, where $C = 0.06\sigma_u$ = constant

Very often we consider $B = \psi t$, where ψ lies between the values 3 and 8. Again $t = \pi \cdot m$, where $m \simeq 0.3 - 75$

$$P = C \cdot \psi \cdot t^2 = 0.06\ \sigma_u \cdot \psi \cdot \pi^2 \cdot m^2$$

$$= 0.6\ \sigma_u \cdot m^2 \cdot \psi.$$

$$P = \frac{75 \cdot N}{v}$$

where N = power in H.P. and v is the velocity at the contact point in metres/sec.

Let us now take a generalised case shown in Figure 3.15(b), where height of the tooth = $l = \gamma t$ and thickness at the root = $a = \beta \cdot t$.

$$\text{From this we get } P = \frac{\beta^2 . Bt\sigma_b}{6\gamma} = y \cdot \sigma_b \cdot (Bt) \qquad (3.27)$$

where $y = \dfrac{\beta^2}{6\gamma}$.

This y is called Lewis' factor and is generally greater than 0.06. It depends on (i) pressure angle and (ii) the number of the tooth.

For 20° pressure angle, $Y = 0.154 - \dfrac{0.912}{Z}$

and for 14-15° pressure angle, $Y = 0.124 - \dfrac{0.684}{Z}$

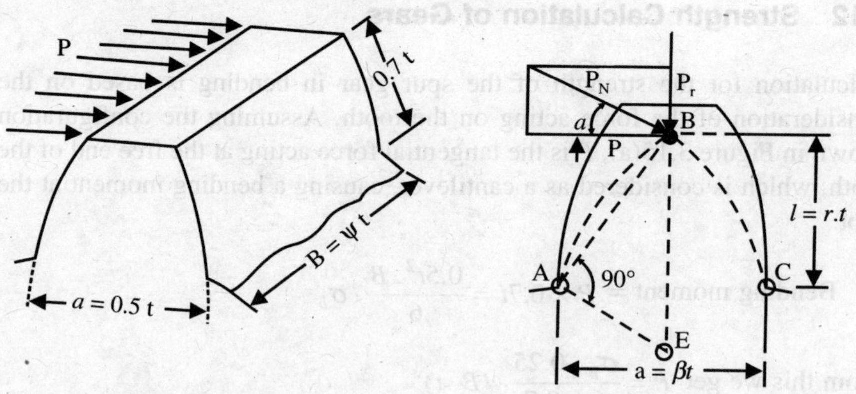

Figure 3.15(a) **Figure 3.15(b)**

Since the teeth function under contact phenomenon, it is possible to design gear tooth based on maximum contact shear stress (τ_{con}). Maximum value of τ_{con} is denoted as

$$|\tau_{\text{con}}| = 0.145\sqrt{\frac{P_n \cdot E_{eq}}{B \cdot \rho_{eq}}} \tag{3.28}$$

$$P_n = \frac{P}{\cos\alpha} \text{ and } \frac{1}{\rho_{eq}} = \frac{2(Z_2 \pm Z_1)}{m \cdot Z_1 \cdot Z_2 \sin\alpha} \tag{3.29}$$

where α = pressure angle (14° or 20°), Z_2, Z_1 = number of teeth in mating gears and m = module in mm. The ± sign indicates exttrnal or internal gearing respectively.

$$P = \frac{75N \cdot 60 \times 100}{\pi \cdot d_1 \cdot n_1} = \frac{75N \cdot 60 \times 100}{\pi \cdot m \cdot Z_1 \cdot n_1},$$

when d_1 is in mm and N is power in H.P.

Equivalent Modulus $E_{eq} = \dfrac{2E_1E_2}{E_1 + E_2} \simeq E$, when both the mating gears are of same material. Here E_1, E_2 represent Youngs modulus for gear and pinion respectively. $E_{eq} = E = 2.1 \times 10^6$ kg/cm²

$$|\tau_{con}|_{\max} = 0.45\sqrt{\frac{75N \times 6000 \times 2.1 \times 10^6 \times 2(Z_2 + Z_1)^2}{\pi n_1 \sin\alpha\cos\alpha\, Bm^2\, Z_1^2 Z_2^2 \cdot 2}} \tag{3.30}$$

But $\qquad 2\sin\alpha\cos\alpha = \sin 2\alpha$

$$\text{Centre distance} = A = \frac{d_2 + d}{2} = \frac{m(Z_2 + Z_1)Z_1}{2Z_1}$$

$$= \frac{mZ_1}{2}\left(\frac{Z_2}{Z_1} + 1\right)$$

$$= \frac{mZ_1}{2}(i + 1) \tag{3.31}$$

$$\text{For internal gearing } A = \frac{mZ_1}{2}(i - 1) \tag{3.32}$$

Thus

$$\frac{m^2 Z_1^2}{4} = \frac{A^2}{(i+1)^2}$$

$$\therefore \quad |\tau_{\text{con}}|_{\max} = 0.145\sqrt{\frac{75.6000\,(2.1 \times 10^6)\, N(i \pm 1)^2\,(Z_2 \pm Z_1)}{\pi \sin 2\alpha\, n_1\, A^2 \cdot B \cdot Z_2}} \tag{3.33}$$

The above equation can be further simplified as under

$$|\tau_{con}|_{max} = 0.145\sqrt{\frac{75\times 6000\times 2.1\times 10^6}{\pi \sin 40°}}\cdot\sqrt{\frac{N(i\pm 1)^2}{n_1 B\left(\frac{Z_2}{Z_1}\right)}\left(\frac{Z_2}{Z_1}\pm\frac{Z_1}{Z_1}\right)}$$

$$= \left(\frac{100,000}{A}\right)\sqrt{\frac{N(i\pm 1)^3}{n_1 B\cdot i}} \tag{3.34}$$

As such $n_1 = i\cdot n_2$ where n_2 is number of revolution of the gear, then

$$|\tau_{con}|_{max} = \frac{100,000}{Ai}\sqrt{\frac{N}{n_2}\frac{(i\pm 1)^3}{B}} \tag{3.35}$$

Assuming permissible value of $|\tau_{con}|_{max}$, and taking into account the known technical data N, n_2 and i, the values of A and B can be determined. In fact the value of B obtained from Lewis' equation may be utilised to fix up suitable centre distance between the gears.

3.13 Rigidity of Grinding Wheel-workpiece System in an Internal Grinder[87]

The degree of vibratory motion or vibration marks in an internal grinding operation depends, in the first instance, upon the grinding wheel-workpiece system. Hence the rigidity plays a predominant role in all technological aspects of precision engineering. Here is presented a simplified analysis of the dynamic behaviour of the system in an undamped condition where excitation is inevitably derived from the grinding wheel imbalance. Dynamic compliance, the inverse of dynamic rigidity, has been the criterion for evaluating the dynamic behaviour. The static rigidities of the workpiece, the grinding wheel spindle and the contact rigidity have been experimentally determined by applying a known magnitude of static force and observing corresponding deflection with the help of micron dial gauges. The waviness of the ground surface was magnified and traced for the analysis also with the notations as shown below:

Notations	*Quality*	*Unit*
A	Cross sectional area	cm^2
E	Young's modulus of elasticity for steel	kg/cm^2
g	Acceleration due to gravity	cm/sec^2
I	Moment of inertia	cm^4

K	Static rigidity	kg/micron
L	Length	cm
P	Static force	kg
q	Intensity of pressure on contact surface	kg/cm^2
t	Time	sec
U	Average unbalanced force	gm-mm
x_{max}	Maximum peak to peak waviness	micron
δ	Deflection	micron
ω_n	Natural frequency	rad/sec
π	Density of steel	gm/cc

Subscripts

b	bearing (front support)
c	contact of grinding wheel and workpiece
t	grinding wheel and spindle combined
f	workpiece
s	grinding wheel spindle alone
w	grinding wheel alone

Let us assume the grinding wheel and the workpiece as two equivalent isolated masses at the end of a massless elastic cantilever beam vibrating due to their own excitation. The equivalent masses at the end of cantilever beams have been separately worked out for analysis. When the grinding wheel and the workpiece are coupled or brought in contact, a relative vibratory radial motion is developed between them which causes waviness on the ground surface. The amplitude of the vibration marks or the waviness depends mainly on (i) the rigidity of the grinding wheel-workpiece system and (ii) the magnitude of the grinding wheel imbalance at a particular dressing and grinding condition.

The grinding wheel-workpiece system is dynamically represented, in the present work, by a very simplified mathematical model for the purpose of analysis. The dynamically represented system has static rigidities of the grinding wheel spindle and the workpiece, the contact rigidity, the equivalent masses of the grinding wheel and the work-piece as the variables in an undamped condition where excitation is mainly due to grinding wheel imbalance. Dynamic rigidity of the system has been evaluated by introducing differential equations.

The static rigidities of the grinding wheel spindle, the workpiece and the contact rigidity are determined experimentally by applying known magnitude of static forces with the help of a specially designed loading device, and measuring corresponding deflections by micron dial gauge.

A large workpiece diameter and a very small grinding wheel were purposely selected in this particular experiment to observe pronounced waviness on the inner ground surface of the workpiece. The specimen workpiece was thoroughly polished before the experiment in order to eliminate the initial waviness present in the workpiece. The grinding operation is done dry with an axial feed and small radial depth of cut. Magnified tracings of the radial displacements or the vibration marks at different axial positions of the workpiece are randomly selected to measure the maximum peak to peak waviness. With the help of the derived expression for the average dynamic compliance and the maximum value of peak to peak waviness, a statistical assessment of the grinding wheel imbalance present during grinding is made.

3.13.1 *Mathematical model*

The grinding wheel workpiece system is dynamically represented by a very simplified model as shown in Figure 3.16. Neglecting the low operating frequency of the workpiece, the equation of motion of the masses due to F0sin <o0t excitation is of the form:

$$\left.\begin{aligned} m_1\ddot{x}_1 + k_t x_1 + k_c(x_1 - x_2) &= F_0 \sin \omega_0 t \\ mj\ddot{x}_2 + k_j x_2 + k_c(x_2 - x_1) &= 0 \end{aligned}\right\} \quad (3.36)$$

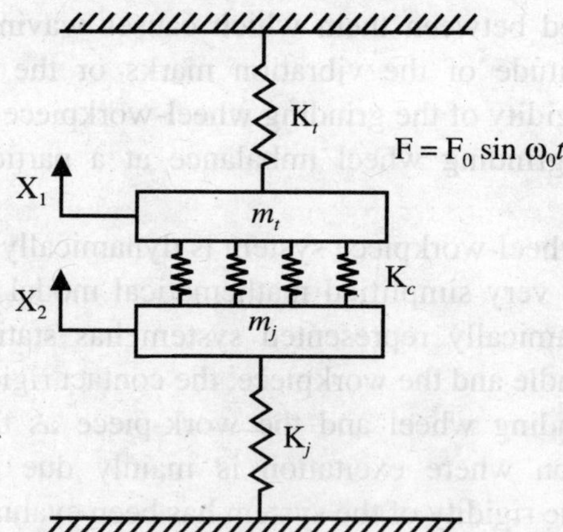

Figure 3.16 Vibration model of the system (K_t = 0.155 kg/micron, K_c = 1.4 kg/micron m_t = 3.62 × 10^{-5} kg-sec^2/cm, m_j = 1.35 × 10^{-6} kg-sec^2/cm ω_n = 2000 rad/sec)

Grinding wheel and the spindle may be thought of as a cantilever beam carrying a single isolated load at the end. Considering the first mode of vibration, the natural frequency of the spindle is given by the relation:

$$\frac{1}{(\omega_{ns})^2} = \frac{m_s L_s^4}{E_s I_s (1.875)^4}$$

where $m_s = \rho_s A_s / g$

Next, considering the spindle massless and having a load (W_w) at the end equal to the weight of the grinding wheel, the natural frequency is:

$$\frac{1}{(\omega_{nw})^2} = \frac{W_w}{g K_t}$$

where $\frac{1}{K_t} = \frac{1}{K_s} + \frac{1}{K_b}$

By Dunkerley's approximation the natural frequency of the grinding wheel and the spindle combined is:

$$\frac{1}{\omega_{nt}^2} = \frac{1}{\omega_{ns}^2} + \frac{1}{\omega_{nw}^2} = \frac{m_t}{K_t}$$

Or $$\frac{m_s \cdot L_s^4}{E_s \cdot I_s (1.875)^4} + \frac{W_w}{g K_t} = \frac{m_t}{K_t}$$

Therefore

$$M_1 = \frac{m_s \cdot K_t \cdot L_s^4}{E_s \cdot I_s (1.875)^4} + \frac{W_w}{g} \tag{3.37}$$

Equivalent mass, m_t, is the isolated mass at the end of a massless cantilever beam.

Similarly, considering the workpiece as a uniform cantilever fixed at one end

$$\frac{1}{\omega_{nJ}^2} = \frac{m L_J^4}{E_J I_J (1.875)^4} = \frac{m_J}{K_J}$$

where $m = \frac{P_J A_J}{g}$

Therefore, equivalent mass of the workpiece

$$m_J = \frac{m \cdot K_J L_J^4}{E_J I_J (1.875)^4} \tag{3.38}$$

With reference to Eqn. (10.20), let $x_1 = 0 = \dot{x}_1$ at $t = 0$ and $x_2 = 0 = \dot{x}_2$ at $t = 0$.

Applying Laplace transformation to Eqn. (10.20) and solving them simultaneously,

$$x_1(S) = \frac{F_0\omega_0^2\{m_J S^2 + (K_J + K_c)\}}{(S^2 + \omega_0^2)[\{m_t S^2 + (K_t + K_c)\}\{m_J S^2 + (K_J + K_c)\} - K_c^2]}$$

$$x_2(S) = \frac{F_0 K_c \omega_0}{(S^2 + \omega_0^2)[\{m_t S^2 + (K_t + K_c)\}\{m_J S^2 + (K_J + K_c)\} - K_c^2]}$$

The relative displacement of the system per kg of the exciting force is given by:

$$\frac{x_1(S) - x_2(S)}{F_0} = \frac{-\omega_0\,(m_1 S^2 + K_J)}{(S^2 + \omega_0^2)[\{m_t S^2 + (K_t + K_c)\}\{m_J S^2 + (K_J + K_c)\} - K_c^2]} \quad (3.39)$$

The denominator, when facton&ca, gives the values of two natural frequencies of the system, viz, ω_{n1}, and ω_{n2} apart from the operating frequency, ω_0. In order to study the complete solution of the relative displacement function, we may rewrite Eqn. (3.39) as

$$\frac{x_1(S) - x_2(S)}{F_0} = \frac{\omega_0}{m_t}\,\frac{(S^2 + \omega_{n1})}{(S^2 + \omega_0^2)\,(S^2 + \omega_{n1}^2)\,(S^2 + \omega_{n2}^2)} \quad (3.40)$$

Inverse Laplace transformation (for $t > 0$) after partial fraction gives a general solution of the relative displacement of the system per kg exciting force of dynamic compliance.

$$\frac{x_1 - x_2}{F_0} = a_0 \sin \omega_0 t + a_1 \sin \omega_{n1} t + a_2 \sin \omega_{n2} t \quad (3.41)$$

where $$a_0 = \frac{(\omega_{nJ}^2 - \omega_0^2)}{m_t(\omega_{n1}^2 - \omega_0^2)\,(\omega_{n2}^2 - \omega_0^2)}$$

$$a_1 = \frac{-\omega_0(\omega_{n1}^2 - \omega_{nf}^2)}{\omega_{n1}\, m_t(\omega_{n1}^2 - \omega_0^2)\,(\omega_{n1}^2 - \omega_{n2}^2)}$$

$$a_2 = \frac{-\omega_0(\omega_{nf}^2 - \omega_{n2}^2)}{\omega_{n1}\, m_t(\omega_{n2}^2 - \omega_0^2)\,(\omega_{n1}^2 - \omega_{n2}^2)}$$

3.13.2 *Static Rigidities of the Elements*

The static rigidities of the grinding wheel spindle, workpiece, and the contact rigidity were determined by applying static force with the help of a spring loaded, jack type of loading device and then measuring the deflection by micron dial gauge as shown in Figure 3.17 (a, b, c). The loading device had detachable brackets to suit the surfaces where the applications of static force was envisaged. The static force was applied by the compressions of two identical and calibrated springs. Proper care was taken in centering the spindle, workpiece and to align the axis of loading device and the dial gauge.

Contact Rigidity

The grinding wheel and the workpiece were brought in contact and diametrically opposite static forces were applied. The dial gauge indi-cated the bulging out of workpiece at the contact point only. As the bending force was' neutralised there was no bending deflection as shown in Figure 10.10. Load versus deflection graphs were plotted for different axial positions of contact and hyperbolic nature of the curves was observed which can be given by the relation.

$$P = C\delta^{1/m} \tag{3.42}$$

In order to determine the coefficient c and the index $1/m$, the curves were replotted on a log-log graph and the straight line nature was verified. The slope and they-axis intercepts of these lines give the values of $1/m$ and c respectively.

The value of c and m were found in the ranges 0.89-1.1£ and 0.884-0.837 respectively. On the basis of the computed values of c and m, a general formula $P = \delta^{1/0.85}$ may be given, without any appreciable error, for the specimen taken in this particcular case C›nsequently, the contact rigidity may be expressed as

$$K_c = P^{0.15} \tag{3.43}$$

Variation of contact rigidity with applied force as shown in Figure 10.11 increases with increase in the applied force. This is, probably, due to the fact that the grinding wheel does noi have a smooth surface, and consequently there are contacts of a number of grains which cause higher pressure intensity-and thereby more deflections. Conversely, at higher static force, there is more surface contact and less pressure intensity causing less deflection thus higher rigidity.

Figure 3.17(a) Measurement of spindle and support deflections with applied load.

Figure 3.17(b) Measurement of contact point deflection with radial load.

Figure 3.17(c) Measurement of workpiece deflection with applied load.

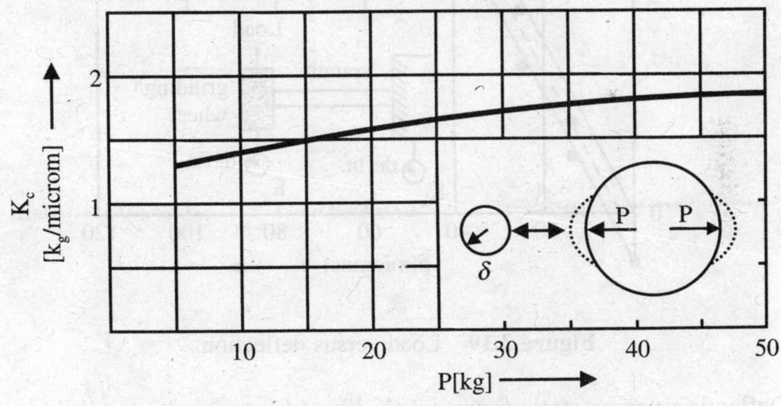

Figure 3.18 Contact rigidity with applied load.

Spindle Rigidity

Figure 3.10(a) shows micron dial gauge mounted at the front support of the spindle and deflections from the front support versus the applied force were plotted as shown by K_b (Figure 3.12). Up to 7 kg of the applied force the nature of the graph was a straight line passing through the origin and afterwards relatively more deflections were observed as the force increased. This indicates that the support was tending to be less and less rigid as the force increased. This is one of the reasons for the internal grinders being subjected to smaller grinding force. The deflections observed in the front bearing support led to the conclusion that the rigidity of the spindle bearing must also be considered in conjunction with the rigidity of the spindle.

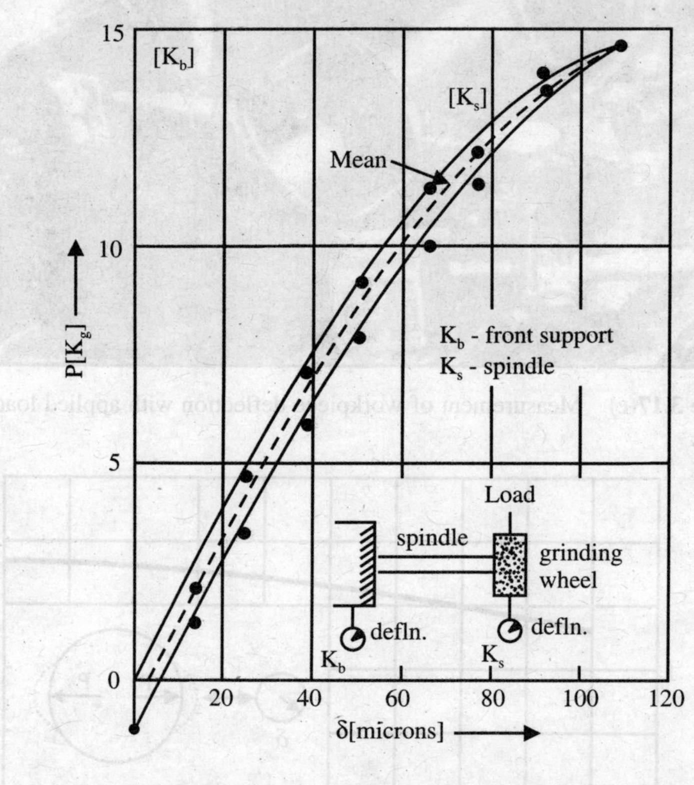

Figure 3.19 Load versus deflection.

Deflection versus static force graph, like a hysterisis loop, while loading and unloading the spindle was plotted as shown by K_s (Figure 3.12). Hence, a mean dotted curve was plotted to assess the rigidity of the spindle. This mean curve was a straight line passing through the origin only up to 10 kg of applied force and the rigidity of the spindle started dropping, on further application of force, which clearly indicates that very low grinding force is permissible in internal grinding.

Workpiece Rigidity

Static force was applied at different axial positions as shown in Figure 3.10(c) and straight line nature of graphs passing through the origin was found. The slope of these straight lines gave the static rigidities. Variation of rigidities along the axial position was plotted as shown in Figure 3.13 and Table 3.1 (columns 2 and 3). It is known that the rigidity of cantilever type of a job varies inversely with the cube of the length and the nature of the graph shown agrees with it.

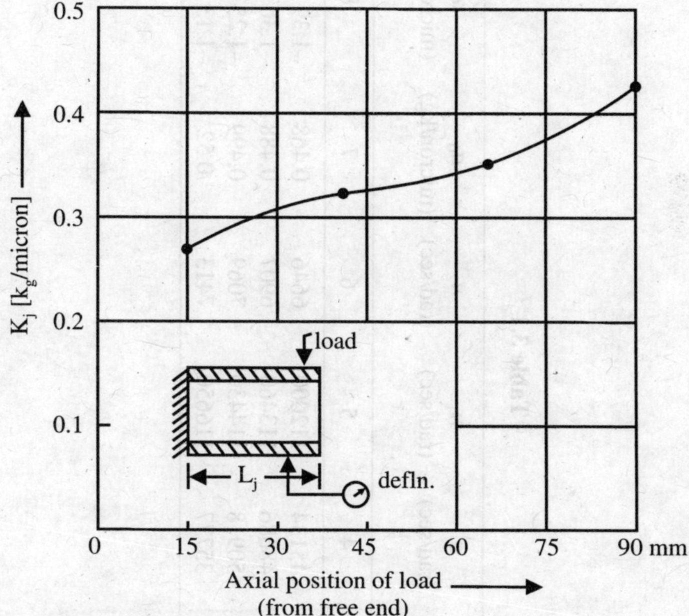

Figure 3.20 Workpiece static rigidity along axial position.

3.13.3

The magnified image of the Talyrond record was traced in a number of segments and random selection of portions of these magnified tracings were taken for measuring peak to peak deflections.

3.13.4 *Analysis of the Experimental Results*

Based on the experimental evaluations of the static rigidities of the elements in the system, a sample calculation is carried out with a simplifying assumption that the grinding force does not exceed 7 kg. The operating

Table 3.1

SI No.	Contact position from the free end (cm)	K_j (kg/micron)	n_j (rad/sec)	n_1 (rad/sec)	n_2 (rad/sec)	a_0 (micron/kg)	a_1 (micron/kg)	a_2 (micron/kg)	U_t (gm-mm)
1	2	3	4	5	6	7	8	9	10
1	15	0.275	45134	112000	6646	0.468	-1.39×10^{-4}	–0.139	5.6
2	40	0.32	48686	113460	6907	0.488	-1.30×10^{-4}	–0.139	4.9
3	65	0.35	50918	114430	7069	0.499	-1.25×10^{-4}	–0.139	6.1
4	90	0.42	35777	116650	7413	0.52	-1.13×10^{-4}	–0.139	5.9

frequency is a determined parameter as the speed of the driving motor and the spindle stepped up by means of pulleys being known. It is seen that K_j increases as we move from free end of the workpiece to the fixed end. Hence, ω_{nj} for different contact point of the grinding wheel no longer remains constant. The moment ω_{nj} changes, the natural frequencies of the system are affected and correspondingly amplitudes a_0, a_1, a_2 also vary.

The values of ω_{nj}, ω_{n1}, ω_{n2} and a_0, a_1, a_2 for different contact points are given in Table 3.1 (columns 4-9). The amplitude a_1, is negligibly small and the first natural frequency of the system is so high that the net effect of a_1 sin $\omega_{n1}t$ on the dynamic compliance may be safely omitted. Hence eqn. (3.42) can very well be modified as:

$$\frac{x_1 - x_2}{F_0} = a_0 \sin \omega_0 t + a_2 \sin \omega_{n2} t \tag{3.44}$$

$$\text{or, dynamic rigidity} = \frac{F_0}{(x_1 - x_2)}$$

$$= 1/(a_0 \sin \omega_0 t + a_2 \sin \omega_{n2} t) \tag{3.45}$$

From the foregoing, it is apparent that

$$[(x_1 + x_2)/F_0]_{avg} \le (a_0 + a_2)/2 \tag{3.46}$$

In other words, the system rigidity, on an average, will be at least equal to $= 2/(a_0 + a_2)$ or

$$\frac{2m_t\omega_{n2}\,(\omega_{n1}^2 - \omega_0^2)\,(\omega_{n1}^2 - \omega_{n2}^2)\,(\omega_{n2}^2 - \omega_0^2)}{[\omega_{n2}(\omega_{n1}^2 - \omega_{n2}^2)\,(\omega_{nj}^2 - \omega_0^2) - \omega_0(\omega_{nj}^2 - \omega_{n2}^2)\,(\omega_{n1}^2 - \omega_0^2)]}$$

From $[(x_1 - x_2)/F_0]_{avg}$ and x_{max}, an average amount of unbalanced force, in gm-mm, already present in the grinding wheel may also be statistically apprehended

$$U_t \frac{\omega_0^2}{g} = \frac{x_{max}/2}{[(x_1 - x_2)/F_0]_{avg}}$$

$$\text{Therefore,} \qquad U_t = \frac{x_{max}\, g}{(a_0 + a_2)\,\omega_0^2} \times 10^4 \tag{3.47}$$

3.13.5 *Discussion*

(i) It appears reasonably difficult to determine the deflection of the grinding wheel-workpiece contact point free from bending deflection by the application of unidirectional force. Hence, equal but diametrically opposiie forces were applied with grinding wheel and the workpiece in contact to neutralise the effect of bending and only the bulging of the workpiece was determined. The validity of this simplification is, however, not known. But the literature survey reveals that Prof. A.P. Sokolovsky, his school of associates, and D. N. Reshetov have greatly emphasized the hyperbolic nature of the load versus deflection graphs which are quite in agreement with the present experimental results at the contact point of the grinding wheel and the workpiece. They proposed the empirical relation $\delta = Cq^m$ of which eqn. (3.42) is a further simplified form.

(ii) It is seen from the experimental results that the workpiece is remarkably less rigid compared to the theoretically calculated values. The reasons are partly explained by Z.M. Levina that only a minor part of the joint area accepts the pressure owing to the presence of waviness and flatness deviations in actual machine tool joints and partly due to the fact that the workpiece taken in this case was of quite a large diameter and only 3 mm of its length was held in the grip of the jaws. Hence, in actual practice, the end condition cannot be taken as fixed one and any theoretical calculation of the workpiece rigidity may lead to confusing results if otherwise not corrected.

(iii) Grinding machine in the present work has been considered rigid enough compared to the grinding wheel-workpiece system. As a matter of fact, there will be excitations imposed on the system due to machine vibrations also and this necessitates a further rigorous analysis.

However, utmost care was taken to avoid the effect of excitations imposed on the system while doing the experiment.

(iv) The vibration energy may be considerably dissipated by providing visco-elastic dampings on the spindle and taking a damped system into operation The mathematical model here does not include the effects of waviness already present on the workpiece before the experiment and the fluctuations in the cutting force while grinding.

However, it is difficult to account for fluctuating grinding force. The relative displacement between the grinding wheel and the workpiece is a periodic phenomenon and hence the grinding force would also fluctuate. A close examination of the cutting action in internal grinding has, necessarily, become imperative.

(v) The computed value of U_t, as given in Table 3.1 (column 10), lies in the range 5-6 gm-mm imbalance and so an average value will indicate a close approximation of the imbalance that may be present in the grinding wheel.

Thus it is possible to conclude that.

(a) The contact rigidity increases with the applied load at the contact point of the grinding wheel and the workpiece.
(b) The rigidity of the spindle bearing (support) must also be considered in conjunction with the rigidity of the spindle.
(c) The rigidities of the spindle and the support are very low and tney drop abruptly after a certain load of small magnitude. This is why internal grinding involves a very small magnitude of grinding force and depth of cut.
(d) The maximum peak to peak waviness on the workpiece gives a fairly good statistical assessment of the imbalance in the grinding wheel with the help of the derived expression for average dynamic compliance.
(e) The amplitude associated with the second harmonic in the expression for dynamic compliance eqn. (3.41) is negligibly small and the net effect of this harmonic on the system may be safely omitted.

CHAPTER 4

Further Studies On Kinematics

4.1. Transmission in Stepped Regulation

From drive (prime-mover) to spindle we go through transmission links having multiplying factors. Systems of regulation obtained can be divided into two

1. Stepped regulation ; and
2. Stepless regulation.

In (1) we utilize A.C. motors with constant speed and then some devices to multiply the speeds. Drives are :

(a) cone pulley drive,
(b) change gear drive, and
(c) gear boxes.

In (2), we utilize, in modern machine tools, either

(a) variable speed electric drive, or
(b) mechanical friction drive.

Transmission in the System of Stepped Regulation

(a) CONE PULLEY DRIVE. In Figure 4.1 is shown the sketch of a cone pulley drive. In regard to drive mechanism a steady reversion to the individual electric motor from the cone pulley has taken place, during the last 50 years. Nevertheless, the cone pulley drive possesses the advantages in regard to cheapness and simplicity and it will always survive and be in use cn very many types of machine tools. *Advantages*: (1) cheap and simple from production and maintenance point of view.

Disadvantages: (1) big in dimensions; (2) provides small range of stepped speeds (3 or 4 different steps); (3) As the velocity of. the belt decreases, the power decreases. So this cannot be used in very slow speeds, as torque is low and requires large pulley and wide belt which tells upon efficiency of the system; and (4) can't change speed automatically.

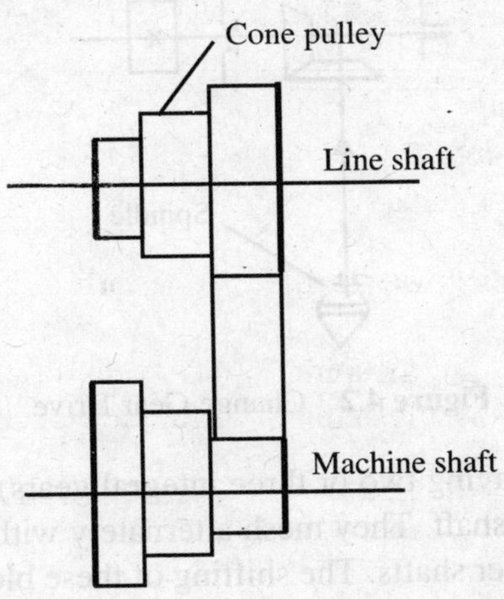

Figure 4.1 Cone Pulley.

(b) CHANGE GEAR DRIVE. A typical example of change gear drive has been shown in Figure 4.2. The advantages and disadvantages of this system are as under. *Advantages*: (1) simplicity; (2) low cost: (3) safety in use and maintenance; (4) accurate speed ratio. *Disaavantages*: (1) low range ol speed regulation. Gear ratio i lying between $l/4 \leq i \leq 1$ or $1/5 \leq i \leq 2$; (2) more time is taken for speed changing in this type of transmission; (3) impossible to practice automatic change of speed, in a case like this.

(c) GEAR BOX DRIVES: There are five different types of mechanical devices for transmission.

(i) *Gear box with sliding gears.* Typical exam pie of a gear box with sliding gears has been clearly shown in Figure 4.3. It is almost an unavoidable solution of the problem of obtaining variable speeds in a gear box of the machine tool. Sliding gear

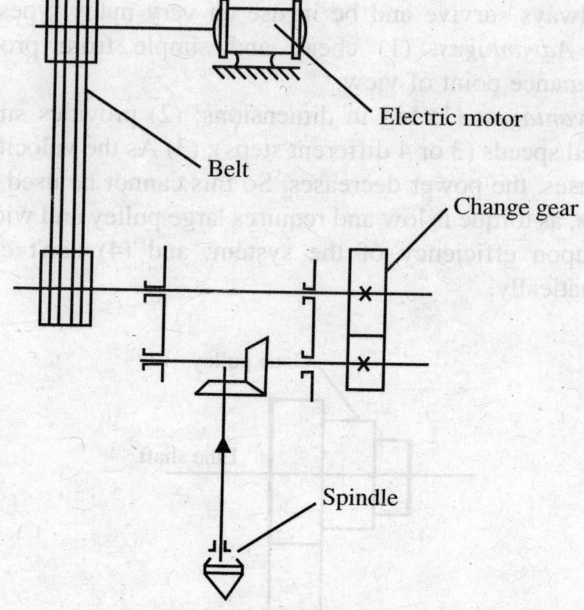

Figure 4.2 Change Gear Drive.

blocks (having two or three integral gears) are usually slided on a splined shaft. They mesh alternately with different gears fixed on the other shafts. The shifting of these block gears can be done by various gear shifting mechanisms shown in Chapter 11. *Advantages*: (1) simplicity in operation; (2) provides long range of speed ragulation; (3) accurate speed ratios; (4) good for constant changes of speeds as in automatic or semi-automatic lathes.

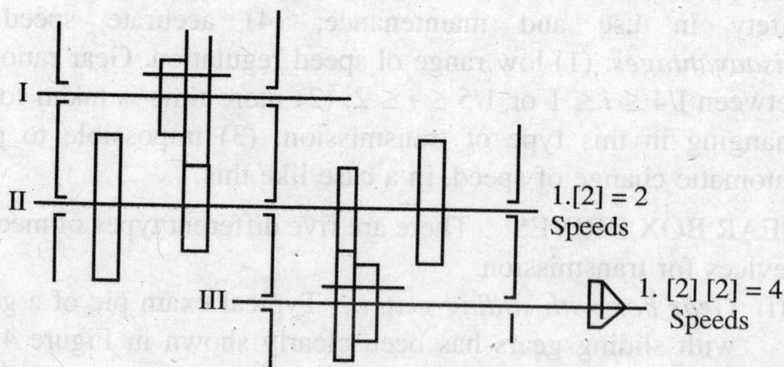

Figure 4.3 Gear Box Drive

(ii) *Gear box with clutches.* Such a gear box arrangement is shown in Figure 4.4 where different speeds are obtained by using a number of clutches and by shifting the clutches and sliding gears.

The clutches used in a system like this are usually ordinary friction or electromagnetic clutches. *Advantages*: (1) the gear box becomes compact, and (2) automatic shifting is possible. Disadvantages: (1) expensive; (2) friction losses considerable, if friction clutch is used (efficiency in friction clutch drive is about .4 to .5).

(iii) *Gear box with tumbler gears.* This system of gear box is mostly used for slow speeds. A typical example is its use in apron of a lathe and for gear changing in screw cutting machine. In Figure 4.5 has been shown the complete arrangement of a tumbler gear box by line diagram. Here we can have three different positions, namely forward, neutral and reverse.

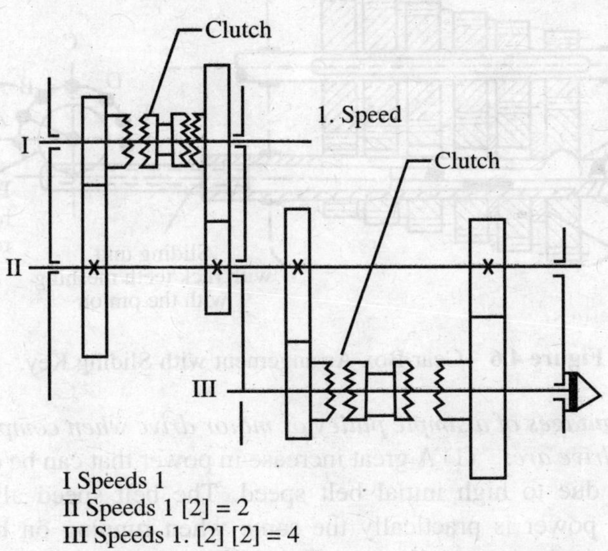

Figure 4.4 Gear Box with Clutches.

(iv) *Gear box with gears on sliding key.* This is an improvement in the design over the design of sliding gear. Here the gears to be meshed afe all placed on a sliding key. The sliding key can be

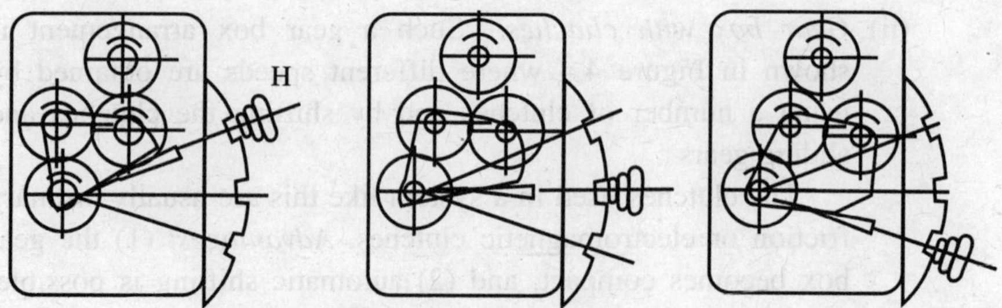

Figure 4.5 Tumbler Gears in Three Positions (Forward, Neutral and Reverse).

pushed in by moving the handle H to the required stud. (See Figure 4.6).

(v) *Gear box with arrangements combining with two or more of the above types*. Any of the above can be used in conjunotion with a variable speed motor to simplify the design without restricting the speed range.

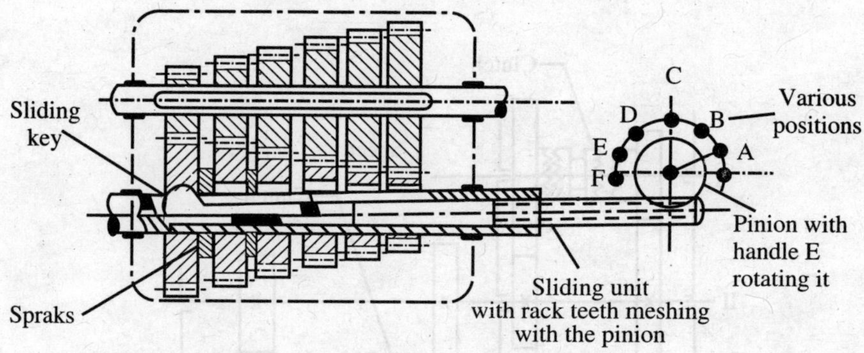

Figure 4.6 Gear Box Arrangement with Sliding Key.

The advantages of a simple pulley or motor drive when compared with a cone-pulley drive are: (1) A great increase in power that can be delivered to the machine due to high initial belt speed. The belt speed always being constant, the power is practically the same when running on high or low speeds, whereas the cone-pulley acts inversely in this respect. That is as the diameter of the work increases, for a given cutting speed, the power decreases. (2) Speed changes being made by levers, any speed is quickly obtained. (3) The machine can be driven direct from the line shaft or motor and no light restricting countershaft is required. Floor space is also often saved. (4) Longer life of the driving belt as the belt shifting is eliminated.

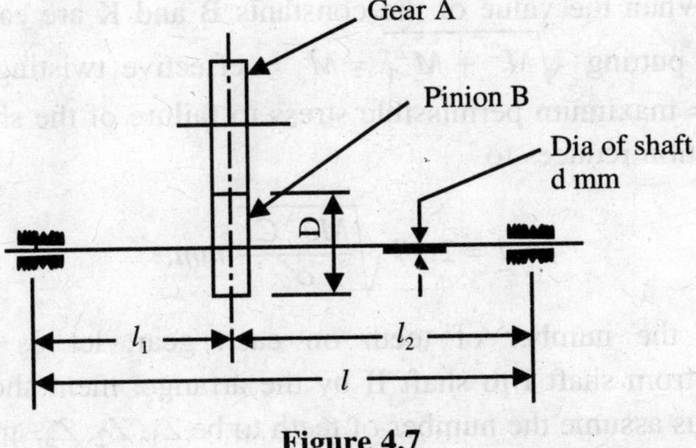

Figure 4.7

Design of Intermediate Shafts of All Geared Headstock

Let us assume a simplified sketch of the intermediate shaft with gear mating as shown in Figure 4.7. The design can be effected by taking into account the failure of the shaft in bending and endurance.

$$d = 2.17 \sqrt[3]{\frac{\sqrt{(B.M)^2 + (K.Mt)^2}}{\frac{\sigma'_e}{C_s}}} \text{ in mm,} \tag{4.1}$$

where M–bending moment on the shaft in kg.mm = $P \frac{l_1 . l_2}{l}$

(where P→tangential force acting on the pitch diameter line of the gear train).

$$M = \frac{P . l_1 l_2}{l} = \frac{M_t}{D} . \frac{2 l_1 l_2}{l}$$

M_t–turning moment on the shaft = 975000 $\frac{N}{n}$ kg. mm.

(where N = power transmitted to shaft in k watt when the shaft is running at a speed of n r.p.m.).

B and K – coefficients characterizing the cycles of alternating stresses and stress concentration in the case of bending and in case of torsion respectively:

σ'_e – endurance limit of material under the condition of bending wi:h symmetrical cycle of stress in kg/mm.^{2}

C_s–safety factor.

When the value of the constants B and K are each equal to 1, then putting $\sqrt{M^2 + M^2_1} = M_e$ = effective twisting moment and σ'_e = maximum permissible stress to failure of the shaft, the above equation reduces to

$$d = 2.17 \sqrt[3]{\frac{Me \cdot Cs}{\sigma'_s}} \; mm. \tag{4.2}$$

To find out the number of teeth on each gear, let us represent the transmission from shaft I to shaft II by the arrange: ment shown in Figure 4.8. Also let us assume the number of teeth to be Z_1, Z_2, Z_3, and Z'_1, Z'_2, Z'_3 respectively.

If the transmission ratios are represented by i_1, i_2 and i_3, then

$$i_1 = \frac{Z_1}{Z'_1} \quad i_2 = \frac{Z'_2}{Z} \quad i_3 = \frac{Z_3}{Z'_3} \qquad \text{condition 1}$$

and condition 2 :

$$Z_1 + Z'_1 = Z_2 + Z'_2 = Z_3 + Z'_3 = 2\,Z_0 \text{ (say)},$$

so that

$$Z_1 + \frac{Z_1}{i_1} = 2\,Z_0$$

or

$$Z_1 = 2\,Z_0 \frac{i_1}{i_1 + 1},$$

and

$$Z'_1 = 2\,Z_0 \frac{1}{i_1 + 1}, \text{ and so on.}$$

Suppose, $i_1 = \dfrac{A_1}{B_1}$, $i_2 = \dfrac{A_2}{B_2}$ and $i_3 = \dfrac{A_3}{B_3}$,

where A_1 = pitch circle diameter of gear having Z_1 teeth and so on.

$A_1 = mZ_1$ where m–module which is kept same for all these gears.

$$A_1 + B_1 = A_2 + B_2 = A_3 + B_3$$

From the point of view of interference and minimum wear

Z_{min} = minimum number of teeth on any gear ≥ 17 provided the transmission ratio is upto 1.

For all other transmission ratios $Z_{min} \geq 40$

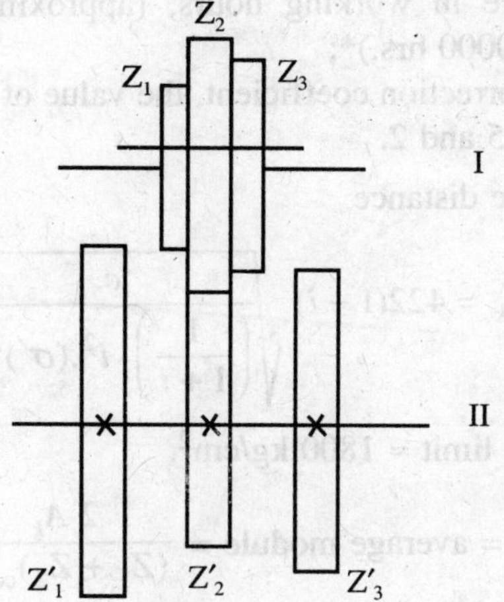

Figure 4.8

Also $2Z_0$ under all conditions should be ≤ 120.

The design of gear from the point of view of strength should be done on the basis of Lewis formula about which a student is asked to refer to any book on machine design.

Method of Designing Bevel Transmission

With reference to the figure shown in 4.9

Z_1 = number of teeth of bevel no. 1

$$= \frac{(Z_1 + Z_2)_{cond}}{1 + i} \cos \delta_1 \tag{4.3}$$

where i = transmission ratio;

$(Z_1 + Z_2)_{cond}$ – summation of the conditional number of teeth (assuming the gear to be equivalent spur); and

δ_1 – angle as shown

$$Z_2 = Z_1 \times i \tag{4.4}$$

Coefficient of performance [48] taking into account the life of the bevel pairs

$$= C = \frac{N}{n_2} \cdot (n_1 h)^{1/3} \cdot K_s, \tag{4.5}$$

where N = power transmitted in H.P.;

n_2, n_1 = r.p.m. of bevel no. 2 and no. 1 respectively;

h = life in working hours, (approximately estimated upto 20000 hrs.)*;

K_s = correction coefficient, the value of which varies between 1.5 and 2.

Conditional centre distance

$$A_k = 422(1+i)\sqrt[3]{\frac{c}{\left(\frac{1}{1+i}\right)\cdot i^2.(\sigma')^2}}, \tag{4.6}$$

where σ′ – endurance limit ≈ 1800 kg/cm^2.

$$m_{av} = \text{average module} = \frac{2\,A_k}{(Z_1+Z_2)_{\text{cond}}}, \tag{4.7}$$

where A_k is the distance shown in the sketch in Figure 4.9

$$m = \text{face module} = m_{av} + \frac{b \sin \delta_1}{Z_1} \tag{4.8}$$

$$\frac{L}{b} = \frac{m \times Z_1}{2l \sin \delta_1} \tag{4.9}$$

The peripherial velocity of the bevel transmission (assuming h equal to

$$20000 \text{ hrs.}) = v = \frac{m_{av}\cdot Z_1 \cdot n_1}{20000\,*} \text{ m/sec.} \tag{4.10}$$

Value of $(Z_1 + Z_2)_{\text{cond}} = f(i)$, *i.e.* a function of i:

i	$(Z_1 + Z_2)_{\text{cond}}$
1–3	100
3–5	150
5–8	200

Problem 1: Design the headstock gear box of a turret lathe having arrangement for nine spindle speeds ranging from 30 to 1000 r.p.m. Only layout and not a full drawing of ihe gear box should be shown. The H.P. of the machine may be taken as 6 and minimum number of teeth on a gear 25.

---

* The bevel pairs, as used in machine tools can very easily be estimated to have a life equal to 10000–20000 working hours or in some special capes even more. The ratio $\frac{L}{b}$ is usually equal to or greater than t.

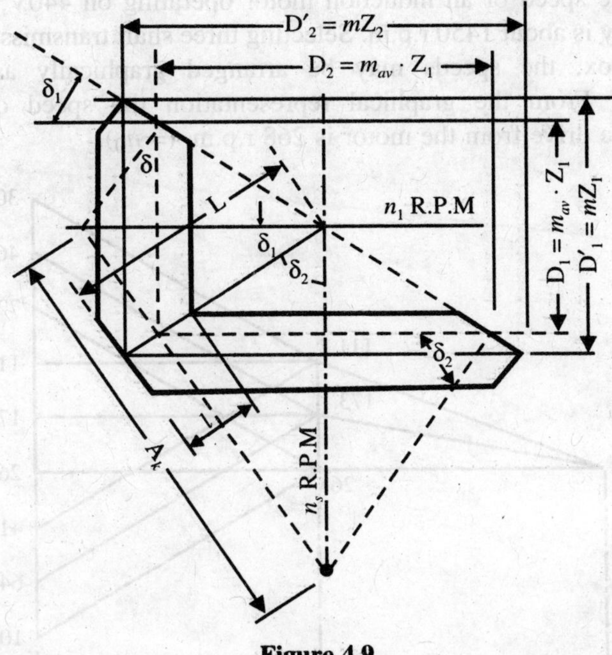

Figure 4.9

(1) Represent the speeds graphically ;
(2) Draw the structure diagram ;
(3) Show the layout of gear box and connection to motor; and
(4) Draw the deviation diagram for the available speeds.

Solution : Let n_1, n_2 ... n_9, be the 9 spindle speeds. For minimum loss in average speed, the speed should be arranged in geometric progression. Now if φ = ratio in geometric progression, then

$$\frac{n9}{n1} = \varphi^{9-1} = \varphi^2$$

$$\text{or } \varphi = \left(\frac{1000}{30}\right)^{1/8} \approx 1.55.$$

Therefore

$n_1 =$ 30 r.p.m.
$n_2 =$ 30 × 1.55 = 46.5 r.p.m.
$n_3 =$ 72 r.p.m.
$n_4 =$ 111.7 r.p.m.
$n_5 =$ 173.1 r.p.m.
$n_6 =$ 268 r.p.m.
$n_7 =$ 416 r.p.m.
$n_8 =$ 645 r.p.m.
$n_9 =$ 1000 r.p.m.

Now the speed of an induction motor operating on 440V, 50 cycle, 3 phase supply is about 1450 r.p.m. Selecting three shaft transmission drive for the gear box, the speeds may be arranged graphically as shown in Figure 4.10. From the graphical representation the speed of the shaft receiving the drive from the motor is 268 r.p.m. (= n_0).

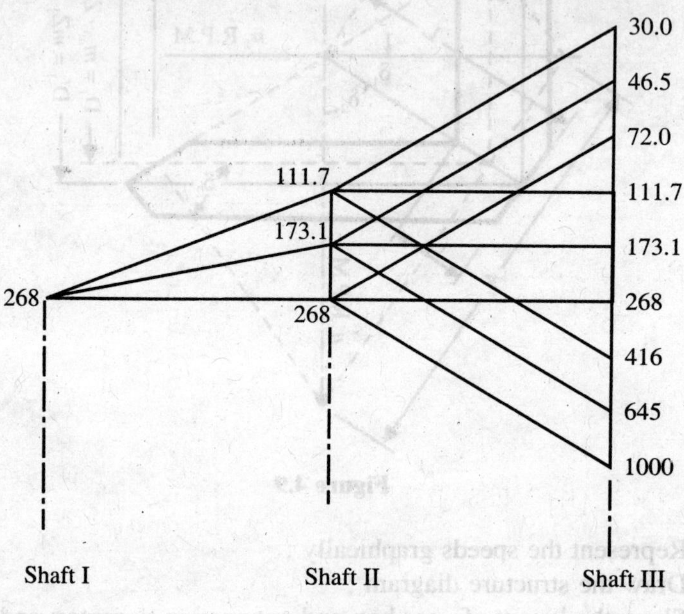

Figure 4.10

Assuming the motor shaft pulley of 75 dia., the diameter of the pulley on shaft I

$$D = \frac{1450}{268} \times 75 = 406.25 \text{ mm}.$$

Gear arrangements have been shown in diagram in Figure 4.11. For the reduction of speed from 268 to 111.7 r.p.m. number of teeth on minimum sized gear = 27, say.

$$G_1 = 27,$$

then $$G_4 = \frac{268}{111.7} \times 27 = 65.$$

As the modules of gears are kept the same,

$$\frac{G_5}{G_2} = \frac{268}{173.1} = 1.548 \qquad G_5 + G_2 = 65 + 27 = 92$$

$$G_2 = 36 \text{ teeth and } G_6 = 56 \text{ teeth}$$

also, $\quad G_3 + G_6 = 92 \quad$ and $\quad G_3 \doteq G_6$

(since speed reduction = 1)

$$G_3 = G_6 = 46 \text{ teeth.}$$

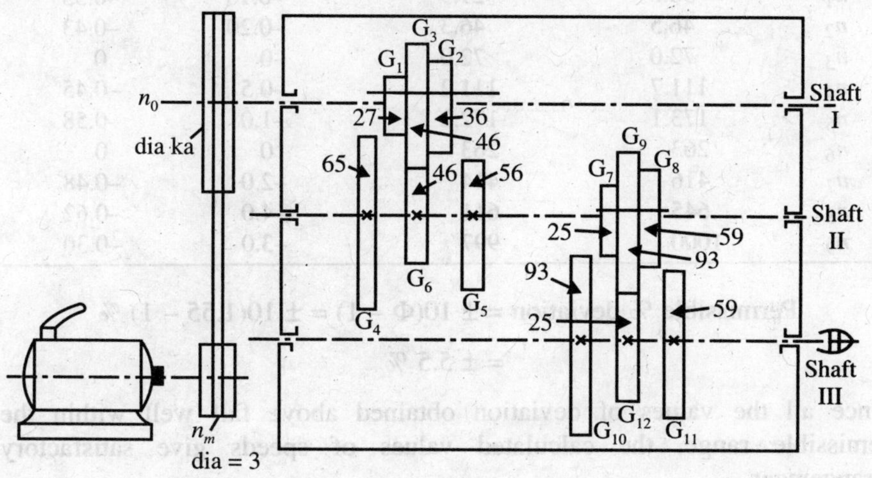

Figure 4.11

For shaft II to shaft III:

$$\frac{G_7}{G_{10}} = \frac{72}{268} = 0.268$$

If $G_7 = 25$ teeth, $G_{10} = 93$ teeth.

In the same way as before we get

$$G_8 = G_{11} = 59 \text{ teeth}$$
$$G_9 = \quad 93 \text{ teeth}$$
$$G_8 = \quad 25 \text{ teeth}$$

Structural diagram :

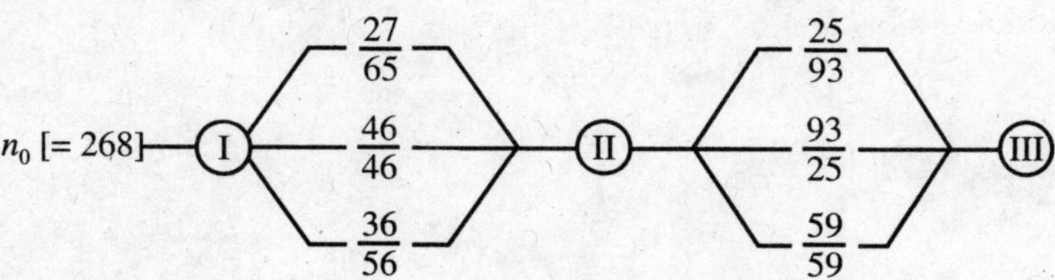

All the gears have the same module ($m = 3$)

Speed	*Calculated value of speed in rpm* (n_c)	*Obtainable value of speed in rpm* (n_{obt})	*Difference* ($n_{obt} - n_c$)	*% Deviation*
n_1	30.0	29.9	–0.10	–0.33
n_2	46.5	46.3	–0.20	–0.43
n_3	72.0	72.0	–0	0
n_4	111.7	111.2	–0.5	–0.45
n_5	173.1	172.2	–1.0	–0.58
n_6	263	263	0	0
n_7	416	414	–2.0	–0.48
n_8	645	611	–4.0	–0.62
n_9	1000	997	–3.0	–0.30

$$\text{Permissible \% deviation} = \pm\, 10(\Phi - 1) = \pm\, 10(1.55 - 1)\ \%$$

$$= \pm\, 5.5\ \%$$

Since all the values of deviation obtained above fall well within the pemissible range, the calculated values of speeds give satisfactory arrangement.

CHAPTER 5

Stepless Regulation In Machine Tools

5.1. Classification

The system of stepless regulation is more common now-a-days in modern machine tools. By such system we can have practically no loss in speed and can arrange speed adjustment without stopping the machine. It reduces time losses and is therefore very useful in automatic machine tools. Stepless regulation can be effected by any one of the following methods :

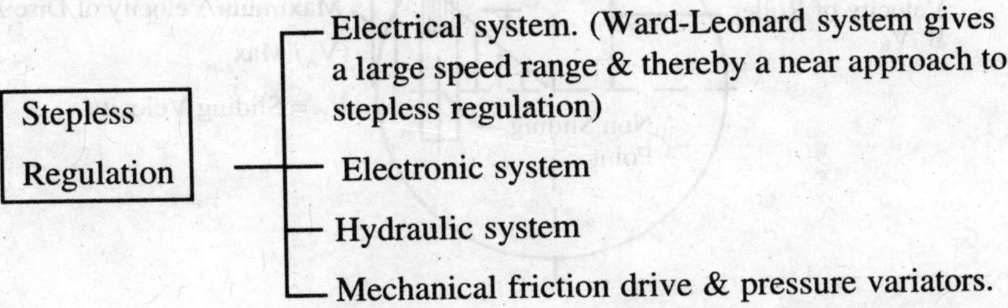

The electrical systems have been dealt with separately in Chapter 12 of this book. The hydraulic drive in machine tool avoids speed fluctuation. Stepless regulation of speed can be obtained by hydraulic mechanism up to a range of 1000. (Maximum speed can be obtained near about 60 m/min while the minimum speed is limited by instability). Hydraulic cylinder and piston drive are extensively used to get straight line motion. The principle of regulation by the help of hydraulic motor is explained in Chapter 13 of this book. Grinders, boring machines, gear grinding, shaping, planing machines, etc., are usually provided with hydraulic stepless regulation.

In this chapter the author has dealt only with the mechanical friction drives and pressure variators.

5.2 Mechanical Faction Drive

1. *Two rolling bodies.* Here in this particular case (Figure 5.1), the two rolling bodies rotating about their axes are made to remain in contact with the help of the preloaded spring force. The contact between the two bodies can be either line or point (if the roller is of B' type). In many cases the two rolling members A and B may be made to be in contact by the help of their weight or centrifugal force depending upon the constructional feature. Such systems cannot obtain clean rolling friction. Such machanism sliding with velocity V_0 has been shown in the velocity diagram (Figure 5.1).

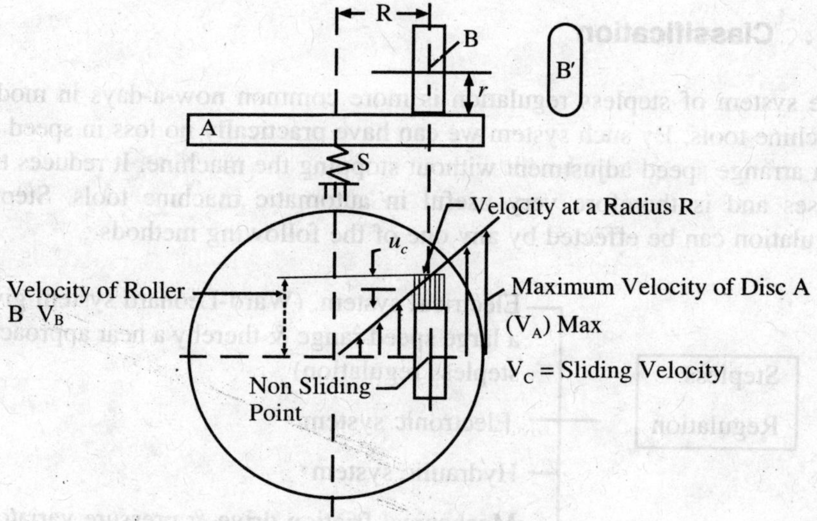

Figure 5.1 Two rolling bodies.

Speed Ratio obtained in this case $= i = \dfrac{r}{R}$. By changing the position of the roller, i.e. by changing the distance R we can change the regulation. We can get in a case like this a regulation range $\leq$ 2–2.5

1 (a) *Friction Drive with Two Discs and One Roller*
This mechanism in its simplest form is shown in Figure 5.2.

Ratio of speed obtained : $i = \frac{r}{R_2} = \frac{R_1}{r} = \frac{R_1}{R_2}$. Usually we can have a regulation up to 4.

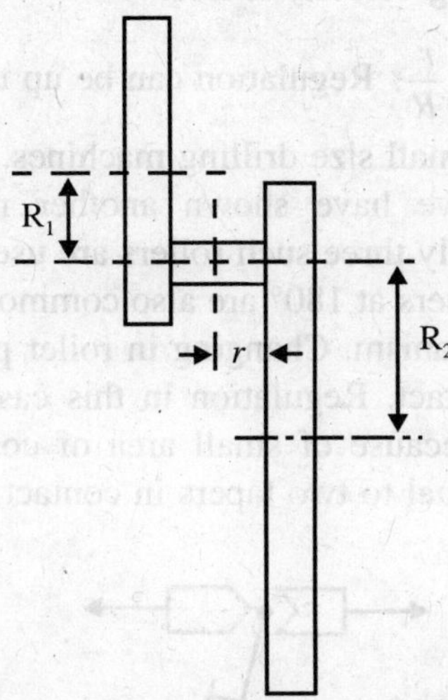

Figure 5.2 Two discs and one roller.

Disadvantage. Bigger sliding or friction loss. To remedy this, we can reduce area of contract. It may be possible to reduce the area to a point contact. We can change the design of the disc or roller contour, so that the amount of sliding is less.

2. Another type of stepless regulation used in light duty

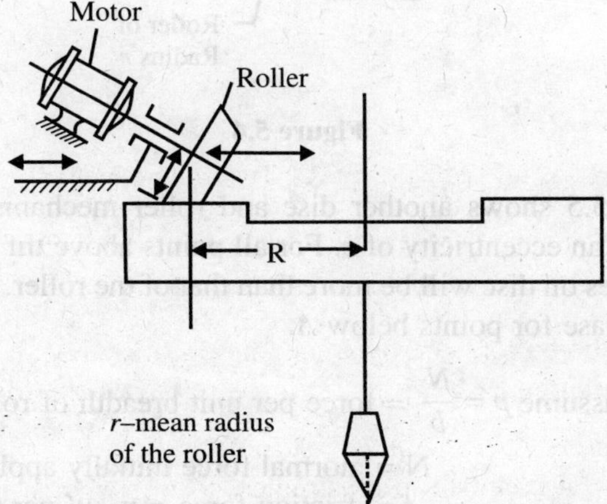

Figure 5.3

machine tool and mostly in feed mechanism, can be obtained by the arrangement shown in Figure 5.3.

In comparison to variators shown in Figs. 5.1 and 5.2 in this case the loss due to sliding velocity is less.

Speed Ratio : $i = \frac{r}{R}$. Regulation can be up to 3.5

This is used in small size drilling machines.

In Figure 5.4 we have shown another mechanism for stepless regulation. Usually three such rollers are used at an angular distance of 120°. Two rollers at 180° are also common. Rollers are controlled by shifting mechanism. Changing in roller position means changing the point of contact. Regulation in this case ≤ 8–10. This reduces friction losses because of small area of contact. Theoretically this mechanism is equal to two tapers in contact.

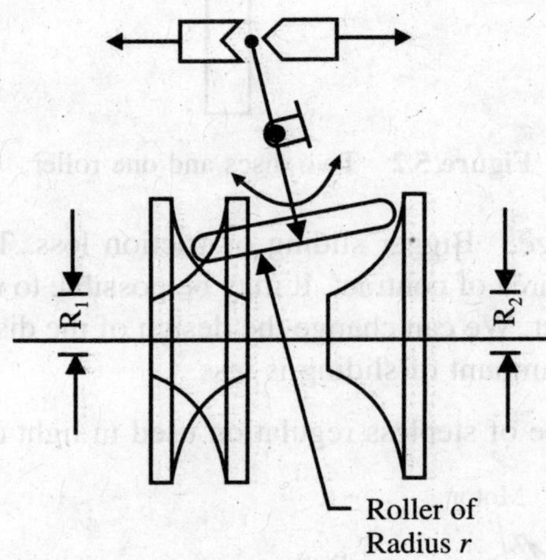

Figure 5.4

3. *Figure* 5.5 shows another disc and roller mechanism with contact point at an eccentricity of e. For all points above thf point marked A, velocities on disc will be more than that of tne roller. The reverse will be the case for points below A.

Let us assume $p = \frac{N}{b}$ = force per unit breadth of roller;

N = normal force initially applied; and

f = friction force = p . μ' per unit breadth,

where μ' = coefficient of friction; and
M = outside torque = $M_1 - M_2$

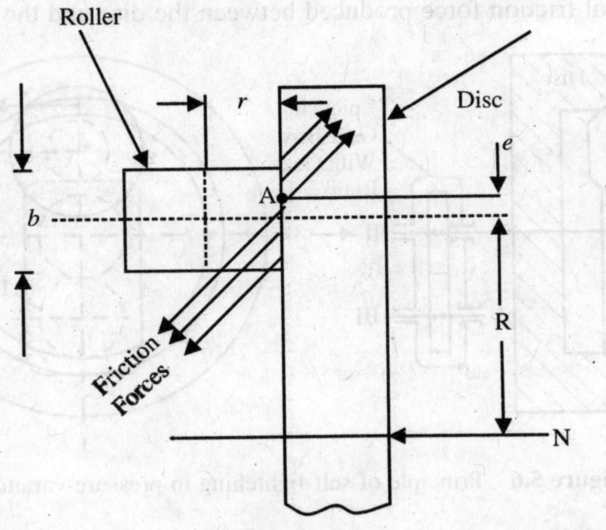

Figure 5.5

where M_1 = first friction torque, caused by forces acting above A.
M_2 = second friction torque, caused by friction forces acting below the point A.

$$M_1 = f\left(\frac{b}{2} - e\right) \times \left[R + e + \left(\frac{b}{2} - e\right) \times \frac{1}{2}\right]. \qquad (5.1)$$

$$M_2 = f\left(\frac{b}{2} + e\right) \times \left[R + \frac{e}{2} - \frac{b}{4}\right]. \qquad (5.2)$$

Add up equations 5.1 and 5 2 to get a quadratic equation for e:

$$e = \Psi\ (M/p\mu', R, b)$$

where Ψ denotes mathematical function. That is to say, e will be a function of three parameters, $\frac{M}{p\mu'}$, R and b.

$$i = \text{Ratio of speed} = \frac{r}{R + e}$$

It is in fact needless to mention here that in all the notations above, e is a variable which depends upon the speed of outside rolling as well as on R which is again a variable parameter.

4. Principles of Self-tightening. Figure 5.6 shows a mechanism of drive with self-tightening arrangement. The initial pressure produced by the weight of the power disc is very small. Let us assume that P = initial friction force produced between the disc and the roller.

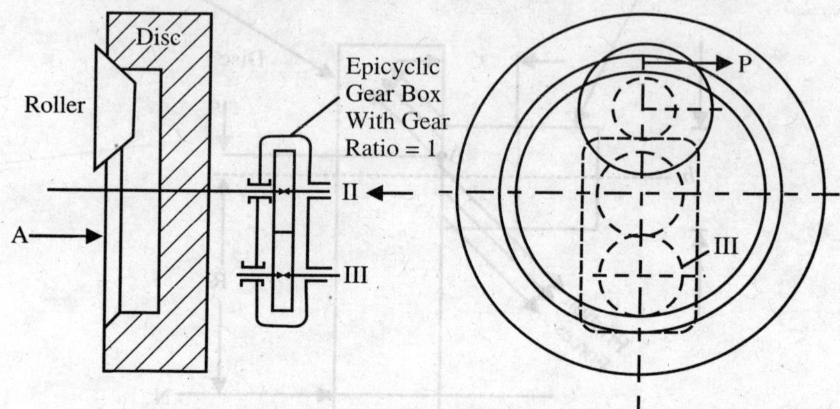

Figure 5.6 Principle of self-lightening in pressure variator.

The main feature of the above design is that the entire frame containing the two gears along with disc can rotate freely about shaft III. As the motion is transmitted to the spindle III through the friction variator, the force P is developed between the roller and the disc. This P multiplied by leverage takes care of outside torque. Now in the vertical position P is generated by the weight of the box. Now as the outside torque is increased, there is a tendency of the gear box to rotate about shaft III and thus the pressure between roller and disc is increased, so consequently the force P. Thus the self-tightening is effected.

5. *Application of Pressure Variators.* The scheme consists of two internally tapered discs, with a number of balls in between them as as shown, in Figure 5.7 (a) and (b). Also there are rollers arranged, as shown, on worm, and their positions can be actuated by means of a worm shaft. Details are shown below in Figures 5.7 (a), (b) and (c). axial as well as transverse thrusts. Each ball has got three-point contacts. Two contacts are effected with two discs and one with the roller. When in central position, the axis of rotation of balls is parallel to the central spindle. If the positions of rollers are changed, then the axis of rotation of balls changes as shown in Figure 5.7 (c). It is always parallel to the axis of rollers, and this has been found experimentally. So the ratio of motions is altered.

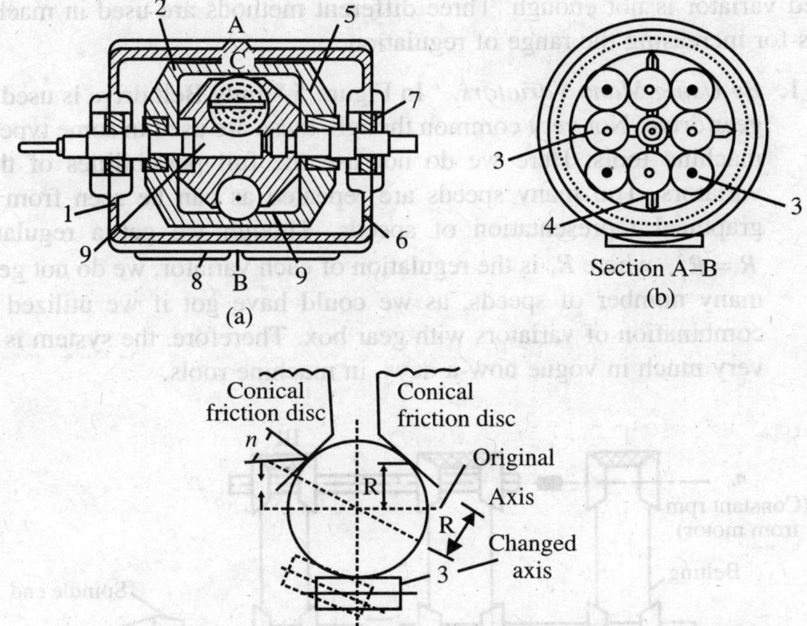

2,5.–Variators with conical friction surface
3.–Rollers fitted inside worm wheel
4.–Worm Wheel
7.–Self Regulated Bearings
8.–Hardened Balls
9.–Worm Shaft
1.6.–Housing made in two halves

Figure 5.7

We use a number of rollers in actual design and install them by means of frames of worm gears, and position the rollers by means of worm shaft. *Advantages*. (1) This design is very compact, i.e. the dimension of the whole system is within 8' dia. for 10 balls. (2) Small friction losses, since theoretically we have point contacts. *Disadvantages*. 8 to 10 H.P. is the limit of power capacity. If the speed increases wear becomes too much and so we must limit the diameter of balls.

5.3 Methods of Increasing the Range of Regulation in Modern Machine Tools

All these variators can only provide a narrow range of regulation say 8 to 10. But in machine tools we require speed and feed range of 100 to 150. So one speed variator is not enough. Three different methods are used in machine tools for increasing the range of regulation.

1. *By Using Many Variators.* In Figure 5.8 only Belt drive is used, no gear drive. Not very common though, these are used in some types of machine tools. Here we do not use the full possibilities of three variators. Too many speeds are repeated as can be seen from the graphical representation of speeds. Though we get a regulation $R = R_v^3$, where R_v is the regulation of each variator, we do not get as many number of speeds, as we could have got if we utilized the combination of variators with gear box. Therefore, the system is not very much in vogue now-a-days, in machine rools.

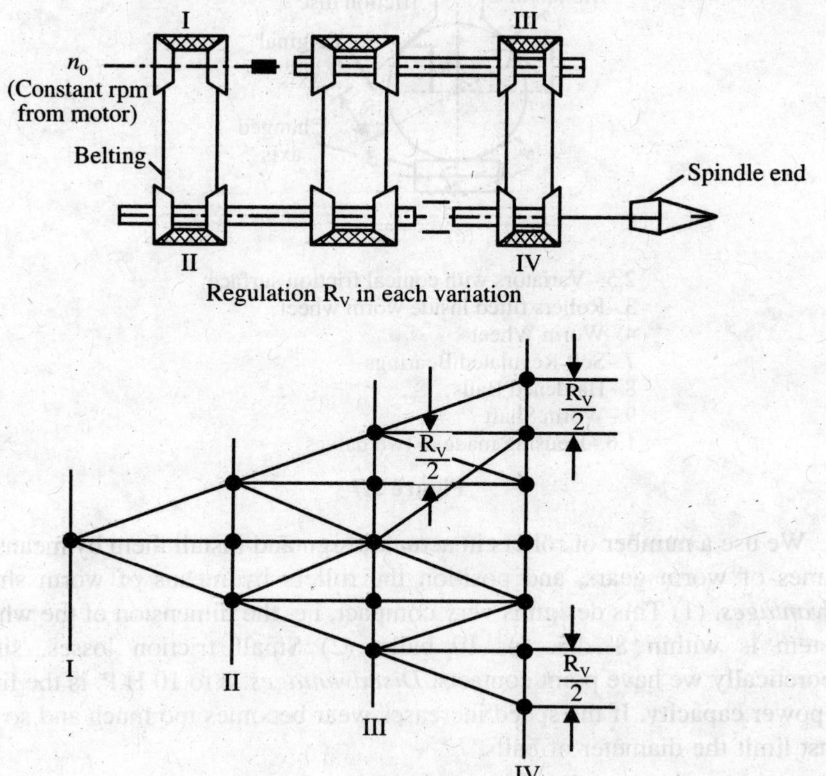

Figure 5.8 Combination of variator.

In the case shown, we can analyze the speeds as follows:
Shaft I: has got one specific speed;
Shaft II: has a range of speeds;

Shaft III: any point on shaft 11 gives an infinite number of speeds, within a range, in shaft III. Or in other words, shaft III can have any speed within the range.

2. *Combination of Variators with Gear Box.* The next improvement is the combination of variators with gear box as shown in Figure 5.9.

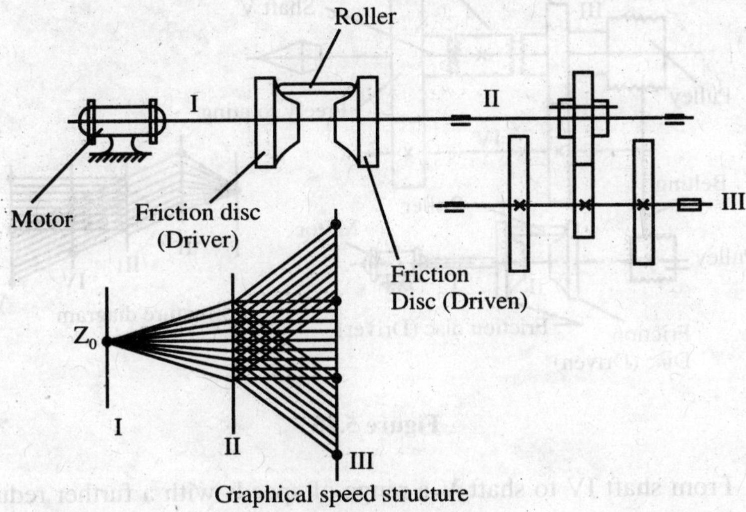

Figure 5.9.

Shaft I-can have only one constant speed;
Shaft II-a range of speeds;
Shaft II-three differerent speeds, reduced by the gear ratios corresponding to any point on the second shaft.

In the 3rd shaft all extended values of speeds obtained are different as can be seen from the graphical structure. The continuity is the best theoretical figure, otherwise there will be either repetition or break, when both are undesirable.

Practical difficulties. If the initial range is big, then big ratio cannot be achieved as the gear box has its limitations. So in this combination, we use special gear box.

Method used very commpnlv in a Back gear Lathe. In Figure 5.10 is shown the method commonly adopted in a back gear drive of lathe machine.

Shaft I-has one speed;
Shaft II-has a range of speeds;
Shaft III-has a range of speeds.

From shaft III–(1) Direct-a range of speeds to shaft V.
(2) to shaft IV a range of speeds, with a reduction.

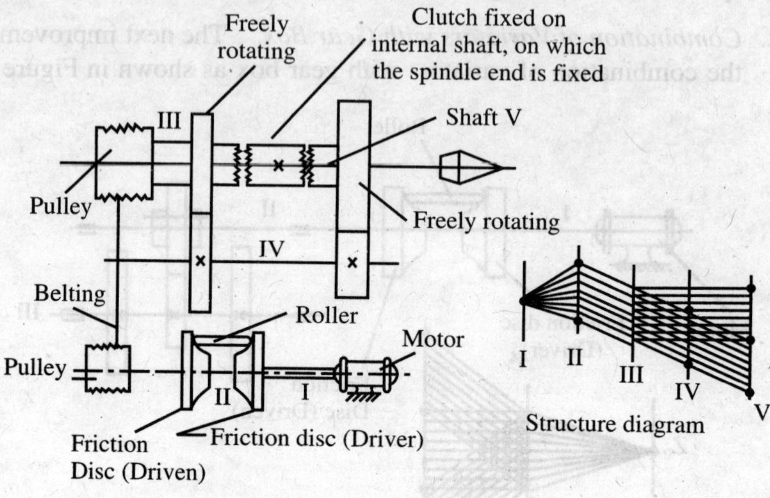

Figure 5.10

From shaft IV to shaft V-a range of speeds with a further reduction.

3. *Variators with Epicyclic Mechanism*. Figure 5.11 shows the variator with epicyclic mechanism, as used in a machine tool. This is used only in feed drive. Let us assume n_0–speed of the motor, n_1–r.p.m. of gear Z_1, n_2–r.p.m. of gear having Z_2 number of teeth, n–spindle r.p.m. and n_0–r.p.m. of epicyclic casing also.

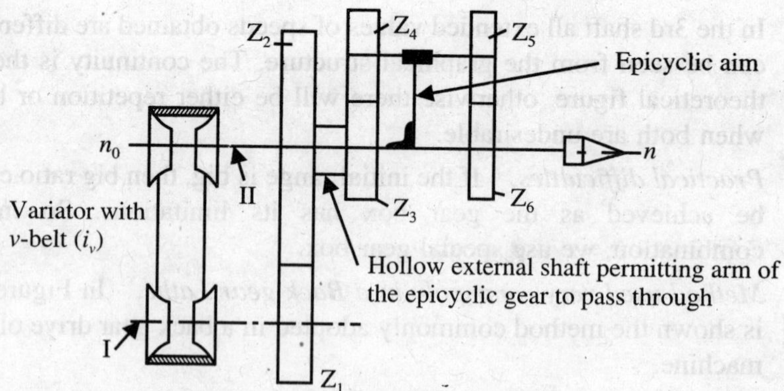

Figure 5.11 Variator with epicyclic mechanism.

$$n_2 = n_0 \cdot i_0 \cdot \frac{Z_1}{Z_2}, \quad \text{where } i_v = \text{ratio ot variator}$$

$$= n_0 \cdot i_v i_2, \quad \text{where } i_2 = \frac{Z_1}{Z_2} = \text{constant ratio}$$

$$\frac{n - n_0}{n_0 \, (i_v \, . \, i_2 - 1)} = i_1; \tag{5.3}$$

here Z_1 and Z_2 are the number of teeth of the gears, shown in the sketch.

$$n = n_0 \, . \, i_1 \, . \, i_2 - n_0 \, . \, i_1 + n_0 = n_0 \, [i_1 \, . \, i_2 \, . \, i_v - (1 + i_1)] \tag{5.4}$$

$i = f(n, i_v, i_2)$, where f denotes function.

$n = n_0 \, (c_1 \, . \, i_v + c_2)$,

where c_1 and c_2 are constants.

From this expression, it is easy to find out values of n:

Put i_v maximum, to get $n_{max.}$

Put i_v minimum, to get $n_{min.}$

Theoretically, it is a very good combination, but practically there are some disadvantages:

(1) Epicyclic mechanism is complicated to manufacture and cannot be used in high speeds.

(2) Low efficiency, if high range is provided.

This may be used in speed drive, if the speed is low and efficiency is not of much consideration.

5.3 Semi Toroidal Drive[153]

Continuously variable speed and power transmission can be obtained by using half-toroidal drive power roller as shown in Figure 5.11(a). The speed ratio can be changed by the control of altitude angle. The ideal speed ratio is given by equation as given below:

$$e_s = \frac{\omega_3}{\omega_1} = \frac{r_1}{r_2}$$

Where ω_1 is the angular speed of rotation of the input disk or driver and ω_3 is the angular speed of output (Driver) disk. The half Toroidal power roller is rotating with angular speed of ω_2.

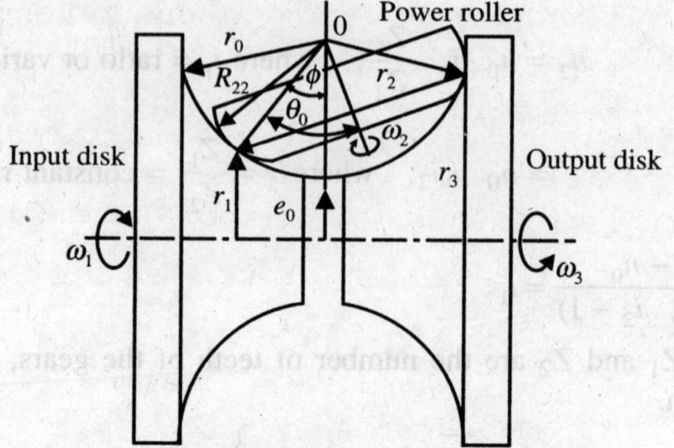

Figure 5.11(a) Half Toroidal drive.

Referring to the Figure 5.11(a)

$$r_1 = r_0(1 + K_0 - \cos \phi)$$

$$r_3 = r_0[1 + K_0 - \cos (2\theta_0 - \varphi)]$$

$$K_0 = \frac{e_0}{r_0}$$

In the above equation e_0 is the minimum radius of the input torso and r_0 is the radius of the power roller, which is a half toroidal roller.

5.4 Considerations in Variator Calculation

(1) Contact stresses (based on the Hertz's equation) for linear contact $q_{max} \leq (0.25 - 0.30)\ H_B\ kg/mm^2$,

where H_B = Brinell hardness number and

q_{max} = permissible contact stress

(2) Friction forces (provide high coefficient of friction).

(3) Bearing stresses $\sigma_{per} = 2000\ kg/cm^2$; σ_{per} means permissible value of the bearing stress.

Torque is due to μ′ × (normal force), where μ′ –coefficient of friction. To increase the torque with the same normal load, μ′ has to be increased.

Material used: (a) Low *C–Cr* steel; *C*–0.15, Cr 1.5–2. (b) C.J. It is not very good against bearing stresses, (c) Hardened steel, plastics (textolites), ferroids, etc., are also used as variator materials. Hardened steel can have μ′ up to 0.2.

5.5 Friction Loss in Friction Variators due to Loss in Sliding Velocity

In continuation of variator shown in Figure 5.5, it is necessary to analyse the friction loss in details. Pressure on the sliding surface and its relation with sliding velocity can be considered for finding out the magnitude of work lost in friction.

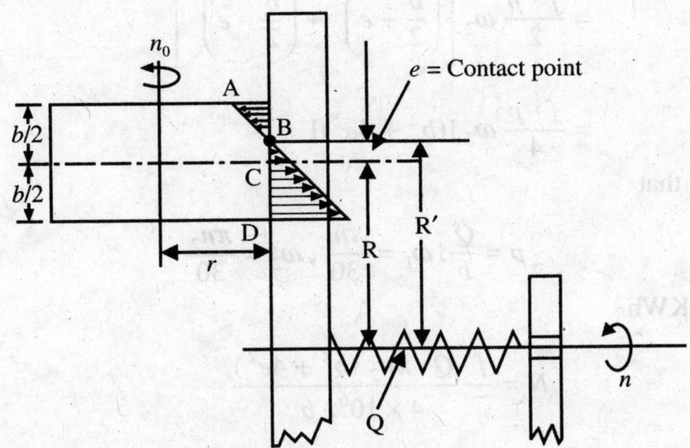

Figure 5.12 Analysis of variators

In the part BD, shown in Figure 5.12, where $V_1 > V_2$

$$dA_1 = f \cdot p\, dx\, (r\omega_1 - R\omega_2) \tag{5.5}$$

where dA_1 – Elemental work done due to friction

R – Radius of the disc at the point of rolling obtained by coordinate x, measur d from B to D

r – Radius of the roller at the same point

ω_1, ω_2 – Angular velocity of the roller and the friction disc respectively

p – Pressure on unit length of contact face

f – Coefficient of friction.

Analogically for the section BA, where $V_2 > V_1$ and $x < 0$ elemental work done due to friction

$$dA_2 = fp \cdot dx\, (R\omega_2 - r\omega_1)$$

$$\therefore \quad dA_2 = -fp \cdot dx\, (r\omega_1 - R\omega_2) \tag{5.6}$$

Generally

$$R = R' - e$$

and $$r\omega_1 - R\omega_2 = r\omega_1 - R'\omega_2 + x\omega_2$$

and $$r\omega_1 - R'\omega_2 = 0 \text{ (at the point of sliding)}$$

Total work due to friction

$$A = f \cdot p\left[\omega_2\left(\int_0^{(b/2+e)} x\,dx + \int_0^{(b/2-e)} x\,dx\right)\right] \tag{5.7}$$

$$= \frac{f \cdot p}{2}\,\omega_2\left[\left(\frac{b}{2}+e\right)^2 + \left(\frac{b}{2}-e\right)^2\right]$$

$$= \frac{f \cdot p}{4}\,\omega_2\,[(b^2 + 4e^2)] \tag{5.8}$$

We know that

$$p = \frac{Q}{b};\ \omega_1 = \frac{\pi n_1}{30},\ \omega_2 = \frac{\pi n_2}{30}$$

Power in KWh

$$N = \frac{f \cdot Q \cdot \pi n_2\,(b^2 + 4e^2)}{4 \times 10^6 \cdot b}$$

$$= \frac{f \cdot Q \cdot \pi b n_2}{4 \times 10^6}\left[1 + \left(\frac{2e}{b}\right)^2\right] \tag{5.9}$$

Here $$e = R' - R$$

5.6 Principles of Self-locking in Variator

Figure 5.13 shows the detailed diagram of the stepless self tightening pressure variator shown earlier in Figure 5.6. Forces acting on the contact point of the cone and the sizes and other notations are clearly shown in the sketch with reference to this figure:

$$P \times CF + P_1\,\frac{mZ_2}{2} \geq Q \cdot OC$$

$$P(a\cos\gamma + R) + P_1\,\frac{mZ_2}{2} \geq Q \cdot a\sin\gamma \tag{5.10}$$

where m is module and Z_1 and Z_2 are the gear teeth respectively

$$P_1 \times \frac{mZ_1}{2} = P \times R \tag{5.11}$$

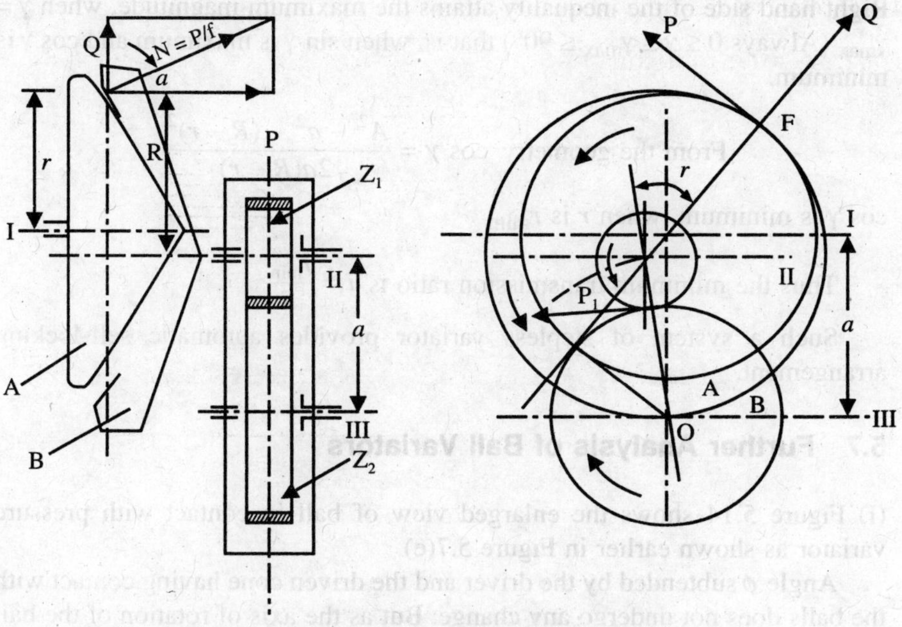

Figure 5.13 Stepless variator with self-locking system.

$$Q = N \cos \alpha = \frac{p}{f} \cos \alpha \tag{5.12}$$

$$a = \frac{m\,(Z_1 + Z_2)}{2} \tag{5.13}$$

$$\therefore \qquad P(a \cos \gamma + R) + PR\,\frac{Z_1}{Z_2} \geq \frac{P}{f} \cos \alpha \cdot a \sin \gamma$$

or,
$$m\,\frac{(Z_1 + Z_2)}{2} \cos \gamma + R + R\,\frac{Z_2}{Z_1} \geq \frac{m\,(Z_1 + Z_2)}{2f} \sin \gamma \cos \alpha \tag{5.14}$$

For all relative positions of the friction cone

$$R\left(1 + \frac{Z_2}{Z_1}\right) \geq \frac{m\,(Z_1 + Z_2)}{2} \cdot \left[\frac{\sin \gamma \cos \alpha}{f} - \cos \gamma\right] \tag{5.15}$$

or in other words

$$R \geq \frac{mZ_1}{2}\left[\frac{\sin \gamma \cos \alpha}{f} - \cos \gamma\right] \tag{5.16}$$

Right hand side of the inequality attains the maximum magnitude, when $\gamma = \gamma_{max}$ (Always $0 \le \gamma \le \gamma_{max} \le 90°$) that is, when sin γ is maximum and cos γ is minimum.

$$\text{From the geometry } \cos\gamma = \frac{A^2 - a^2 - (R-r)^2}{2a(R-r)}$$

cos γ is minimum, when r is r_{mim}

Thus the minimum transmission ratio is $i = \frac{r_{min}}{R}$

Such a system of stepless variator provides automatic self-locking arrangement.

5.7 Further Analysis of Ball Variators

(I) Figure 5.14 shows the enlarged view of ball in contact with pressure variator as shown earlier in Figure 5.7(e).

Angle ϕ subtended by the driver and the driven cone having contact with the balls does not undergo any change. But as the axis of rotation of the ball tilts to the position x' x' at an angle with original axis of rotation x x, we find that the speed ratio changes.

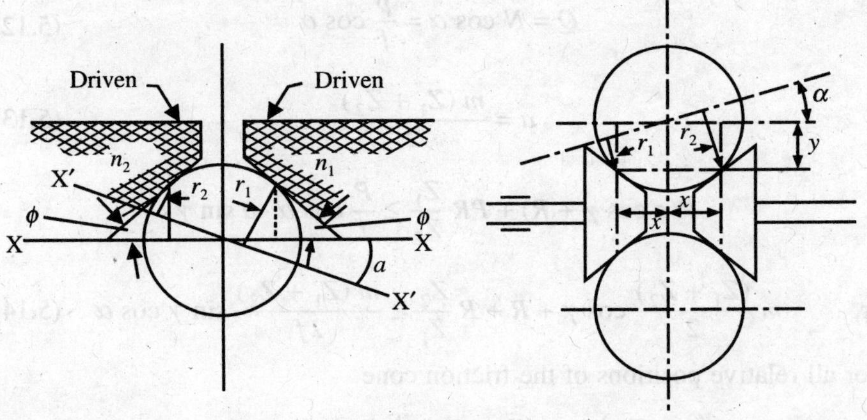

Figure 5.14(a) Figure 5.14(b)

$$i = \frac{n_2}{n_1} = \frac{r_2}{r_1} = \frac{\cos(\phi + \alpha)}{\cos(\phi - \alpha)}$$

Here $r_2 = r\cos(\phi + \alpha)$

$r_1 = r\cos(\phi - \alpha)$

where r = distance of the line of contact from the original axis. Since angle ϕ is constant we can get the following conditions:

Angle	α	0°	90°	90° – ϕ	90° + ϕ
Speed ratio	i	1	–1	0	∞

$\alpha = 90° + \phi$ is of purely theoretical interest.

(II) In Figure 5.14(b) has been shown another case of ball variator with exterior cones as driver and driven elements. Let us assume α as the angle of tilt of axis of rotation of ball.

Here speed ratio $$i = \frac{r_2}{r_1}$$

$$r_1 = (Y - x \tan \alpha) \cos \alpha$$

$$r_2 = (Y + x \tan \alpha) \cos \alpha$$

$$\therefore \quad i = \frac{y + x \tan \alpha}{y - x \tan \alpha}$$

In such cases transmission ratio is not directly proportional to α.

CHAPTER 6

Machine Tool Guides

6.1 Classification of Guides used in Machine Tool

In this chapter we shall deal with the trends in the design of machine tools guide ways, the main purpose of which is to produce the accuracy of motion. The classification of guides generally used in machine tools can be done as under:

In the integral construction the material of the guide is the same as the material of the bed. But since the resistance of guides depends on the surface hardness, it is usually better to use two different materials for table and bed guides with different hardness. If the table guide is made of softer material, it will receive uniform wear in course of use, whereas if the bed guide is of softer material, wear over the entire length will not be uniform. If both are of same hardness, initial friction will be more. In usual practice, the bed guides are made harder than the sliding guides of the table.

In Figures 6.1 and 6.2 the machine tool guides have been classified into various categories depending upon the directions of motions of sliding, shapes, material, construction, frictional behaviours, etc. The main purpose of the guide is to produce accurate directional motion of the guided parts. The choice of a guide for a particular machine tool must be made with proper attention to various factors enumerated in the subsequent paragraphs.

6.2. Wearing of Guides

Machine tool guides may develop wear or surface breaking because of the following few important reasons :

1. Bad quality of working surfaces.

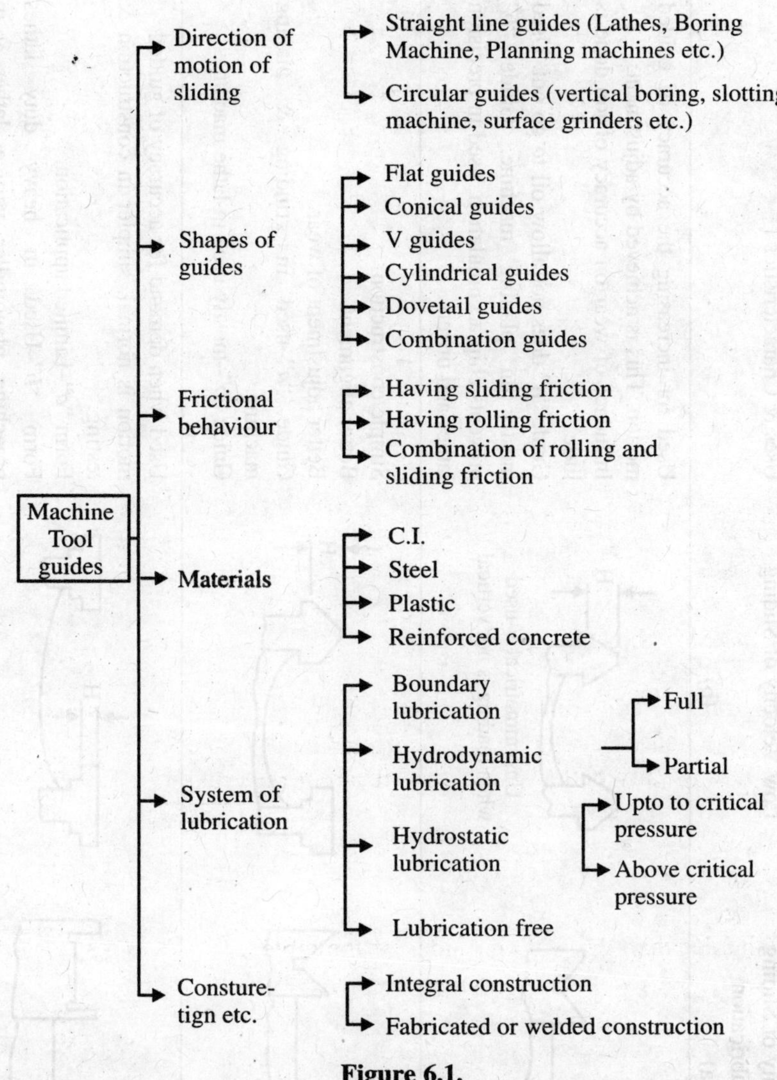

Figure 6.1.

2. Uneven pressure on the guiding surface because of the cutting forces and the turning moments resulting therefrom.
3. Abrasive actions of chips, dirts, etc. entrapped between the working surfaces (quite common in the case of open guides); this can be prevented by the use of plastic or metallic coverings, as are usually common in some grinding machines.
4. Difference in the quality characteristics (mechanical and chemical properties) of the working materials.

Conditions of Working

Types	High Velocity of Sliding Good Lubrication (a)	Low Velocity of Sliding (b)	Uses & Characteristics
V-Shaped	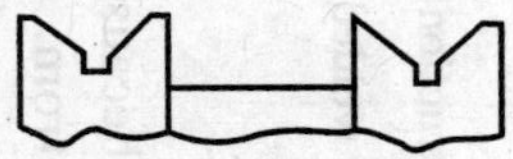	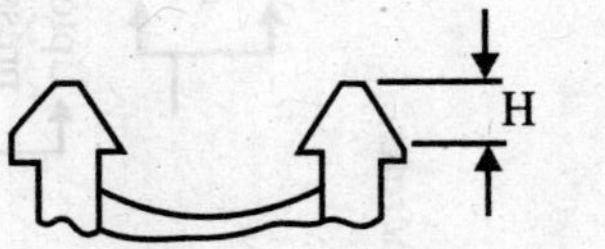 H Unsymmetrical V-used when loading is not vertical	Used for increasing the accuracy of guided motion. This is achieved by adjustment. Influence of wear on accuracy of job done is little. Guide "*a*" does not allow oil to go out. Used mostly in planing machine. Guide "*b*" prevents chip accumulation, used in precision lathes and turrets.
Combination	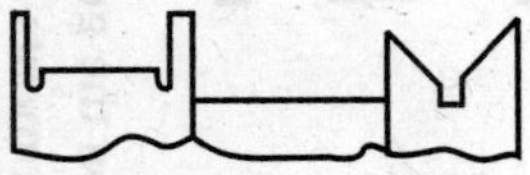	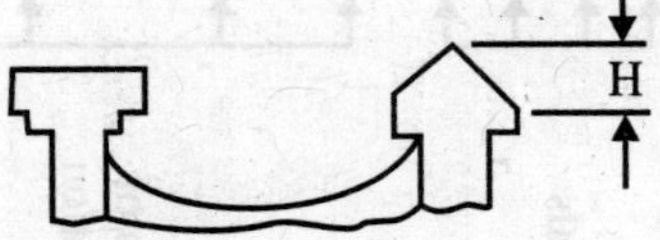 H	Simple construction. Better alignment. Better adjustment of wear. Guide "*a*"–used in grinding & planing machine. Guide "*b*"–mostly used in lathe machines
Straight or Flat	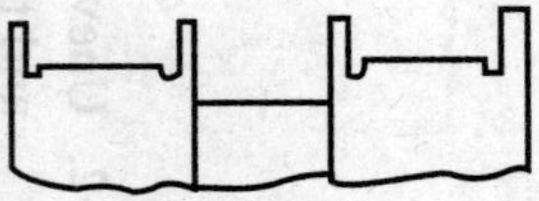	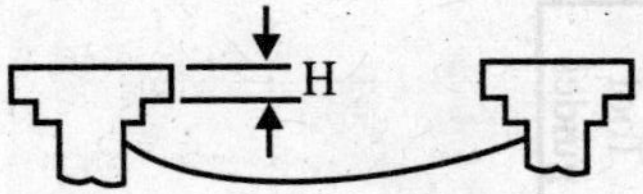 H	Used when demand for accuracy of guided motion is normal, simpler in construction & testing. Form "*a*"–Limited application. Form "*b*"–Used in heavy duty lathes broaching, planomiller, vertices lathes & in columns & cantilever of milling machines, etc.

Dovetail	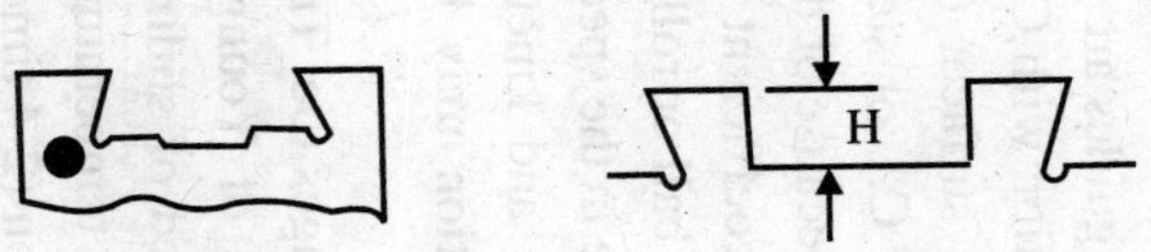 H	Easy to adjust. But does not work properly, when the overturning moment is considerably great. Form "*a*"–used in milling machines. Form "*b*"–In lathes supports.
Straight & dovetail	 H	This is better than the last type so far as overturning moment is concerned. Widely used in transverse slides of cradial drill, verticle lathes etc.
Cylindrical	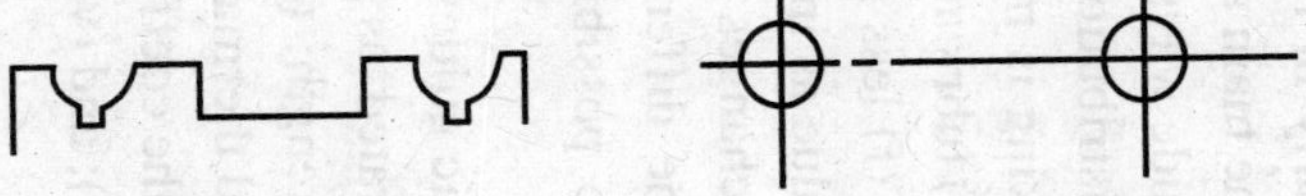	Used in presses, broaching & honing machines. High resistance to wear can be achieved by using "Steel-C.I" combination.

Figure 6.2.

5. Insufficiency in the working characteristics of the thrust bearing at the end of the central spindle of the circular table, (see Figure 6.4a) causing an increase of pressure on the circular guides.

In about 66 per cent of the cases, the wear may be caused because of the improper and insufficient lubrication while in about 45 percent of the cases it may be due to the damage caused to lubrication system.

6.3. Guide Materials

The usual materials for guides are cast iron and steel. The various specifications of such conventional materials have been shown in Chapter 7. The material specifications vary depending upon the type of the machine tool.

The plastic guides are now-a-days being used more extensively in the heavy machine tools, particularly in heavy boring machine, vertical lathe, large planing machines, etc. The main advantages of the plastic guides are as follows: (a) less wear of the guide ways, when used in combination with C.I. or steel; (b) more uniform distribution of pressure on guide surface; (c) possibility of the surface breaking is much less than in case of C.I. or steel guides ; (d) easy fabrication; (e) lubrication may not be needed because of the porosity of the guide material; (f) less friction against sliding (coefficient of friction has almost the same value as in rolling friction in usual ball or roller guides). Coefficient of friction changes very slightly with change in the speed of sliding; and (g) because the difference between the static and kinetic friction is not very great, the possibility of stick slip vibration may be minimized.

As against the above, plastic guides have serious disadvantages too. The disadvantages may be enumerated as follows : (i) low value of Young's modulus, less hardness and strength; (ii) limited range of speed of sliding (usually 60–70 m/min); (iii) bad thermal conductivity; (iv) more temperature deformation; (v) high value of the coefficients of linear expansion (2-4 times more than that of iron and steel); and (vi) humidity also has severe effects on plastic guides.

A large number of experiments have been carried out into the working characteristics of the plastic guides, by A.S. Lapiduce, (S8J G. A. Levit,[33] M. N. Tsirlin,[31] U. N. Sokolov[47] and others of the Machine Tools Research Institute of U.S.S.R., Moscow. Figure 6.8 shows a typical fabrication of plastic guide.

6.4. Temperature Deformation of Guides

If the temperature generated by friction is high, the guides undergo thermal deformation. To take out the heat generated, it is necessary to flood the guiding surface with lubrication oil.

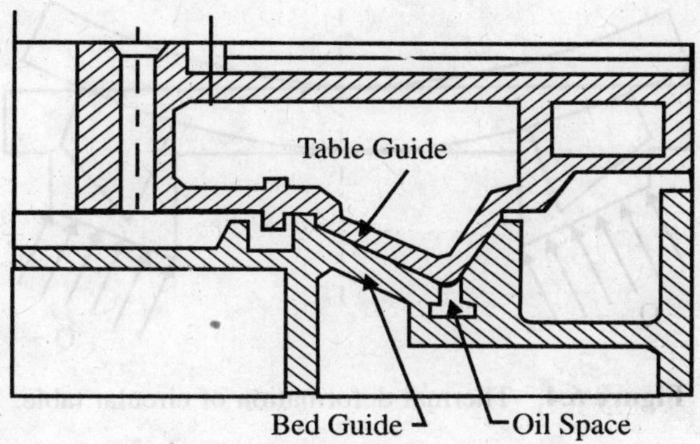

Figure 6.3.

The amount of heat transferred to the lubrication oil may be calculated from the formula given below:

$$Q = \frac{psv\,f}{427} \times 3600, \text{ K. calories/hr}$$

where Q = total heat generated in the guide by friction in *K*. calories;
p = intensity of pressure on the guides in kg/m^2;
s = area of the working surface in m^2;
v = speed of sliding in m/sec;
f = coefficient of friction;
with reference to Figure 6.3.

If t_m is the absolute temperature, to which the oil is raised, because of the frictional beat transfer, then t_m may be obtained from the equation[47]

$$t_m = \left(\frac{Q + C \cdot V \cdot \gamma \cdot t_0}{K_n + K_0 + C \cdot \gamma\, V} + t_0 \right) \tag{6.1}$$

and in the above equation

t_0 = temperature of the surrounding, in °C;
Q = frictional heat generated in *K*. calories per hr;
C = specific heat of oil in *K*. cal/kg. °C;
γ = specific weight of oil in kg/cm^3;
V = volume of oil circulated in cm^3/hr;
t_0 = temperature of oil entering the guide in °C;

K_n = overall heat transfer coefficient from the guide and walls of table to the surrounding medium in K cal/hr. °C;

K_0 = overall heat transfer coefficient from the guide and walls of the bed to the surrounding medium in K. cal/hr. °C.

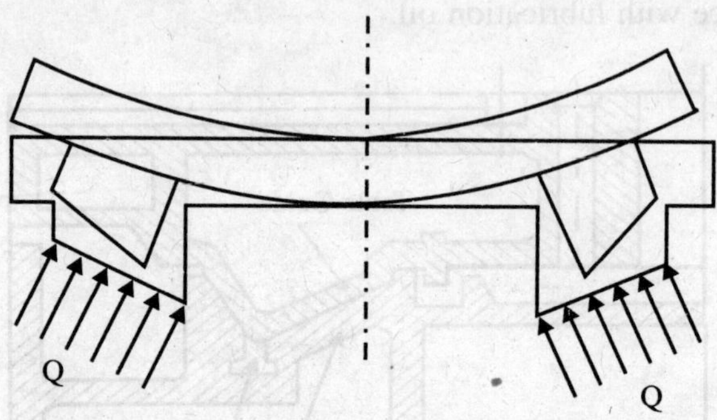

Figure 6.4 Thermal deformation of circular table.

Now K_n and K_0 can be found out separately by the equations of heat balance. In his paper on calculation of the temperature of the oil film in the circular guide of vertical lathe, Sokolov (1956)[47] has shown methods of calculating K_n and K_0.

The temperature t_m to which the oil is raised, causes thermal deformation of the guide. Assuming the table to be a compact rigid body, the circular table of a vertical lathe will be expected to take the shape of a concave curve. This type of temperature deformation of the table considerably affects the accuracy of the work done in the machine. The magnitude of such deformation may become much more than the deformation taking place because of improperly clamping the job on work table. Figure 6.4 shows the thermal deformation of the circular table of a vertical lathe. The deformation can be calculated from the equation (7.6) given in Chapter 7.

Because of the low thermal conductivity of the plastic guide, there are more possibilities of severe thermal deformation of the guide. This limits the range of working temperature of a plastic guide. The normal range of temperature, within which the plastic guides work satifactorily, is 20°C–90°C.

6.5. Liquid Friction in Guides

Usually horizontal circular tables having circular guides are supported on a central spindle, which transmits a portion of the load on the table to the thrust bearing and can be adjusted in the vertical direction by the help of the hydraulic arrangement or some other mechanical methods. In U.S.S.R. machine tools produced by the vertical lathe manufacturing factory (named after Sedin) have hydraulically adjustable thrust bearing.

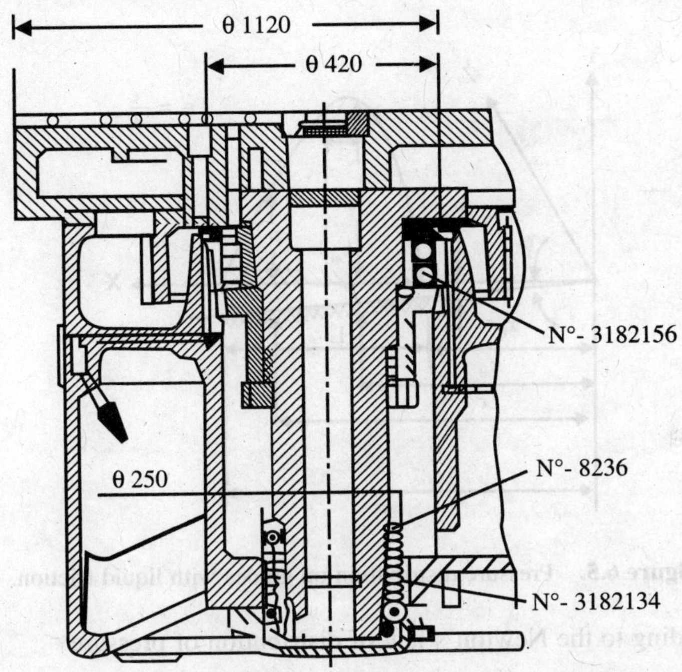

Figure 6.4(a) Showing the horizontal table guided on the bed of a vertical boring machine through central spindle supported on bearings No. 8236 and 3182184. These bearings can be preloaded to relieve the guides of the bed from excessive pressure.

To reduce the kinetic friction between the guiding surfaces to a minimum, it may be necessary to use hydrodynamic lubrication. The upward force generated in the lubrication film under the; dynamic condition partially balances the external load on ihe guide. By external load we mean $G = G_1 + G_w + G_v$ where G_t = weight of the table, G_w = weight of the workpiece on the table and G_v = force contribution because of the cutting force and the overturning moment produced by the eccentricity of tae cutting force. If the entire value of G is balanced by the upward force or the lifting force (as it is popularly known) of the lubrication film, then full hydrodynamic lubrication will be effected and under such condition, the table and the bed will be separated by a film of lubricant of considerable thickness. Due to the hydrodynamic lubrication, in a straight flat guide shown in Figure 6.5, a lubrication wedge is formed in-between the guiding surfaces. This will give rise to a parabolic type of distribution of the guide pressures, the maximum pressure being exhibited at a point, where the thickness of the oil film is hm while the minimum thickness of the oil film is only h_0. (See Figure 6.5).

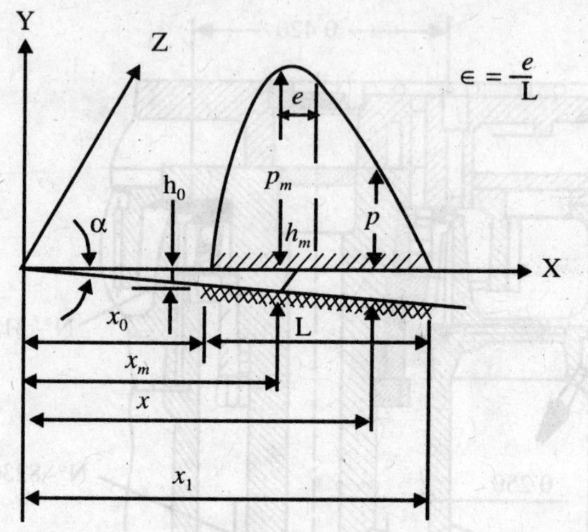

Figure 6.5. Pressure distribution on guides with liquid friction.

According to the Newton's law of distribution of pressure

$$\frac{\partial}{\partial x} \cdot \left(h^3 \frac{\partial p}{\partial x} \right) + \frac{\partial}{\partial z} \cdot \left(h^3 \cdot \frac{\partial p}{\partial z} \cdot \right) + 6\mu v \cdot \frac{\partial h}{\partial x} = 0 \tag{6.2}$$

where h = thickness of the oil film at a point x where the pressure is equal to p;

μ = absolute coefficient of viscosity of oil;

v = velocity of sliding.

Since h_m is the thickness of the oil film at a point where p is maximum, it can be easily said' that $\partial_p/\partial_z = 0$ assuming infinite breadth of the guide.

$$\therefore \qquad h^3 \frac{\partial p}{\partial x} + 6\mu v \frac{\partial h}{\partial x} = 0$$

$$\therefore \qquad h^3 \frac{\partial p}{\partial x} + 6\mu v h = C_1 = \text{constant}$$

when $h = h_m$ p is maximum and hence $\frac{\partial p}{\partial x} = 0$

$$\therefore \qquad C_1 = 6\mu v h_m$$

$$\frac{\partial p}{\partial x} = 6\mu v \cdot \left(\frac{h_m}{h^3} - \frac{1}{h^2} \right) \tag{6.3}$$

Since $h = \alpha \cdot x$, $h_m = \alpha \cdot x_m$

$$\frac{dp}{dx} = \frac{6\mu v}{\alpha^2} \cdot \left[\frac{x_m}{x^3} - \frac{1}{x^2} \right] \tag{6.4}$$

Integrating we have

$$p = \frac{6\mu v}{\alpha^2} \cdot \left[-\frac{1}{2} \cdot \frac{x_m}{x^2} + \frac{1}{x} + C \right]$$

Putting the boundary conditions : $p = 0$, when $x = x_0$,

$p = 0$, when $x = x_1$,

we get

$$C = -\frac{1}{x_0 + x_1} \text{ and } x_m = \frac{2x_1 \, x_0}{x_0 + x_1}$$

Thus,

$$p = \frac{6\mu v}{\alpha^2} \cdot \frac{(x_1 - x)(x - x_0)}{(x_0 + x_1)\, x^2}. \tag{6.5}$$

We may say that $p = AW$, where A is a constant, independent of co-ordinates, that is equal to $\frac{6\mu v}{\alpha^2}$ and W is a factor depending on co-ordinates.

Also.

$$\frac{dp}{dx} = A \frac{dW}{dx}. \tag{6.6}$$

Comparing equation (6.5) with equation (6.4) we get

$$\frac{dW}{dx} = \frac{x_m}{x^3} - \frac{1}{x^2}$$

If we substitute

(i) $\frac{h_0}{x^0} = \alpha$

(ii) $m = \frac{x_0}{L}$

(iii) $M = \frac{m^2 (1 - \xi)\, \xi}{(2m + 1)(m + \xi)^2}$

(iv) $\xi = \dfrac{x - x_0}{L}$

in the above equations, then the equation (6.5) can be reduced to the simple expression

$$p = \frac{6\mu v \cdot L}{h_0^2} M. \tag{6.7}$$

From this expression we can find out the total force of the lubrication film of unit breadth by integrating p with respect to x between the limits x_0 and x_1.

If the total force is equal to P_1, then it can be expressed as

$$p_1 = \frac{6\mu v \cdot L}{h_0^2} \cdot K, \tag{6.8}$$

where

$$K = m^2 \left(\log_e \frac{m+1}{m} - \frac{2}{2m+1} \right).$$

Modifying this equation to suit guides having a finite breadth B and assuming parabolic pressure distribution, Shiebel (1934)[44] has shown that the total resultant force

$$P = \frac{5\mu v}{h_0^2} \cdot K \cdot \frac{L^2 B}{1 + \left(\frac{L}{B}\right)^2} \tag{6.9}$$

In the case of straight guides, as in the case of planing machines or grinding machines slideways, where the length is very great as compared to the breadth, above equation may profitably, be reduced to

$$P = \frac{5\mu v \cdot B^3 \cdot K}{h_0^2}. \tag{6.10}$$

It is apparent from the above equation that the length has got no influence on the magnitude of the total force. The value of h_0 in the above equation may be found from the following expression

$$h_0 = \Delta E_1 + \Delta E_2 + \Delta E_3 + \ldots,$$

where ΔE_1, ΔE_2, ΔE_3 etc., are the partial errors (in geometrical forms)-micro and macro errors because of surface roughness, weaviness, non-parallelity, and so on. From the equation

$$K = m^2 \left(\log_e \frac{m+1}{m} - \frac{2}{2m+1} \right),$$

a characteristic curve for K can be plotted (see Figure 6.6). For $m \approx 1$ wc get the maximum value of K. The maximum value of K as read from the curve is 0.0267 and accordingly the maximum total force on the guide can be equal to

$$F = \frac{0.133\mu v \cdot B^3}{h_0^2} \tag{6.11}$$

In the working regime of machine tools, it is better to have hydrodynamic lubrication, since the resultant upward force of the hydrodynamic lubricant film balances a fraction of the load on the table and thereby reduces the wear on guide and increases its working life.

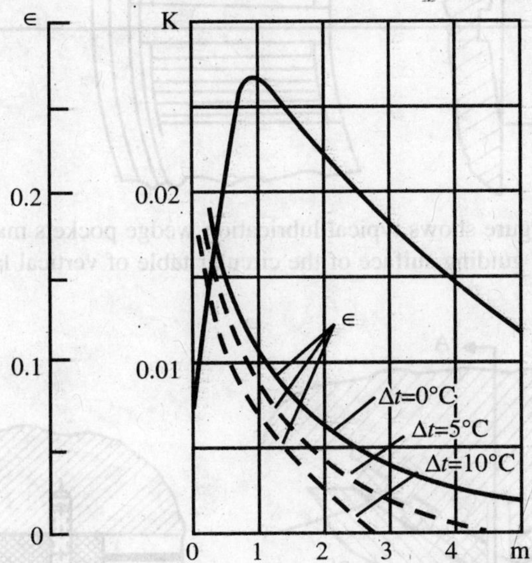

Figure 6.6 The curve showing the variation of the factor K with the factor "m".

To provide hydrodynamic lubrication, it is therefore necessary to machine on the rubbing surface of the guide several pockets in the shape of wedge complementary to the lubrication wedge,

The distribution of lubrication oil in such cases are controlled by special electrical system, where two or three positioned valves are operated'by electomagnet and time relays. Generally two electrical motors are needed, one for running the supply pump and the other for operating the valve pusher.

6.6. Kinetic Friction and Stick-slip Vibration

At low speed of sliding, the coefficient of friction is high, but as the speed increases, the value of the kinetic friction comes down.

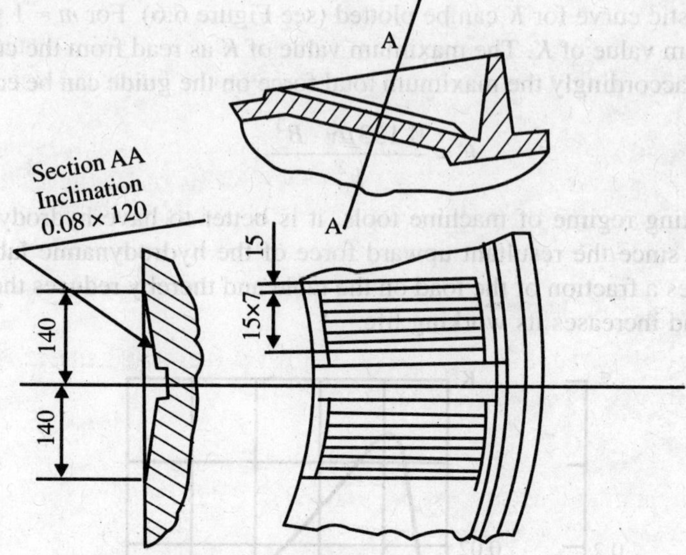

Figure 6.7 The figure shows typical lubrication wedge pockets made by scraping on the guiding suiface of the circular table of vertical lathe.

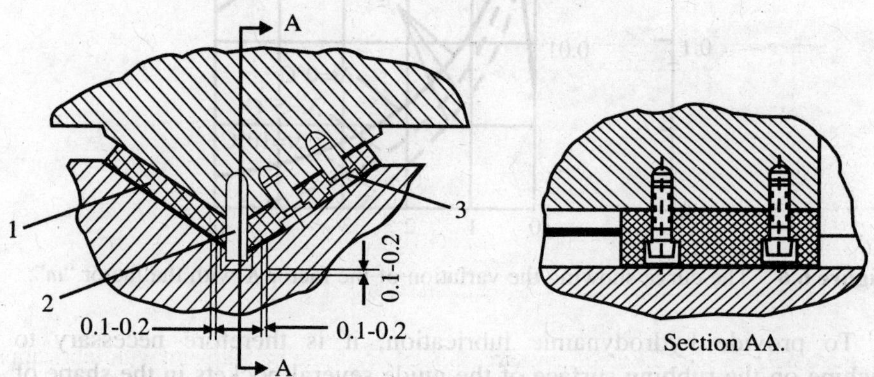

Figure 6.8 Shows typical methods of fabricating the plastic guides.

At low sliding speeds, the phenomenon of stick-slip is more prominent. The amplitude of such vibration may be regarded to be proportional to the difference between the coefficients of friction at the end of stick and at the end of slip. Because of the occurrence of stick-slip in the sliding mechanism of heavy machine tools, it may be necessary to have more power input to the machine at the start.

Not much work has been done to study the stick-slip behaviour of the guides at very low speed of sliding. In the Machine Tools Research

Organisation of U.S.S.R., the work is being carried out to study such characteristic behaviour of the machine tool guides, under the condition of low sliding speed.

Rolling friction instead of sliding friction considerably reduces the difference between the static and kinetic coefficients of friction and thereby reduces the possibility of stick-slip vibration. Plastic material because of high internal friction also reduces the possibility of stick-slip vibration to a great extent.*

6.7. Specimen Calculation for Guides having Lubrication Wedge, as shown in Figure 6.7

Figure 6.9 shows a typical pressure distribution, found in actual practice with either straight guide mechanism or circular table mechanism. The inclination a is made to provide lubrication wedge, necessary for hydrodynamic lubrication.

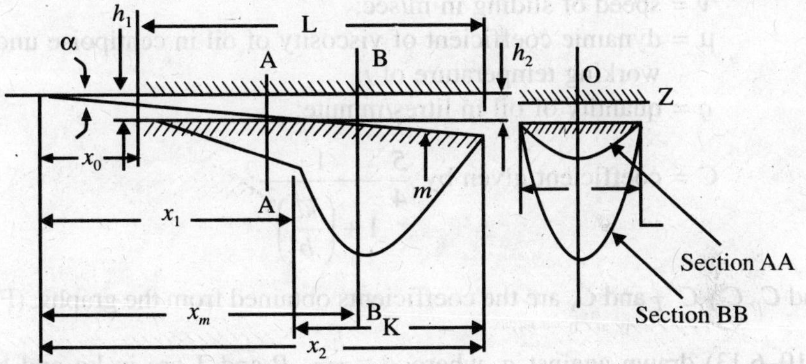

Figure 6.9 Distribution of pressure on guide of finite breadth and with lubrication wedge pocket

The design of such guide requires the following calculations:[32]

(a) Hydrodynamic force:

$$P = \frac{1}{981 \times 10^4} \cdot \frac{2\mu v l^2 b}{3h_1^2} \cdot C \, . \, C_v; \tag{6.12}$$

---

*Minimization of stick-slip vibration, increases the sensibility of the guided motion. It is mainly with this purpose that the sliding friction is being superseded by the rolling friction.

(b) Frictional force:

$$T = \frac{1}{981 \times 10^4} \cdot \frac{\mu \cdot v \cdot l \cdot h}{h_1} \left(C'_T + C''_T \, C \right) \quad (6.13)$$

(c) Coefficient of friction:

$$f = \frac{T}{P} = 1.5 \frac{C'_T + C''_T \, C}{C \, . \, C_p} \, \frac{h_1}{l}; \quad (6.14)$$

(d) Quantity of oil flowing in guides:

$$q = \frac{60}{10^3} \cdot \frac{v \cdot h_1 \, l^2}{b} \, C \, C_v; \quad (6.15)$$

where, l = length, in figure, in mm;
b = finite breadth, in figure, in mm;
kl = length of the inclination, in mm;
α = angle of inclination;
h_1, h_2 = minimum and maximum thickness of oil film in mm;
v = speed of sliding in m/sec;
μ = dynamic coefficient of viscosity of oil in centipoise under the working temperature of t;
q = quantity of oil in litres/minute;
C = coefficient given by $\frac{5}{4} \frac{1}{1 + \left(\frac{kl}{b} \right)^2}$;

and C_p C'_T C''_T and C_v are the coefficients obtained from the graphs, (Figures 6.10–6.13) drawn against a, where $a = \frac{h_2}{h_1}$. P and T are in kg and kg.mm respectively. If i is the number of such lubrication pockets (all around the perimeters, in case of circular table or along the length in the case of straight line table mechanism), then P and T should be multiplied by i.

That is to say

$$P = \frac{1}{981 \cdot 10^4} \cdot \frac{2\mu \cdot v \cdot Pb}{3h_1^3} \cdot C \cdot C_p \quad (6.16)$$

and

$$T = \frac{i}{981 \cdot 10^4} \cdot \frac{\mu \cdot v \cdot lb}{h_1} \left(C'_T + C''_T \cdot C \right) \quad (6.17)$$

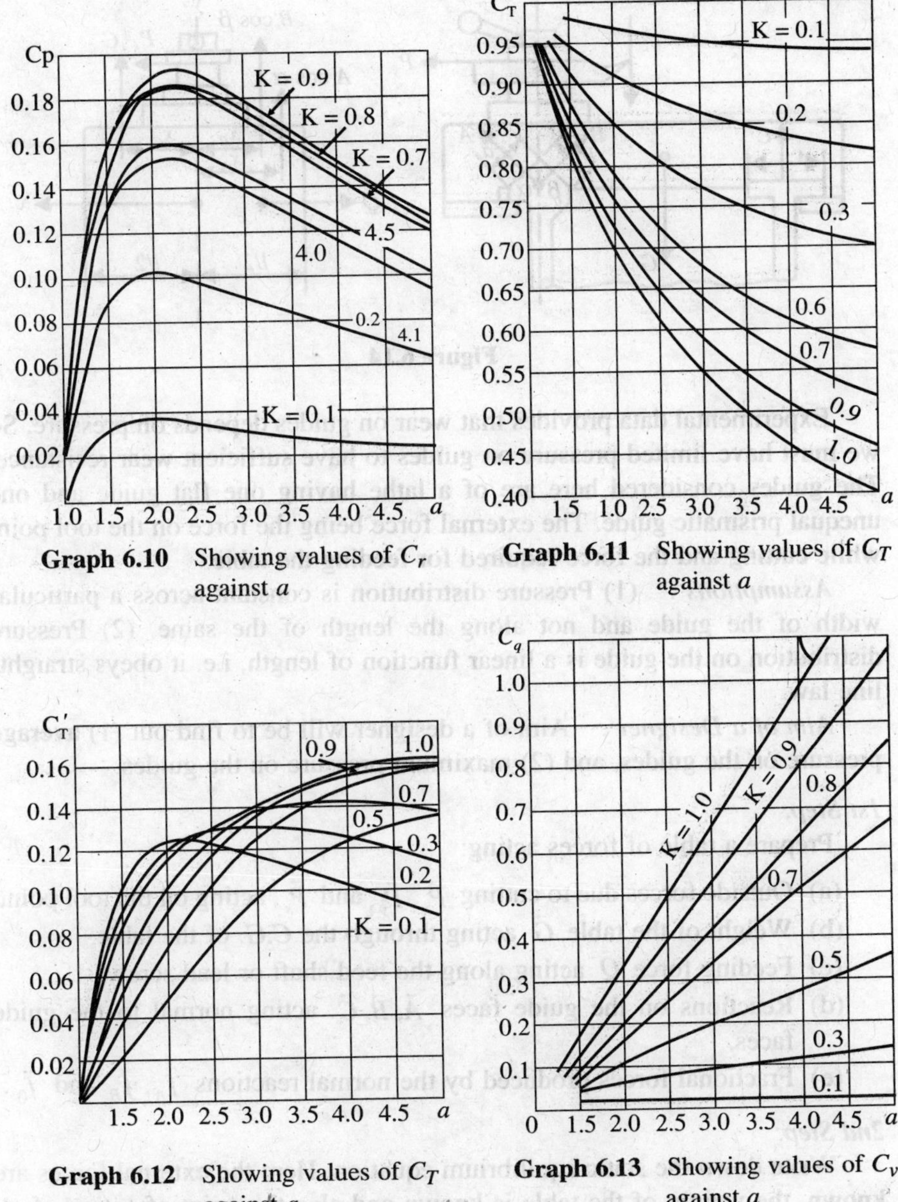

Graph 6.10 Showing values of C_r against *a*

Graph 6.11 Showing values of C_T against *a*

Graph 6.12 Showing values of C_T against *a*

Graph 6.13 Showing values of C_v against *a*

6.8. Methods of Calculating Pressure on Guides

In Figure 6.14 we have shown the sketch for calculating the pressure distribution in guides.

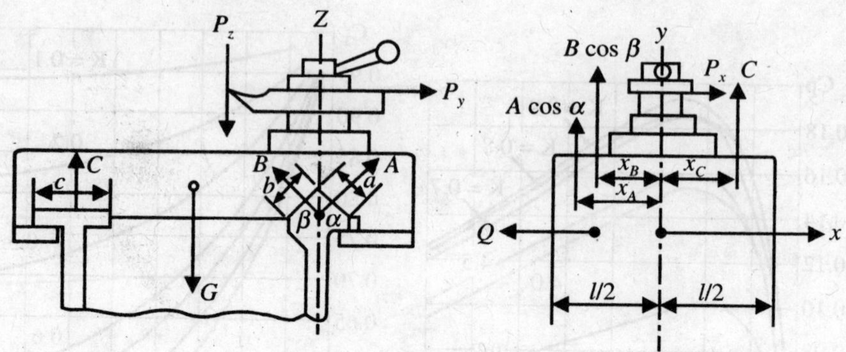

Figure 6.14

Experimental data provides that wear on guides depends on pressure. So we must have limited pressure on guides to have sufficient wear resistance. The guides considered here are of a lathe having one flat guide and one unequal prismatic guide. The external force being the force on the tool point while cutting and the force required for feeding the table.

Assumptions : (1) Pressure distribution is constant across a particular width of the guide and not along the length of the same. (2) Pressure distribution on the guide is a linear function of length, i.e. it obeys straight-line law.

Aim of a Designer : Aim of a designer will be to find out (1) average pressure on the guides, and (2) maximum pressure on the guides.

1st Step:

Prepare a table of forces acting:

(a) Outside forces due to cutting $\vec{P}_x$, $\vec{P}_v$, and $\vec{P}_s$, acting on the tool point.

(b) Weight of the table $\vec{G}$ acting through the *C.G.* of the table.

(c) Feeding force $\vec{Q}$ acting along the feed shaft or lead screw.

(d) Reactions on the guide faces $\vec{A}, \vec{B}, \vec{C}$ acting normal to the guide faces.

(e) Fractional forces produced by the normal reactions $\vec{f}_A$, $\vec{f}_B$ and $\vec{f}_0$.

2nd Step:

Write down the static equilibrium equation. Here the external forces are known, the weight of the table is known and also the point of action of all these forces. The unknown quantities, naturally, are $\vec{A}, \vec{B}, \vec{C}, \vec{Q}$ and the point of application of the first three forces, i.e. X_A, X_B and X_C (in the diagram in Figure 6.14). Hence there are seven unknown quantities, for which we require seven independent equations. For static equilibrium

$$\sum X = 0 \quad (1) \qquad \sum M_z = 0 \quad (4)$$

$$\sum Y = 0 \quad (2) \qquad \sum M_v = 0 \quad (5)$$

$$\sum Z = 0 \quad (3) \qquad \sum M_z = 0 \quad (6)$$

3rd Step :

From the 1st four equations we can find out the unknown quantities $\vec{A}, \vec{B}, \vec{C}$ and $\vec{Q}$

4th Step :

To find the average pressures on any one of the guides, we have:

$$P_A = \frac{A}{a \cdot l}$$

$$P_B = \frac{B}{b \cdot l}$$

$$P_C = \frac{C}{c \cdot l}$$

where a, b, c—widths of the guides; and

l—length of the guides

But usually in machine tools we get non-uniform distribution of pressure and, therefore, it is necessary for us to calculate the maximum pressure. This is essential while calculating for the wear of guides.

5th Step :

Equation due to elastic deformation : When the number of unknown quantities is more than the number of static equilibrium equations, it is essential to find out other relations with consideration to deformations.

Moment of the reaction about Y axis:

$$M_v = M_{\text{I}} + M_{\text{II}},$$

where M_{I}— moment of reaction acting on the first guide ; and

M_{II}— moment of reaction acting on the second guide.

This will become the seventh equation necessary for solving all the seven unknown quantities.

6th Step :

Find out the Maximum pressure on guide : As per our second assumption and considering only one face of the v-guide, we can represent the pressure distribution diagram as shown in Figure 6.15, where N is the midpoint of the rectangular pressure distribution in the diagram, while G represents the centre of gravity of the total presssure diagram.

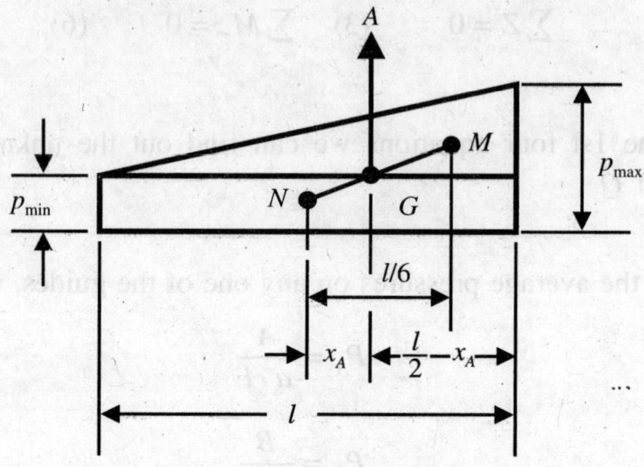

Figure 6.15 Pressure distribution on guide.

$$\text{Total force} = \frac{p_{\max} + p_{\min}}{2} \times l \cdot a,$$

where a—width of the face of guide under consideration.

$$\frac{\left(\dfrac{p_{\max} - p_{\min}}{2}\right) l}{p_{\min} \cdot l} = \frac{x_A}{\dfrac{l}{6} - x_A}$$

and $p_{\min} = 2p_{A\nabla} - p_{\max}$, where $p_{A\nabla}$ – average pressure;

$$\therefore \qquad p_{\max} = \frac{p_{A\nabla}\ (6x_A + 1)}{1} \qquad (6.18)$$

Out of this equation four different cases may arise namely:

Case I if $x_A = 0$ $\quad p_{\max} = p_{A\nabla}$.

Case II if $x_A = \dfrac{l}{6}$ $\quad p_{\max} = 2p_{A\nabla}$.

In the first case the figure of distribution will be a rectangular one, or, in other words, this will give the best possible and uniform distribution of pressure. In case II the distribution of pressure will be of triangular nature and therefore presents the worst condition of distribution-the extreme condition of distribution.

Case III $0 < x_A < \frac{l}{6}$ This is a real case. That is to say in practice in all machine tool guides we expect a conditional distribution of this nature.

Case IV $x_A > \frac{l}{6}$. In this case the full length of the guide will not be in contact. The slide may be inclined at an angle. A case may arise when we shall get a distribution as shown in Figure 6.16 which can never be allowed in actual practice.

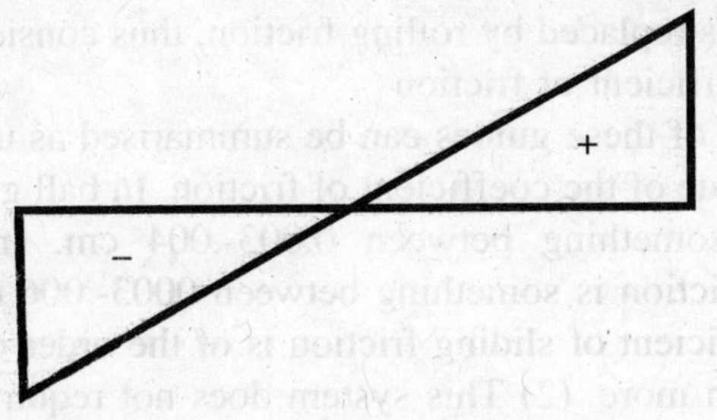

Figure 6.16 Distribution of guide pressure if $x_A > \frac{l}{6}$.

Recommendations. The following recommendations can be mads regarding the permissible maximum pressure on guides. (The recommendations are valid for C.I. guides only). For low velocity of sliding (lathe, milling and similar machines):

$$p_{\max} = 25 \text{ to } 30 \text{ kg/cm}^2.$$

For the guides of special purpose machines, working in the heavier regime of metal cutting, the above value of $p_{\max}$ is diminished by 25 per cent approximately. For guides of heavy machine tools (planer, planomiller, etc) $p_{max} = 10$ kg/cm^2, when the velocity of sliding is low. With high velocity of sliding in a heavy machine tool $p_{max} \leq 2\text{-}4$ kg/cm^2.

In practice, very often the grinding machines permit an average value of guide pressure $p_A \approx 0.5$ to 1 kg/cm^2. For grinding machines only, the maximum permissible pressure is not specified. Instead only the average pressure on guide is specified.

N.B. If the guides are C.I.-Steel, almost the same specifications hold good. If the guides are made of steel only, the values given above may be increased by 20-30 per cent. *Standardised values of the relative dimensions.* The dimension H in Figure 6.2 can be selected from the following series in *mm*;

4, 5, 6, 8, 10, 12, 16, 20, 25, 32, 40, 50, 65, 80, 100, 125, 160, 200, 250, 320.

For fiat guides, H should be in the range of the series 8 to 100 mm. For triangular or V guides, it should be selected from 6 to 320 mm. For dovetail guides, H should be selected from the series 4 to 80 mm. Once the value of H is chosen, the other relative dimensions can be obtained.

6.9. Guides having Rolling Friction

Now-a-days we are using more and more the ball and roller guides in machine tools. The increased use is due to the fact that in such guides, the sliding friction is replaced by rolling friction, thus considerably reducing the value of the coefficient of friction.

Advantages of these guides can be summarised as under :

(1) Low value of the coefficient of friction. In ball guides the coefficient of friction is something between 0.003-.004 cm. In roller guides ihe coefficient of friction is something between 0003-.006 cm, while in sliding guides the coefficient of sliding friction is of the order of 0.15 to 0.20 or in some cases even more. (2) This system does not require any lubrication of the guides. (3) Wear of the ball or roller guides is much less than ordinary sliding guides. (4) Durability is high. (5) Permissible higher speed; and (6) Accurate displacement of the slides. Of course the increased cost of ball and roller guides is a definite disadvantage.

Ball and roller guides are usually used in machine tools, which require (a) high velocity, (6) accurate displacement of the guided motion, (c) durability, and (d) less wear.

It should be noted in this connection that the possibility of stick-slip vibration occuring in such a guide is also remote. The guided surface must be finished to a high surface quality, since here we have got point contact or line contact instead of an area contact in sliding guides and therefore any small error in waviness, will greatly affect the accuracy of the guided motion. The rolling friction guides are always used in combination with steel tables and beds.

For ball guides. The maximum pressure on the guide can be calculated from the formula of contact stress:

$$p_{max} = 91.8 \sqrt[3]{\frac{P}{k^2 \cdot d^2}} \text{ kg/cm}^2 \qquad (6.19)$$

where P— load on in each ball in kg;

d— diameter of the ball in mm ;

$$k = \left(\frac{1-\mu_1^2}{E_1} + \frac{1-\mu_2^2}{E_2} \right) \frac{\text{mm}^2}{\text{kg}},$$

where E_1 and E_2–modulus of elasticity of the 1st order for the material of the ball and guide respectively in kg/mm^2; and μ_1, μ_2– values of the Poisson's coefficient for the materials respectively.

For Roller guides. Maximum contact stress in such case can be calculated from the formula:

$$p_{\max} = 0.798\sqrt{\frac{q}{k \cdot d}} \approx 0.8\sqrt{\frac{q}{k \cdot d}}\ \frac{\text{kg}}{\text{mm}^2} \tag{6.20}$$

where q—load per unit length of the roller in kg/mm;
d—diameter of the roller in mm ;

$$k = \left[\frac{1-\mu_1^2}{E_1} + \frac{1-\mu_2^2}{E_2}\right]\frac{\text{mm}^2}{\text{kg}} \text{ (as in case of ball guides).}$$

We may also write

$$p_{\max} = 80\sqrt{\frac{q}{10 \cdot k \cdot d}} = 25.2\sqrt{\frac{q}{k \cdot d}}\ \frac{\text{kg}}{\text{cm}^2}$$

where q—is in kg/cm.
d—in mm.
k—mm^2/kg.

Usual material of guide, in such cases, is steel, for which the permissible maximum contact stress p_{max} = 30000 – 35000 $\frac{\text{kg}}{\text{cm}^2}$. For C.I. this value is little lower, but it is seldom used. The recirculating ball guides are now very much in use, particularly in some grinding machine table, honing machines, etc.

Combined guides. The main object of using sliding guides is to have accuracy of guided motion and that of using rollers is to reduce friction. The two aspects can be combined as shown in Figure 6.17.

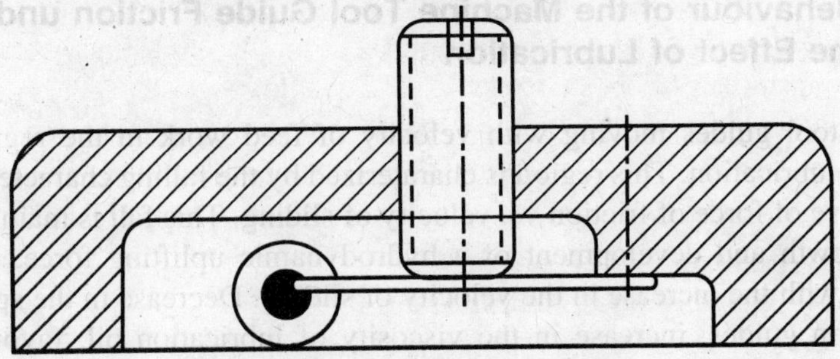

Figure 6.17 Combination Guide.

Some universal grinders do use this design. By adjusting the screw down, we put pressure on leap spring and the reaction is transferred to the roller, so that the sliding friction is decreased. By adjusting the screw properly we can divide the full weight of the table between the rolling and the sliding surfaces. Generally we use four rollers below the table.

In using guides having combination of rolling and sliding friction, if we have a total normal load of N, this load is distributed into two N_1 and, N_2 acting normal to the sliding contact and rolling contract respectively. If the coefficient of sliding friction is denoted by f and rolling friction by λ (cm), then the total force of friction F is given by:

$$F = f\,N_1 + \frac{\lambda}{r} N_2, \tag{6.21}$$

where r—radius of roller.

If the coefficient of proportionality of normal load is denoted by K, then

$$K = \frac{N_1}{N_1 + N_2} = \frac{N_1}{N}.$$

Under such conditional substitution, the overall force of friction F can be represented in the form

$$F = \left[K\,f + \frac{\lambda}{r}(1 - K) \right] N. \tag{6.22}$$

And consequently the overall coefficient of friction can be represented by

$$f_{ov} = Kf + \frac{\lambda}{r}(1 - K).$$

Depending upon the value of $K(0 \le K \le 1)$, the value of the overall coefficient of friction f_{av} oscillates between maximum value of coefficient of sliding friction at one extremity and the same of rolling friction at the other extremity.

6.10. Behaviour of the Machine Tool Guide Friction under the Effect of Lubrication

Machine tool guides moving with velocity of feed work in the region of boundary lubrication. This region is charaterized by the falling characteristics of the curve of force of friction vs. velocity of sliding. This fall is mainly due to the growth and development of a hydrodynamic uplifting force, which increases with the increase in the velocity of sliding. Decrease in the specific pressure on guides, increase in the viscosity of lubrication oil, favourable distribution of load on the guides-all these factors increase the steepness characteristics of the force of friction as the sliding speed increases and decrease the value of critical velocity, wherefrom starts the regime of liquid friction. The author himself found out experimentally the characteristic relationship between the coefficient of kinetic friction of machine tool guides and the velocity of sliding under different types of lubricants and specific pressures on guides.

Various curves are plotted to show the characteristic relationship between coefficient of friction and velocity of sliding under different types of lubricating oils and specific pressures on the guides (Figures 6.18, 6.19 and 6.20). The values of the coefficient of static friction f_n, was 0.25 in the case of industrial–45 oil, 0.24 to 0.26 in the case of Avtol-18, and 0.19 in the case of anti-stickslip oil. The lower value was obtained when the specific pressure was 2 kg per cm^2, and the higher value when the specific pressure was 0.5 kg per cm^2.

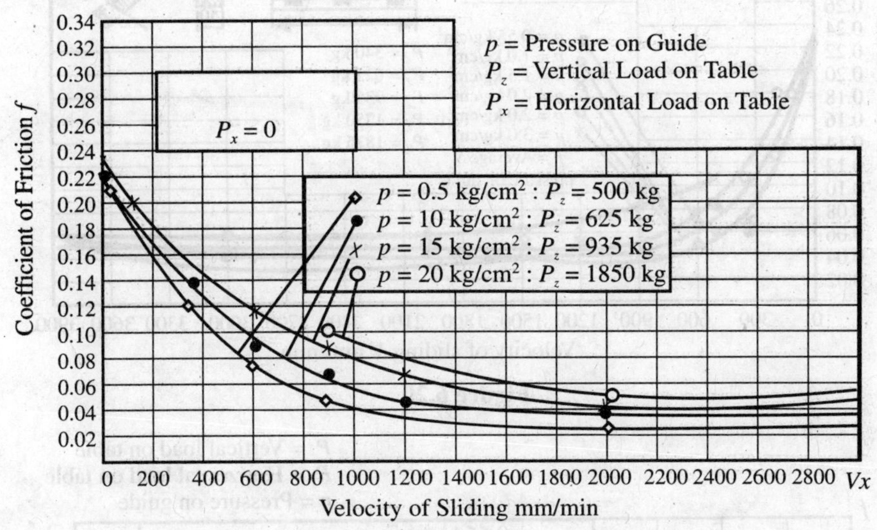

Figure 6.18

The coefficient of friction, f_0, under very low speeds of sliding (say 6 mm per min), was 5 per cent less than the coefficient of static friction.

The lubricating oils used have the following coefficient of viscosity:

(1) AvtoI-18 $\mu = 0.075$ kg. sec/m^2 at $t = 20°C$
(2) Industrial 45, $\mu = 0.021$ kg. sec/m^2 at $t = 20°C$
(3) Antistick slip oil μ. = 0.0068 kg. sec/m^2 at $t = 20°C$.

In Figures 6.21 and 6.22 we have made a comparative study of the behaviour of different lubricating oils.

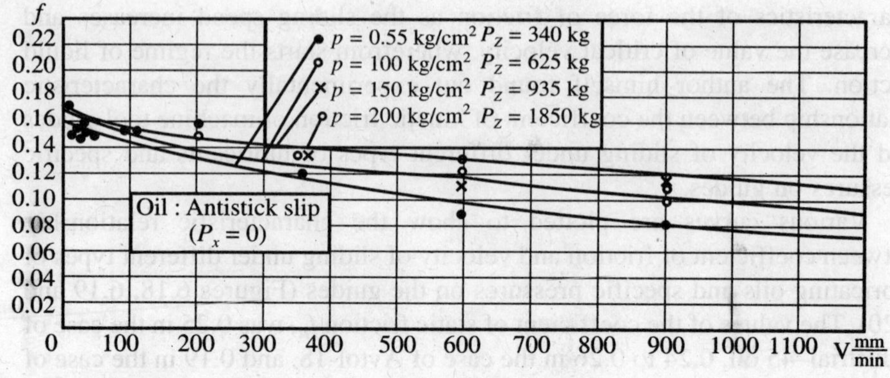

Figure 6.19

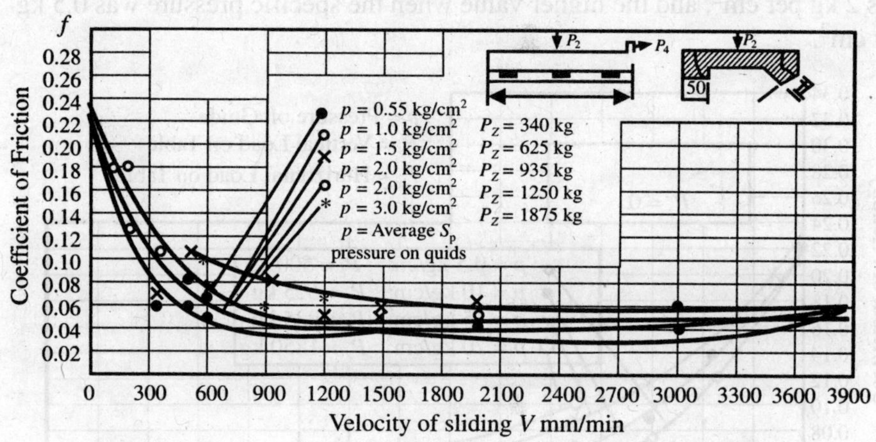

Figure 6.20

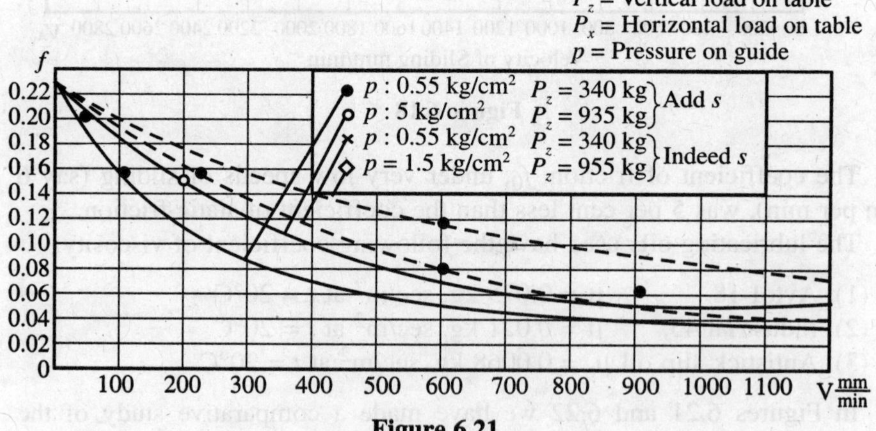

Figure 6.21

6.11. Fundamental Types of Circular Guides

1. *Flat Ring Guides.*

   Simplest in manufacturing. Used in machines having larger diameter table and in machines where guides are not expected to participate in sharing the major part of the radial load.

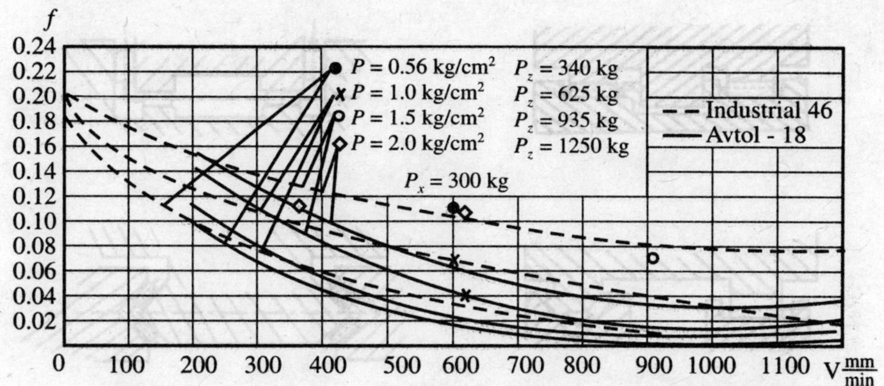

Figure 6.22 Comparative study of different lubricating oils.

2. *Conical Guides.*

   Construction is relatively simple. Used where guides are required to share a portion of the radial thrust also. Angle of the cone is usually 30°. It permits free deformation of the table under temperature.

3. *V-Guidcs.*

   Complicated in construction. Takes radial as well as axial load. Also prevents overturning moment. Such shape can save lubrication oil to a greater advantage.

4. *Spherical or Ball Guides.*

   Construction gives easy rotation, so necessary for accurate angular displacement, for example, in rotary table of a jig boring machine.

6.12. Accuracy and Wear of Machine Tool Guides

Accuracy of machine tool guides depends upon the technology of manufacture and conditions of working. Correct selection of material should depend on optimizing the working conditions under which the operations are done on the machine tools.

The rate of wear of the guides under normal condition lies within the range of 0.03 to 0.12 mm/year. But under severe conditions of working, it can be sometimes even more than 0.3 mm/year. Theoretical and experimental investigations done by A.S. Pronikov show that the accuracy of the product and vibration characteristics of the machine tools are influenced by the amount of wear of the guides and the form of wear follows a certain pattern which could be determined by the method of Pronikov.[1]

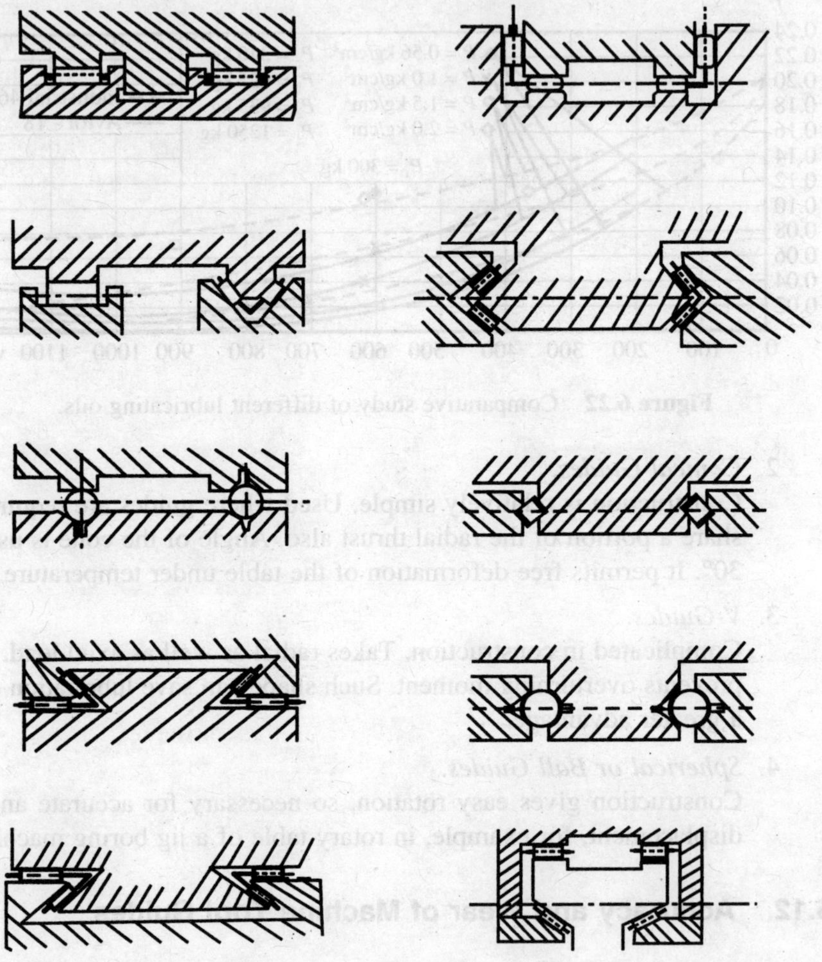

Figure 6.23 Various types of ball & roller guides for straight line motion.

Normally the guides of the bed are harder than the guides of the carriage or table. This is because the guides of the bed should undergo less wear than the mating parts.

6.13. Hardness of the Different Guide Materials

Material	*Hardness*
Grey cast iron	40–50 Rc.
Mehanite casting iron	45–53 Rc.
C-steel (case hardened)	56–62 Rc.

Machine tool guides are being made now-a-days of hardened steel, having hardness ranging from 52-58 R_e For heavy duty machine tool, as for example, the guides of the vertical boring machine, the plastic guides are very often used. The variety of plastic material for machine tool guide is textollite.

Heat-treatment of the guides can be done by flame hardening process. But now-a-days flame hardening process is more and more being replaced by induction hardening.

Problem I: You have been asked to design straight flat guides for precision cylindrical grinding machine. The guides are to work under lubrication. The lubrication chosen is antistickslip oil with polar additives having a coefficient of kinetic viscosity μ = 0.007 kg. sec/m^2. The minimum thickness of oil film under the dynamic condition arising out of microerror and macroerror of the surface is 0.01mm. If the maximum sliding velocity of the carriage is limited to 3 m/min and the maximum load on each guide is not to exceed 750 kg, find out the dimension of the guide (length and breadth) assuming that the maximum intensity of pressure on the guide does not exceed 0.5 kg/cm^2.

Solution : Maximum hydrodynamic force (= P_1) is given by:

$$P_1 = \frac{5\mu \cdot v}{h_0^2} \cdot K \cdot \frac{L^2 B}{1 + \left(\frac{L}{B}\right)^2}$$ as Per notations in the text.

If $L \gg B$, then $\left(\frac{L}{B}\right)^2$ is very great compared to 1.

$$P_1 = \frac{5\mu \cdot v}{h_0^2} K \cdot B^2.$$

But $P_1 \rightarrow P_{\max}$ when $K \rightarrow 0.027$

$$P_{\max} = \frac{0.135}{h_0^2} \cdot \mu \cdot v \cdot B^2;$$

$$B = \left[\frac{P_{max}\ h_0^1}{0.135\mu \cdot v}\right]^2 = \left[\frac{750 \times 10^{-10} \times 60}{0.135 \times 0.007 \times 3}\right]^2 = 0.1167 \text{ metres}$$

That is to say the width of the guide B

= 117 mm (approxim).

Maximum intensity of pressure $= 0.5 \dfrac{\text{kg}}{\text{cm}^2} = 5000 \dfrac{\text{kg}}{\text{m}^2}$

and P = 750 kg.

$$L = \frac{750}{5000 \times 0.117} = 1.285 \text{ metres or } L = 1285 \text{ mm.}$$

Note: It has been presumed here that the weight of the carriage and the vertical component of the cutting force combined together is negligible as compared to the maximum hydrodynamic force generated by this film of lubrication oil.

In problem like the one illustrated above, if we are provided with the weight G of the sliding carriage or table, then we can very easily find out the limiting velocity at which the liquid lubrication will take place. Such value of limiting velocity is given by

$$V_{lim} = \frac{G \cdot h_0^2}{0.135\ \mu \cdot B^2}$$

Usually in machine tool design h_0 is taken to be 0.01 to 0.02 mm. But in the case of heavy machine tools with longer contact surface h_0 = 0.06 – 0.1 mm.

6.14. Design of Guides under Hydrostatic Lubrication

Hydrostatic slides, in contrast to hydrodynamic slides, demand an external fluid pressure to support the load on the machine. Though recent in its application, it is fast replacing conventional slides in the field of heavy duty machine tools, requiring precision in the quality manufacture. The main advantages can be summarised as under: (1) hydrostatic slides offer extreme rigidity; (2) load capacity in such slides is totally independent of the velocity of sliding, and (3) such a system is characterized by a very small frictional drag. With continuity and for steady flow of incompressible fluid we can write the Reynold's equation in the form :

$$\frac{\partial}{\partial x}\left(\frac{h^2}{\mu} \cdot \frac{\partial p}{\partial x}\right) + \frac{\partial}{\partial z} \cdot \left(\frac{h^2}{\mu} \cdot \frac{\partial p}{\partial z}\right) = 6U \frac{\partial h}{\partial x}$$

The above equation is valid for journal bearing having the following nomenlatures: h = thickness of the oil film, x, y, z = the directions of three reference planes, μ = coefficient of viscosity of the lubricant, U = velocity of the journal, and p = pressure in oil.

The same equation will suit the machine tool slides moving under liquid friction, but U will replace the velocity feed of the table'on the guides.

For hydrostatic bearings:

$$\frac{\partial}{\partial x}\cdot\left[\frac{h^3}{u}\cdot\frac{\partial p}{\partial x}\right]+\frac{\partial}{\partial x}\cdot\left[\frac{h^3}{\mu}\cdot\frac{\partial p}{\partial z}\right]=0$$

With reference to the Figure 6.24 if p-pressure in the direction of the flow and the flow or consumption of oil over a length L and breadth unity is Q, then[44].

$$p=\frac{12\mu\cdot Q\cdot l}{h_0^3\,L} \tag{6.23}$$

where P – pressure in the working surface of the distribution canal;

l – width (as shown in Figure 6.24) through which the oil follows;

L – length of the guide;

h_0, h – minimum and maximum oil film thickness (as shown in Figure 6.24) respectively.

The total load that can be carried by the hydrostatic pressure of the fluid is given by

$$P_{ov}=\frac{p\cdot L\cdot(a+b)}{2} \tag{6.24}$$

If U_0 = velocity of sliding between the two successive layers of the oil film when $Y = 0$, then the coefficient of kinetic friction of the slides is given by[22]

$$f=\frac{2\mu\cdot U_0}{h_0 p} \tag{6.25}$$

Ideally $f \approx 0$ since $U_0 \to 0$. But in practice, it is not so, and as a result we get "hydrostatic-hydrodynamic condition".

The total energy lost in the system = energy lost by pumping oil to the accumulator up to the required preessure + energy lost in friction.

If this total energy lost is denoted by E_{loss} then we express it as follows:[72]

$$E_{loss}=\frac{p^2\cdot h_0^3\,L}{3\eta\mu\cdot(b-a)}+\mu\cdot U_0^2\left[\frac{b-a}{h_0}+\frac{a}{h}\right]L, \tag{6.26}$$

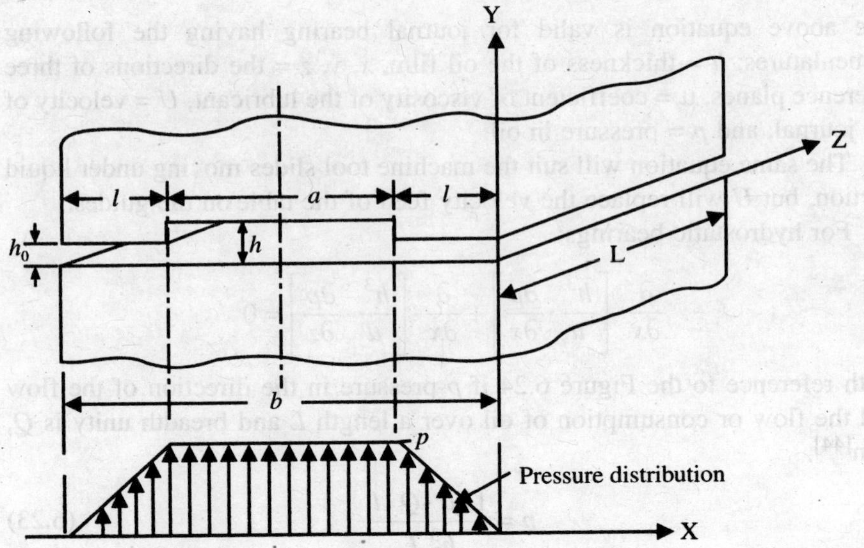

Figure 6.24 Guide with hydrostatic lubrication.

where η–efficiency of the pump.

Substituting equality $P_{ov} = \dfrac{p \cdot L\,(a+b)}{2}$ and representing the design parameter $\dfrac{a}{b} = x$, the above equation is reduced to:

$$E_{\text{loss}} = \frac{4p_{ov}^2 \cdot h_0^3\ L}{3\eta\mu \cdot L^2 b^3\ (1+x)^2\ (1-x)} + \frac{\mu b U_0^2\ L}{h_0} - \frac{\mu b x U_0^2 L}{h_0} + \frac{\mu U_0^2 bx \cdot L}{h}$$

$$E_{\text{loss}} = \frac{1.333\ P_{ov}^2 \cdot L}{\eta\mu L^2 b^2\ (1+x)^2\ (1-x)} + \frac{\mu \cdot bU_0^2\ L}{h_0} - \frac{\mu bxU_0^2 L}{h_0} + \frac{\mu U_0^2 bxL}{h}$$

Differentiating E_{loss} with respect to x, which is a variable parameter we can get the optimum solution for x, which gives minimum E_{loss}.

$$\frac{dE_{\text{loss}}}{dx} = \frac{1.333\ P_{ov}^2 h_0^3 L}{\eta\mu\ L^2 b^3} \cdot \frac{3x-1}{(1-x^2)^2\ (1+x)} + \frac{\mu U_0^2 b \cdot L}{h} - \frac{\mu U_0^2 b \cdot L}{h_0}$$

$$\therefore \quad \frac{(1-x^2)^2\ (1+x)}{3x-1} = \frac{1.333}{\eta} \left(\frac{P_{ov} \cdot h_0^2}{\mu \cdot L \cdot U_0 b^2} \right)^2 \cdot \frac{1}{\left[1 - \dfrac{h_0}{h} \right]}$$

Neglecting the magnitude of h_0/h in comparison, we get the equality

$$\frac{(1-x^2)^2\,(1+x)}{3x-1}=\frac{1.333}{\eta}\cdot\left(\frac{P_{ov}\cdot h_0^2}{\mu L U_0 b^2}\right)^2. \tag{6.27}$$

When P_{ov}, ≈ 0 then $\dfrac{(1-x^2)^2\,(1+x)}{3x-1}=0$ or in other words the value of $x=\dfrac{a}{b}=1$. Similarly when $U_0=0$, $\dfrac{(1-x^2)^2\cdot(1+x)}{3x-1}=\infty$ which will give a solution $x = 0.333$. Therefore we can conclude that $0.333 \le x \le 1$.

By plotting x against the $f(A)=\left(\dfrac{P_{ov}\cdot h_0^2}{\mu\cdot L U_0 b^2}\right)^2\times\dfrac{1.333}{\eta}$ we get a characteristic curve which runs asymptotic with the horizontal base line at $x = 0.333$.

The usuallly attained value of $f(A)$ in machine tools, make us select x ranging from 0.9 to 0.7 or near about. That is, in most machine tools:[22]

$$0.7 \le x \le 0.9. \tag{6.28}$$

This parameter is very important from the view-point of a machine tool designer. The coefficient of friction in hydrostatic slides is usually equal to zero or less than its value under full hydrodynamic condition. In actual machines using hydrostatic slides, the weight of the table, weight of the job, and the contribution of average cutting force add up themselves to be balanced by the external pressure of the oil.

If the pressure of the hydrostatic film p increases above the value satisfied from the equality below, then we get actual condition of floating :

$$p=\frac{2\,(G_i+G_w+P_u)}{L\,(a+b)} \tag{6.29}$$

where G_i— weight of the table;

G_w— weight of the work piece;

P_z— contribution of the cutting- force (average only acting vertically downwards).

Considerable a.nount of researches is being done in this important field of machine tools by PERA, U.K. and ENIMS, U.S.S.R. Commercial machines with hydrostatic lubrication system in slides have already been marketed. But much more developments in this field have to be effected to result in its wider applicability in modern machine tools.

6.15 Type of Hydrostatic Slides

Hydrostatic slides can be of two types:

(i) open slideways which cannot prevent separation of the main mating surfaces; and
(ii) closed slideways capable of withstanding considerable tilting moments.

The simplest hydrostatic thrust bearing is shown in Figure 6.25.

The oil is pumped at constant supply pressure p_z through a throttle valve with resistance R_1 into the bearing pocket. The throttle is regulated to control the oil flow and develop the required bearing pressure p_b. In case of full hydrostatic lubrication, the upper slide is separated from the bottom by an oil film of thickness *h*. Oil while flowing through the lubricating gap h overcomes another resistance R_2 and the pressure in the pocket drops from p_b, to atmospheric pressure at the point where it emerges from the bearing into atmosphere. Thus, for a constant supply pressure p_z, p_b depends on two resistances R_1 and R_2 R_1 can be regarded as constant under the given conditions and R_2 depends on the thickness of the gap. When gap *h* is reduced, R_2 increases and as a result pressure p_b increases. This determines the rigidity of hydrostatic slideways.

If it is assumed that the bearing pressure p_b drops linearly from edges of the oil pocket to the edge of the bearing, then the upward lift calculated on the basis of volumes of truncated pyramid should equal the external load *W* applied on the bearing. Hence, we can write

$$W = \frac{p_b}{6}[(2L + l)B + (2l + L)b] \qquad (6.30)$$

Type of compensator

The normally used compensators which provide resistance Rt to the oil flow are of two types:

(i) In the form of a diaphragm with a small hole known as 'orifice restrictor' (with a very small ratio of length to diameter).
(ii) In the form of channels known as 'capillary restrictor' (with a large ratio of length to diameter).

Levit and Lure[58][59][60] have studied the comparative suitability of the above two types of throttle and have concluded that capillary restrictors are more suitable because the oil film thickness and hence the bearing stiffness are independent of the oil viscosity. The other advantage of channel type

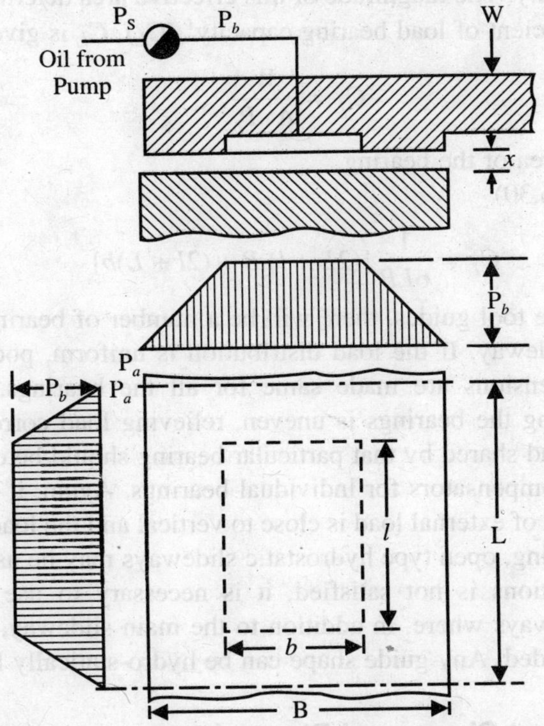

Figure 6.25 Hydrostatic pocket calculation diagram.

throttle is that the diameter being larger, there is very little possibility of throttle clogging even when the filtration of the lubricating oil is not satisfactory.

6.16 Hydraulic Load Relief[60]

In case of hydraulic load relief, the bearing pocket pressure should be such that the upper slide does not float over the lower one for any type of loading. The fundamental difference between hydrostatic lubrication and the hydraulic load relieving system is that in the later case, there is direct contact between the mating slideways.

If the rubbing surfaces of the slideways are ideally flat, no oil should leak from the oil pockets and thearea over which the oil pressure is distributed is equal to the area of the oil pockets. But since the mating surfaces have inherent roughness, certain amount of oil will invariably escape and the effective area of pressure distribution will be much greater than the area of the oil pockets only. The magnitude of this effective area determines, what is known as 'coefficient of load bearing capacity' (C_L). C_L is given by

$$C_L = \frac{W}{A \cdot P_b}$$

where A is the area of the bearing.

From eqn. (6.30)

$$C_L = \frac{1}{6LB}[(2L + l)\,B + (2l + L)b] \tag{6.31}$$

In actual machine tool guides, there will be a number of bearings along the length of the slideway. If the load distribution is uniform, pocket pressure and pocket dimensions are made same for all the bearings. If the load distribution among the bearings is uneven, relieving load corresponding to the functional load shared by that particular bearing should be developed by using separate compensators for individual bearings. Again, if the direction of the application of external load is close to vertical and the load is expected to be nonfluctuating, open type hydrostatic slideways may be used. If one of the above conditions is not satisfied, it is necessary to use closed type hydrostatic slideways where, in addition to the main slideway, an auxiliary slideway is provided. Any guide shape can be hydro-statically lubricated.

6.17 Oil Pocket Shape and Dimensions

The possible shapes of oil pockets which can be commonly used in a hydrostatic slideway are as follows:

(i) In the form of groove or grooves running along the bearing length.
(ii) In the form of groove or grooves running across the bearing length.
(iii) In the form of closed channel.

The coefficient C_L, which is a measure of lead bearing capacity, is used as a criterion to intercompare among the various types of oil pockets. While the maximum value of C_L (near to unity) is the requirement in some designs, a compromise between C_L and the ratio A_p/A (A_p = pocket area, A = bearing area) is the criterion of design in many other applications. To make C_L higher, the pocket dimensions have to be increased which, on the other hand, makes the functioning of the slideways very much risky in the event of lubrication failure. So, the pocket shape which maximises C_L with lower value of A_p/A is regarded as the optimum. Levit and Lure have reported that the best results are obtained by making the pockets in the form of longitudinal parallel grooves in each bearing.

A serious disadvantage of transverse grooves is the narrow bridging pieces at the ends due to small slideway width. While scraping the oil pockets, these sections may be accidentally cut below the slideway surface, thereby causing excessive oil leaks and a reduction in the hydrostatic pressure. A similar effect may be produced due to uneven wear of the slideways.

For a given input pressure closed type slideways ensure a greater rigidity of the oil film than open-type slideways.

To attain high rigidity, it is necessary to make the clearance h as small as possible. With high quality scraping a minimum design clearance of 15 to 25μ can be maintained which provides a rigidity of the order of 100 kgf per micron.

The coefficient of kinetic friction as reported by Hanovitch, in case of open type slideways, is given by,

$$f = \frac{2\mu v_0}{h_0 p_b} \tag{6.32}$$

where v_0 = velocity of sliding between two successive layers of the oil film

h_0 = minimum oil film thickness

μ = kinetic coefficient of viscosity

If $v_0 \rightarrow 0$. f should be equal to 0, but in practice it is not so and we get hydrostatic-hydrodynamic condition.

6.18. Gas Film Lubrication[62]

Though air was thought to be a slideway lubricant in as early as 1854, it is only since 1950 that the study of gas film lubrication has been noticeably accelerated. Air-lubricated slideways are advantageous because they are protected from contamination and because of the low viscosity of air a machine tool slide can be stopped almost instantaneously in the desired position by switching-off the air-supply. The slideways become virtually wear-proof and because of negligible friction forces, power losses are appreciably small.

The main difference between the behaviour of gas and liquid films is that there is a difference in the viscosity of liquid and gas and furthermore the gas is compressible. However, at low sliding speeds as in the case of machine tools, gas film density remains nearly constant. Most gas lubricating films are-laminar and have negligible fluid inertia. In addition, they are usually isothermal.

Aerostatic slideways can have two types of entry resistancesr (a) simple diaphragm type and (b) annular diaphragm type.

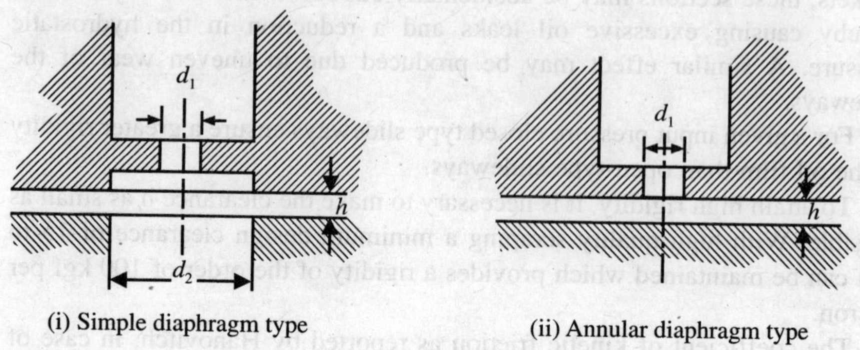

Figure 6.26 Different types of entry resistance.

The difference between these two types is that in the former type, there is a recess cut after the restrictor which ensures a constant cross-section for the flow being independent on the film thickness. On the other hand, in case of annular diaphragm type, the cross-section changes with the film thickness. It has been experimentally established that performance of aerostatic slideways with simple diaphragm type is better than with the other type of entry resistance.

Figure 6.27 shows the calculation diagram of the simplest aerostatic slideway with longitudinal flow. Air is supplied at a constant supply pressure p_s. In the supply line the velocity of air is small and often negligible. This air now passes through the circular restrictor with area $\dfrac{\pi d^2}{4}$. This restricted passage may be through either an orifice, or a capillary or a flow control valve. The flow now emerges into the recess of height $h + \delta$ where th'pressure is p_b, before flowing into the lubricating film of thickness h. In doing so, the air has to overcome a resistance and hence the pressure drops from pb in the bearing pocket to the atmospheric pressure p_b at the point where it emerges from the gap into atmosphere or on the edges of each bearing. Length L of each bearing is normally many times greater than B and consequently the influence of adjoining sections on airflow can be neglected. For a constant supply pressure p_s, the bearing pressure p_b depends on the entry resistance and the resistance of the lubricating gap. Normally the entry resistance is constant depending on the particular design of the restrictor. But the resistance offered by the lubricating gap depends on the film thickness. When the film thickness is reduced, the bearing pressure pb increases and so does the bearing load. Thus increase in pb at a reduced film thickness h determines the rigidity of aerostatic slideway.

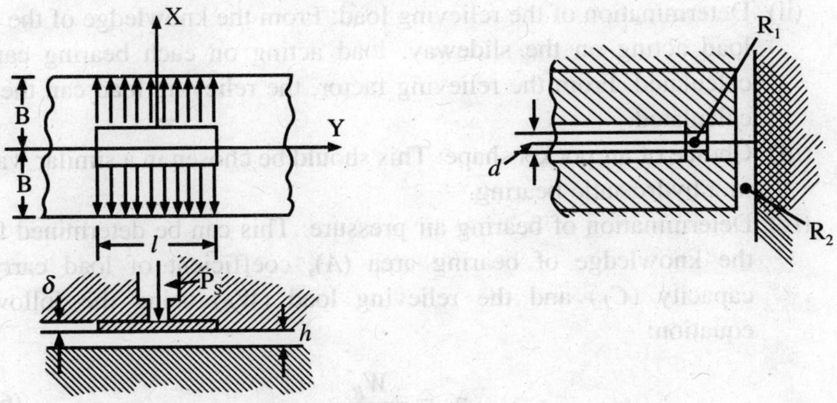

Figure 6.27 Calculation diagram.

Load carrying capacity of the aerostatic slideway can now be cal-culated on the basis of the pressure diagram and afterwards deducting the load due to atmospheric pressure.

Flow requirement

Weight flow through the restrictor orifice (Q_0) can be calculated by the following approximate equation.

$$Q_0 = 0.76.a. \sqrt{\frac{(p_1 - p_b)\, p_b}{T_0}} \tag{6.33}$$

where a = area of the orifice

p_1 = air pressure before orifice

p_b = bearing pressure

T_0 = absolute temperature of supplied air which usually falls a little below the ambient temperature.

6.19 Design Procedure for Aerostatic Slideways

For designing an aerostatic slideway, usually the load, the required stiffness, and the supply air pressure are given. The following calculation sequence is recommended:

(i) Determination of the relieving factor: This depends upon the tolerable frictional power that will be lost during slideway traversing. Normal value is 0.6 to 0.75.

(ii) Determination of the relieving load: From the knowledge of the total load acting on the slideway, load acting on each bearing can be calculated. From the relieving factor, the relieving load can then be calculated.

(iii) Choice of air pocket shape: This should be chosen in a similar way as in a hydrostatic bearing.

(iv) Determination of bearing air pressure: This can be determined from the knowledge of bearing area (A), coefficient of load carrying capacity (C_L) and the relieving load (W_R) using the following equation:

$$p_b = \frac{W_R}{A \cdot C_L} \tag{6.34}$$

Value of C_L depends on the type of air-pocket as shown.

Table 6.1

Type of air pocket	$\frac{A_p}{A}$	C_L (Experimental values)
Two parallel longitudinal narrow grooves	0.05	0.26
Three parallel transverse narrow grooves	0.06	0.45
Closed rectangular channel	0.09	0.48

(v) Determination of the pocket dimensions: The important pocket dimensions are the equivalent length (1) and breadth (b) of the air pocket. The equivalent length of the pocket is determined from the equivalent length of the slideways, number of bearings and the minimum clearance between the bearing edge and the pocket end (which is normally 0.3 of mideway width). The equivalent pocket width can then be determined from the equation of load bearing capacity.

(vi) Determination of air film thickness: This can be determined from the required value of bearing rigidity.

(vii) Determination of the area of the restrictor orifice: It should not be normally more than 0.5 mm in diameter.

(viii) Determination of the compressor capacity: Knowing the. orifice area and the supply air pressure, the air flow rate can be deter-mined. From the air flow rate and the supply air pressure, compressor capacity can be evaluated.

6.20 Influence of Hardness of Material on Guide Wear

Though hardening one or both the elements of sliding pair has often been practised to increase wear resistance, it has been frequently observed that harder materials wear more. It has been reported by Angus that surfaces of hardness below 160 Brineil has a good wear resistance under condition of boundary lubrication because such surfaces contain large amount of ferrite. Thus, hardness itself is not a criterion for wear resistance particularly under controlled conditions. However, where the ingress of dirt and metallic particles is unavoidable, hardening of surfaces is advantageous.* Nelsson, however, reported that the hardness ratio should be in the range of 2 to 3 and the slide whose length is shorter should be harder. Pal and Basu have confirmed this and have observed that total wear is reduced, in such case, by 30 per cent. Hardening, however, becomes increasingly ineffective with increased sliding distance.

6.21 Effect of Surface Preparation

The major ways in which the guideway surfaces are normally prepared are surface grinding, cup grinding and push scraping. Nelsson has found that scraped surfaces result in minimum wear whereas the cup-ground surfaces suffer from the highest wear. The combined wear of the cup-ground/periphery ground pair is the highest whereas scraped/scraped pair or cup-ground/scraped pair has minimum and almost equal amount of wear. The main criterion in this respect is the fullness ratio of the surfaces. The order of increasing wear is the same as the order of decreasing fullness ratio. The added advantage of scraped surface is that it has ability to hold lubricants. Nelsson further concluded that with the increase in sliding distance, influence of surface preparation on wear reduces.

Opitz has recommended different grinding procedures for the mating surfaces when both of them are ground.

*Lapidus recommended that if one of the sliding pair is to be left unhar-dehed, it should be the lower one. Opitz recommended the hardness ratio of the mating surfaces in the range of 1.1 to 1.2 and it does not matter which surface is harder.

Alekseer has suggested that in order to increase the wear resistance the sliding surfaces should be burnished.

Chrome-plating of the slideway surfaces also reduces wear by 3 to 4 times. A deposit of 0.0015 to 0.002 inch of chrome by flash chroming gives the optimum result. When cast iron is used, coating must be sound since the plating will follow the surface pattern exactly and will not fill in the pits. Since the chrome has natural porosity, easy lubrication is also ensured. The recommended hardness of the coating is from 68 to 72 Rockwell C.

6.22 Effect of Micro-Structure and Chemical Composition

Angus reported that undcrcooled graphite is the worst structure for wear and scoring resistance, particularly in the condition of boundary lubrication. This particular structure is often formed on or little within the surface when chills or densers are used and the way to get rid of this is to provide adequate machining allowance.

Regarding effect of chemical composition on wear, it is reported that for minimum wear, carbon content should be about 3.2 per cent.

6.23 Effect of Surface Pressure

It is usually recommended that mean specific pressure in the sliding surface, a large fraction of which is due to the cutting forces, should not exceed 7 kg/cm^2. However in most machine tools, this value is significantly lower. For light duty precision machine it may be as low as 0.5-1.0 kg/cm^2.

6.24 Seizure and Tearing of Guides

High actual pressure on certain areas of contact increases the danger of seizure and tearing. The abnormal high pressures are caused by (a) residual elastic and thermal deformation of main castings, (b) low accuracy machining and fitting of rubbing surfaces and (c) non-uniform contact in reciprocating straight line slideways due to non-uniform wear of the mating surfaces. Contamination of the slideways by the waste products of machining operation is another cause of seizure and tearing. Presence of hard particles of metal and abrasives in the oil film between the rubbing pair causes (a) formation of 'oxide-free' areas which are more prone to seizure, (b) increase of contact str esses at the edges of snatches produced by work-hardened metal chips, (c) a break in the oil film due to longitudinal indentations on the guide surface and consequent reduction of hydrodynamic lifting force and increase of pressure on the slides and (d) absorption of the lubricant in case of insufficient lubrication.

To reduce tearing and seizure, the following practices have been recommended:

(i) Use of rubbing pair resistant to seizure (cast iron against hardened steel, plastics against cast iron etc.).
(ii) Generous lubrication with the circulatory oil flow and fine filtering of the oil.
(iii) The slideway shape should be of raised shoulder type with reliable slideway protection using seals and scrapers.
(iv) To ensure reliability of the lubricating mechanism.

6.25 Error Contribution to the Job Due to Longitudinal Wear on Lathe Guides

Lapidus has shown that during the period of normal wear, the bed guide of a machine tool exhibits a linear relationship between the maximum wear and the period of working. But along the length of the guide, the wear is not uniform. It follows a mixed parabolic curve. If the wear was uniformly distributed over the entire slideway length, there would be little effect on the working accuracy of a machine tool, e.g. lathe. Because of maximum wear in the middle of the bed, a barrel-shaped workpiece would result, even though the workpiece and the bed are sufficiently stiff. The vertical deviation of the cutting edge from its theoretical path due to guide wear has a relatively small influence upon the working accuracy compared to the horizontal deviation.

6.26 Contact Deformation of Guides

A major part of the total elastic displacement in a machine tool which affects its operating characteristics is due to contact deforma-tions in the slideways. Contact deformations in plane surface in sliding contact are determined mainly by the compression of surface irregularities whereas in case of large areas, deformation of microwaves also has a major role to play. Levina[66] reported that if σ is the menn pressure in joint in kg/cm^2 and δ is the mean contact deformation in microns, then.

$$\delta = C\sigma^m$$

Values of C' and m for typical combinations of finished slideway surfaces are shown in the table 6.2.

Table 6.2

VALUES OF *C'* AND *m*

Type of finish	C	m
Scraped-Ground	0.8 to 1.0	0.4
Scraped-Scraped (No. of spots/inch^2 = 15 to 18)	0.8 to 1.0	0.5
Scraped-Scraped (No. of spots/inch^2 = 10 to 12)	1.3 to 1.5	0.5
Scraped-Scraped (No. of spots/inch^2 = 20 to 25)	0.3 to 0.5	0.5

If the loading is eccentric, there is a tilt (ϕ) in the joint given by

$$\phi = \frac{KM}{J} \text{ micron/cm}$$

where J = Moment of inertia of the joint section, cm^4

M = Torque due to eccentric loading in kg-cm

K = Coefficient of contact compliance in micron cm^2/kg

$$= \frac{d\delta}{d\sigma}$$

With large nominal contact area as in the case of actual slideways, contact deformation (δ) is given by

$$\delta = K\sigma$$

where K = contact compliance coefficient

= 1.0 (for σ = 3 kg/cm^2 and slideway face width $\leq$ 100 mm)

= 2.0 (for σ = 3 kg/cm^2 and slideway face width $\leq$ 200 mm)

If σ exceeds 3 kg/cm^2, K values are reduced by 40 to 50%. For bad quality slideway K values are increased by 50 to 70%.

6.27 Contact Deformations in Guides with Clearance

In mating slideways, an optimum clearance is always provided to ensure accuracy of travel with minimum friction loss. J. Pic[61] Pal and Basu[62] have obtained theoretical relationships for determining contact pressure in the slideways having various shapes with clearance. Following the calculation diagram in Figure 6.28, the following equation may be used for determining the contact length in case of flat slideways loaded with torque.

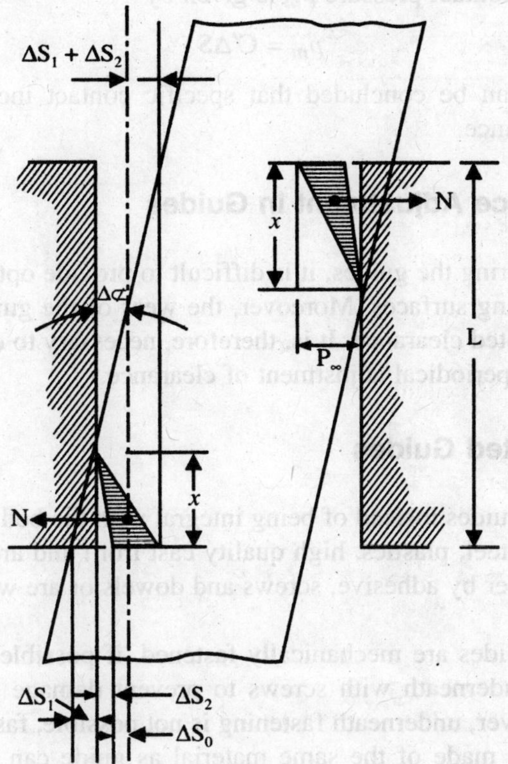

Figure 6.28 Calculation diagram.

$$X^3 - \frac{3L}{2} X^2 - \frac{2X}{C} + \frac{L}{C} = 0 \tag{6.34}$$

where X = contact length, cm

L = length of the slideway, cm

C = a characteristic parameter given by $\dfrac{C'.\Delta S_0.b}{6M}$, $\dfrac{1}{\text{cm}^2}$

where C' is contact stiffness of slideway material, ΔS_0 the clearance, b the width of the slideway and M the loading torque

After determining X from the above equation, maximum contact deformation is obtained from the formula

$$\Delta S = \frac{X \cdot \Delta S_0}{2L - 4X} \tag{6.35}$$

and the specific contact pressure p_m is given by

$$p_m = C'\Delta S \tag{6.36}$$

From these, it can be concluded that specific contact increases with the increase in clearance.

6.28 Clearance Adjustment in Guides

While manufacturing the guides, it is difficult to provide optimum clearance between the mating surfaces. Moreover, the wear of the guides would alter the initially adjusted clearance. It is, therefore, necessary to equip the guides with devices for periodical adjustment of clearance.

6.29 Fabricated Guides

Sometimes, the guides instead of being integral with the bed are made in the (form of strips (steel, plastics, high quality cast iron) and are secured to the cast iron bed either by adhesive, screws and dowels or are welded if the bed material is steel.

When the guides are mechanically fastened, if possible they should be fastened from underneath with screws to prevent damage to the slideway surface. If, however, underneath fastening is not possible, fastening from the top using screws made of the same material as guide can be used. Screw heads are cut and flush ground with the guide surface. In order to increase the transverse rigidity and lateral load-bearing capacity mechanically secured guides usually have an integral key.

6.30 Error Estimation in Guide Design

With reference to the Figure 6.29, *B* is the spacing of the guides and *H* is the height of the workpiece (measured from the centre to the interface of the guide as shown). The total guide forces are P_n on the flat guide and P_t on the '*V*' guides with non-uniform angles.

Width of the '*V*' guide faces are a and *b* respectively. From the analysis we get

$$\beta = 90 - \alpha,\ P_3 = P_Y \text{ and } P_1 = P_Z - P_{II} \tag{6.36}$$

$$P_Z\left(\frac{B-D}{2}\right) - P_Y H = P_{II}.B \tag{6.37}$$

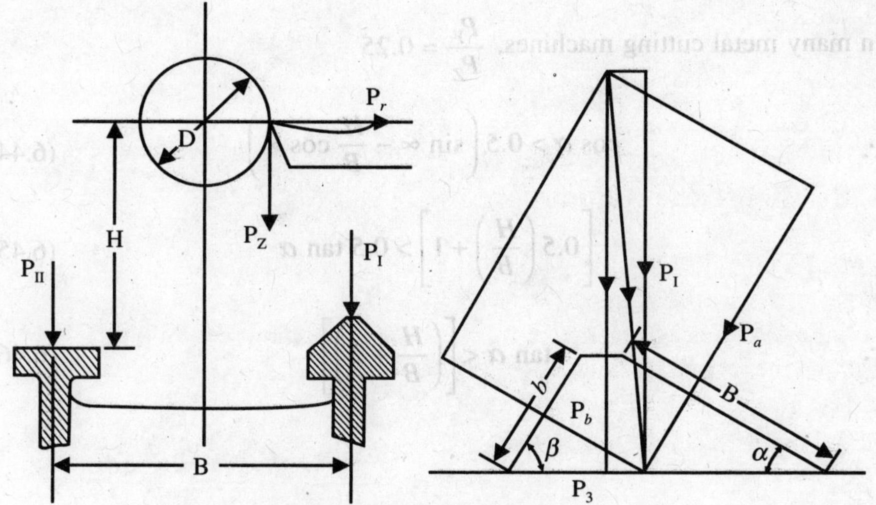

Figure 6.29 Distribution of guide pressure.

$$\left.\begin{aligned} P_{\mathrm{II}} &= P_Z\left(\frac{B-D}{2B}\right) - P_Y \cdot \frac{H}{B} \\ P_{\mathrm{I}} &= P_Z\left(\frac{B+D}{2B}\right) + P_Y\,\frac{H}{B} \end{aligned}\right\} \tag{6.38}$$

and

Again $P_a = P_I \cos \alpha - P_3 \sin \alpha$,

$$= P_z\left(\frac{B+D}{2B}\right)\cos\alpha - P_Y\left(\sin\alpha - \frac{H}{B}\cos\alpha\right) \tag{6.39}$$

Similarly

$$P_b = P_I \sin \alpha + P_3 \cos \alpha \tag{6.40}$$

$\therefore$

$$P_b = P_z\left(\frac{B+D}{2B}\right)\sin\alpha + P_3\left(\cos\propto + \frac{H}{B}\sin\alpha\right) \tag{6.41}$$

Thus P_a, b = mathematical function of D and $\propto$. Tilting of the table takes place if P_a is negative, which cannot be allowed and Assuming $D \to 0$.

$$P_a = \frac{P_Z \cos\alpha}{2} - P_Y\left(\sin\alpha - \frac{H}{B}\cos\alpha\right) \tag{6.42}$$

For no tilting :

$$\frac{P_Z}{2}\cos\alpha > P_Y\left(\sin\alpha - \frac{H}{B}\cos\alpha\right) \tag{6.43}$$

In many metal cutting machines, $\dfrac{P_Y}{P_Z} \simeq 0.25$

$$\therefore \qquad \cos \alpha > 0.5 \left(\sin \propto - \frac{H}{B} \cos \propto \right) \tag{6.44}$$

$$\left[0.5 \left(\frac{H}{B} \right) + 1 \right] > 0.5 \tan \alpha \tag{6.45}$$

$$\therefore \qquad \tan \alpha < \left[\left(\frac{H}{B} + 2 \right) \right] \tag{6.46}$$

CHAPTER 7

Design of Beds, Tables and Columns

In this chapter we shall deal with the various considerations for designing the beds, tables and columns of a machine tool. Materials used for these elements of machine tools are many, each having its own characteristic features. The author discusses only the salient points in the design of beds, tables and columns and the universally used materials for their construction.

7.1. Various Types of Beds Used in Machine Tools-Their Construction and Design Features

The purpose of the bed is to provide true relative positions of all units in machine tools. Some of these units may be integral with beds while some may slide. Depending on this we can have on the machine tool bed (a) fixed connecting surfaces, and (b) guides. The main repuiremcnts of machine tool bed are: (i) Rigidity, which is an important consideration for providing true positions of various units; (ii) Accuracy; (beds should be so designed as to have predetermined initial accuracy). But this is not enough. The accuracy must be maintained even over a considerably long period; and (iii) wear resistance of guides and other moving surfaces.

The materials of construction and their composition and heat treatment must, therefore, be decided on the above fundamental basis. It is only very common in machine tools to have guides integral to the bed casting. But sometimes separate guides are also fabricated on the bed or on the table. The latter is assuming greater importance in the recently designed machine tool. The main factors that predominate in the macnine tool bed are the material and the design.

Usual materials for machine tool bed are C.I. of the various compositions. The table below shows the typical composition of C.I. used in beds and bed guides of different types of machine tools.

Table 7.1

Material for Beds and guide C.I. Types of Machine Tool	*Total carbon (combin-ed+free)*	*Si*	*Mn*	*Cr*	*Ni*	*Strength* σ_B *kg/cm²*	*Brinell hardness number* H_B
Light machine tool	3.4	1.75	0.55	–	0.75	2300	200
Heavy machine tool	3.2	1.00	0.80	0.3	1.25	2800	200
Extra duty machine tool	2.85	1.75	0.80	–	1.50	3100	240

The Brinell hardness numbers, specified above, relate to the guiding surface of the bed only, attained after the usual heat treatment process. Mechanite or modified cast iron is also used in many cases as material for the bed of a machine tool. Chemical composition is almost the same, only the percentage of *C* and *Si* being slightly less. But the difference lies in the method of inocculation. Ferrosilicon, calcium silicides, etc. are normally used as inocculants. Mechanite castings give improved structure and improved machanical properties with a uniform distribution of fine graphite flakes which reduces friction on guides. The free graphites serve the purpose of lubricant.

All C.I. are good for absorbing vibration. But the disadvantage lies in the fact that it is difficult to produce intricate castings without defects or cracks. In the case of compli-catedly shaped castings we shall' have to use other material or C.I. with alloying elements like Ni, Cr, Mo, etc. Advantages of alloy casting can be described as under: (*i*) very high surface hardness even without heat-treatment and therefore the resistance to wear increases; and (*ii*) alloy castings usually have small internal stresses and this is very important in the case of precision machine tools.

The residual stresses are usually relieved in course of time. Ageing or seasoning for'stress relieving is done either naturally or artifically in the heat treatment furnaces at a temperature ranging from 500°C to 600°C, the rate of soaking being 1 hour per inch section. After the heat has been completely soa'ced in, it is allowed to cool slowly in the furnace. For alloy C.I. the temperature is slightly higher.

But for precision machine tools natural seasoning of two to five years is required. Since annealing of grey iron at 730°C to 823°C, followed by slow cooling, reduces tensility, it is always better to go through the process of natural ageing.

Alloy castings are very expensive. As a general rule the weight of the bed is aproximately 50 per cent of the weight of the machine tool itself and the material cost is approximately half the cost of a finished machine >tool bed. That is to say the bed material cost is approximately 25 per cent of the total cost of a finished machine tool. Thus to keep the cost of a machine tool well within the reach of a buyer, we must not go for expensive bed material until and unless it is unavoidable.

Steel is also one of the many materials of machine tool bed. The choice is dependent on the fact that it can be welded. Usually in machine tool we use medium carbon steel having C = 0.3 to 0.5, σ_B = 800 kg/cm^2 and hardness approximately 156 BHN. as cast. Reinforced concrete is recently being used for single and heavy duty machine tools as, for example, for tool grinders. Not much material is available about the characteristics of reinforced concrete as a bed material, but research is being carried out in many parts of the world.

Bed design. The beds should bs designed in such a way as to give maximum rigidity against torsion and bending. A few usual sections of the beds have been shown in the sketches in Figure 7.1. The various sections have been compared on the basis of maximum bending moment and twisting moment that they can successfully withstand. In such comparison three assumptions have been made, *viz.* that (*a*) cross-sectional areas of the different sections shown in Figure 7.1 are the same; (*b*) maximum bending stress permissible is the same for all the sections; and (c) similarly maximum shear stress permissible is the same for all the sections.

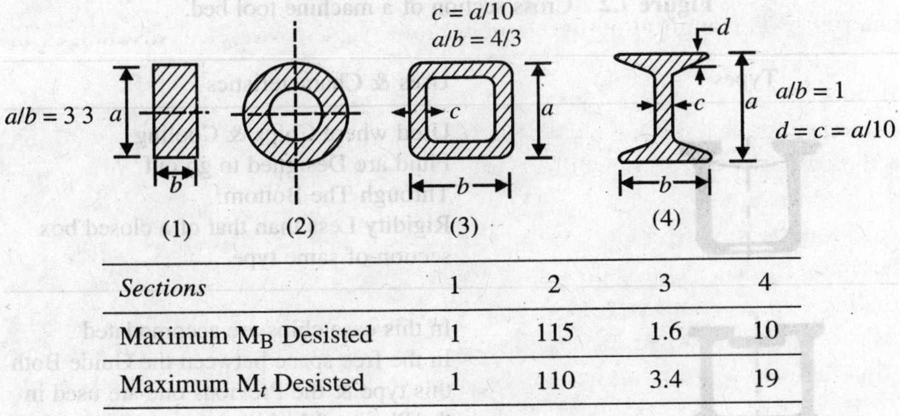

Sections	1	2	3	4
Maximum M_B Desisted	1	115	1.6	10
Maximum M_t Desisted	1	110	3.4	19

Figure 7.1 Comparison Between Different Bed Sections M_B–Bending moment, M_t–Twisting moment.

The comparison at once shows that box is the most suitable section from the point of view of maximum bending moment and maximum twisting moment that it can resist. But in machine tool beds in practise it is not possible to provide complete box section from the point of view of ease in manufacturing though it is the best from the point of view of rigidity. Any deviation from the box section will reduce the rigidity and something must be provided to reduce such losses in rigidity. This may be done by allowing few ribs either straight or at an angle.

Usual cross-section of medium and small-duty lathe has been shown in Figure 7.2. In such cases of design of bed section a hight rigidity has to be provided against high deflection and that will prevent vibration. Ribs are provided in the hollow sections for improving upon the rigidity. They may be rectangular or angular in construction. In Figure 7.2 (*a*)a few typical horizontal bed sections with their relative characteristics have been shown.

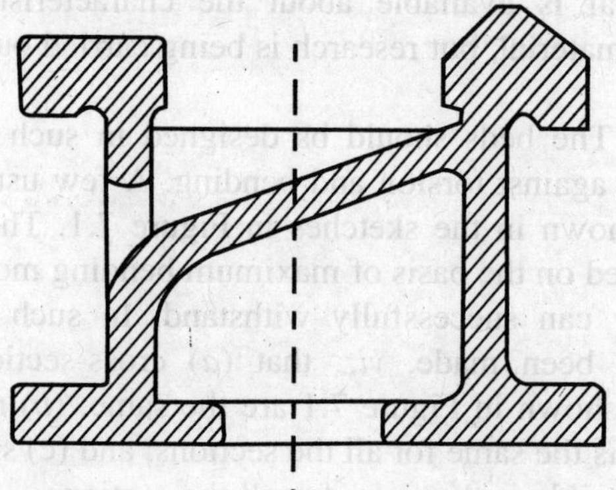

Figure 7.2 Cross section of a machine tool bed.

Types	Uses & Characteristics
	Used where Chips & Cooling Fluid are Designed to go out Through The Bottom. Rigidity Less than that of a closed box section of same type.
	In this case chips are accomodated In the free space between the Guide Both this type & the Previous one are used in the Planing Machine etc.

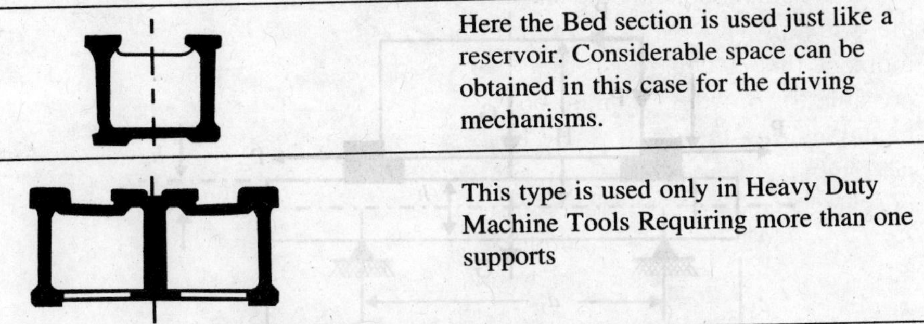

Figure 7.2(a) Typical horizontal bed sections.

DESIGN OF TABLES AND CARRIAGES

7.2. Determination of the forces acting on the horizontal table or a vertical boring machine

In such a machine tool the circular table moves over the circular guides of the bed and carries the load due to the weight of the workpiece as well as due to cutting forces. Some forces also result from the reactions of the holding clamps. With reference to Figure 7.3, let us assume that P_x, P_y, P_z, or three component of the cutting farce in kg;

G_w – weight of workpiece in kg;
d_o – diameter of the work in cm.
d – diameter of table in cm
G_t – weight of table in kg;
h_k, h_n and H – distances (shown in sketch) in cm.

(a) Let $P = G_w + P_x\ kg$ = total external forces acting vertically down (7.1)

Weight of the table G_t is assumed to be uniformly distributed, the intensity of pressure in guide being q.

(b) Turning moment M_k obtained due to the action of P_{cl} is given by the following expression:

$$M_k = Z\ .\ P_{cl},\times (h_k + h_n)\ \text{kg . cm},$$

where Z is the number of clamps used, and P_{cl} is the clamring force.

(c) Intensity of the uniformly distributed load

$$q = \frac{4G_t}{\pi d^2}\ \text{kg/cm}^2. \tag{7.2}$$

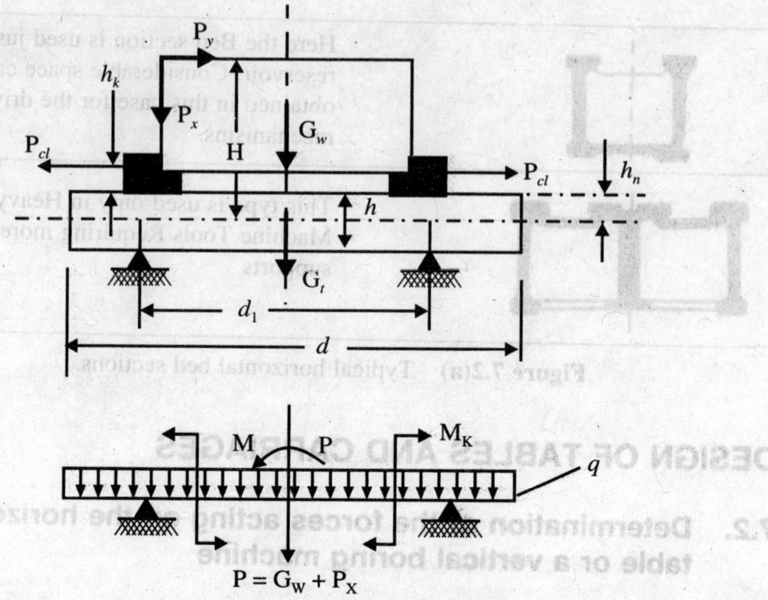

Figure 7.3.

(d) Moment M calculated from the cutting force:

$$M = \sqrt{(0.5\,P\alpha\, d\theta \mp P_v H)^2 + (P_s \cdot H)_s^2} \tag{7.3}$$

Sign – for turning operation; and

Sign + for facing operation.

If σ – intensity of pressure on the guide, then

$$\sigma = \frac{Gw + Px + Gt}{\pi d_1 \times b}, \tag{7.4}$$

where b–width of the guide in cm.

If the guides are working under hydrodynamic lubrication, then calculations should follow the steps given below:

(a) Hydrodynamic force for each lubrication wedge pocket = F

$$= \frac{1}{981 \times 10^4} \cdot \frac{2\mu \cdot v \cdot l^2 \cdot b}{3h^2{}_1} C \cdot C_P.$$

where $$C = \frac{5}{4} \cdot \frac{1}{1 + \left(\dfrac{kl}{b}\right)^2},$$

where kl is the length of inclination of wedge pocket in mm

l– length each lubrication wedge pocket in mm;

C_p– constant dependent on "a" which is equal to $\frac{h_1}{h_2}$,

b– width of the guide in mm;

μ– coefficient of viscosity in centipoise ;

v– speed of sliding in m/sec.

The value of C_p is determined from graph given in Figure 6.10. hx and Aa are maximum and minimum oil film thicknesses respectively in mm.

(It has been assumed that hydrodynamic condition is obtained by providing wedge pockets).

(b) If the number of lubrication wedge pocket = i, then the total uplifting force is given by $Q = iF$. If the job is nonsymmetrical, then the job may be acted upon by a moment M arising out of the eccentricity of the cutting force, then critical lifting force [29] $= Q_{cr} = iF \mp \frac{4M}{d}$,

where d– average dia of the guide in cm ; and

M– turning moment (overturning moment) in kg.cm.

Here the critical uplifting force Q_{cr} tries to relieve the guide of excessive pressure.

(c) Specific pressure on the guide

$$q_{cr} = (G_t + G_w + P_x - Q_{cr}) \times \frac{1}{\pi \cdot d} \text{ kg/cm}^2$$

assuming the breadth "b" to be unity.

If the guides are working under boundary lubrication, then conditions of the expressions will be completely different.

(d) Taking into account the inclination of the table θ in radians (due to elastic and thermal deformation which can be determined depending upon the form)

σ_{max} = (intensity of pressure on the guides)

$$= (G_t + G_w + P_x - Q_{cr}) \times \frac{1}{\pi d b} + \frac{\theta \cdot b}{2 K_{sl}} \cdot \frac{\text{kg}}{\text{cm}^3}, \quad (7.5)$$

where b–width of the guide and K_{sl}–coefficient of slenderness in $\frac{\text{cm}^3}{\text{kg}}$ (If a certain force acting on a body produces a stress equal to σ and deformation δ, then the coefficient of slenderness ratio can be written as $K_{sl} = \frac{\delta}{\sigma}$ which is inverse of rigidity).

Table 7.2[29]

Material of guide	*Width of guide b in mm*	*Value of K_{sl} for calculation purposes, in cm³/kg*
C.I. – C.I.	200 – 300	$(4–5)\ 10^{-4}$
C.I. – C.I.	300 – 450	$(5–6)\ 10^{-4}$
Plastic C. – I.	2 to 2.5 time more than in case using C.I. – C.I. pair	

Permissible value of the specific pressure $\sigma_{derm} \approx$ 0–20 kg/cm². It is very difficult to practically measure the value of K_{sl} since the amount of contact deformation is too small. The inclination angle θ of the work table due to thermal deformation can be written as [47]

$$\theta = \frac{\varepsilon\,(\theta_u - \theta_1)\cdot d_1}{2h} \text{ radian,} \tag{7.6}$$

where d_1 – average diameter of the guide in mm;
h – height of the table in mm;
ε – coefficient of linear expansion (for CI.

$(1 \text{ to } 1.1) \times 10^{-5}$ 1/°C);

θ_u and θ_l are the temperatures of the upper and lower layer of the table in degrees centigrade.

Ratio $\frac{h}{d}$ *of the Work table of a Vertical Lathe and the Number of* Ribs. As a general rule, used by machine tool designers, if diameter of the table is 2000 – 4000 mm, the average ratio $h/d \approx 0.12$ and the number of radial ribs in such case ≈ 0.12. If d = 5000 – 8000 mm, $h/d \approx 0.08 – 0.09$ and the number of radial ribs in such a case is 12 – 16. In case of vertical boring machine the height of work H that can be accommodated can be obtained from the following table.

Table 7.3

Maximum dia. of the Job in mm.	*Average value of the ratio H/d*
2000	0.87
2500	0.75
3000–3200	0.62
4000 – 4500	0.57

Assumptions made in forming the table of data above:[29]

1. In making the above table it has been assumed $P_y = 0.5\ P_z$ which is a reasonable assumption.
2. Weight of workpiece for the table diameter 2000 – 4000 mm ≤ 1000 kg.
3. Maximum value of P_z while machining jobs having considerable height does not exceed 1/3 of total permissible load on the the table as guaranteed in the pamplet.

7.3. Column Design of a Milling Machine and Maximum Deflection Error in a Milling Machine

In Figure 7.4 we have given the schematic diagram of a milling machine with the vectors showing the three dimensional components of the cutting forces acting on the cutter. With the

I_0 = polar moment of inertia of the column

notations shown in the sketch, the deflection of the tool in the direction X, Y and Z are δ_x, δ_y and δ_z respectively. Then

$$\delta_x = P_x\left(\frac{l^3}{3EI_{vv}} + \lambda \cdot \frac{1}{GA} + \frac{(a+y)^2 \cdot l}{G \cdot I_v}\right), \tag{7.7}$$

where I_{vv}– moment of inertia of the cross-section of the vertical column about the axis yy (same as in Figure 7.5;

λ– coefficient of shear distribution for the cross-section under bending in the plane XY;

A– area of the cross-section;

I_o– polar moment of inertia of the cross-section, and

G– modulus of rigidity for the meterial of the column.

In the equation 7.7 the first term represents the deformation of the cutter due to cantilever action, the second term denotes the shear deflection, and the last term represents the torsional deflection of the cutter along the plane XY. Similarly we can find out the magnitudes of δ_y and δ_z. Machine tool designers will usually take $l \geq 1.5\ h$. The cross-section of the column is the same as in the case of the drilling machine, shown in Figure 7.5. Since in a milling machine the value of P_x is maximum, the deflection of the cutter in the direction X will also become maximum. So for the purpose of designing from the point of view of accuracy of the job milled we must equate 8, to the amount of deflection permissible. Usually the magnitude of this permissible deflection will be given as a necessary data to the designer.

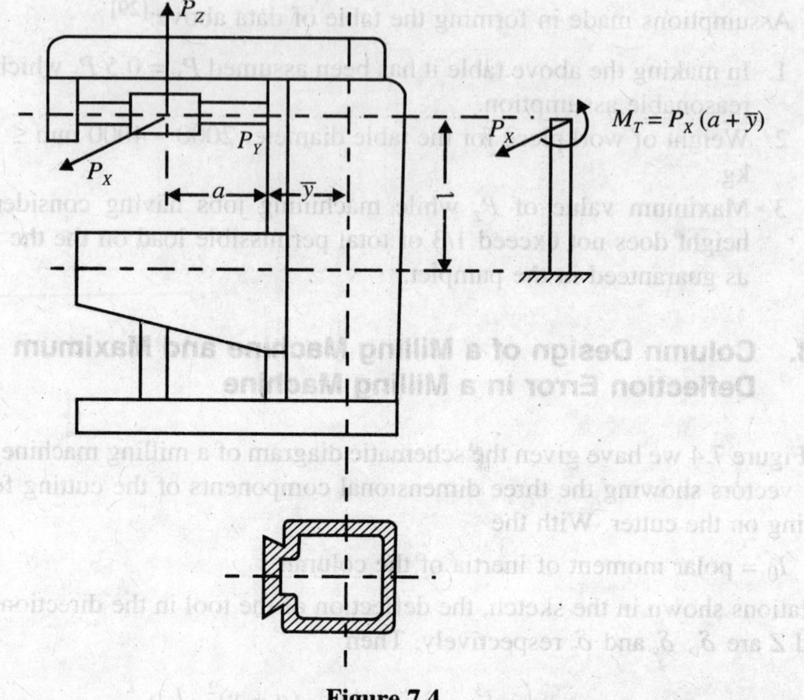

Figure 7.4.

Stresses on the two extreme fibres of the sections can be found out from the bending and direct stresses as in the case of column design of a drilling machine.

7.4. Column Design of Drilling Machine

In Figure 7.5 is shown the diagram of vertical drilling machine with the force or thrust P acting on the drilling during the operation. It is but essential to consider the suitability of cross-section of the pillar from the point of view of maximum per-missible angle of inclination of the column under the dynamic condition, since such inclination resulting from the deforma-tion tells upon the accuracy of the drilled hole. It has been assumed that: (*i*) maximum distance between the bracket supporting the drill head and the table, when the table is in the lowest position = l (measured in between neutral axes); (*ii*) vertical length of the bracket H; (*iii*) I_{xx} and I_w are the moment of inertia of the cross-section about xx and yy axis respectively; (*iv*) θ = angle of inclination of the column as a result of the deformation.

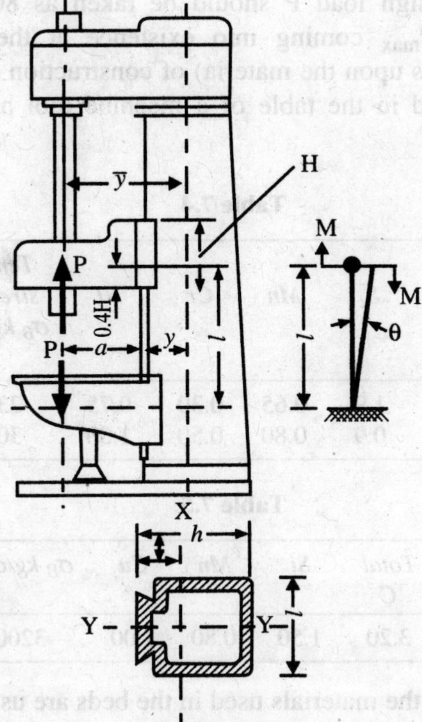

Figure 7.5.

Then $$\theta = \frac{P\,(\overline{y} + a)\,.\,l.}{E \cdot I_{xx}} \tag{7.8}$$

For box type of section maximum permissible value of $\theta^{[29]} = 25/1000$ in radians. Of course there are many varieties of drilling machines ensuring permissible 6 as 1 mm per 1000 mm. ⌊utting the permissible value of θ in the above equation we can find out approximate size of section which will have a moment of inertia I_{xx} about xx as obtained from above. Stresses in the cross-section can be found out from the rela-tion below:

$$\sigma_{\overline{y}} = \frac{P\,(\overline{y} + a)\,.\,\overline{y}.}{I_{xx}} + \frac{P}{A} \tag{7.9}$$

$$\sigma_{h-\overline{y}} = \frac{P\,(\overline{y} + a)\,(h - \overline{y})}{I_{xx}} + \frac{P}{A} \tag{7.10}$$

where A = area of the cross-section;

$$\sigma_{\overline{y}}, \sigma_{h-\overline{y}} \leq \sigma_{Permissble}.$$

The value of the design load P should be taken as 80 per cent of the maximum load of P_{max} coming into existence at the steady dynamic condition σ_{per} depends upon the materia) of construction of the column.

CI. materials used in the table of a machine tool have the following compositiont [56]:

Table 7.4

CI. for tables Type of machine	*Total C*	S_t	*Mn*	*Cr*	*NI*	*Tensile strength σ_B kg/cm²*	*Brinell Hardness Number H_B*
Light duty	3.4	1.9	0.65	0.30	0.75	2300	200
Heavy duty	2.9	0.9	0.80	0.50	1.50	3000	200

Table 7.5

Type of elements	*Total C*	*Si*	*Mn*	*Cu*	*σ_B kg/cm²*	*Brinell Hardness H_B*
Column, pillars, etc.	3.20	1.50	0.80	1.00	3200	225

Heat treatment of the materials used in the beds are usually done to have high hardness of the surface layers. Surface hardening process can be done much more rapidly than the ordinary hardening. Usually it can be done by any one of the processes named below:

(a) flame hardening;
(b) induction hardening; and
(c) electrolytic hardening.

All these processes can be fully or partly automatized.

7.5. Stiffness and Natural Frequency of Machine Beds[78]

Normally a surface grinding machine bed is made of a box section in the form of two beams connected by ribs and walls with large opening. This can be likened to a simulated form shown in Figure 7.6, as two beams separated by an elastic layer. The bed is mounted on a rigid concrete foundation, which is regarded as flexible solid.

Let Y_1 be the bending of the top beam at points $l > x > 0$ (where l is the bed length, and x is the coordinate of the point); Y_2 is the bending of the bottom beam at the same point, b is the width of the bearing surface. Bending in such structure having uniform loading.

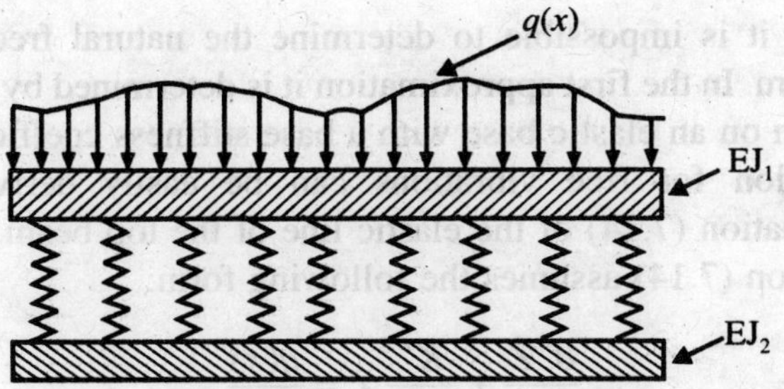

Figure 7.6 Simulated condition of a grinding machine bed for calculation purpose.

$q(x)$ on it can be determined by the equation:

$$EJ_1 \frac{d^4 y_1}{dx^4} = q_1(x) - K.b.\,(y_1 - y_2) \tag{7.11}$$

$$EJ_2 \frac{d^4 y_2}{dx^4} = q_2(x) + Kb.\,(y_1 - y_2) \tag{7.12}$$

Subtracting Eq. (7.12) from Eqn. (7.11) and assuming that the elastic layer is compressed:

$$y_1 - y_2 = Z;\; q_1(x) - q_2(x)\frac{J_2}{J_1} = q(x)$$

thus

$$EJ_1 \frac{d^4 Z}{dx^4} = q(x) - Kb\left(\frac{J_1}{J_2} + 1\right) Z$$

Designating

$$K' = Kb\left(\frac{J_2}{J_1} + 1\right) \tag{7.13}$$

$$EJ_1 \frac{d^4 Z}{dx^4} = q(x) - KZ \tag{7.14}$$

where E is the Young's modulus of material and J_1, J_2 moment of inertia of the beams respectively.

For static loading conditions Eqn. (7.14) is solved by the Krylov method

$$V(\xi) = C_1 V_1(\xi) + C_2 V_2(\xi) + C_3 V_3(\xi) + C_4 V_4(\xi)$$

where $V_1(\xi)$ .... $V_4(\xi)$, are tabulated Krylov functions and C_1 .... C_4 are integration constants depending upon boundary conditions.

At present it is impossible to determine the natural frequency of the described system. In the first approximation it is determined by the vibrations of the top beam on an elastic base with a base stiffness coefficient K.

The equation for free vibrations can be easily derived from the differential equation (7.14) of the elastic line of the top beam. For dynamic loading condition (7.14) assumes the following form:

$$\frac{\partial^4 Z}{\partial x^4} + \frac{K'}{EJ_1} Z = \frac{q(x)}{EJ_1} \tag{7.15}$$

Since in free vibration, the top beam is loaded only by inertia forces, load $q(x)$ should be replaced by those forces. Designating the mass of a unit length of the top beam by p, the mass of element dx will be $p\ dx$, and the inertia force

$$dP_{in} = -\rho dx \frac{\partial^2 Z}{\partial t^2} = q(x)\, dx$$

thus

$$q(x) = -\rho \frac{\partial^2 Z}{\partial t^2} \tag{7.16}$$

Substituting the value of $q(x)$ in equation (7.15) we obtain

$$\frac{\partial^4 Z}{\partial x^4} + \frac{\rho}{EJ_1} \frac{\partial^2 Z}{\partial t^2} + \frac{K'}{EJ_1} Z = 0 \tag{7 17}$$

The equation (7.17) describes the vibrations of the top beam under the following conditions:

(i) When the unit length mass of the bottom beam is at least one order lower than that of the top beam.
(ii) When the moment of inertia of the bottom beam is at least one order greater than that of the top beam.

The first condition applies to the present case being considered. The second condition applies where the bed forms a solid elastic body on a separate rigid foundation. Partial solutions of the vibration equations can be written in Fourier series:

$$\left.\begin{aligned} \alpha(x, t) &= f(x) \sin \omega t \\ \dot{\alpha}(x, t) &= \omega f(x) \cos \omega t \\ \ddot{\alpha}(x, t) &= -\omega^2 f(x) \sin \omega t \end{aligned}\right\} \tag{7.18}$$

And after substituting relationship (7.18) in equation (7.17) and after manipulation we obtain

$$\frac{d^4 f(x)}{dx^4} - M^4 . f(x) = 0 \tag{7.19}$$

where

$$M = \sqrt[4]{\frac{1}{EJ_1}(p\omega^2 - K')} \tag{7.20}$$

By introducing a dimensionlcss quantity $\xi = Mx$ as the variable, the general equation (7.19) is obtained in the following form:

$$V(\xi) = C_1V_1(\xi) + C_2V_2(\xi) + C_3V_3(\xi) + C_4V_4(\xi)$$

where $V(\xi)$ are hyperbolic-trigonometrical functions of the following values:

$$V_1(\xi) = \frac{1}{2}(\text{ch } \xi + \cos \xi)$$

$$V_2(\xi) = \frac{1}{2}(\text{sh } \xi + \sin \xi)$$

$$V_3(\xi) = \frac{1}{2}(\text{ch } \xi - \cos \xi)$$

$$V_4(\xi) = \frac{1}{2}(\text{sh } \xi - \sin \xi)$$

Usually the sidewalls of a machine bed are sufficiently thick to permit the assumption that the ends of the top beam are fixed. It can therefore be considered that the linear and angular displacements at the ends are zero, that is, $C_1 = C_2 = 0$. The equations necessary to determine the integration constants C_3 and C_4 are obtained by equating to zero the linear and angular displacements, when

$$\xi_1 = M_1 l \tag{7.21}$$

since
$$\frac{dv}{dx} = M\frac{dv}{d\xi}$$

We obtain

$$\left.\begin{array}{l} C_3V_3(\xi) + C_4V_4(\xi) = 0 \\ C_3V_2(\xi) + C_4V_3(\xi) = 0 \end{array}\right\} \tag{7.22}$$

To ensure that the solution of the system of equations (7.22) is other than zero, the determinant of the system should equal to zero:

$$D = \begin{vmatrix} V_3(\xi) & V_4(\xi) \\ V_2(\xi) & V_3(\xi) \end{vmatrix} = U$$

Expanding to series

$$\frac{1}{4}(\text{ch}\,\xi + \cos\xi)^2 - \frac{1}{4}(\text{sh}^2\,\xi - \sin^2\xi) = 0$$

or

$$\text{ch}\,\xi\cos\xi - 1 = 0 \tag{7.23}$$

The roots of the above equation are

$$\xi_0 = 0 \qquad \xi_1 = 4.730 \qquad \xi_2 = 7.8532$$

$$\xi_{1,2} \simeq \left(\frac{2i+1}{2}\right)\pi \tag{7.24}$$

where i is the harmonic number.

Taking into account Eqn. (7.20) and (7.21) we have

$$\frac{\xi_1}{L} = \sqrt[4]{\frac{1}{EJ_1}(p\omega_i^2 - K')} \tag{7.25}$$

Solving the expression (7.25) for o>t and taking into account well known ratio $f = \dfrac{\omega}{2\pi}$ cycles/sec, the following formula is obtained for determining the natural frequency of the top beam in the construction shown diagramatically in Figure 7.6

$$f = \frac{1}{2\pi}\sqrt{\frac{EJ_1}{p}\frac{\xi_1^4}{l^4} + \frac{K'}{p}} \tag{7.26}$$

Numerical calculations show that the first natural frequency wholly depends on the base stiffness and the mass per unit length of the top beam, which can be calculated from tlfe following simplified formula:

$$f_0 = \frac{1}{2\pi}\sqrt{\frac{K'}{p}} \tag{7.27}$$

CHAPTER 8

Design of Power Screws of Machine Tools

8.1 Types and Classification

The main purpose of the power screw of a machine tool is to provide accurate and uniform motion of the executive units of a machine tool. For example in the case of a lathe machine the purpose of power screw is to provide the motion of the carriage to slide on the bed; similarly in the case of milling machine it is required for the guided movement of the work table.

Power screws are classified on the basis of accuracy of motion to be executed and the particular machine operation. Apart from this, there is another way of classification depending upon the geometrical parameters of the screws, viz. its shape, size and form the table below gives a reasonable classification of the power screws:

In the table below the power screws have been classified according to its accuracy in manufacture.

The design of such power screw requires the following preliminary steps:

1. design for specific pressure on the screws;
2. design for elastic stability, considering it to be an axially loaded column;
3. design for maximum bending moment and deflection;
4. design for the twisting moment;
5. design based on combined principal stresses.

In order to calculate all the design factors, it is absolutely necessary, right at the beginning, to decide upon the material (hardness and strength), extent of heat treatment and standard thread profile and other geometrical

Table 8.1

Types	*Tolerance on pilch on av. dla. in (1μ = 0.001 mm)*	*Application in machine tools*
I	± 2μ	Used in very precision machine, such as, a jig boring machine.
II	± 4μ	High precision lathes, Thread Grinders, etc.
III	± 8μ	Precision Machine tools, such as Thread millers, Lathes, Milling machines.
IV	± 20μ	Milling Machines, and also used in any feed mechanism.
V	± 40μ	Machines not requiring any accuracy, machine saws etc.

parameters. Usually a lead screw is made from steel having hardness number varying between 50 and 56 R_e. The nut is usually made up of brass or bronze.

The approximate permissible intensity of pressure on the screws is shown bejow:

Screw material	*Nut material*	*Permissible intensity of pressure*	*Machines used*
Steel	Bronze	30 kg/cm^2	Thread milling, Screw cutting, Lathes, etc.
Steel	Bronze	120 kg/cm^2	Feed Mechanism of Milling machine.
Steel	Cast Iron	80 kg/cm^2	,,

Lead screw is made of high carbon steel, when high accuracy is needed. Normally we use alloy steel containing Cr, Mn and carbon 1 to 1.2 per cent M.S. with high surface finish is also used in many cases. The nuts can be manufactured in the form of steel bushing or by centrifogal casting with a lining of bronze. This is to reduce the cost of the material.

Forms of threads: Usual forms of threads are as shown in Figure 8.2. The included angle of the thread is denoted by 2β.

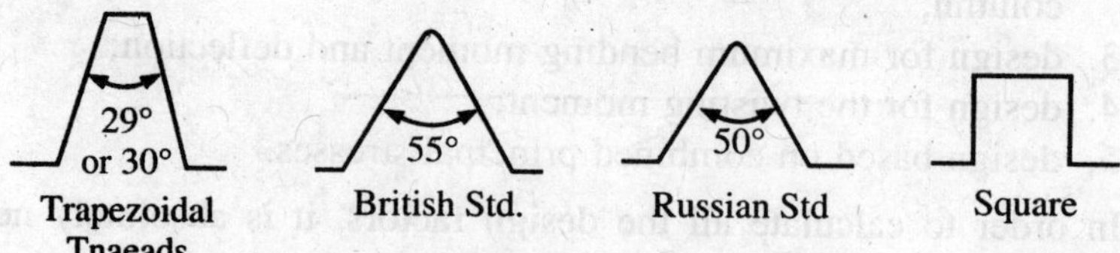

Figure 8.2.

By reducing the angle P, the friction can be reduced and therefore square thread would have been an ideal one. But for technical difficulties, for example, difficulty of using a half nut with square form, difficulty of manufacture, etc. it is not used in machine tool lead screw.

8.2 Design Calculations

With reference to Figure 8.3(a) and using the notations :

h – depth of the thread

p – pitch of the screw thread

L – length of the nut

q_0 – average axial pressure (intensity of stress) distributed over the contact area

R – radius (effective) of the screw thread we can write:

$$q_0 = \frac{Q}{2\pi . R \cdot \frac{L}{p} \cdot h} \tag{8.1}$$

where $q_0 \leq (100, 120) \frac{\text{kg}}{\text{cm}^2}$ for steel material.

If the material is C.I., the value of q_0 is reduced to 50 per cent of the above mentioned value.

The deflection caused in the lead screw will produce error of pitch. With reference to Figure 8.3 (a), (b) & (c) and assuming the lead screws to be a beam loaded axially the error in pitch caused due to the axial load Q can be expressed by

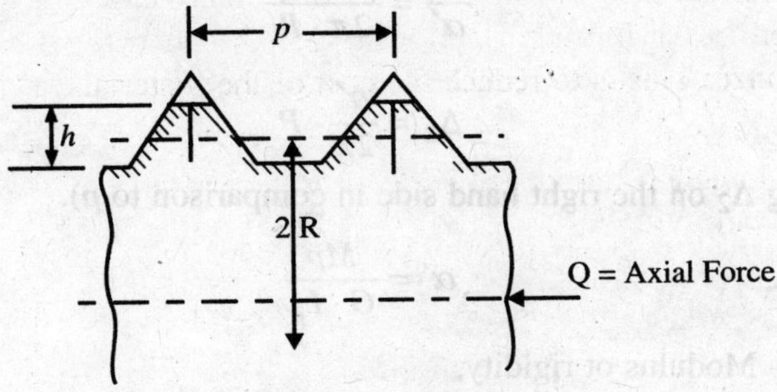

Figure 8.3 (a)

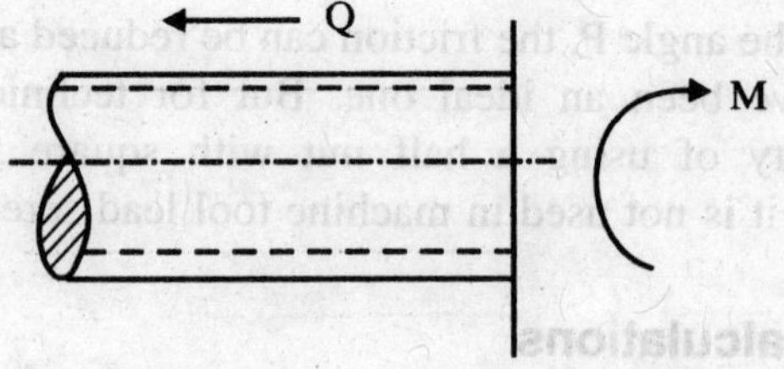

Figure 8.3 (b)

the formula

$$\Delta_1 = \frac{Q \cdot p}{E \cdot A} \tag{8.2}$$

where E – Young's Modulus of the material of the screw,
and A – cross-sectional area of the screw.

Under the effect of Torsion the error in pitch will be Δ_2 as shown in Figure 8.3(c)

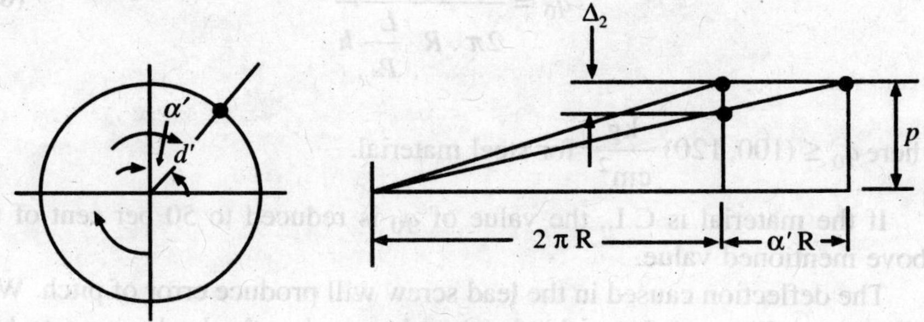

Figure 8.3(c)

$$\frac{\Delta_2}{\alpha'} = \frac{P - \Delta_2}{2\pi \cdot R}$$

or,

$$\Delta_2 = \frac{\alpha'}{2\pi} \cdot P$$

(Neglecting Δ_2 on the right hand side in comparison to p).

But

$$\alpha' = \frac{Mp}{G \cdot I_p}$$

where G – Modulus ot rigidity,
M – Twisting moment
I_p – Polar moment of inertia and
α' – the angle of twist in radians.

The efficiency η of the screw transmission can be calculated from the basic conception of effort and work done.

$$\eta = \frac{Q \cdot p}{2\pi \cdot M} \tag{8.3}$$

Therefore, $M = \dfrac{Q \cdot p}{2\pi \cdot \eta}$

Substituting the value of M in the above equation we get

$$\Delta_2 = \frac{Q \cdot p^3}{4\pi^2 \cdot \eta I_p \cdot G} \tag{8.4}$$

But $G \approx \dfrac{4}{10}$ for steel material and $I_n = \dfrac{\pi}{32} \cdot D^4 = A \cdot \dfrac{D^2}{8}$

$$\therefore \quad \Delta_2 = \frac{Q \cdot p}{E \cdot A} \cdot \frac{p^2}{2\eta \cdot D^2} = \Delta_1 \cdot \frac{p^2}{2\eta \cdot D^2} \text{ (approximately)} \tag{8.5}$$

Where D = effective dia. of the screw thread.

The total error in pitch

$$= \Delta = \Delta_1 + \Delta_2 = \Delta_1 \left(1 + \frac{p^2}{2\eta \cdot D^2}\right) \tag{8.5}$$

But the ratio $\dfrac{p}{D}$ is usually small and as such the computed value of $\dfrac{p^2}{D^2}$ is still smaller and may be neglected in comparison and therefore it is enough to calculate only Δ_1. Only in the case of multistart thread Δ_2 is also considerable and hence its effect cannot be neglected.

8.3. Strength of Lead-screw

The total stress $\quad \sigma_0 = \sqrt{\sigma^2 + 4\pi^2}$ (8.6)

where $\quad \sigma \dfrac{Q}{A}$ = direct stress,

and $\quad \tau$ = shear stress $= \dfrac{16M}{\pi \cdot D^3}$

$$\sigma_0 \le \sigma_{\text{permissible}} \le \frac{\sigma_e}{K_s}$$

where σ_e – yield stress for the material of the screw, and
K_s – stress concentration factor (usually taken as 3 or 3.5).

Buckling load in the case of lead screw can be calculated by considering it to be centrally loaded column and applying the Euler's theory of buckling.

The efficiency of the screw can be calculated from the commonly known formula:

$$\eta = \frac{\tan\alpha\left(1 - \tan\alpha\dfrac{f_B}{\cos\beta}\right)}{\tan\alpha + \dfrac{f_B}{\cos\beta}} \tag{8.7}$$

where α – angle of helix.,
f_B – coefficient of friction, and
2β – angle of the thread.

But this equation does not give the correct value of η when the screw assembly is working under the effect of lubrication. In such cases the viscous frictional force of the lubricant has also got to be taken into account. The procedure for obtaining η in such cases can be mentioned as under:

1. Find out the Viscous force of friction:

$$T_0 = 10^{-7} \cdot \mu \cdot \frac{v_c}{\delta} \cdot F \cdot K_n \tag{8.8}$$

where T_0 – viscous frictional force;
μ – dynamic coefficient of viscosity of the lubrication oil in centipoise;
v_c – velocity of sliding on the average diameter of the thread profile in mm/sec.

$$\left(\text{if } N - RPM,\ v_c = \frac{\pi \cdot D \cdot N}{60 \cdot \cos\alpha}\right);$$

F – Total transverse sectional area of the thread in mm^2;
δ – Transverse gap between the thread profiles of the nut and the screw (average clearance) in mm; and
K_n – correction coefficient (K_n =1.5 – 2. Value of K_n will be more as the diameter D increases above $D > 50$ mm).

2. Calculate the conditional coefficient of friction:[11]

$$f_y = 0.18\left(1 - \sqrt[3]{v_c}\right) \text{ where } 0 < v_c < 0.35\,\frac{M}{\text{sec}} \tag{8.9}$$

If $$v_c \geq 0.35 \frac{M}{sec}, \text{ then } f_y \approx 0.05$$

3. Obtain the overall coefficient of friction f_u by the formula :[11]

$$f_B = \frac{f_v + \frac{T_0}{Q} \cdot \cos \beta \cdot \cos \alpha}{1 + \frac{T_0}{Q} \cdot \sin \alpha} \tag{8.10}$$

when $T_0/Q \geq 0.02$

If $\frac{T_0}{Q} < 0.02$, then f_B can be taken as f_v.

4. Calculate now the efficiency from the formula:

$$\eta = \frac{\tan \alpha}{\tan \alpha + \frac{f_B}{\cos \beta}} \left(1 - \tan \alpha \cdot \frac{f_B}{\cos \beta}\right).$$

8.4. Ball Recirculating Power Screw Assemblies

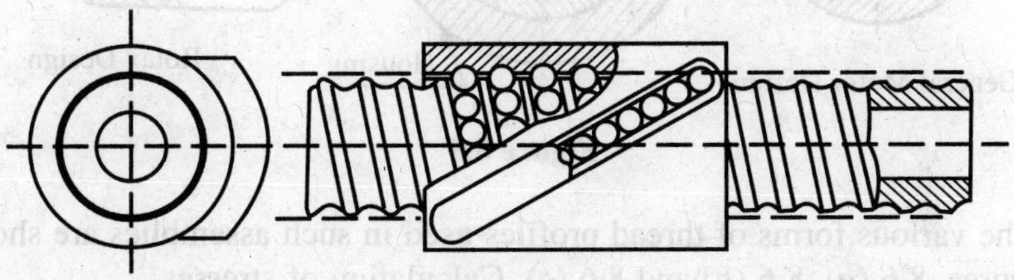

Figure 8.4

Now-a-days in precision machines and machines having programme control systems we are more and more using the Ball recirculating screw Nut assembly. These assemblies are characterized by high efficiency (independent of the velocity of rotation, axial loading or lubrication) minimum loss due to friction (coefficient of friction as low as 0.004 cm), minimum error due to backlash; no substantial reduction in overall rigidity of the system. The efficiency curves are shown in Figures. 8.7(*a*), 8.7 (*b*) and (8.8).

The efficiency of such assemblies as shown in sketches

(Figures. 8.4 and 8.5) can be calculated by the same equation quoted before, *viz.*

$$\eta = \frac{\tan \alpha}{\tan \alpha + \dfrac{f_B}{\cos \beta}} \left(1 - \tan \alpha \cdot \frac{f_B}{\cos \beta}\right).$$

where $f_B = \dfrac{fr}{R_B}$

where f_r – coefficient of rollingfriction

and R_B – radius of balls with which the working threads are filled up. The threads are filled up with balls which keep on circulating thereby replacing sliding friction by rolling friction. The balls are fed back to the groove of the first thread either by means of external tube or canal or internal canals formed on an insert (see the sketches Figure 8.5).

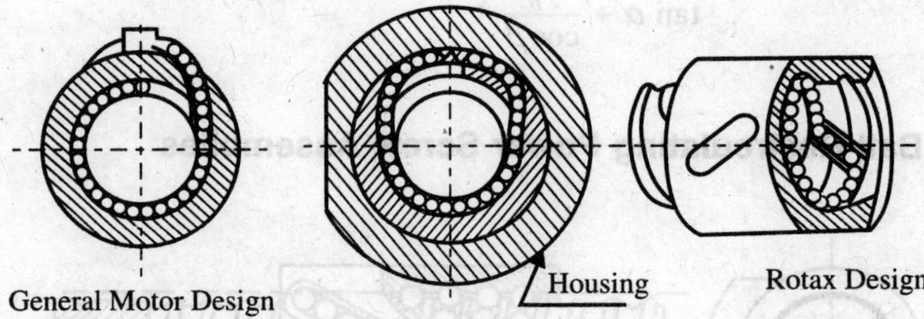

Figure 8.5

The various forms of thread profiles used in such assemblies are shown in Figures. 8.6 (*a*), 8.6 (*b*) and 8.6 (*c*). Calculation of stresses

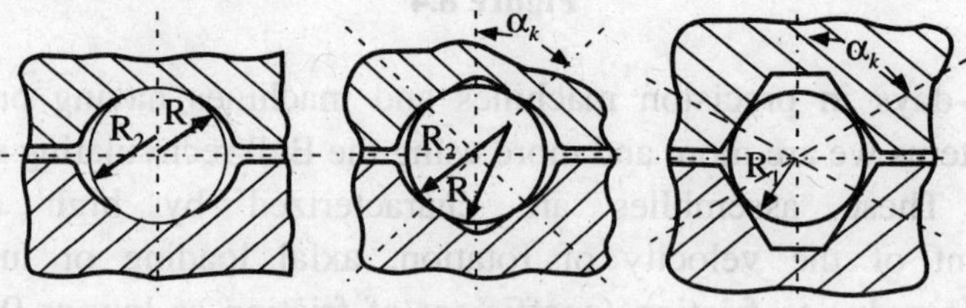

Figure 8.6(a)

and hence the load-bearing capacity in each one of these cases can be done by using the simplified form of Hertz equation. For the besf construction it is recommended that the angle of contact α_k should be near about 60° and the best profile is the gothic arc one. Usually the ratio R_2/R_1 should be between 1.05 to 1.1, where R_1 is the radius of Ball and R_2 is the radius of the thread profile. The material of the screw and nut should be hardened chromium steel [1]SAE 8626 and [2]SAE 4615 having the Hardness number 58–62 Rockwell C. Allowable permissible contact stress is

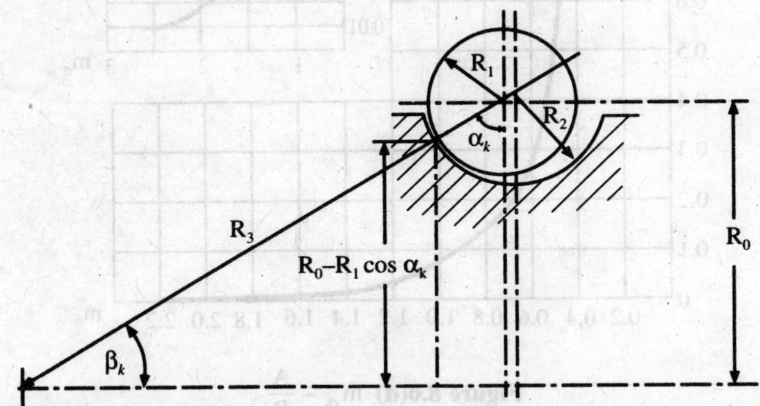

Figure 8.6(b) Semicircular Profile.

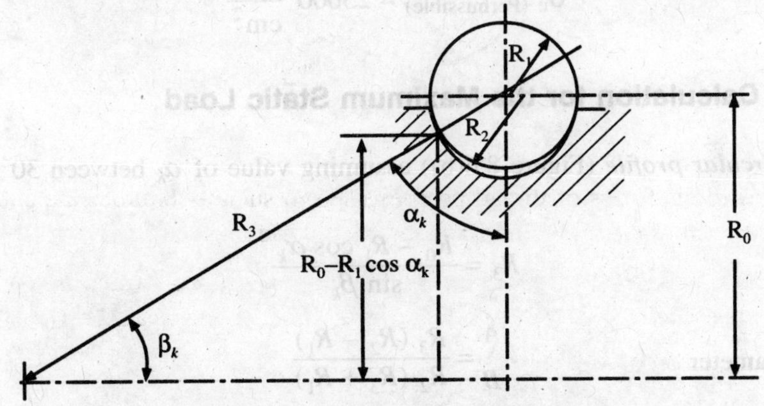

Figure 8.6(c) Gothic Arc Profile.

---

[1]C–0.18 to 0.23%. Mo–0.7 to 0.9%. Si–0.2 to 0.35%. Ni-O.4 to 0.7%. Cr–0.4 to 0.6%. Mo–0.15 to 0.25%. $H_{RC} = 60$.

[2]C–0.17 to 0.24%. Mn–0.4 to 0.7%. Si–O.2 to 0.35%. Ni-1.5 to 2.0%. Mo–0.2 to 0.3%. $H_{RC} = 60$.

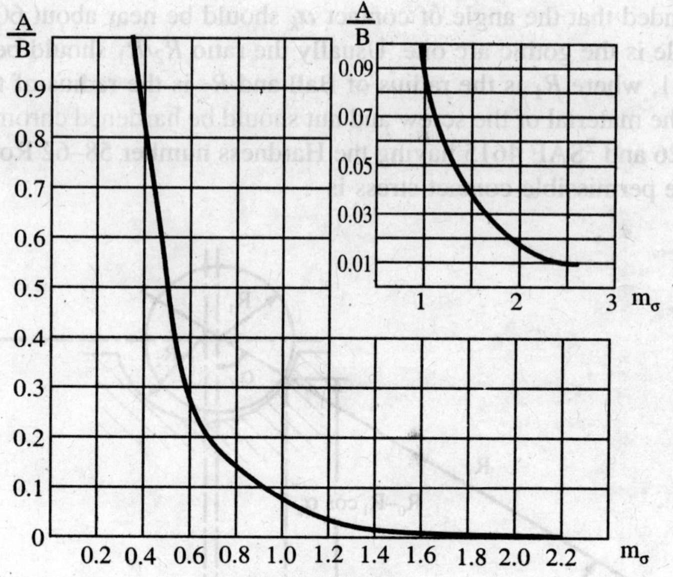

Figure 8.6(d) $m_\sigma - \frac{A}{B}$

$$\sigma_{e\ (\text{Permissible})} \approx 23000\ \frac{\text{Kg,}}{\text{cm}^2}$$

8.5. Calculation for the Maximum Static Load

Semicircular profile (Figure 8.6(b)) assuming value of α_k between 30° and 60°

$$R_3 = \frac{R_0 - R_1 \cos \alpha_k}{\sin \beta_k}$$

(!) parameter
$$\frac{A}{B} = \frac{R_3\ (R_2 - R_1)}{R_2\ (R_3 + R_1)}$$

(!) Essential parameter required for using the equation for contact stress on the profile.

$$\sigma_c = \text{contact stress} = m_\sigma \sqrt[3]{\frac{P \cdot E^2\ (R_2 - R_1)^2}{(R_2 \cdot R_1)^2}}\ \frac{\text{Kg}}{\text{cm}^2}$$

where
$$E = \frac{2E_1E_2}{E_1 + E_2}$$

Here E_1, E_2 represent elastic modulus of the balls and screw respectively.

The value of constant m_σ depends on parameter $\frac{A}{B}$ and can be found out from the Figure 8.6 (d)

Load on each ball normal to the contact point – P, can be found out by putting $\sigma_c = \sigma_c$ (Permissible).

The number of balls with which each thread is filled up

$$n \approx \frac{2\pi.\ R_0}{D_B}$$, where DB –diameter of balls. DB

Total axial load that the screw can withstand

$$Q = i.\ P. \cos \beta_k$$

where i = number of threads through which the balls recirculate.

Trapezoidal profile

With reference to the tropezoidal profile

$$\sigma_c = 0.388\ \sqrt[3]{\frac{P \cdot E^2}{R_1^2}}$$

P can be found out by putting $\sigma_c = \sigma_c$ (Perm)

$$Q = i.\ n.\ P \cos \beta_2$$

Gothic Arc Profile

Calculation same as that in case of Semicircular profile.

$$E - \text{Young's modulus} = \frac{2E_1E_2}{E_1 + E_2}$$

where E_1 and E, are the two different values of the modulus of elasticity when the contacting materials (balls and screw or nut) are different.

Example : To find the total axial load (static) to which Ball recirculating power screw assembly can be subjected if D_B = 4 mm, R_0 = 19 mm. Allowable contact stress $\sigma_{c(\text{derm})}$ = 23000 kg./cm^2. Assume n = 4 (number of threads)

Answer: $$\sigma_c = 0.388\ \sqrt[3]{\frac{PE^2}{R_1^2}}$$

Assuming balls and the screw to be of the same material having

$$E = 2.1 \times 10^6 \text{ kg/cm}^2, \quad \sigma_2 = 11400 \sqrt[3]{P}$$

$$P = \left(\frac{23000}{11400}\right)^3 = 8.3 \text{ kg and } n \approx \frac{314 \times 38}{4} \approx 30$$

If the angle of contact a_k is assumed to be 45° then the total axial load

$$Q = i.n.p. \cos \beta_k = 4.30\ 8.3.\ 0.707 \text{ kg}$$
$$= 704 \text{ kg.}$$

Gothic Arc Profile

In the same way the problem can be solved for the Gothic arc profile. Let us say that the radius of the profile = R_2 = 2.1 mm and $\alpha_k = \beta_k$

$$R_3 = \frac{R_0 - R_1 \cos \alpha_k}{\sin \beta_k} = \frac{19 - 2 \cdot 0.707}{0.707} = 24.9 \text{ mm}$$

$$\text{Parameter } \frac{A}{B} = \frac{R_3\,(R_2 - R_1)}{R_3\,(R_3 + R_1)} = \frac{24.9\,(2.1 - 2.0)}{2.1\,(24.9 + 2.0)} = 0.044$$

From Figure No. 8.6 (*d*) $m_\sigma = 1.38$ when $\frac{A}{B} = 0.044$

$$\therefore \qquad \text{Contact stress } \sigma_c = 1.38 \sqrt[3]{\frac{P.(2.1.10^6)^2 \cdot (0.21 - 0.20)}{0.20^2 \cdot 0.21^2}}$$

$$\sigma_c = 8694\ P1/3$$

from this it follows that the total axial force on the screw

$$Q = i.n.\ P \cos \beta_k = 4.30.\ 20.0.\ 0.707 = 1696 \text{ kg.}$$

8.6. Efficiency of the Ball Recirculating Power Screw

The efficiency can be calculated from equation given in (8.7), only by putting $f_B = \frac{2fr}{D_B}$ where $D_B = 2r_B$ = diameter of the balls. But two cases may arise:

1. When the rotating screw makes the nut move axially in which case

$$\tan \alpha = \frac{S}{\pi . D_{\text{av (screw)}}}$$

$D_{\text{av (screw)}}$ = Effective dia of the screw thread (joining the contact points where the balls make contact).

2. When the rotating Nut makes the screw move axially. In this case tan

$$\alpha = \frac{S}{\pi . D_{av\,(Nut)}}$$

$D_{av(Nut)}$ = Effective dia. of the threads of the Nut.

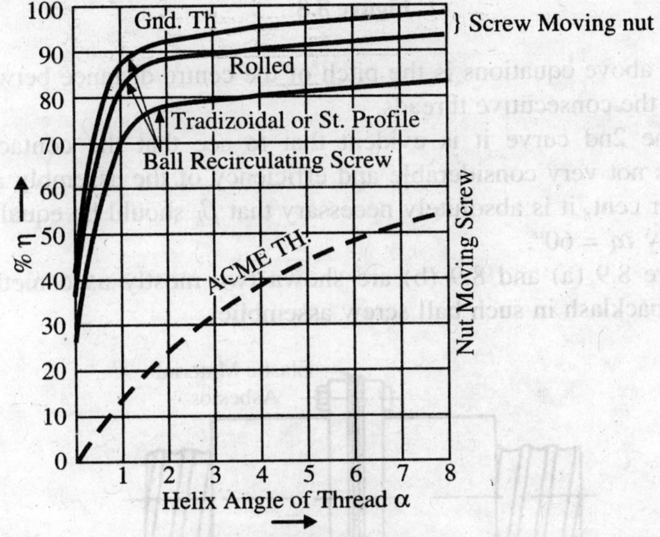

Figure 8.7(a)

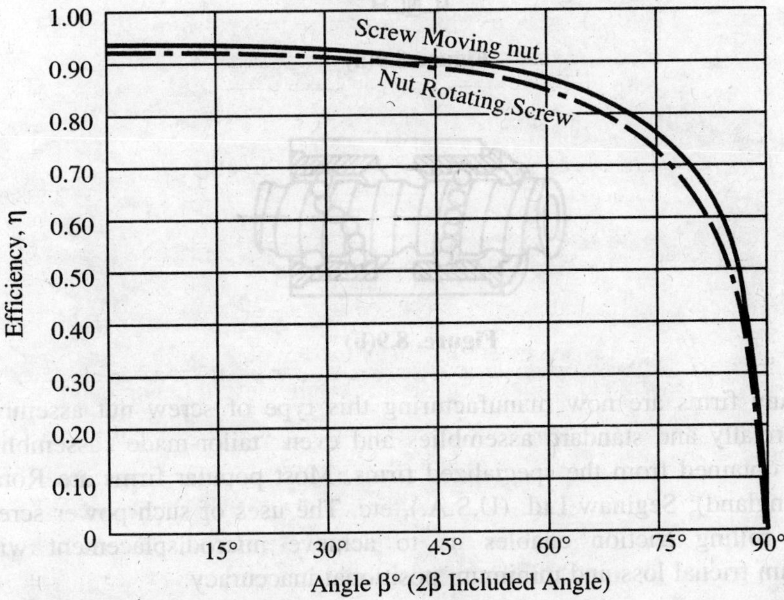

Figure. 8.7(b) Efficiency of Recirculating Power Screw Assembly.

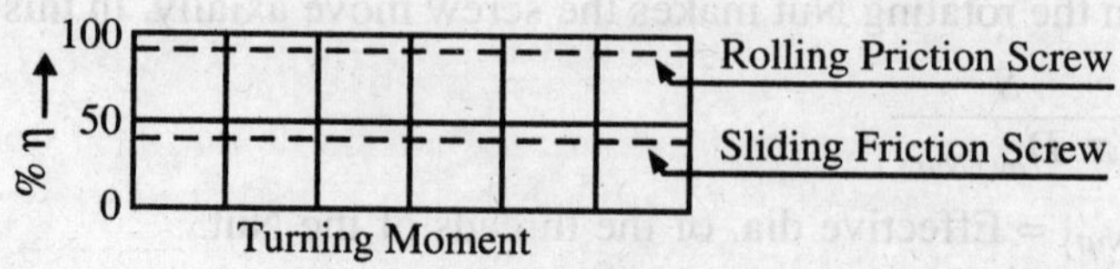

Figure 8.8

S in the above equations is the pitch of the centre distance between the two balls in the consecutive threads.

From the 2nd curve it is evident that to see that the contact stress developed is not very considerable and efficiency of the assembly also not below 90 per cent, it is absolutely necessary that β_k should be equal to 30°. That is to say $\alpha_k = 60°$.

In Figure 8.9 (a) and 8.9 (b) are shown two mostly used methods of eliminating backlash in such ball screw assemblies.

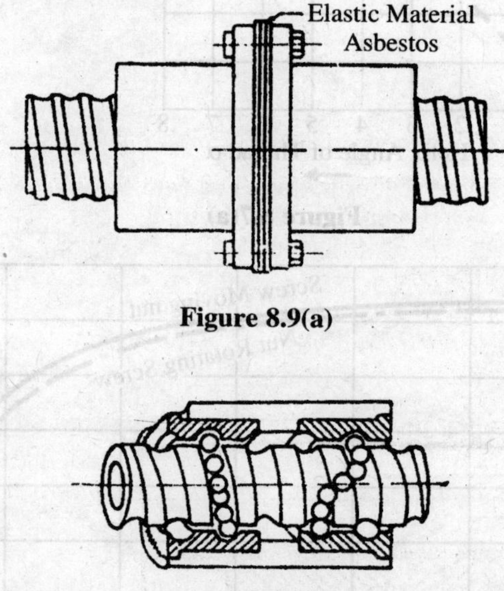

Figure 8.9(a)

Figure. 8.9(b)

Many firms are now manufacturing this type of screw nut assembly commercially and standard assemblies and even "tailor-made" assemblies can be obtained from the specialized firms. Most popular firms are Rotax Ltd. (England); Seginaw Ltd. (U.S.A.), etc. The uses of such power screw having rolling friction enables us to achieve microdisplacement with minimum fricnal loss and minimum positional inaccuracy.

8.7. Compensation of 'Backlash' in Ordinary Sliding Screw Assemblies

There are various methods available today and used extensively in machine tools for compensating the errors in lead screws. Some of them are enumerated below:

Ordinary method of adjustment of clearance: The axial force acting on the lead screw, in the case of a table feed mechanism, can very easily be found out from the frictional forces developed. And in such determination, we shall have to take into account the weight of the table, weight of workpiece fixed on the table, as well as, the vertical and horizontal component of the cutting forces acting on the job. Any "backlash" or the axial play of the lead screw, resulting from the clearances in between the thread profiles of the nut and the screw, can be adjusted by the method shown in Figure 8.10. Here, as the sketch shows, we use double nuts, one of the twin nuts being fixed to the housing, while the other can be regulated. The adjustable nut can be moved in or out depending upon the axial play of the screw, by using the pair of locknuts. Once this nut is regulated to the required amount, it can be locked in the said position with the help of the lock nuts. As is evident, in this case, we can only regulate periodically and the automatic compensation is also not possible by this method.

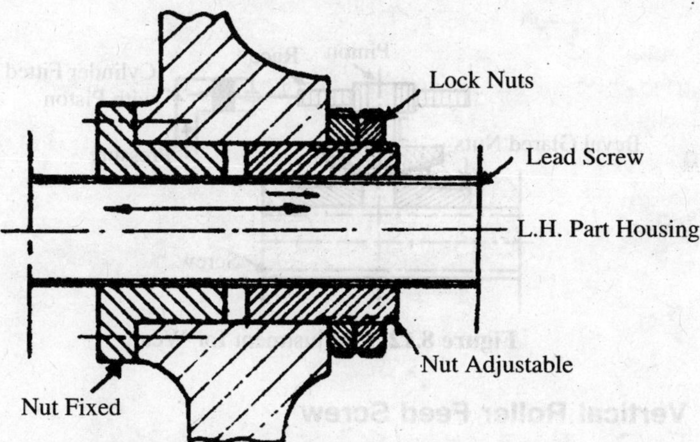

Figure 8.10. Method of Wear Compensation.

For automatic compensation of clearances. In Figure 8.11 and Figure 8.12 are shown two methods for automatic compensation of clearances due to wear. In these the nuts are regulated either by a spring force or by a bevel gear, rotated with the help of a rack and pinion arrangement. The pressure in the oil cylinder moves the rack and thereby effects the rotation of the bevel gear and the rotation of the bevel means rotation of the bevel geared twin nuts. Thus the clearance can be com-pensated. In the arrangement shown in Figure 8.11, any large increment of external axial force may upset the preloading of the spring loaded bridge, but that is not possible in Figure 8.12. In both the cases, twin nuts must be encased in a housing, to which the sliding unit is fixed.

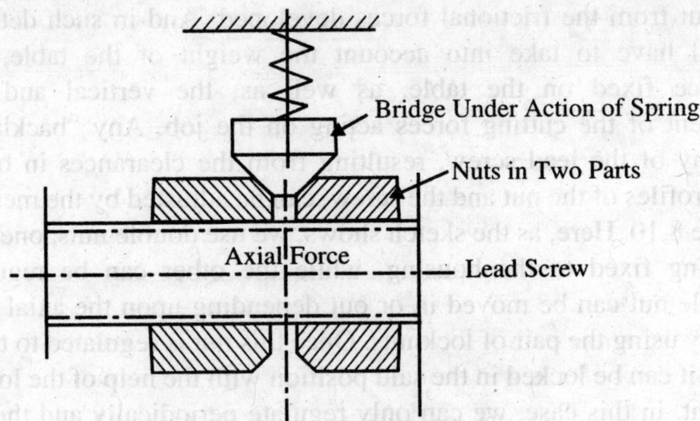

Figure 8.11. Wear Adjustment.

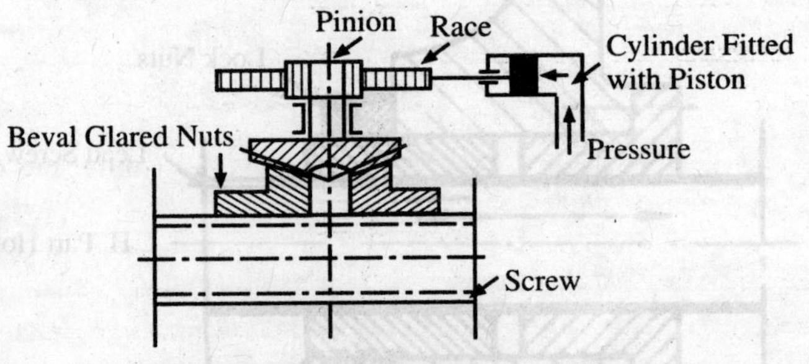

Figure 8.12. Adjustment for Wear.

8.8. Vertical Roller Feed Screw

In Figure 8.13 we have shown only a skeleton sketch of a vertical roller feed screw, as used in modern machine tool.

Here the nut is replaced by three cylindrical rollers having rack teeth of the same profile as the, screw thread. These three rollers are placed-in the main housing of the nut at an angular distance of 120°. But in the vertical

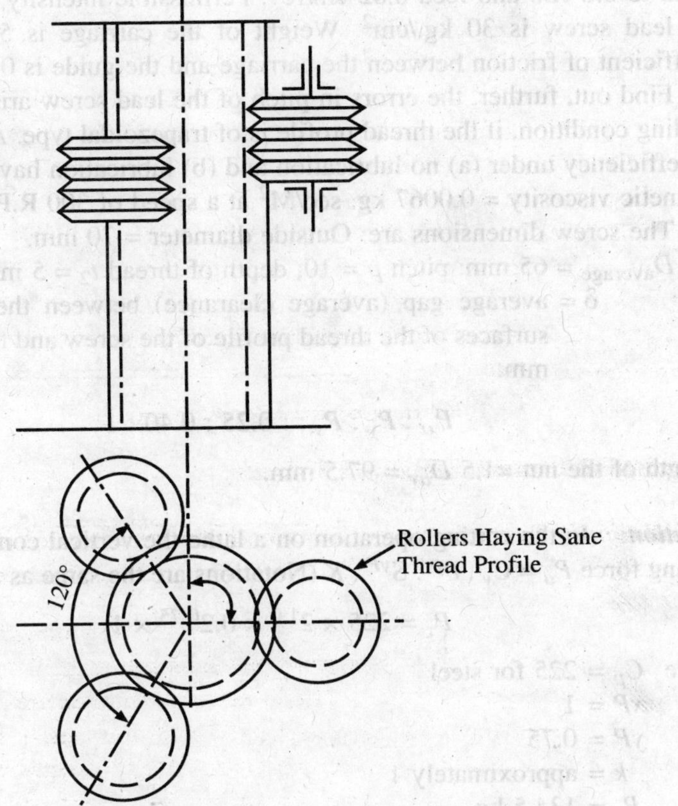

Figure 8.13. Vertical Roller Feed Screw.

plane their positions are offset by an amount equal to one-third of the pitch of the screw. The main housing of the nut does not rotate; but only slide in the vertical direction. Inside the housing, a proper frame is made for positioning of the rollers, which should rotate freely about their cwn axes. Since, in this particular arrangement shown, the contact area is being reduced to three points theoretically, we can expect full replacement of sliding friction by rolling friction. But, in real working, certain amount of sliding friction will always be accompanying the rolling friction. However, the coefficient of friction, in this case, is very many times less than in the conventional sliding screw assembly, resulting in higher efficiency and ltss vibratory motion of the unit moved by the nut.

Problem. Find out the specific pressure on a lead screw assuming a m.s. material being cut by H.S.S. tool on a lathe machine using a depth of cut equal to 0.2 cm. and feed 0.02 cm/rev. Permissible intensity of pressure on the lead screw is 30 kg./cm^2. Weight of the carriage is 50 kg. and the coefficient of friction between the carriage and the guide is 0.20.

Find out, further, the errors in pitch of the lead screw arising out of the loading condition, if the thread profile is of trapezoidal type. Also determine the efficiency under (a) no lubrication and (b) lubrication having coefficient of knetic viscosity = 0.0067 kg. sec/M^2 at a speed of 300 R.P.M.

The screw dimensions are: Outside diameter = 70 mm.

$D_{average}$ = 65 mm: pitch p = 10; depth of thread t_2 = 5 mm.

δ = average gap (average clearance) between the non-working surfaces of the thread profile of the screw and the nut = 0.182 mm.

$$P_\alpha : P_y : P_s : : 0.25 : 0.40 : 1$$

Length of the nut =1.5 D_{av} = 97.5 mm.

Solution. In the cutting operation on a lathe the vertical component of the cutting force $P_a = C_v . t^{xp} . S^{yP} . K$ (Notations are the same as in Chapter 2.).

$$P_z = 225 \times 2^{1.0} \times 0.2^{0.75} \times 1$$

Since C_p = 225 for steel

xP = 1

yP = 0.75

k = approximately 1

P_z = 134.5 kg.

P_x = 025 P_x = 33.62 kg.

Total axial force on the screw $Q = P_x + f(G + P_z)$

where f = coefficient of friction

G = weight of the carriage

P_x and P_z = are the components of the cutting force in the x and z direction respectively

$$\therefore \quad Q = 33.62 + 6.2\ (50 + 134.5) = 70.52 \text{ kg.}$$

Contact area = π . D_{av} . t_2 × (Number of threads in contact)

$$\pi \cdot 65.5 \cdot \frac{97.5}{10} = 99.5 \text{ sq. cm.}$$

specific pressure on the screw $q = \dfrac{70.52}{99.5} = 0.709$ kg/sq. cm.

This is well within the permissible value of 30 kg./cm^2.

Now the error in pitch $= \Delta_1 + \Delta_2 \cdot \dfrac{p^2}{2\eta.D}$ (Notations are given in this chapter).

In single start thread $p \angle\angle$ D, hence, $\dfrac{p^2}{2\eta} \cdot D \cdot \Delta_1$ becomes negligible and hence will not affect the accuracy of calculation

$\therefore$ Error in pitch $\Delta_1 = \dfrac{Q \cdot p}{A.E}$

where Q = total force;
p = pitch of thread;
A = area of cross-section of the screw;
and E = Young's modulus of elasticity (= 2.1×10^6 kg/cm^2)

Substituting the values:

$$\Delta_1 = \frac{70.52 \times 10}{\frac{\pi}{4}(6.5)^2 \cdot 2.1 \times 10^6 \times 10} = 1.012 \times 10^{-6} \text{ cm}$$

With no lubrication the efficiency of the screw $\eta = \dfrac{\tan \alpha}{\tan(\alpha + \rho)}$

where $\alpha = \dfrac{P}{\pi.D}$ and $\tan \rho = \dfrac{f_B}{\cos \beta}$

where f_B = 0.22 to 0.24, $\beta = 15°$
η = 0.163 or 16.3%.

With lubrication. Efficiency of the screw $\eta = \dfrac{\tan \alpha}{\tan(\alpha + \rho)}$

where $\tan \rho = \dfrac{f_B}{\cos \beta}$

and $f_B = \dfrac{f_v + \frac{T_0}{Q}\cos \beta \cos \alpha}{1 + \frac{T_0}{Q}\sin \alpha}$ when $\dfrac{T_0}{Q} \geq 0.02$

otherwise $f_B \approx f_y$

where f_y is given by: $f_v = 0.18\,(1 - \sqrt[3]{V})$

when $V_c \leq 0.35 \dfrac{M}{\text{sec}}$

and $f_v = 0.005$ when $V_c > 0.35 \dfrac{M}{\text{sec}}$

$$V_c = \text{velocity of sliding} = \frac{\pi \cdot D \cdot N}{60 \cos \alpha}$$

From the calculations $V_c = 1.022 \dfrac{M}{\text{sec}}$.

$$F = \pi . D_{av} . t_2 \approx \pi . 65 \times 5 \times 10^{-6} = 0.999 \times 10^{-3} M^2$$

$$i = \text{Number of threads} = \frac{97.5}{10} \approx 10$$

$$T_0 = \mu . \frac{V_c}{\delta} . F . k_n . i \; Kg$$

$$= \frac{0.0067 \times 1.022 \times 0.999 \times 10^{-3} \times 2 \times 10}{0.182 \times 10^{-3}}$$

$$= 0.752 \text{ kg}$$

$$\frac{T_0}{Q} = \frac{0.752}{70.5} = 0.01066 \quad \text{As } V_c = 1.022 \frac{M}{\text{sec}} \; f_v = 0.05$$

Again since $\dfrac{T_0}{Q} < 0.02$, $f_B = f_v = 0.05$

$$\therefore \qquad \tan \rho = \frac{f_B}{\cos \beta} = \frac{0.05}{\cos 15} = 0.0510$$

$$\eta = \frac{\tan \alpha}{\tan (x + \rho)} = 0.487 \text{ or } 48.7\%$$

The result shows that the efficiency of a sliding screw assembly has been increased by the application of lubrication (of viscosity 0.0067 kg. sec./m^2) from 16.3 percent to 48.7 per cent when running at 300 R.P.M.

8.9. Distribution of Load Between the Threads of Nut in the Power Screw with Sliding Friction

This considers the distribution of load based on elastic deformation and shear following the traditional method of N.E. Zukovsky as early as 1902. The deduction, as obtained from his works, has been presented here:

Assumptions

(1) Angle of helix of the screw-nut is very small and threads are horizontal

(2) N = load on 1st thread
N_1 = load on 2nd thread
N_3 = load on 3rd thread

(3) G = Modulus of rigidity of material
S_1 = Length of thread of nut
S = Length of thread of screw
F = Area of cross-section of thread at the root of screw
F = Area of the cross-section of thread at the root of the nut.

With reference to the Figure 8.14(a)

$$aC = \frac{SN}{FG} + \frac{S_1 N}{F_1 G_1} \tag{8.11}$$

$$bD = \frac{SN_1}{FG} + \frac{S_1 N_1}{F_1 G_1} \tag{8.12}$$

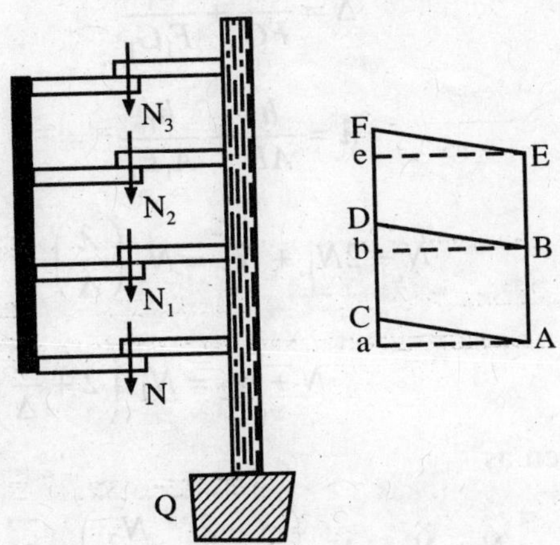

Figure 8.14(a)

Total tensile force acting on AK

$$= T = N_1 + N_2 + N_3 + ... + N_k$$

Tensile strain (BA) on screw thread

$$= \frac{Th}{AE}$$

Compressive strain (CD) on the nut

$$= \frac{Th}{A_1 E_1}$$

where h = Distance between the consecutive threads
E, E_1 = Young's modulus of the screw and the nut respectively
A, A_1 = Area of cross-section of the screw and nut (tranverse)

$$aC - bD = T \frac{h}{AE} + \frac{h_1}{A_1 E_1}$$

$$(N - N_1)\left(\frac{S}{FG} + \frac{S_1}{F_1 G_1}\right) = T\left[\frac{h}{AE} + \frac{h_1}{A_1 E_1}\right] \tag{8.13}$$

$$\left.\begin{aligned} N - N_1 &= T \cdot \frac{\lambda}{\Delta} \\ N_1 - N_2 &= (T - N_1)\frac{\lambda}{\Delta} \end{aligned}\right\} \tag{8.14}$$

where
$$\Delta = \frac{S}{FG} + \frac{S_1}{F_1 G_1}$$

and
$$\lambda = \frac{h}{AE} + \frac{h_1}{A_1 E_1}$$

$\therefore$
$$N - 2N_1 + N_2 = N_1\left(\frac{\lambda}{\Delta}\right)$$

$$N + N_2 = N_1\left(2 + \frac{\lambda}{\Delta}\right) \tag{8.15}$$

This can be written as

$$N + N_2 = N_1\left(2 + \frac{\lambda}{\Delta} - \frac{N_2}{N_1}\right) + N_2$$

$$\frac{N_1}{N} = \frac{1}{2 + \frac{\lambda}{\Delta} - \frac{N_2}{N_1}}$$

Similarly
$$\frac{N_2}{N_1} = \frac{1}{2 + \frac{\lambda}{\Delta} - \frac{N_3}{N_2}} \text{ and so on}$$

$$\frac{N_1}{N} = \frac{1}{\left(2+\frac{\lambda}{\Delta}\right) - \frac{1}{\left(2+\frac{\lambda}{\Delta}\right) - \frac{1}{\left(2+\frac{\lambda}{\Delta}\right)}}} \qquad (8.16)$$

If $\qquad \dfrac{N_1}{N} = q$, then $\dfrac{N_2}{N_1} = \dfrac{N_3}{N_2} = q = \dfrac{1}{2+\frac{\lambda}{\Delta} - q}$

$\therefore \qquad q^2 - \left(2+\dfrac{\lambda}{\Delta}\right)q + 1 = 0$

$$q = \left(1+\frac{\lambda}{\Delta}\right) \pm \sqrt{\left(1+\frac{\lambda}{2\Delta}\right)^2} - 1,\ q < 1 \qquad (8.17)$$

From this expression, it is evident that the load distributed on successive threads goes on diminishing, resulting in an asymptotic nature of the curve. In fact after a limited number of thread, the load sharing is negligible.

If $\qquad Q = 100$ kg. and $q = 0.7$

$\therefore \qquad N$ = load carried by first thread (nearest to load application point)

$= 0.3\ Q = 30$ kg

Load carried by second thread $= N_1 = qN = 0.7 \times 0.3\ Q$

$= 0.21\ Q = 21$ kg

$N_2 = 0.21Q \times 0.7 = 0.147\ Q = 14.7$ kg

$N_3 = 10.29$ kg

and so on

Then the curve between load carried and number of threads can be plotted as under

$\therefore \qquad q = \left(1+\dfrac{\lambda}{2\Delta}\right) - \sqrt{\left(1+\dfrac{\lambda}{2\Delta}\right)^2 - 1}$

We know that $\qquad N - N_1 = T \cdot \dfrac{\lambda}{\Delta} = (Q - N)\dfrac{\lambda}{\Delta}$

$$N - qN = (Q - N)\frac{\lambda}{\Delta}$$

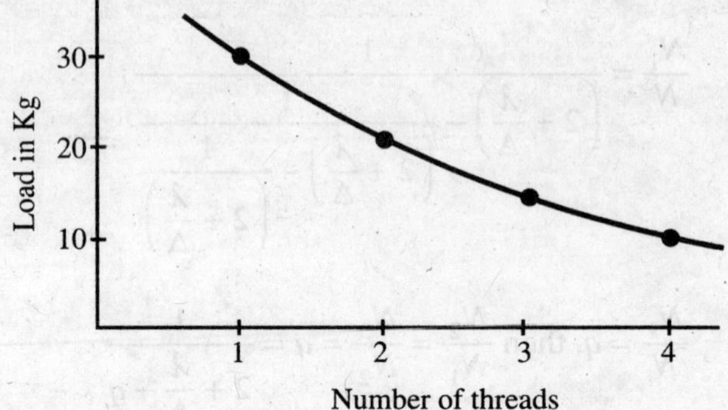

Figure 8.14(b)

$$\therefore \qquad N = \frac{\dfrac{\lambda}{\Delta} Q}{1 - q + \dfrac{\lambda}{\Delta}}$$

$$N + Nq + Nq^2 + \ldots\ldots = Q$$

$$Q = \frac{N}{1 - q}$$

or $$N = (1 - q)Q \tag{8.18}$$

8.10. Load Distribution on the Threads of the Nut of Ball Recirculating Screw Assembly

The distribution of load on the threads of nut in an ordinary screw nut assembly was first investigated by Prof. N.E.Zukovsky[146] of the Moscow Higher Technological Institute in the year 1902. But in this work Prof.Zukovsky for the sake of simplicity assumed that the nut had infinite number of threads having helix angle of 90° and that the threads worked only in shear.

Though for the time being the work of Zukovsky remained unnoticed in countries outside U.S.S.R., it is about 30 years later that E.Jacquet[147] (1931) and Madushka11481 (1936) in their respective solutions related to the distribution of load on the threads of a bolt nut assembly, used the principles of Zukovsky. The entire work of Zukovsky in this field was published in the form of two papers. Zukovsky did experiments with a nut having 10 number of threads to show that the maximum load is taken up by the first thread (near the load applying point) and that the loads on the subsequent threads go on decreasing according to a nonlinear function. Maduschka in his work experimented with finite number of threads and plotted the load distribution curve. In his work Maduschka considered the elastic deformations of the threads . taking into the effect of bending (due to pressure on the flanks of the thread) and shear.

B.G.Galerkin[149] of Leningrad found out the elaborate method of solving the problem of distribution taking into consideration bending, shear deflection and the radial elastic displacements of the threads.

V.B.K.uklin[150] in (1957) in "VestnikMashinostroenie" (U.S.S.R.), gave a more accurate method for the determination of the load per unit thread of a nut in a screw-nut assembly taking into account the effect of the frictional force between the threads acting along the flanks in addition to the factors mentioned above.

The character of the load distribution between the threads of nut in a ball recirculating screw assembly presents a still more complicated problem because of the contact deformation caused by the balls on the surface of the thread profiles at the contact points. The problem can be simplified by neglecting the contribution of the frictional force. The elimination of this from the calculation is justified, since the coefficient of friction in a ball recirculating screw assembly is much less than in ordinary screw assembly.

8.10.1 *Mathematical Approach*

On the basis of the Schematic diagram of Zukovsky (Figure 8.15) we can write the following equation connecting the deformations of the screw and the nut:

$$\Delta_1 + \Delta_2 = [\delta_1(z) + \delta_2(z)] - [\delta_1(0) + \delta_2(0)]$$

where

$$\delta_1(z) + \delta_2(z)$$

summation of the deflection of the middle point of the thread of the screw with respect to the coincident point on the nut, in the section Z;

$$\delta_1(0) + \delta_2(0)$$

same as above but in the section O;

The above equation can be enlarged to the following form:

$$\Delta_1 + \Delta_2 = (\delta u_1 + \delta u_2) + (\Delta D_1 + \Delta D_2) + (\delta c_1 + \delta c_2) + (u_1 + u_2)\, tg\, \beta \quad (18.19)$$

The above equation contains the following characteristic notations

Δ_1 – axial displacement of the body of the screw in the given section under tension;

Δ_2 – axial displacement of the body of the nut under compression;

δu_1 – deformation of the point under consideration on the thread of the screw in the section under bending;

δu_2 – deformation of the point coincident to the above on the thread of the nut in the same section in bending;

ΔD_1 – contact deformation at the point of contact on thread of the screw caused by the ball . measured in the axial direction , in the section;

ΔD_1 – contact deformation at the point of contact on the thread of the nut caused by the ball measured in the axial direction, in the scetion;

δc_1, δc_2 – axial displacements of the points of contact of the screw and nut respectively resulting from the shear deflection;

u_1 – radial elastic displacement at the given point of the thread of the screw;

u_2 – radial elastic displacement at the given point of the thread of the nut;

2β – included angle of the thread profile.

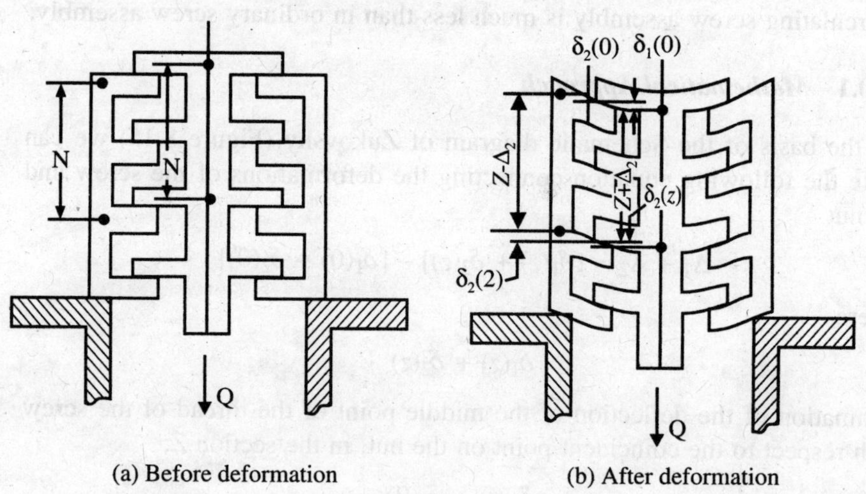

(a) Before deformation (b) After deformation

Figure 8.15

The solution of the above equation gives the nature of distribution of the load per thread of a ball circulating nut. for solving the above equation it is necessary to determine . before all. the absolute value of each of the factors mentioned above δc, δu and u can be expressed in simplified form using Golerkin's method.

8.10.2 *Axial displacement resulting from the deformation due to bending*

Let us assume that the normal force acting on the surface of the thread profile of the screw or the nut at the point of contact of the ball is equal to P. for trapezoidal thread profile (as shown in Figure 8.19)

$$M_z = P \cos \beta \, [(x - b) - b \, t \, g^2 \, \beta];$$

and the moment of inertia of the section :

$$J_x = \frac{2}{3} x^3 \, tg^3 \, \beta w;$$

Where. w = distance between the points of contacts of the successive balls on the surface of the thread profile.

Accordingly

$$\delta u = \int_b^c \frac{[P \cos \beta (x - b) - b \, tg^2 \, \beta](x - b)}{\frac{2}{3} x^3 \, tg^2 \beta \, w \, E} dx,$$

where,

$1 \, (x - b)$ = Bending moment caused by unit force.

E = Young smodulus of elasticity;

By simplifying we obtain

$$\delta u = \frac{3}{2E} \frac{P \cos \beta}{2E \, tg^3 \, \beta \, w} \left[\int_b^c \frac{x^2 - 2bx + b^2}{x^3} dx - tg^2 \, \beta \int_b^c \frac{bx - b^2}{x^3} dx \right]$$

From the above by integration

$$\delta u = \frac{3}{2E} \frac{P}{w} \frac{\cos \beta}{tg^3 \, \beta} \left[\left\{ l_n \frac{c}{b} + \frac{2b}{c - b} - \frac{b^3}{2(c^2 - b^2)} \right\} \right.$$

$$\left. - tg^2 \, \beta \left\{ - \frac{b}{c - b} + \frac{b^2}{2(c^2 - b^2)} \right\} \right] \tag{8.20}$$

Considering the thread profile to be a metric one we have

$$b = \frac{t_0}{2}, \, C = \frac{7}{8} t_0, \, a = \frac{t_0}{8} \text{ and } 2\beta = 60°$$

Putting these values in the above equation:

$$\delta u = \delta u_1 = \delta u_2 = Cu \frac{p}{E} \tag{8.21}$$

where,

Cu = constant coefficient, the value of which is equal to 20.1 and

$$p = \frac{P}{w}$$

$$E = \frac{2E_1 E_2}{E_1 + E_2}$$

where

E_1 = Young's modulus for the material of the screw or the nut;

E_2 = Young's modulus for the material of balls;

For the purpose of simplicity let us assume that the screw, nut as well as the balls are of the same material.

8.10.3. *Vertical Displacement of the point M resulting from the shear deflection*

We have

$$\delta c = \delta c_1 = \delta c_2 = \frac{K_c}{G} \int_b^c \frac{P \cos \beta}{2x\, w\, tg\, \beta}\, dx = \frac{K_c}{2G} \frac{P}{w} \frac{\cos \beta}{tg\, \beta}\, l_n \frac{c}{b}$$

where

K_c = Dimensionless coefficient, depending on the distribution of the shear stresses on the surface of the conical section.

G = Modulus of rigidity $= \dfrac{E}{2(1+\mu)}$.

where

μ = Poisson's coefficient.

It can be said approximately that $K_C = \dfrac{6}{5}$ and μ = 0.3 from that

$$\delta c = \delta c_1 = C_c \frac{p}{E},$$

where

C_c = Constant coefficient having a value equal to 1.34 .

8.10.4. *Constant deformation*

We have

$$\Delta D_1 + \Delta D_2 = 2.08 \sqrt[3]{\left\{\frac{1}{D}\left(\frac{1-\mu_1^2}{E_1}+\frac{1-\mu_2^2}{E_2}\right)^2\right\}} \cos\beta \times P^{2/3}$$

Where μ_1 and μ_2 are the poisson's ratios for the materials of the screw or the nut and ball respectively,

It may be stated that

$$\Delta D_1 + \Delta D_2 = C_1 P^{2/3} \qquad (8.24)$$

Where

$$C_1 = 2.08 \cos\beta \sqrt[3]{\left\{\frac{1}{D}\left(\frac{1-u_1^2}{E_1}+\frac{1-u_2^2}{E_2}\right)^2\right\}}$$

And

$D = d_w$ where d_w - diameter of the bail.

8.10.5 *Radial elastic displacement of the point M*

Assuming it a thick tube:

$$\mu_1 = \frac{r_1 (1-\mu)}{E_1} \sigma_{av} \qquad (8.25)$$

$$U_2 = \frac{r_0}{E_2}\left(\mu_2 + \frac{R_e^2 + r_0^2}{R_e^2 - r_0^2}\right)\sigma_{av} \qquad (8.26)$$

Where,

R_e = equivalent outside radius of the mil,
d_0 = $2r_0$, outside diameter of the thread of the nut,
d_1 = $2r_1$, inner diameter of the thread of the nut,
σ_{av} = average stress on the surface,
S = lead of the thread.

$$\sigma_{av} = \frac{P \sin\beta}{\omega S} = \frac{p}{S}\sin\beta$$

$$U_1 \, tg\,\beta = r_1(1-\mu_1)\frac{\sin\beta \, tg\,\beta}{S}\frac{p}{E_1} \qquad (8.27)$$

$$U_2\, tg\, \beta = r_v \left(\mu_2 + \frac{R_e^2 + r_0^2}{R_e^2 - r_0^2} \right) \frac{\sin \beta\, tg\, \beta}{S} \frac{p}{E_2} \tag{8.28}$$

Equation of load distribution

Putting the values obtained in eqns. 8.20, 8.21, 8.22, 8.24, 8.27 and 8.28 in the fundamental equations we obtain:

$$\Delta D_1 + \Delta D_2 = \left[C_u \frac{p}{E_1} + C_c \frac{p}{E_1} r_1 (1 - \mu_1) \frac{\sin \beta\, tg\, \beta}{S} \frac{p}{E_1} \right]$$

$$+ \left[C_u \frac{p}{E_2} + C_c \frac{p}{E_2} + \frac{r_0}{E_2} \left(\mu_2 + \frac{R_e^2 + r_0^2}{R_e^2 - r_0^2} \right) \frac{p}{S} \sin \beta\, tg\, \beta \right] + C_1 P^{2/3} \tag{8.29}$$

The above equation may be written in the form

$$\Delta D_1 + \Delta D_2 = \lambda_1 \frac{p}{E_1} + \lambda_2 \frac{p}{E_2} + C_1 P^{2/3} \tag{8.30}$$

where

$$\lambda_1 = C_u + C_c + \frac{r_1}{S} (1 - \mu_1) \sin \beta\, tg\, \beta.$$

$$\lambda_2 = C_u + C_c + \frac{r_0}{S} \left(\mu_2 + \frac{R_e^2 + r_0^2}{R_e^2 - r_0^2} \right) \sin \beta\, tg\, \beta.$$

The value of Δ_1 and Δ_2 can be expressed (vide Figure 8.16) in the following forms:

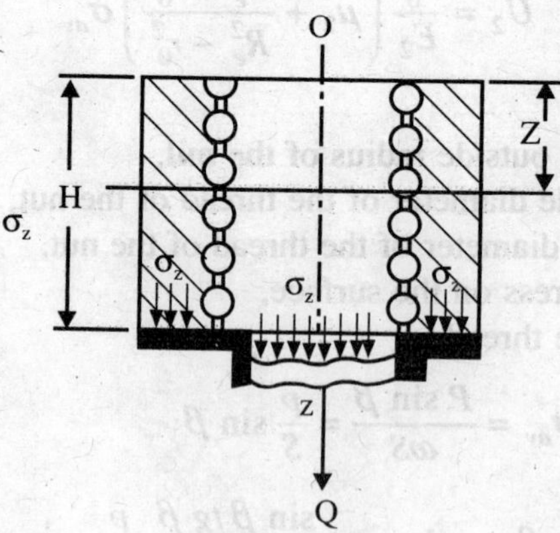

Figure 8.16

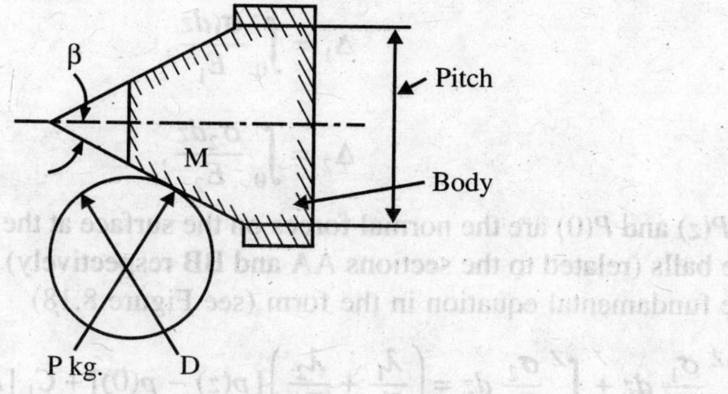

Figure 8.17

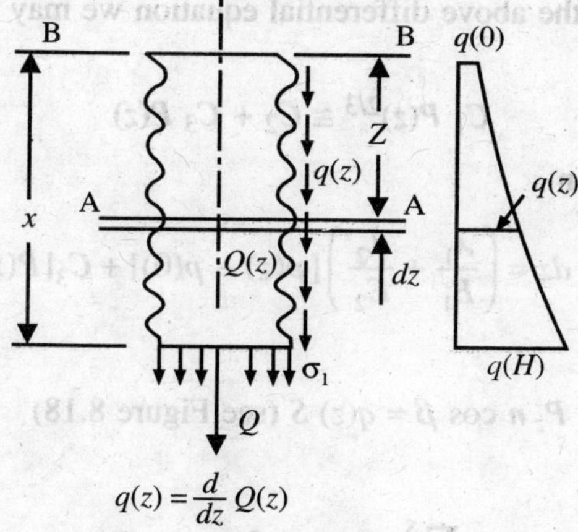

Figure 8.18

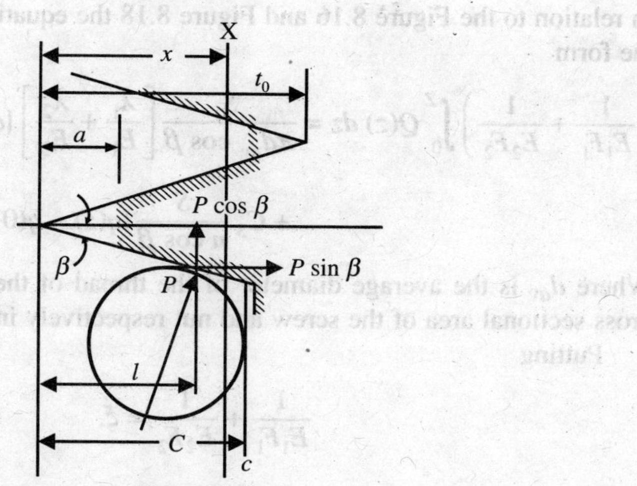

Figure 8.19

$$\Delta_1 = \int_0^z \frac{\sigma_1 dz}{E_1}, \tag{8.31}$$

$$\Delta_2 = \int_0^z \frac{\sigma_2 dz}{E_2}, \tag{8.32}$$

If $P(z)$ and $P(0)$ are the normal forces on the surface at the point of contact of the balls (related to the sections AA and BB respectively) then we may write the fundamental equation in the form (see Figure 8.18)

$$\int_0^Z \frac{\sigma_1}{E_2} dz + \int_0^Z \frac{\sigma_2}{E_2} dz = \left(\frac{\lambda_1}{E_1} + \frac{\lambda_2}{E_2}\right)[p(z) - p(0)] + C_1 \, [P(z)^{2/3} - P(0)^{2/3}] \tag{8.33}$$

While simplifying the above differential equation we may assume upto first approximation

$$C_1 \, P(z)^{2/3} \cong C_2 + C_3 \, P(z)$$

Under this condition

$$\int_0^Z \frac{\sigma_1}{E_1} dz + \int_0^Z \frac{\sigma_2}{E_2} dz = \left(\frac{\lambda_1}{E_1} + \frac{\lambda_2}{E_2}\right)[p(z) - p(0)] + C_3[P(z) - P(0)] \tag{8.34}$$

Again

$$P_z \, n \cos \beta = q(z) \, S \text{ (see Figure 8.18)} \tag{8.35}$$

And

$$p\sum_{S=0}^{S} \omega \cos \beta = q\,(z)\, S \tag{8.36}$$

In relation to the Figure 8.16 and Figure 8.18 the equation can be written in the form

$$\left(\frac{1}{E_1F_1} + \frac{1}{E_2F_2}\right)\int_0^Z Q(z)\, dz = \frac{S}{\pi d_{av} \cos \beta}\left[\frac{\lambda_1}{E_1} + \frac{\lambda_2}{E_2}\right]\{q(z) - q(0)\}$$

$$+ C_3 \frac{S}{n \cos \beta}\{q(z) - q(0)\} \tag{8.37}$$

Where d_{av} is the average diameter of the thread of the nut. F_1 and F_2 are cross sectional area of the screw and nut respectively in the section AA.

Putting

$$\frac{1}{E_1F_1} + \frac{1}{E_2F_2} = \xi, \tag{8.38}$$

$$\left(\frac{\lambda_1}{E_1}+\frac{\lambda_2}{E_2}\right)\frac{S}{\pi d_{av}\cos\beta}+\frac{C_3 S}{n\cos\beta}=\gamma \tag{8.39}$$

We obtain

$$\xi\int_0^Z Q(z)\,dz=\gamma[q(z)-q(0)] \tag{8.40}$$

$$\xi Q(z)=q(z)'\gamma \tag{8.41}$$

Repeating the operation and considering the relation

$$q(z)=\frac{d}{dz}Q(z),$$

we obtain

$$q(z)''=\frac{\xi}{\gamma}q(z)=m^2\ q(z)$$

where

$$m=\sqrt{\left(\frac{\xi}{\gamma}\right)} \tag{8.42}$$

The solution of the above equation can be written in the form

$$q(z)=A\text{ sh }m\,z+B\text{ ch }m\,z. \tag{8.43}$$

Putting the boundary conditions (see Figure 8.18)

$Q(z)=0$	$q(z)=0;$	under Z = 0
$Q(z)=\text{Q}$	$q'(z)=m^2Q;$	under Z = H

We obtain

$$A=0;\quad B=\frac{m^2}{m\ sh\ mH}=\frac{Q\ m}{sh\ mH} \tag{8.44}$$

The equation (8.43) can be written in the form

$$q(z)=\frac{Q\ m}{sh\ mH}\ ch\ mZ \tag{8.45}$$

8.10.6 *The Influence of pitch error on the distribution of the load on the ball recirculating nut*

If the pitch of the nut is greater than that of the screw and the error is constant over the entire length, then

$$\Delta_1 = -Z\frac{\Delta S}{S} + \frac{1}{E_2F_2}\int_0^Z Q(z)\,d(z);$$

$$\Delta_2 = \frac{1}{E_1F_1}\int_0^Z Q(z)\,d(z);$$

where ΔS = inaccuracy in the pitch.

The fundamental equation of load distribution in al ball recirculating nut under such condition can be written in the form

$$-Z\frac{\Delta S}{S} + \frac{1}{E_2F_2}\int_0^Z Q(z)\,d(z) + \frac{1}{E_1F_1}\int_0^Z Q(z)\,d(z) = \frac{S}{\pi\, d_{av}\cos\beta}$$

$$\times\left[\frac{\lambda_1}{E_1} + \frac{\lambda_2}{E_2}\right]\{q(z) - q(0)\} + \frac{C_c S}{n\cos\beta}\{q(z) - q(0)\} \qquad (8.46)$$

While comparing the nature of load distribution on the screw nut assembly under rolling friction with that under sliding friction, the author found (with absolute values of the assemblies used in practice in Machine handling) that the load distribution is more uniform in the case of ball recirculating nut than in the case of ordinary nut.

8.11. Analysis of Axial Load and Contact Rigidity of a Recirculating Ball Screw

We have already seen that

$$\sigma_c = m_\sigma \sqrt[3]{\frac{PE^2(R_2 - R_1)^2}{R_1^2 R_2^2}} \qquad (8.47)$$

$\frac{A}{B}$ may also be written as approximately equal to

$$\frac{A}{B} = \left(1 - \frac{R_1}{R_2}\right)\left(1 - \frac{R_1}{R_0}\cos\alpha_k\right) \qquad (8.48)$$

Similarly $m_\sigma \approx \left(1.32 - 3.49\frac{A}{B}\right)^2$. For the range of screw diameter from 15 mm to 100 mm, m_σ normally lies between 0.03 and 0.1.

From the load equation, we may write

$$P = \left[\frac{|\sigma_{con}|_{perm}}{m_\sigma}\right]^3 \frac{R_1^2 R_2^2}{E^2\,(R_2 - R_1)^2} \qquad (8.49)$$

where $|\sigma_{con}|_{perm}$ = permissible value of contact stress, normally lies between 20,000 and 25,000 kg/cm^2.

Assuming helix angle of the thread as λ, we may write for axial thrust

$$Q = PZ \sin \alpha_k \cos \lambda$$

where Z = total number of balls recirculating in i grooves

$$= i \cdot n = i \cdot \frac{\pi R_0}{R_1}, \text{ neglecting } \lambda$$

In general $\alpha_k = 45°$ and $\lambda = 2$ to $2\frac{1}{2}$. Hence the above load equation can be written in the form

$$Q = 0.7Z \left[\frac{|\sigma_{con}|_{perm}}{m_e} \right]^3 \frac{R_1^2 R_2^2}{E^2 (R_2 - R_1)^2} \tag{8.50}$$

assuming cos λ value approximately equal to unity.

In case of grooves in screw or nut having semi-circular or gothic arc profile average value of $\frac{R_2}{R_1} = 1.05$ and for usual material of recirculating ball screw (approximate chemical composition identical to hardened chromium steel SAE 8626 and SAE 4613 having hardness value between 58 and 62 Rockwell C), the value of $F_1 = 2.1 \times 10^6$ kg/cm^2.

$$Q = \pi \times 0.7 \times \frac{iR_0}{R_1} \left[\frac{|\sigma_{con|perm}}{m_e} \right]^3 \cdot \frac{R_1^2 \cdot R_2^2}{(R_2 - R_1)^2} \tag{8.51}$$

In deducing the above expression it has been assumed that the pattern of distribution of load is uniform amongst successive threads of the nut in contact.

8.12 Evaluation of the Rigidity

The contribution of contact deformation and hence contact rigidity is greater than any other contribution and hence a designer prefers to design recirculating ball screw only on the basis of the contact rigidity. The contact deformation is given by the equation

$$\delta = m_\delta \sqrt[3]{\frac{P^2 (R_2 - R_1)}{E_2 R_1 R_2}} \tag{8.52}$$

$$m_\delta \simeq \left(1.41 - 1.17\frac{A}{B}\right)^2 \tag{8.53}$$

$$P = \frac{Q}{Z \sin \alpha_k \cdot \cos \lambda} \tag{8.54}$$

$$\delta = \frac{m_\delta}{Z^{2/3}}\left(\frac{R_2 - R_1}{R_1 R_2}\right)^{1/3} \frac{1}{E^{2/3}} \frac{Q^{2/3}}{(\sin \alpha_k\ cos\ \lambda)^{2/3}} \tag{8.55}$$

δ_0 = total component of the contact deformation along the screw axis, neglecting the effect of the helix angle of the thread which is negligibly small = $2\delta \sin \alpha_k$.

Contact rigidity = $J = \dfrac{dQ}{d\delta} =$

Rate of change of Q in relation to rate of change of δ

$$= (110 \times 10^3)\,\frac{3}{2}\,Z\delta_0^{1/2}$$

Substituting δ_0 by axial load equation above and simplifying we get

$$J = 3300\sqrt[3]{QR_1Z^2},\ \text{where } R_1 \text{ is in cm and } Q \text{ in kg.}$$

$$\simeq 1500\sqrt[3]{QR_1Z^2},\ \text{where } R_1 \text{ is in mm and } Q \text{ in kg.} \tag{8.29}$$

Again $$Z = i \cdot \pi \frac{R_0}{R_1}$$

$$J \simeq 1500\sqrt[3]{Q\,R_1 i^2 \pi^2 \left(\frac{R_0}{R_1}\right)^2} \tag{8.56}$$

$$\simeq 1500\sqrt[3]{\frac{Q}{R_1} i^2 \pi^2 R_0^2}$$

$$= 3300\sqrt[3]{Q\,i^2\,\frac{R_0^2}{R_1}}\ \text{ in kg/mm}$$

$$\simeq 2000\sqrt[3]{Q\,i^2\,\frac{D_0^2}{R_1}} \tag{8.57}$$

where D_0 happens to be the pitch diameter of screw in mm and R_1 in mm, Q being in usual kg measure.

8.13 Sensitivity Analysis

The calculation of contact rigidity is subjected to certain error, depending upon the error contribution in the measurement of thrust and tolerances in diameter of the screw and radius of the ball.

$$J \simeq 2000 \sqrt[3]{\frac{Q\, i^2\, D_0^2}{R_1}}$$

or $$J \simeq K\, Q^{1/3}\, D_0^{2/3}\, R_1^{-1/3}, \text{ where } K = 2000 i^{2/3}$$

Using partial derivatives.

$$\Delta J = \frac{\delta J}{\delta Q}\Delta Q + \frac{\delta J}{\delta D_0}\Delta D_0 + \frac{\delta J}{\delta R_1}\Delta R_1 \quad (8.58)$$

$$\Delta J = \frac{K}{3} Q^{1/3} D_0^{2/3} R_1^{-1/3}\left[\frac{\Delta Q}{Q} + \frac{2\Delta D_0}{D_0} - \frac{\Delta R_1}{R_1}\right]$$

$$\frac{\Delta J}{J} = \frac{1}{3}\left[\frac{\Delta Q}{Q} + \frac{2\Delta D_0}{D_0} - \frac{\Delta R_1}{R_1}\right] \quad (8.59)$$

Here $\frac{\Delta J}{J}$ sensitivity of rigidity, in percentage, expressed as decimal

$\frac{\Delta Q}{Q}, \frac{\Delta D_0}{D_0}, \frac{\Delta R_1}{R_1}$ are tne percentage error in the measurement of Q, D_0, R_1 respectively.

Assuming $\Delta Q = \pm 30$ kg and $Q = 600$ kg.

$\Delta D_0 = \pm 0.02$ mm when $D_0 = 40$ mm

$\Delta R_1 = \pm 0.001$ mm when $R_1 = 2$ mm

we get:

$$\frac{\Delta J}{J} = 1/3[10 + 0.2 - 0.01]$$

$$= 3.39\%$$

This shows a typical error analysis, assuming simple laws of cumulative error. The case will be different when each error is randomly distributed.

8.14 A Critical Analysis

In the load distribution on the threads of the nut of ball recirculating screw assembly, Basu[63] compared the nature of distribution of load in a screw-nut pair under rolling friction with that under sliding friction and found (with absolute values of size ranges of assemblies used in practice) that the load distribution is more uniform in the case of recirculating ball nut than in case of ordinary sliding screw nut assembly. This evaluation was done considering (*i*) contact deformation, (*ii*) axial displacement resulting from shear deflection, (*iii*) deformation under bending and (*iv*) radial elastic displacement (considering thick tube analogy). These findings of the author were upheld subsequently by Birger and Arutunyan[64]. Both these two findings were based on theoretical work following the lines of Zukovsky[71]. However in case of recirculating ball screw pair the major component of deformation of a point on the thread profile is due to contact deformation, which predominates over other components of deformation. Subsequently Rodinov[69] in his paper has shown, based on experimental observations, that the distribution of load in ball leadscrew is not uniform compared to that in a sliding leadscrew pair. But his method of experimentation is not adequate. According to Rodinov[69] first turn absorbs about 40% of the load, while Pyasik[67] and Pavlov[68], also contributing to the same opinion as Rodinov, feel that the first turn nearly shares 50% of the total load and that the distribution is not uniform. More recent work on the load distribution pattern on the turns of a ball screw is by Ya. I. Shul'ga1[70], according to whose theoretical observation, it is less uniform than in a sliding leadscrew pair. However he has not been able to completely do away with the major conclusion of Basu[63], Birger and Arutunyan[64]. In fact Shul'ga, in his work, has considered total convergence $2W$ (same as total contact deformation at the two opposing contact points) approximated to a single linear relationship with load, normal to contact point, within ± 30%. But that approximation itself is erroneous and may contribute to wrong finding in regard to load distribution pattern. According to the author from theoretical calculations, it is true that there is non-uniformity but this is much less corresponding to that in a sliding screw pair. This is more so, because of the pattern of contact deformation, and manufacturing errors in pitch and ball diameters. Using the sizes and capacity of the ball leadscrew pair it is found that the sharing of the load by the turns of the nut has less non-uniformity in comparison to that used in sliding friction screw pair having identical size and capacity range.

8.15 Analysis of Preload

Contact deformation may be written in the form:

$$\delta = CP^{2/3} \tag{8.60}$$

where $C = m_\delta \sqrt[3]{\dfrac{R_2 - R_1}{R_1 R_2} \times \dfrac{1}{E^2}}$

= F (major and minor axes of contact area, geometry of the ball nut profile, and equivalent Young's modulus of the ball and nut) F denoting function

$E = \dfrac{2E_1E_2}{(E_1 + E_2)}$ where E_1, E_2 denote the Young's modulus of screw and nut material respectively.

$$P = C'\delta^{3/2}$$

$$J = \text{Rigidity} = \frac{dP}{d\delta} = C' \times \frac{3}{2}\delta^{1/2}$$

$$= C''\delta^{1/2}$$

$$= C''(CP^{2/3})^{1/2}$$

$$\therefore \quad j = CjP^{1/3} \tag{8.61}$$

where $C_j = \dfrac{3}{2C}$

Figure 8.20 shows the characteristic dependence of rigidity J_p corresponding to load P on each ball, normal to the contact point. From this curve it is evident that in order to make a ball lead screw work with steady rigidity $\left(\text{without change in : value : of } \left|\dfrac{dJ_P}{dP}\right|\right)$, it has to be loaded at P or more. But since this value of P is sufficiently large we require an initial amount of preload.

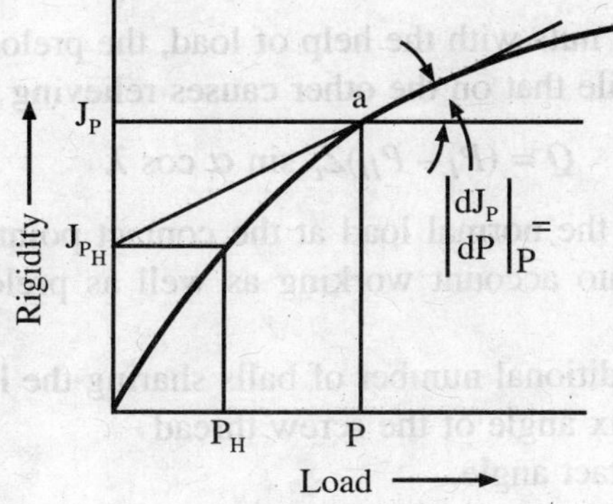

Figure 8.20

$$\left|\frac{dJ_P}{dP}\right|_P = \text{slope of this curve at load } P = \text{constant} = \text{m}$$

$$\therefore \qquad m = \frac{d}{dP}(C_J P^{1/3}) = C_J \times \frac{1}{3} \times P^{-2/3}$$

$$= \frac{C_J}{3} P^{-2/3} \tag{8.62}$$

Referring to Figure 8.15 J_{PH} is the intersection of tangent to rigidity curve at point a produced to meet γ axis.

$$(J_P - J_{PH}) = mP$$

$$C_J(P^{1/3} - P_H{}^{1/3}) = mP = \frac{C_J}{3} P^{1/3} \tag{8.63}$$

$$C_J P^{1/3} - CJP_H{}^{1/3} - \frac{C_J}{3} P^{1/3} = 0$$

$$\therefore \qquad \frac{2}{3} C_J P^{1/3} = C_J P_H{}^{1/3}$$

$$\therefore \qquad \left(\frac{P}{P_H}\right)^{1/3} = \frac{3}{2}$$

$$\therefore \qquad \frac{P}{P_H} = \frac{27}{8} \tag{8.64}$$

Since $P_H = KP$, where K is really the fraction of overall load used as preload, we denote it as preloading factor.

$$\therefore \qquad \text{Preloading factor } K = 0.3 \approx 1/3$$

In preloading by twin nuts with the help of load, the preload on one adds to the working load, while that on the other causes relieving the working load.

Axial load $\qquad Q = (P_I - P_{II}) Z_c \sin \alpha \cos \lambda \qquad$ (8.39)

where P_I and P_{II} are the normal load at the contact points of the two nuts respectively taking into account working as well as preload (as shown in Figure 8.21).

Z_c – conditional number of balls sharing the load

λ – Helix angle of the screw thread

α – contact angle.

P_1 and P_2 – loads shared by successive theards, namely first and and second theard respectively.

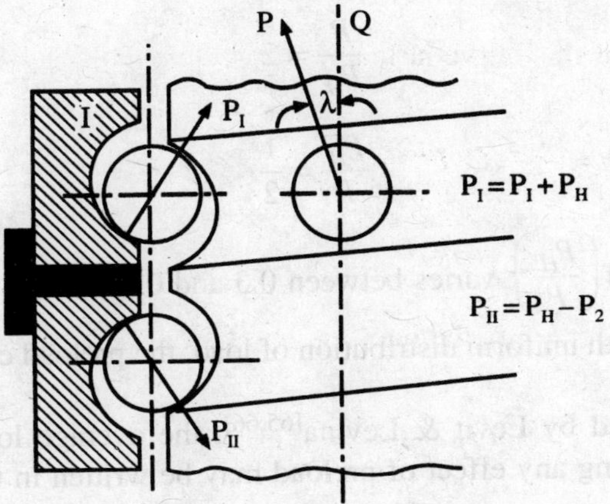

Figure 8.21

According to the author, there is not much variation in the load sharing in successive teeth. Moreover,

$$P_{\text{I}} = P_1 + PH;\ P_{\text{II}} = P_H - P_2$$

From eqn. above,

$$\frac{P_I}{P_H} = \frac{P_I}{P_H} + 1 = 1 + 0.38\frac{P}{P_H} \tag{8.65}$$

assuming 38% of total load P carried by first thread and 30% by the second thread of the nut.

$$\therefore \qquad P_{\text{I}} = P_H + 0.3P \text{ load on one nut}$$

$$P_{\text{II}} = P_H - P_2$$

$$\frac{P_{\text{II}}}{P_H} = 1 - \frac{P_2}{P_H} = 1 - \frac{0.30P}{P_H}$$

$$\therefore \qquad \text{P}_{\text{II}} = \text{P}_H - 0.30P \tag{8.66}$$

From these two above equations also it is possible to conclude that the amount of preload should always be greater than 1/3. Assuming minimum load carried by the last thread as $0.01P$ and the next thread (one before the last) as $0.02P$, we get

$$\left.\begin{aligned} P_I &= P_H + 0.01P \simeq P_H \\ P_{II} &= P_H - 0.02P \simeq P_H \end{aligned}\right\} \tag{8.67}$$

Maximum value of $\dfrac{P}{P_H} = P_I + P_{II} = 2\ P_H$

$$\therefore \qquad \frac{P}{P_H} = 2$$

$$\therefore \qquad \frac{P_H}{P} \simeq \frac{1}{2} \tag{8.68}$$

Thus the value of $\left(\frac{P_H}{P}\right)$ varies between 0.3 and 0.5.

Moreover with uniform distribution of load, the preload coefficient tends to become 0.5.

As formulated by Levit & Levina[65,66], the average load on one ball without considering any effect of preload may be written in the form

$$P_{av} = \frac{2}{5C_1^{3/2} \cdot \Delta}\left[\delta^{5/2} - (\delta - \Delta)^{5/2}\right] \tag{8.69}$$

where $\Delta = \Delta_1 + 0.5\,\Delta d$; Δ_1 being error in pitch measured on contact line of the nut and Arf is error in dimension of the ball.

Again $$P_{max} = \left(\frac{\delta}{C_1}\right)^{3/2}$$

$$\therefore \qquad K_P = \frac{P_{av}}{P_{\max}} = \frac{2\delta}{5\Delta}\left[1 - \left(1 - \frac{\Delta}{\delta}\right)^{5/2}\right] \tag{8.70}$$

When $\delta < \Delta$, the load is taken by only a few balls and then

$$K_P = \frac{2\delta}{5\Delta} \tag{8.70}$$

Z_c = Number of balls taking the load = $Z \cdot K_P$

In this method, advocated by Levit & Levina1[66], only deterministic values (equal distribution) of the error in pitch and error in ball diameter have been taken into account. But the process of manufacturing produces average pitch error distributed according to normal probability due to the manufacturing process capability. So also varies the diameter of the balls often randomly. Pitch error may be anything upto ± 5 microns. Maximum cumulative pitch error of the nut over its complete length should not exceed 5-8 microns. For use in precision machine the overall error in traverse should be within ± 2 microns.

It is a case of two independent random variables, say $f(x_1)$, $f(x_2)$ related to a variable $f(y)$, where $f(y)$ stands for total en or and $f(x_1)$ and $f(x_2)$ are the distribution of Δ_1 and Δd error.

Here 'Monte-carlo' simulation may be applied to find out the total error Δ (in pitch and diameter of ball) taken together. This value could be taken for computation of P_{av}, in equation (8.44). In using 'Monte-carlo' technique, we take recourse to random simulation using random number table. From the final distribution, it is easy to calculate standard deviation, and hence the error in predicted value of Δ, considering specific degree or level of confidence. Such value of Δ refines the value k_p as obtained by Levit and subsequently reported by Levina & Reshetov[66]

8.16 Standard Dimensions of Recirculating Rail Screw Assembly

Based on the theoretical calculations and actual experimentation the various dimensions characterising the dominating parameters of recirculating ball screw assembly with semi-circular profile or Gothic arc profile (shown in Figure 8.22) have been standardised in Russian machine tools practice. Such dimensions are shown in the tabular form below in Table 8.1.

d_e	i	d_l	r_l	r_2	r_3	r_4	d_{ks}	d_{Hs}	d_{Kn}	d_{HN}	C
20	4	2.5	1.25	1.30	0.25	0.2	18.23	19.3	21.77	20.3	0.035
25	5	3.0	1.50	1.56	0.3	0.2	22.88	24.2	27.12	25.4	0.042
30	6	3.5	1.75	1.82	0.4	0.3	27.53	29.0	32.47	30.5	0.049
30	10	6.0	3.00	3.12	0.6	0.4	25.76	28.2	34.24	30.9	0.085
35	6	3.5	1.75	1.82	0.4	0.3	32.53	34.0	37.47	35.5	0.049
35	10	6.0	3.0	3.12	0.6	0.4	30.76	33.2	39.24	35.9	0.08S
40	6	3.5	1.75	1.82	0.4	0.3	37.53	39.0	42.47	40.5	0.049
40	10	6.0	3.0	3.12	0.6	0.4	35.76	38.2	44.24	40.9	0.085
45	8	5.0	2.5	2.60	0.5	0.4	41.46	43.5	48.54	45.7	0.070
45	12	7.0	3.5	3.64	0.7	0.5	40.05	42.9	48.95	46.0	0.099
50	8	5.0	2.5	2.60	0.5	0.-.	46.46	48.5	53.54	50.7	0.070
50	12	7.0	3.5	3.64	0.7	0.5	45.05	47.9	54.95	51.0	0.099
60	8	5.0	2.5	2.60	0.5	0.4	56.46	58.5	63.54	60.7	0.070
60	12	7.0	3.5	3.64	0.7	0.5	55.05	57.9	64.95	61.0	0.099
70	10	6.0	3.0	3.12	0.6	0.4	65.76	68.2	74.24	70.9	0.085
70	16	10.0	5.0	5.20	1.0	0.7	62.93	67.0	77.07	71.5	0.140
80	10	6.0	3.0	3.12	0.6	0.4	75.76	78.2	84.24	80.9	0.085
80	16	10.0	5.0	5.20	1.0	0.7	72.93	77.0	87.07	81.5	0.140
90	12	7.0	3.5	3.64	0.7	0.5	85.05	87.9	94.95	91.0	0.099
90	20	12.0	6.0	6.24	1.2	0.9	81.52	86.4	98.48	91.8	0.170
100	12	7.0	3.5	3.64	0.7	0.5	95.05	97.9	104.95	101.0	0.099
100	20	12.0	6.0	6.24	1.2	0.9	91.52	96.4	108.48	101.8	0.170

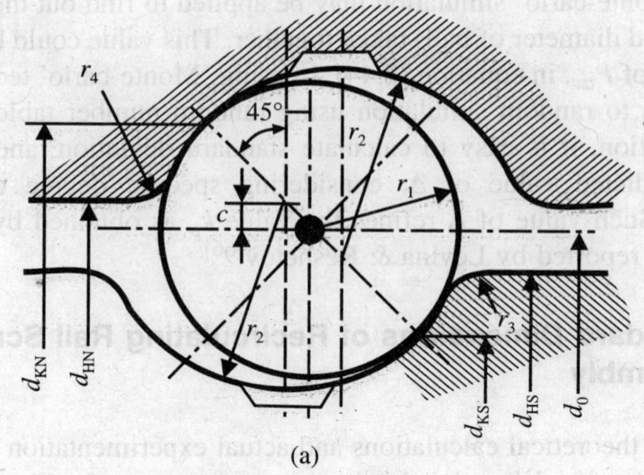

(a)

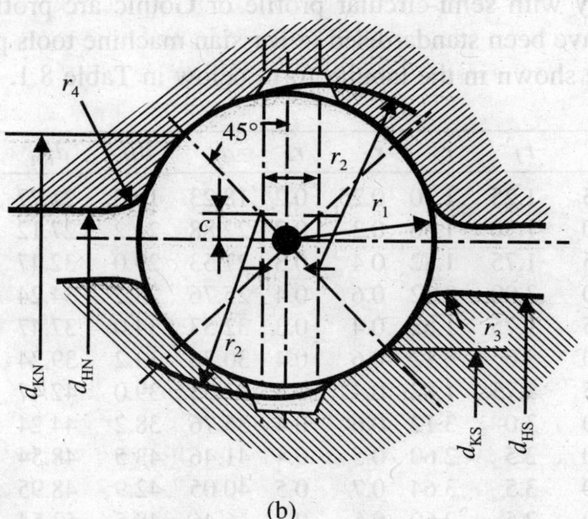

(b)

Figure 8.22 Profile of (*a*) Semi circular (*b*) Gothic arc.

Some empirical method may be used to find out the above values, so as to obtain approximate dimensions

$$d_1 = 0.6t \tag{8.71}$$

Angle of contact $\alpha = 45°$; Helix angle $\lambda \simeq 2°10'$ to $3°30'$

$$C = 0.707\,(r_2 - r_1) \tag{8.72}$$

$$d_{HS} = d_0 - 2[(r_1 + r_3)\cos\,(\mathrm{a} + \mathrm{g}) - r_3] \tag{8.73}$$

$$d_{KS} = d_0 - 1.41r \tag{8.74}$$

$$d_{KN} = d_0 + 1.41r \tag{8.75}$$

$$\gamma^o = \frac{\alpha}{r_1} + \frac{180°}{\pi} \tag{8.76}$$

$$d_{HN} = d_0 + \frac{d_0 + d_{HS}}{2} \tag{8.77}$$

$$r_3 \simeq 0.2r_1 \text{ and } r_4 \simeq 0.15r_1$$

8.17 Calculation for Dynamic Loading

The permissible value of contact stress, for calculating the static stiffness and durability (normally considered 10^7 cycles of stresses as a base)

$$|\sigma|_{static} = [\sigma]$$

Coefficient of durability K based on above cycles of stresses may be evaluated from the following expression

$$K = KQ\sqrt[3]{\frac{60Tnc_i}{10}} \tag{8.78}$$

Here K_Q is coefficient for variability of loading. Assuming that during the dynamic condition of work, the assembly works on some minimum and maximum load values for the same duration of time,

$$K_Q = 0.6 + 0.4\frac{Q_{min}}{Q_{max}} \tag{8.79}$$

where Q_{max} and Q_{min} are maximum and minimum values, respectively of the axial load on the screw. In the absence of preloading; $K_Q \simeq 0.9$, T = calculated life or durability in hours (ormally taken as T = 5000 hours), n = r.p.m. average of the screw and the nut

$$n = \frac{n_{max} + n_{min}}{2} \tag{8.80}$$

C_1 = loading cycles per unit revolution of the screw (or nut). It is approximately denoted by

$$C_1 = 0.5Z_1\left(1 + \frac{r_1}{r_0}\cos\alpha\right) \tag{8.81}$$

where Z_1 = number of working balls in one thread of the screw.

If K (as shown in Eqn. 8.78), is less than 1 i.e.

If $K \leq 1$, then the dynamic load can be calculated from the condition of static loading.

If $K > 1$, the amount of static load has to be divided by K to get the dynanvc load.

$$\left.\begin{aligned} K \leq 1, \text{ then } |Q| &= |Q|_{st} \\ K > 1, \text{ then } |Q| &= \frac{|Q|_{st}}{K} \\ |P| &= \frac{|P|_{st}}{K} \end{aligned}\right\} \tag{8.82}$$

As advocated by Reshetov[88] to sustain the preloading condition in a ball screw assembly having Gothic arc profile, variation from the nominal size of the ball should be given by

$$\Delta d_1 = 2C\sqrt[3]{P_{prc}^2} \tag{8.83}$$

where d_1 = diameter of ball which indicates no clearance in the profile (nominal size);

C = constant depending on m_δ, E and (r_2/r_1);

P_{pre} = amount of preload to which each ball is subjected.

For all sizes for the 'screw-nut' shown in Table 8.1, we may approximately write the equation for contact deformation due to preloading as[84]

$$\Delta d_1 = 2.1\sqrt[3]{\frac{P_{prc}^2}{d_1}} \tag{8.84}$$

$$\delta_{\text{pre}} = 6\sqrt[3]{\frac{P_{prc}^2}{d_1}} \tag{8.85}$$

In the above equations, δ_{pre} and Δd_1 are given in micron and d_1 measured in mm.

CHAPTER 9

Spindle Units in Machine Tools

9.1. Spindles and Their Supports :-Special Features, Material and Construction

The Main objective is to make them run at a high velocity of rotation and maintain high accuracy of the guided motion at all speeds.

Materials : (1) Steel .4 to .45 C; (2) Alloy Steel. Sometimes used, but not so very often because of rigidity consideration. Rigidity of course, depends on Young's modulus.

It usually lies between $2.10^6 - 2.2 \times 10^6$ kg/cm^2. If the spindle is rotated in sleeve bearing, and if we must provide wear resistance characteristics on spindles, we use alloy steel, viz. nickel steel containing .56–.62 C; (3) *C.I.*: Sometimes cast iron is also used, viz. in horizontal boring machines, milling machines and on some heavy lathes. This is particularly used where large diameter spindles are required and rigidity is obviously high. (4) *Steel Casting*: Usually in Machine tools we do not use steel casting because of machining difficulty. In cases of heavy machine tools it may be sparingly used.

9.2. Typical Spindle Ends

Figure 9.1 (a) and (b) show two typical examples of spindle ends to be used for the same purpose.

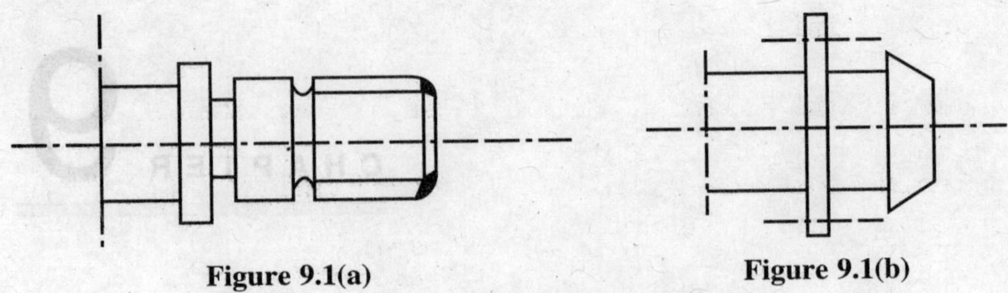

Figure 9.1(a) Figure 9.1(b)

FEATURES

Spindle end–type (a)	*Spindle end–type (b)*
1. Not so good in centering: After some use, it will develop, some clearance.	1. Centering provided from taper
2. Safe clamping of chuck is not possible.	2. Wear do not impair accuracy.
3. Quick stopping is not possible.	3. Clamping of flanges is done by bolts. There are no chances of unclamping when stopping machine from higher speeds.
4. Big cantilever effect causes greater deflection.	4. Smaller length cantilever.
5. Rotation of the Spindle in the reverse direction is not possible.	

These are the reasons that favour the more and more use design (b) and its modified form in modern machine tools. It is made of Cast Iron (spheroidal graphitic C.I.). Spindle ends used in machine tools have been standardized in many parts of the world. A typical spindle unit for a lathe machine is shown in sketch in Figure 9.1 (c).

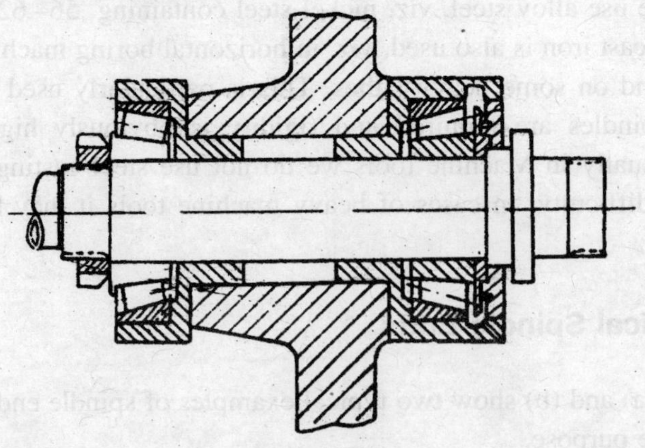

Figure 9.1(c)

9.3. Spindle Supports

Supports may be of two types: (a) sleeve bearings and (b) ball and roller bearings.

Sleeve bearing has some advantages : *Main advantages*: (1) They provide high damping characteristics against vibration required-in grinders, boiing machines, etc. (2) These bearings provide sufficient life under extreme high speeds (only on full fluid lubrication). *Disadvantages*: (1) Usually cost of such bearings is much higher than ball and roller bearings in present mass production. (2) Difficult to provide high accuracy in sleeve bearings. If the speed or load changes, the clearances have to be changed which may not be normally possible (in grinders or in precision machine tools load is almost constant as well as speed and so it could be used).

Types of Sleeve Bearings

(a) *With Radial adjustment for clearances*: Sleeve bearings in Figure 9.2, consists of three segments or parts. One of the parts isusually fixed on the body, while the other parts could be adjusted in the radial directions by screws, springs or hydraulic pressure. Automatic adjustment could be used on hydraulic design. Fixed part is under outside force F, which is the resultant of load and the cutting component Disadvantages: Low rigidity.

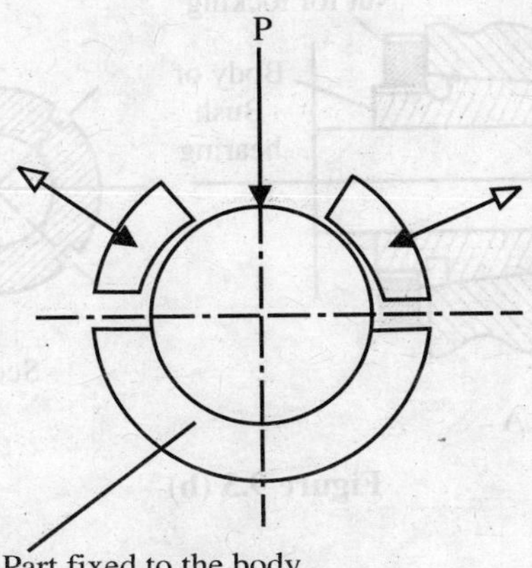

Figure 9.2

(b) *Sleeve bearing with axial. adjustment for clearances*: Depending upon the rotation of the nuts, shown in Figure 9.3. (b) the bush will be pressed, resulting in necessary pre-loading, by its elastic deformation. *Disadvantages*: We cannot provide true cylindrical form inside. But this is used in the design of some brakes. Similar type of design can be used for preloading ball bearing or recirculating ball nut, with slight, modifications.

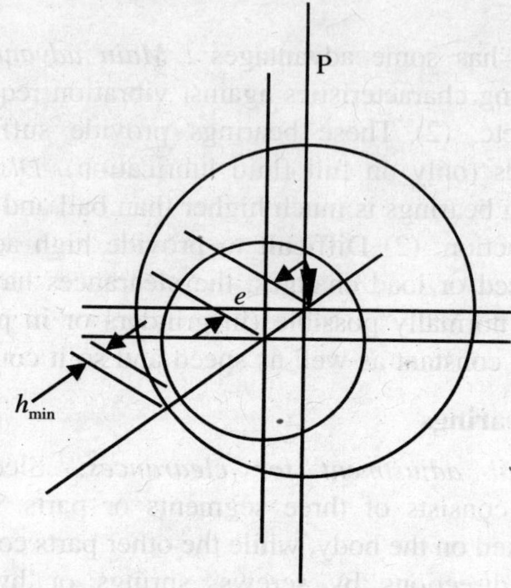

Figure 9.3 (a)

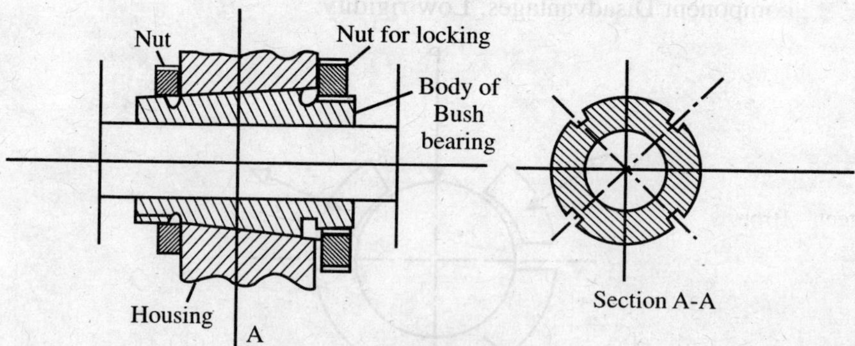

Figure 9.3 (b)

With reference to Figure 9.3 (a),

If P – Radial load on the bearing in kg;

n – speed of the shaft in r.p.m.;

d – diameter of the journal in cm.;

l – length of the bearing in cm.;

p – specific pressure in kg:/cm^2 of diametral projection of the working surface of the bearing;

Δ – diametral gap. (So that the internal diameter of the bearing = $d + \Delta$);

μ – viscosity of oil in centipoise;

h_{min} – Minimum thickness of the oil film ;

C – correction coefficient for the finite length of the bearing and is equal to $C = 1 + \frac{d}{l}$;

maximum value of the

$$\text{radial load} = p_{max} \leq p.d.l. \tag{9.1}$$

where d and l are in cm.

Peripherial velocity of the shaft $\frac{r \cdot d \cdot n}{60 \times 100} \frac{M}{\text{sec}}$. Product of $p.v.$ is a parameter which characterizes the condition of the spindle support

$$\frac{pn}{1910 \cdot l} \leq [p.v]. \tag{9.2}$$

Permissible value of the parameter [p.v.] is dependent on the intensity of lubrication[55]. Right hand side represents the per-missible value of parameter [$p.v$] shown in table below.

value of p dependent on the material of journal and the bearing	p kg/cm^2	$pv \frac{kg.m}{cm^2 - \text{sec}.}$
Under semi-liquid friction		
Steel – C.I-	30	
Steel – Bronze	50	
Steel – Babbit	60	
Hardened steel – Bronze	75	
Hardened steel – Babbit	90	
Hardened steel – Steel	150	
Limititig value under the condition liquid friction		
for C.I.	200	
Babbit	400	
Bronze casting .	650.	
Aluminium alloys	700	
for Shafts and journals of machine tools	5–20	10–25
Electrical motor shafts	10–15	upto–180
Transmission shaft	7-15	10–20

9.4. Calculation on Sleeve Bearing

Due to the work, frictional heat is generated in the bearing:

$$H = \frac{P.f.v.}{427}\ \frac{k.\,cal}{\text{sec}.} \quad \text{where } v \text{ is in metres/sec. (periherial velocity)} \qquad (9.3)$$

f – coefficient of friction in the bearing and is calculated from the formula

$$f = 3.36.10^{-8} \times \frac{d}{\Delta} \times \frac{\mu.n}{p} + 0.55\left(\frac{d}{l}\right)^{1.5} \times \frac{\Delta}{d} \qquad (9.4)$$

A correction coefficient $\alpha\left(\frac{d}{l}\right)^{1.5}$ is introduced only when $\frac{l}{d} < 1$. Otherwise the second term is neglected. From this we briefly say that in a journal bearing working in full hydrodynamic lubrication coefficient of friction is given by:

$$f = F\left(\frac{\mu \cdot n}{p}, \Delta, d, l\right) \qquad (9.5)$$

where F denotes a mathematical function.

If we feed Q litres of oil per min through canal perpendicular to the axis to take away the frictional heat generated.

Then
$$Q = \frac{d^2 \cdot v \cdot \Delta}{10^4 \times 1} \cdot \frac{b}{l} + \frac{9 \cdot 4}{10^6} \cdot \frac{p \cdot h^2}{\mu \cdot a^2} \qquad (9.6)$$

where b – length of the oil hole, in cm.
a – 0.5 $(l - b)$ in cm.
h – gap between the journal and the bearing- at the point where the oil enters the bearing in micron (10^{-3} mm);
P – as before, in kg.

The general equation of thermal equilibrium could now be written in the form:

$$K_0 \times \pi dl\,(t_{b,} - t_a) + \frac{C \cdot Q \cdot \delta}{1000}(t_0 - t_e) = \frac{P \cdot f \cdot v}{427} \qquad (9.7)$$

where t_b, t_a, t_0 and t_e, are the temperature in °C of the bearing, surrounding air, outgoing oil and incoming oil respectively. $C = 0.4 - 0.5$ = specific heat of oil in K. cal/kg. °C, $\delta = 0.87$ to 0.9 = specific weight of the circulating oil in kg/Litre; Q - consumption of oil in cm^3/sec, K_0 = overall heat transfer coefficient from bearing to atmosphere in K. cal/cm^2 °C sec.

9.5. Ball Bearings

Ball and roller bearings are used mainly with the purpose of reducing the friction by replacing sliding action by rolling. Though apparently it seems otherwise, the ball or roller bear-ings exhibit almost same rigidity as the sleeve bearing.

In Figure 9.4 (p. 156) is shown the distribution of lead on a radial ball bearing. If the total load is R and load compressing the balls are P_0, P_1, P_2..., P_n, then

$$R = P_0 + 2\,P_1 \cos\gamma + 2P_2 \cos 2\gamma + \ldots$$
$$2P_n \cos n\gamma, \tag{9.8}$$

where the angle γ's are shown in figure If δ_0, δ_1, δ_2...., δ_n denote the contact deformation under the loading P_0, P_1, P_2 ....., P_n respectively then we can have following two relationships

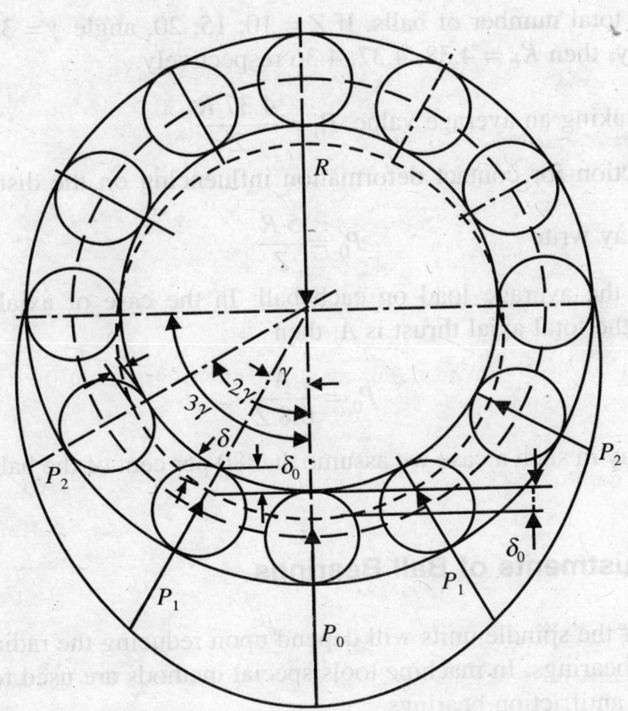

Figure 9.4 Ball Bearing Load Distribution

$$\delta_1 = \delta_0 \cos\gamma;\ \delta_2 = \delta_0 \cos 2\gamma;\ \ldots.\ \delta_n = \delta_0 \cos n\gamma,$$

and $\delta_0 = (\text{constant})\ P_0^{2/3},\ \delta_1 = (\text{constant}) \times P_1^{2/3}$, etc. (9.9)

From the above equations 9.8 and 9.9 we have

$$P_1 = P_0 \cos^{3/2} \gamma;\ P_2 = P_0 \cos^{3/2} 2\gamma;\ ......\ P_n = P_0 \cos^{3/2} n\gamma$$

Putting these values in equation (9.8) we get

$$R = P_0 \left(1 + 2 \sum_{i=1}^{n} \cos^{5/2} i\gamma \right) \tag{9.10}$$

or in other words $$P_0 = \frac{K_b \cdot R}{Z} \tag{9.11}$$

where the constant $$K_b = \frac{Z}{1 + 2 \sum_{i=1}^{n} \cos^{5/2} i\gamma}$$

where Z = total number of balls. If Z = 10, 15, 20, angle γ = 36°, 24°, 18° respectively, then K_b = 4.38, 4.37, 4.36 respectively.

Therefore taking an average value $$P_0 = \frac{4.37\ R}{Z} \tag{9.12}$$

with correction for contact deformation influencing on the distribution for load, we may write $$P_0 = \frac{5\ R}{Z} \tag{9.13}$$

This gives the average load on each ball. In the case of axial thrust ball bearing, if the total axial thrust is A, then

$$P_0 = \frac{A}{0.8\ Z} \tag{914}$$

That is to say in such a case we assume that 80 per cent of the balls only take the thrust.

9.6. Adjustments of Ball Bearings

Accuracy of the spindle units will depend upon reducing the radial and axial play of the bearings. In machine tools special methods are used for reducing the play, in antifriction bearings.

Method 1 : High precision ball and roller bearings should be manufactured. Cost of high precision bearing is about 5-10 times more than that of ordinary bearing. Reducing the play by two times, the cost is approximately increased to five times.

Method 2 : Preloading to eliminate all clearances in bearing and increase their rigidity. Preloading can be done in various ways, but the mostly used method is the one shown in Figure 9.5.

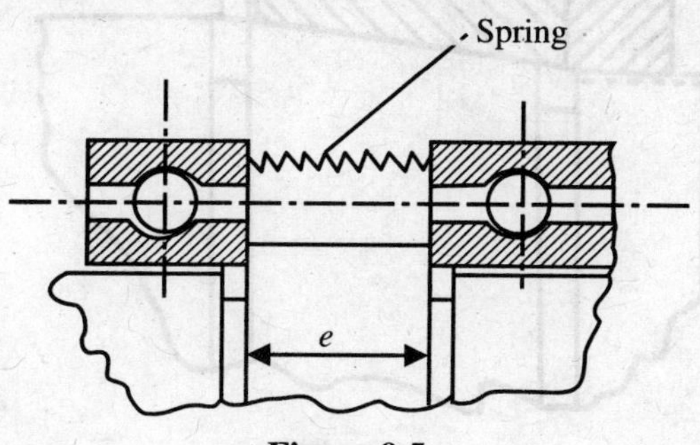

Figure 9.5

Fix the distance between inside rims, but make the distance between outside rims adjustable by means of spring. Axial loading can be transferred to radial loading, if we get angular point of contact, as shown in Figure 9.6. If there is any clearance at the outset, such clearance can be minimized by providing a shift of the races. Provide shifting of the outside race to neglect clearance at the beginning. In assembling we provide opposite directional shifting by spring. Or we could vary this gap *e*, by filler at the outside or at the inside race or at both.

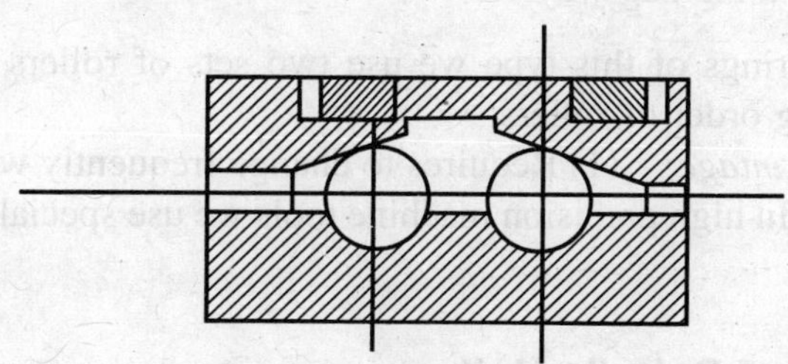

Figure 9.6

9.7. Roller Bearings

There are two main types-(1) cylindrical type and (2) taper roller type.

The cylindrical type provides high rigidity. In machine tools we use such bearings with arrangement for preloading. Such a system of preloading can be best conceived by a look at Figure 9.7.

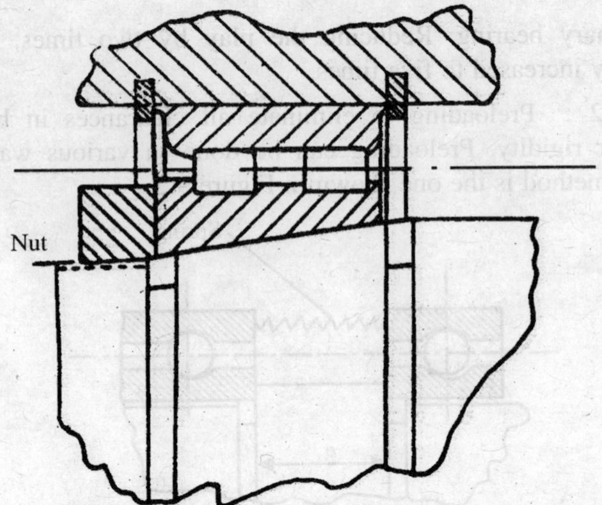

Figure 9.7 Adjustment of Roller Bearing.

When we rotate the nut, the inner race will go in the righ hand direction and there will be elastic deformation of the race in the radial direction. Such deformation will effect the preloading. This system has the following advantages:

1. Initial rigidity is quite high.
2. Outside forces do not influence on preloading.
3. We increase here the number of points of contacts, by disposing the rollers in a zig-zag way.

In roller bearings of this type we use two sets of rollers in a common cage with zig-zag order of rollers.

Main disadvantage; (1) Requires to change frequently when run under very high speed. In high precision machine tools we use special design of this type of bearing.

9.8. Rigidity of Spindle Units

The working characteristic of a machine tool, to a large extent, is related to the accuracy of its spindle rotation, static and dynamic rigidity, allowable speed regulation, heat generated and characteri-stics and durability of the bearings on which the spindle is supported.

The requirement of the rigidity of the spindle may be decided by the normal condition of working of its bearing supports. Rigidity of a spindle unit, supported at its two ends in bearings, should not be less than 25 kg/micron for lathe, automats and other machine tools of normal precision; 50 kg/micron for lathe, boring, grinding and other machine of high precision.

Rigidity of the spindle unit is denoted by

$$J = \frac{P}{Y} \tag{9.15}$$

where P = force in kg, acting at the mid point of the span between two bearing supports

Y = deflection of the spindle under the load in micron

Approximately such rigidity (cqn. 9.15) may be written as:

$$J = 53\,\frac{D^4 - d_1^4}{l^3}\ \text{kg/micron} \tag{9.16}$$

where D = average diameter of the spindle in the span between supports, expressed in mm.

d_1 = average diameter of the internal bore of the spindle, mm

l = length measured between the centrelines of the supporting bearings, mm

Table 9.1 shows the high speed characteristics and regulation of speed of rotation of the spindle where rolling friction bearings are used.

Table 9.1

Type of the Machine Tool	*Parameter dependent on speed = dn_{max} mm/rev per min*	*range* $\frac{n_{max}}{n_{min}}$
Lathe	1,50,000–2,10,000	80–125
Turret and semi-automatics or automatic	1,80,000–2,50,000	10–25
Milling machine	1,50.000–2,00,000	40–50
Boring machine	1,20,000–1,50,000	125–150
Wheel head of internal grinding	5,00,000–6,00,000	1.0

With reference to the spindle deflection diagram shown in Figure 9.8(i) and (ii), we can write down the following equation:

$$\Delta = 1.5\left[\frac{\Delta_l}{\sqrt{m_1}} \pm \frac{1}{K}\left(\frac{\Delta_l}{\sqrt{m_1}} + \frac{\Delta_2}{\sqrt{m_2}}\right)\right] \tag{9.17}$$

Here Δ_1 = radial deflection of the bearing in the front end, mm

Δ_2 = radial deflection of the bearings at the rear end, mm

$K = \frac{l}{a}$ where l is distance between the spindle supports in mm and a is distance between the front end (from bearing support to the face plate) in mm.

m_1, m_2 = number of bearings at the front and rear end respectively.

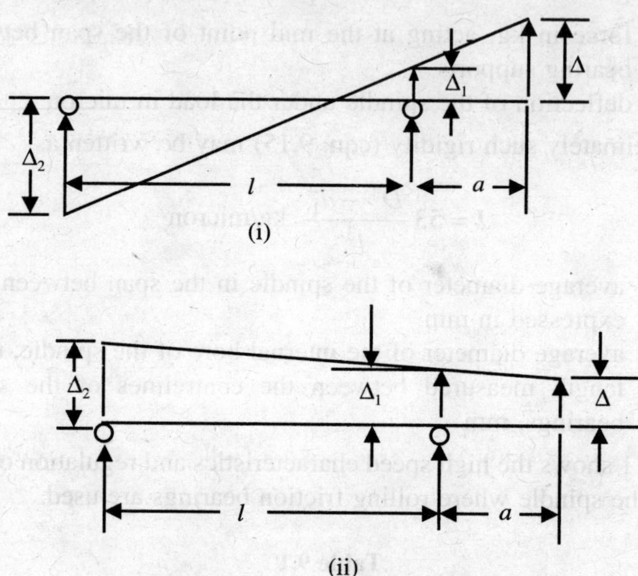

Figure 9.8 Spindle deflection diagram.

9.9. Rigidity of the Rolling Friction Supports

Elastic deformation of rolling friction support consists of (i) elastic radial deformation of the rolling body with the ring (δ'_r) and (ii) contact deformations on tne interface of Journal and inner ring and the outer ring and housing (δ''_r).

Total deformation

$$\delta_r = \delta'_r + \delta''_r \tag{9.18}$$

But
$$\delta'_r = K_1 P^a, \text{ mm} \tag{9.19}$$

where P is load in kg.

The values of K_1, (coefficient) and the power (a) are shown in Table 9.2. The values relate to the optimum conditions of working of antifriction bearings, having diameter ranging from 40 to 100 mm.

$$\delta_r'' = \frac{4PK_2}{db}\left(1 + \frac{d}{D}\right).\ \text{mm} \tag{9.20}$$

where d, D = internal and external diameters of the bearing respectively in mm;

b = width of the bearing in mm;

$K_2 \simeq 0.005$ to $0.025\ mm^3/kg$.

The value K_1 is less, when greater is the accuracy, higher is the preload and the inner ring set on taper.

Table 9.2 Value of the coefficient *K* and constant *a*

Type of bearing	K_1	a
Radial ball bearing	$(0.7–0.00\ 2d) \times 10^{-3}$	$\frac{2}{3}$
Tapered roller bearing	$\frac{0.52}{d} \times 10^{-3}$	1.0
Double row roller bearing with small cylindrical rollers	$\frac{0.40}{d} \times 10^{-3}$	1.0
Single-row roller bearing with small cylindrical rollers	$\frac{0.65}{d} \times 10^{-3}$	1.0

NB: Diameter of hole d is in mm

Rigidity of the radial support, therefore, may be written as:

$$j = \frac{P}{\delta_r' + \delta_r''},\ \text{kg/mm} \tag{9.21}$$

The above expression for the rigidity of the support is obtained considering no clearance in bearing assembly. Changing this clearance we change the load of the bearing.

$$P = iZC_\delta \delta_r f(\rho),\ \text{where}\ \rho = \frac{e}{2\delta_r} \tag{9.22}$$

where Z = number of rolling elements in each row of the bearing

i = number of rows of rollers in multi-row bearings

e = preload, expressed in mm

$f(\rho)$ = interval in load distribution

$$f(e) = \frac{1}{\pi} \int_0^{arc\cos(-\rho)} (\rho + \cos\phi) \cdot \cos\phi \cdot d\phi \tag{9.23}$$

$$
= \begin{cases} \frac{1}{2}\frac{1}{\pi}[\rho\sqrt{1-\rho^2} + \pi - \text{arc cos } \rho] \text{ for } 0 < \rho < 1 \\ \frac{1}{2} \text{ for } \rho \geq 1 \\ 1/4 \text{ for } \rho = 0 \end{cases} \quad (9.24)
$$

d = diameter of the Journal in mm

C_δ = coefficient depending on type of the bearing used in kg/mm

From the above, we may write the rigidity of the roller bearing as

$$
J = \frac{P}{\delta_r} = iZC_\delta f(\rho) \quad (9.25)
$$

9.10. Magnitude of Deflection at the Free end of the Spindle

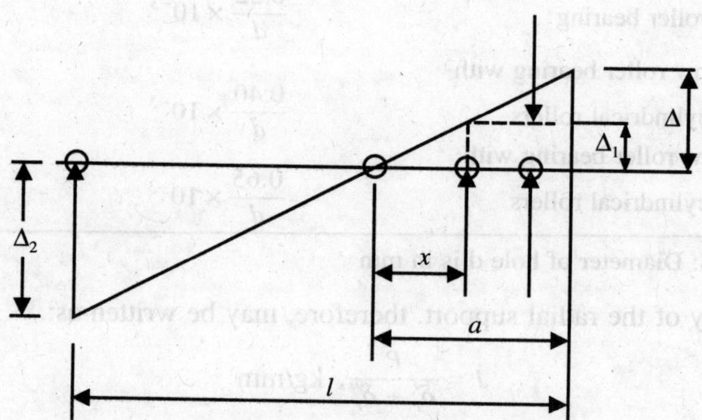

Figure 9.9 Calculation diagram.

$$
\frac{x}{l-x} = \frac{\Delta_1}{\Delta_2} \text{ and hence } x = \left(\frac{\Delta_1}{\Delta_1 + \Delta_2}\right) \quad (9.26)
$$

Again $\frac{\Delta}{\Delta_1} = \frac{a+x}{x} = 1 + \frac{a}{x}$

Therefore $x = \frac{a}{(\Delta_1 / \Delta_2) - 1} \quad (9.27)$

Combining these two equations we get,

$$
\Delta = \Delta_1\left(1 + \frac{a}{l}\right) + \Delta_2 \cdot \frac{a}{l} \quad (9.28)
$$

Often we take $\Delta \le \dfrac{\Delta^*}{3}$

Assuming $$\left(1+\frac{a}{l}\right)\cdot \Delta_1 = \Delta_2 \cdot \frac{a}{l} \simeq \frac{\Delta^*}{6} \qquad (9.29)$$

Where, Δ^* = Maximum permissible radial deflection of the end of the spindle.

Δ_1 and Δ_2 are chosen for bearings with related class of accuracy.

CHAPTER 10

Lubrication and Rigidity in Machine Tools

10.1. Introduction

This Chapter deals with the basis of selecting lubrication for the machine tool on the characteristic behaviour of the moving parts under friction. Though no specific trade marked lubrication oil has been quoted here, the author has enumerated the points to be looked into while making a selection of proper lubricant for a machine tool.

The author has dealt with, in the second part of this Chapter, the meaning and significance of machine tool rigidity about which we hear more than what can be obtained in any text book. The author emphasises specifically the fact that the rigidity though one of the major considerations for designing a machine tool is certainly not the only consideration that a machine tool designer should concentrate whole heartedly on.

In practical purpose dynamic rigidity is more important than static rigidity but uptil now no suitable method is available for obtaining dynamic rigidity value of a machine tool experimentally.

10.2. Steps in Selecting Proper Lubrication Oil

The purpose of lubrication is to eliminate dry friction between two metallic rubbing surfaces. Therefore it minimises wear and surface breaking as well as eliminates the possibility of thermal deformation or stresses induced in the rubbing materials. It increases thus the life of the parts. To carry away the quantity of heat generated by friction we must have sufficient supply of oil, which should be continuously circulated. By using lubrication oil in guides we can permit the use of higher value of specifiic pressure on guides and obtain less heating, less wear, longer life, higher speed and practically less vibratory motion.

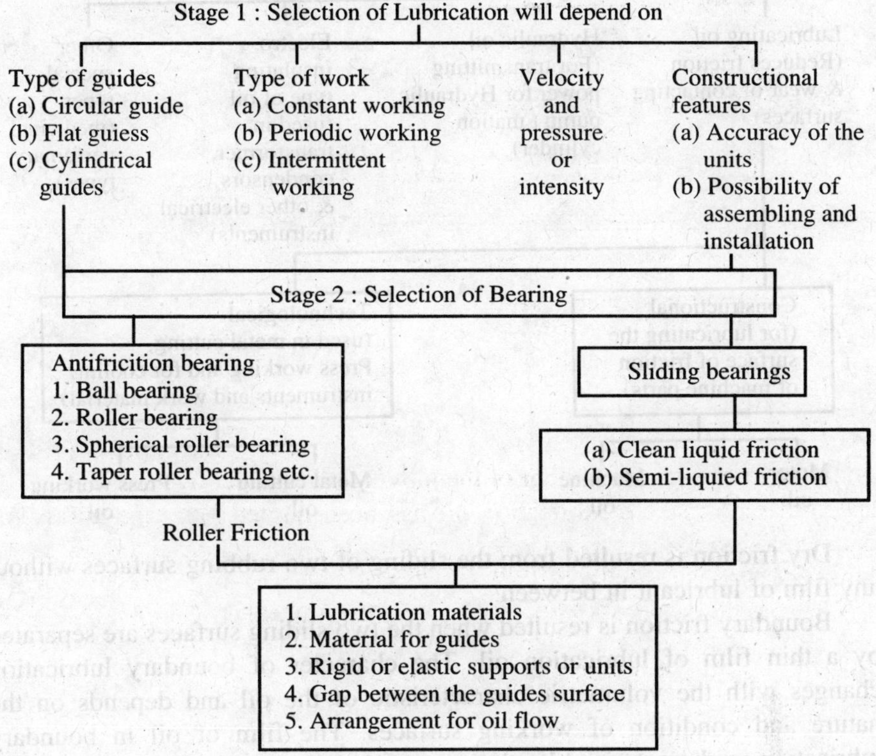

10.3. Frictional Condition of Working

By sliding friction we mean the friction between two surfaces when one undergoes sliding or displacement along the surface of the other, while by rolling friction we mean the friction, resulting by the displacement of one body over another by the help of a moment, vector of which lies on the place of rolling.

From the external conditions of woking, friction can be classified as clean friction, dry friction, boundary friction, semi-dry friction, semi-liquid friction, liquid friction.

Clean friction between the two sliding surfaces depends on the absorbed layer and chemical composition. It is resulted under the considerable plastic deformation of parts. Clean friction is accompanied by molecular adhesion between the particles or grains of the sliding surfaces.

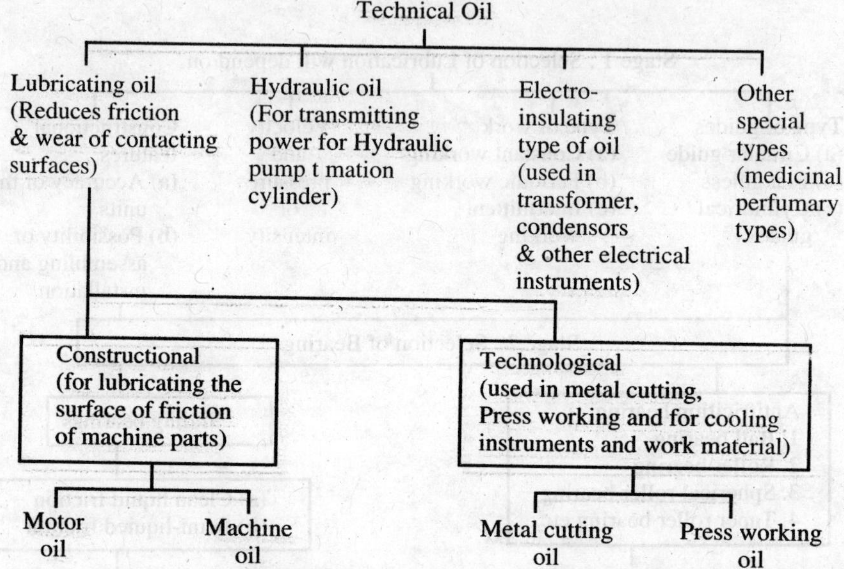

Dry friction is resulted from the sliding of two rubbing surfaces without any film of lubricant in between.

Boundary friction is resulted when the two sliding surfaces are separated by a thin film of lubrication oil. The character of boundary lubrication changes with the volumetric characteristic of the oil and depends on the nature and condition of working surfaces. The film of oil in boundary lubrication is characterized by the presence of absorbed layer with oriented molecular structure.

Semi-dry friction is nothing but transitional friction, that is to say, a combination of dry and boundary friction. It is often referred to as a combination friction. In semi-liquid friction the major portion of the load is taken up by the layer of oil film, while in liquid friction the two sliding surfaces are completely separated by oil and there exists relative shear between each layer of the oil film. In practice, we get only partial or full liquid friction as well as combination friction. Ideal boundary friction is not achieved in practice in any machine tool. It is only possible to obtain boundary friction under laboratory experimental condition.[11] Similarly, to say with confidence, clean friction can be met only when the sliding surfaces are working in vacuum or in inert gas.

The force acting tangentially to the layers of oil film separating the sliding bodies is expressed in the Newton's Law as $$T = \tau \cdot S = \mu \cdot \frac{dv_x}{dy} \cdot S,$$

where μ – dynamic coefficient of viscosity;

$\frac{dv_x}{dy}$ – gradient of the velocity of shears in the lubricant; &

S – area of shear.

The dynamic viscosity oil is usually expressed in the unit $\frac{gm}{cm \cdot sec.}$ and is known as poise. In practice it is useful to have smaller unit equal to 0.01 poise or centipoise :

$$1 \text{ centipoise} = 102.10^{-6} \text{ kg. sec./m}^2.$$

Kinematic viscosity $\upsilon = \frac{\mu}{\rho}$ where ρ is the density in gm/cm^3 or in other words ν is expressed by the unit cm^2/sec, known as stokes. For practice we often use smaller units equal to 0.01 stokes or centistokes. It is imperative that since the value of viscosity changes with the temperature, we should base our consideration, in all usual cases, on normal working temperature, *viz.* 20°C.

Figure 10.1 shows a typical characteristic curve for a machine tool guideways showing the relationship between the coefficient of friction and the velocity of sliding. Let us assume

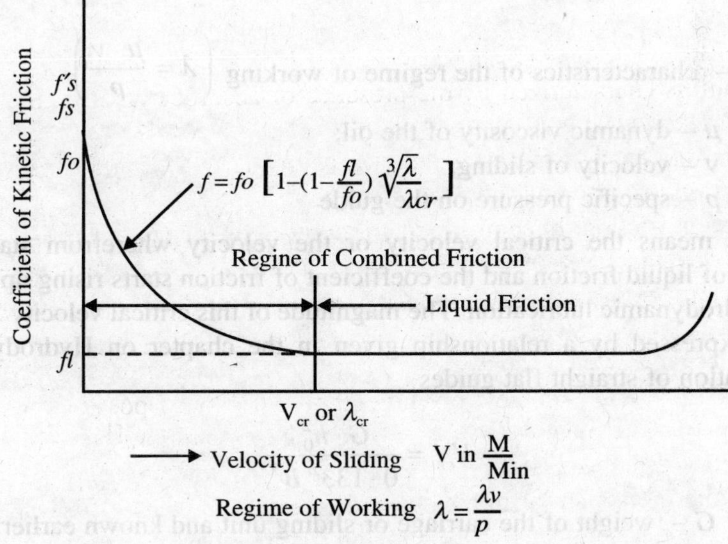

Figure 10.1

f_s – coefficient of static friction when observed during a comparatively smaller period of time of static contact;

f'_s – coefficient of static friction after considerable lapse of time (see Figure 10.2);

f_e – coefficient of kinetic friction when the value of velocity of sliding is the smallest possible;

f_1 – coefficient of kinetic friction corresponding to critical velocity of sliding V_{cr} and critical regime of working λ_{cr}; and

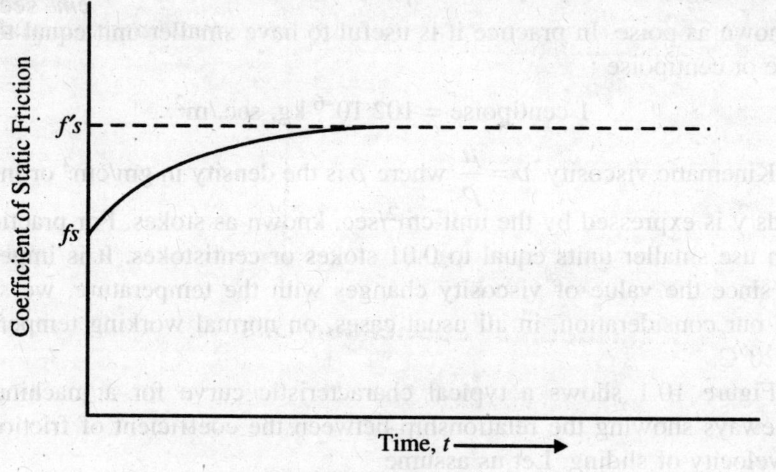

Figure 10.2.

λ – characteristics of the regime of working $\left(\lambda = \dfrac{\mu \cdot v}{p}\right)$

where μ – dynamic viscosity of the oil;

v – velocity of sliding;

p – specific pressure on the guide

V_{cr} means the critical velocity or the velocity wherefrom starts the regime of liquid friction and the coefficient of friction starts rising up due to full hydrodynamic lubrication. The magnitude of this critical velocity V_{cr} has been expressed by a relationship given in the chapter on Hydrodynamic Lubrication of straight flat guides

$$V_{cr} = \frac{G \cdot h_0^2}{0 \cdot 133^{\mu} B^3}$$

where G – weight of the carriage or sliding unit and known earlier;

h_o – minimum thickness of the oil film;

μ – dynamic coefficient of viscosity; and
B – breadth of the fiat straight guides.

In all other cases of guides V_{cr} can be determined experimentally and monograms can be made. For the purpose of calculation h_e, the summation of micro and macro error of the surfaces of contact, can be taken to lie between 0 01 – 0.02 mm. This is in almost all cases of light-and medium-duty machine tool guides. But for heavy machine tools this can be as high as 0.06 to 0.1 mm. Inside the region of combined friction ($V \leq V_{cr}$) the value of kinetic friction corresponding to a particular velocity V or regime of working can be calculated from the empirical formula given by G.A. Levit.[37]

$$f = f_0 \left[1 - K_f \sqrt[3]{\frac{\lambda}{\lambda_{cr}}} \right] \tag{10.1}$$

where $$K_f = 1 - \frac{f_l}{f_e}$$

10.4. Specification of Lubrication Oils

While specifying a suitable lubricating oil for guide or other moving parts we must mention the following properties:

1. Kinematic viscosity in centistokes at 100°C.
2. Ratio between the Kinematic viscosity at 50°C and the same at 100°C.
3. Acid number.
4. Carbon content (0.25 – 0.70%).
5. Ash content (0.005 to 0.25%).
6. Temperature of freezing.
7. Fire temperature.
8. Resistance to foaming,.
9. Contamination.
10. Corrosion and rust prevention.

Various lubricating oils are available in the market, some with SAE specification. It is not our purpose to give here specification and composition of any particular lubricating oil. But it should be noted that by adding certain polar additives to the lubricant as, for example, carbide acid, resin, asphalt and some sulphur compounds, its properties are changed and the lubricant behaves in a much more tamed fashion. That is to say it exhibits certain advantages from the point of view of its characteristic kinetic coefficient of friction.

It is found usually that with such additives the difference between the magnitude of static and kinetic coefficients of friction reduces, thereby minimizing the possibility of the occurrence of stick-slip motion.

In the chapter* on Machine Tool Vibration we have shown such a characteristic curve, and have compared its behaviour with the behaviours of two conventional lubricants.

Typical arrangement of a closed lubrication system for h|bri-cating the gear box of a lathe machine is shown in Figure 10.3.

The lubrication system can be designed for continuous and or automatic lubrication of parts or for discontinuous action. In Figure 10.4 is shown a system of lubrication to be used for providing hydrostatic-hydrodynamic lubrication of machine tool guides.

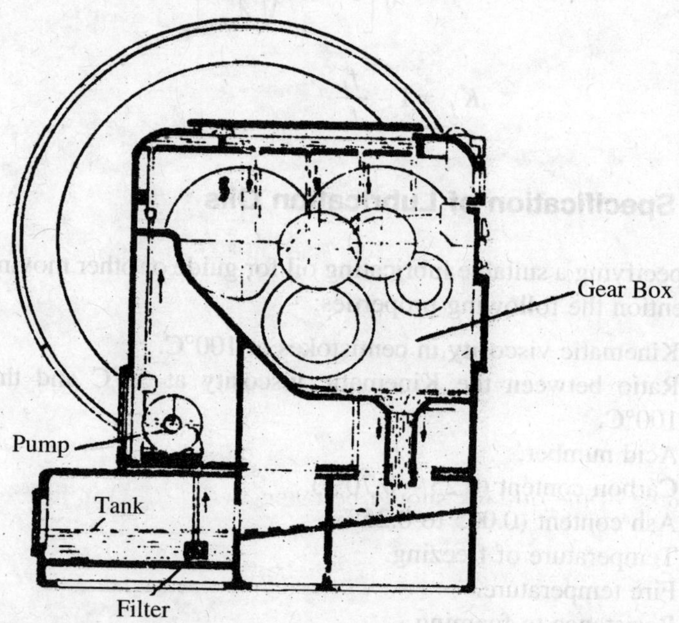

Figure 10.3 Closed Lubrication System.

The consumption or the quantity of lubrication oil, used in the closed (Ref. 4) centralized system, can be approximately calculated from the formula:

$$Q = 14.3 \frac{N(1-\eta)}{C \cdot \gamma \cdot \Delta t} \text{ litres/min.}$$

Where N – power in Kwatts transmitted through the mating pairs, or working pairs;

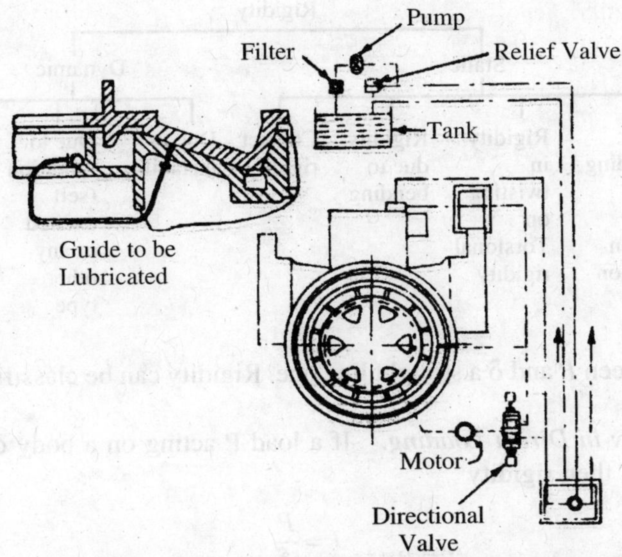

Figure 10.4 Typical Lubrication Arrangement.

η – total efficiency of the lubricated parts or objects;
C – specific heat of the lubricant in K. calories/kg°C. (Usually for mineral oils C ≈ 0.45 K. Cal/kg°C)
γ – specific weight of the lubricant in kg/litres (usually for mineral oils $\gamma \approx 0.88$ kg/litres).
Δt – change in temperature of the lubricating oil (for gear trains and rolling friction bearings $\Delta t \approx (5 - 10)$°C
For sleeve bearings $\Delta t \approx (30 - 40)$°C

Volume of the oil tank should be taken as (5 – 6) Q in litres.

10.5. Rigidity of Machine Tool Units

Rigidity of a part can be defined as the load per unit deformation. That is to say, if under a load P, the deformation that the part undergoes is δ, then the static rigidity $J = \frac{P}{8}$ assuming a linear relationship between P and δ.

The unit by which the rigidity is usually expressed is kg/mm. But if the relationship between the load and the deformation it produces, is not a linear one, then the rigidity will be expressed by the relation $J = \frac{\Delta P}{\Delta \delta}$.

where $\Delta\delta$ is the small change in deformation for the corresponding small change in load P. But in machine tool elements usually we get the average

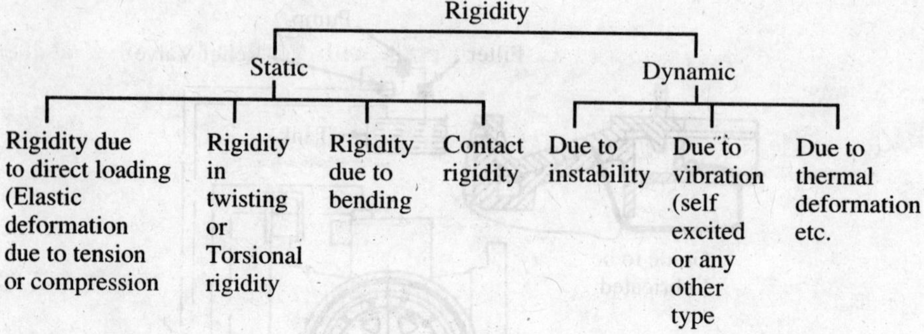

curve between P and δ a straight line one. Rigidity can be classified as shown above.

Rigidity in Direct Loading. If a load P acting on a body causes direct straining δ, then rigidity

$$J = \frac{P}{\delta}. \tag{10.2}$$

The rigidity of a power screw can be calculated from the formula below taking into account the extension in length under the action of axial loading. If Q = axial load on the screw, then

$$\text{strain over length } l \qquad = \frac{\Delta l}{l} = \frac{Q}{\pi/4\, d_{av}^{2}} \Big/ E \tag{10.3}$$

where Δl – extension of length l under Q and E is the Young's modulus of elasticity of the material of the screw, d_{av} – denotes the effective diameter of the screw :

$$J_{direct} = \frac{\pi\, d_{av}^{2}\, E}{4l}. \tag{10.4}$$

Of course, this expression for rigidity does not take into account the rigidity due to contact deformation, which indeed constitutes the major proportion of the overall rigidity nor does it consider the torsional rigidity.

$$\textit{Torsional rigidity: } J_{tor} = \frac{T}{Q} = \frac{GI_{\eta}}{l} = \frac{G \cdot \frac{\pi}{32} \cdot d_{av}^{4}}{l}. \tag{10.5}$$

For the example above : T = torque;
Q = angle of twist;
G = modulus of rigidity of the material;
I_n = polar moment of interia;
l = length of the screw.

Rigidity in Bending

If a shaft is simply supported at both ends with vertical load of P at the middle, then

$$J_{\text{Bend}} = \text{Rigidity due to Bending}$$

$$= \frac{P \times 48\, EI}{P\, l^3} = \frac{48\, EI}{l^3} \tag{10.6}$$

where E – Young's modulus of elasticity for the material;
l – length of the shaft; and
I – moment of inertia of the cross-section of the shaft $\left(= \frac{\pi}{64} d^4\right)$ where d = diameter of the shaft. To generlize, we can say that the rigidity in bending $J_{Bend} = \frac{KEI}{l^3}$, (10.7) where the value of the constant K depends on the condition and position of loading on the beam.

Contact Rigidity

Contact deformation usually does not follow any linear rule and consequently for determining the rigidity we must precisely find out the relationship between contact deformation and the intensity of pressure on contact surface.

Let us follow the symbols:
P – normal load on two bodies at the contact point;
α – contact area;
δ – contact deformation; and
q – intensity of pressure on contact surface.

As shown by A.P. Sokolovsky and emphasized by D.N. Reshetov and J.M. Levina[66], δ and q can be related empirically by the approximate expression

$$\delta = C \cdot q^m \tag{10.8}$$

where m is an exponent, the value of which oscillates between 0.3 and 0.5. This value of m is empirically chosen, based on the experimental results. C is a coefficient depending upon the characteristic of the working surface and materials of the bodies in contact. The above relationship can be simplified by taking first degree approximation, so that $\delta = C.q$. However from equation 10.8, $\delta = C\left(\frac{P}{\alpha}\right)^m = C_1 \cdot P^m$, where C_1 is another constant

$$P = \frac{\delta^{\frac{1}{m}}}{C_1}$$

and the rigidity $$J_{\text{Contact}} = \frac{dP}{d\delta} = \frac{\delta^{\frac{1-m}{m}}}{C_2}, \tag{10.9}$$

where

In a machine the cutting load is resolved into three components:

$$P_x, P_y, \text{ and } P_z.$$

Direction of the application of force	*Direction of measurement of deformation*		
	Along axis y	*Along axis y*	*Along axis y*
		Rigidity	
P_x	J_{xx}	J_{yx}	J_{zx}
P_y	J_{xy}	J_{yy}	J_{zy}
P_z	J_{xz}	J_{yz}	J_{zz}

Rigidity along the axis $y = J_y$

$$\frac{P_y}{J_y} = \frac{P_x}{J_{yx}} + \frac{P_y}{J_{yy}} + \frac{P_y}{J_{yz}} \tag{10.10}$$

$$J_y = \frac{1}{\frac{1}{J_{yx}}\left(\frac{P_x}{P_y}\right) + \frac{1}{J_w}\left(\frac{P_y}{P_y}\right) + \frac{1}{J_{yz}}\left(\frac{P_x}{P_y}\right)} \tag{10.11}$$

where $J_{yz} = \frac{P_y}{\delta_{yz}}$, J_{yx} – deflection along y caused by P_x.

$J_{yy} = \frac{P_y}{\delta_{yz}}$, J_{yy} – deflection along y caused by P_y

$J_{yz} = \frac{P_z}{\delta_{yz}}$, J_{yz} deflection along y caused by P_z.

***Contact rigidity affected by micro-roughness of the surface*[152]**

(i) Micro roughness of the surface generated due to various manufacturing processes, change's the height distribution patterns of the asperities and subsequently the elastoplastic deformations at the interface and hence the coefficient of contact compliance denoted by $\left(\frac{d\delta}{dp}\right)$, opposite of rigidity

(ii) Multimodal probability distribution of asperity heights causes the specific characteristic surface topography, based on random process analysis of various manufacturing processes.

(iii) Prediction regarding functional behavior of the surface, therefore, requires autocorrelation analysis of the surface. Variations in the magnitude of contact compliance with clearances and tilting of slides have also been studied in great details by many researchers.

(iv) According to some researchers, the contact rigidity does not change significantly with the use of lubrication. In fact, some recent work done by the authors has shown that such inference cannot be emphasized strictly particularly when the lubricant (viz. graphite-based grease or MoS_2 mixed with various percentages of base oil) is in the solid lubricant.

Figure 10.5 shows the effects of the varying percentage of solid lubricant (MoS_2) mixed with base oil SAE-20 on contact deformation. All the four graphs are drawn on experimental results [152] based on loading varying from 1 Kg to 5 Kg. For Nylon or filled nylon, as well as for PTFE or Filled PTFE, the optimum concentration of solid lubricant M0S2 is 20% when mixed with Base oil SAE-20.

10.6. Some Errors Affecting Rigidity

In a recirculating ball screw assembly, which is extensively used in programme controlled machine tools, the amount of contact deformation is much higher than in ordinary screw assembly. The bodies of the nut and the screw undergo elastic deformation due to compression and tension when subjected to external load. Besides, there is further deformation due to inaccuracy in manufacture and the force of friction acting along the surface of the thread profile in contact with the nut.

The influence of the frictional force on the rigidity of a recirculating ball screw can be neglected since the coefficient of friction in this case is considerably less than in sliding screw assembly.

The fundamental equation of rigidity of the recirculting ball screw assembly can be expressed in the form

$$\frac{1}{J_{\text{overall}}} = \frac{1}{J_n} + \frac{1}{J_e} + \frac{1}{J_{con}} + \frac{1}{J_M} \qquad (10.12)$$

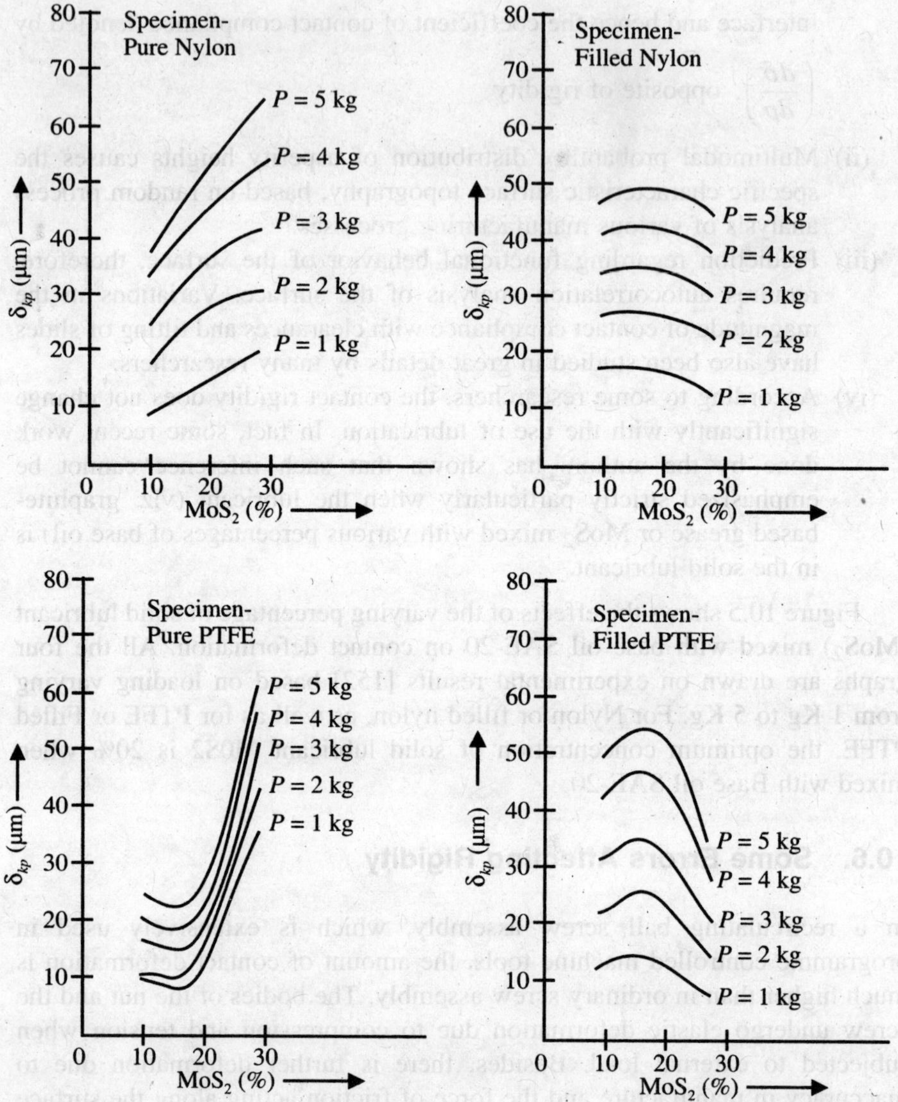

Figure 10.5 Contact deformation against concentration percentage of solid lubricant (MoS_2) in oil.

where $J_{overall}$ is the overall rigidity of the ball screw assembly, J_n – the rigidity of the body of the nut, J_{con} – the contact rigidity, J_{si} – the rigidity resulting from the bending moment acting on the thread, and J_e – the rigidity of the body of the screw or bolt.

In a ball recirculating screw nut, the rigidity of the system varies considerably, depending on the difference in the size of the balls. For example if balls are made within a tolerance of ± 2 micron (1 micron = .001 mm) in diameter, it becomes more rigid than a system using balls having ± 4 micron. The variation in tolerance of the ball size can be considered as listributed equally in groups or according to the law of normal distribution. Automatic control and inspection of balls makes the chances of equal distribution more favourable than the normal distributionof Gauss[10]

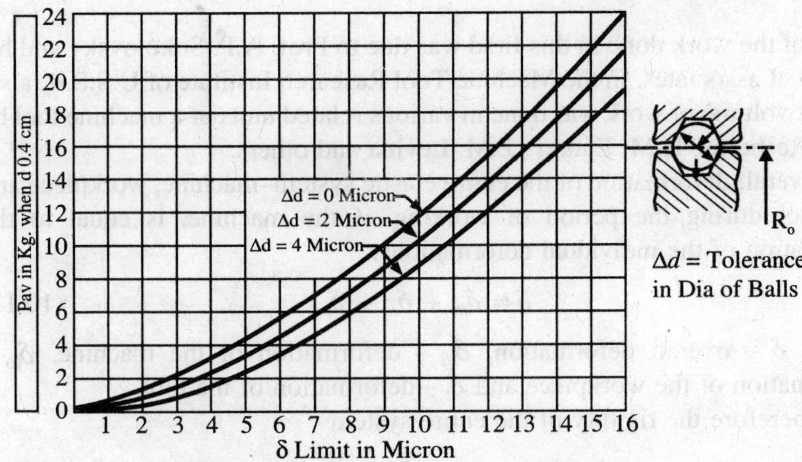

Figure 10.6 P – δ for Trapezoidal Profile.

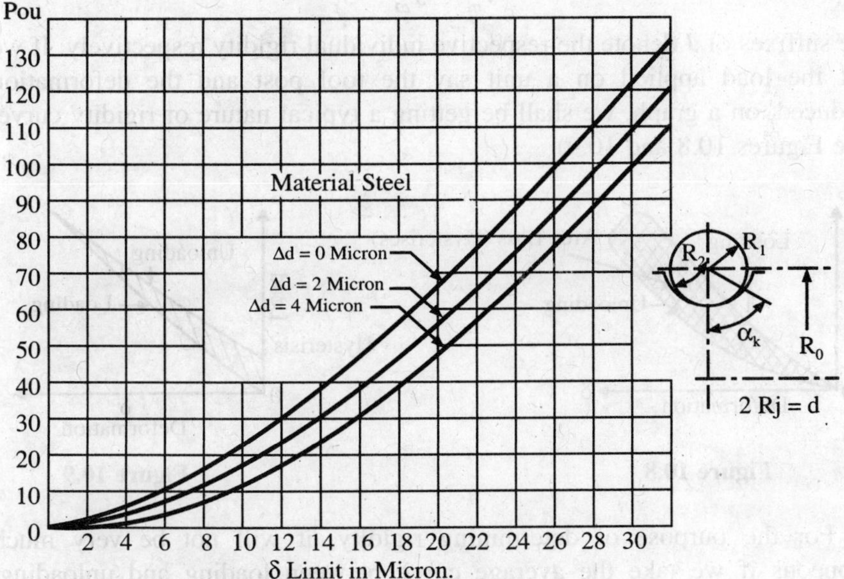

Figure 10.7 P – δ for [Semicircular] Profile.

The figures above show the variation in rigidity depending on the tolerance on the ball sizes for the two different profiles of the ball recirculating screw assembly assuming a range of working, where $P–\delta$ is a linear function. From these two it would appear that for stable working without backlash, it is necessary to have preloaded ball screw.

10.7. Overall Static Rigidity of Machine Tools

Most of the work done in this field was due to Prof. A.P, Sokolovsky and his school of associates. In the Machine Tool Research Institute of U.S.S.R. also, a great volume of work was done in various related units of a machine tool by D.N. Reshetov, H.M. Enikev, Z.M. Levina and others.

Overall deformation of the entire elastic system–machine, workpiece and the tool during the period of working of the machine, is equal to the summation of the individual deformations:

$$\delta = \delta_m + \delta_\omega + \delta_t, \tag{10.13}$$

where δ – overall deformation, δ_m – deformation of the machine, δ_ω – deformation of the workpiece and δ_t – deformation of the tool.

Therefore the rigidity of the entire system

$$J = \frac{1}{\frac{1}{J_m} + \frac{1}{J_\omega} + \frac{1}{J_t}} \tag{10.14}$$

The suffixes of J denote the respective individual rigidity respectively. If we plot the load applied on a unit say the tool post and the deformation produced, on a graph, we shall be getting a typical nature of rigidity curve. (See Figures 10.8 and 10.9).

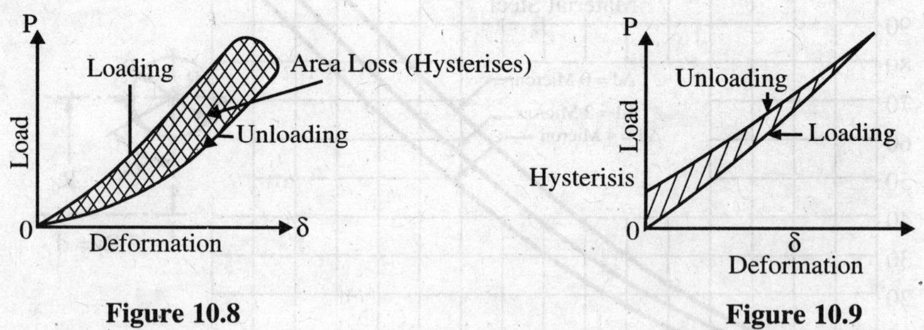

Figure 10.8 **Figure 10.9**

For the purpose of determining rigidity, it will not be very much erroneous if we take the average curve between loading and unloading.

Taking the case of a lathe work, the magnitude of rigidity is dependent on the diameter of the work-piece. The formula given in equation 10.7 shows that the rigidity in bending is proportional to $\frac{d^4}{l^3}$, but $\frac{d^4}{l^3} = \left(\frac{d}{l}\right)^3 d$.

Therefore (1) greater the value of $\frac{d}{l}$ ratio, more the magnitude of rigidity; (2) greater the diameter of the work material, more the rigidity.

Experiment shows that for a lathe machine, having the height of the centre more than 20 cm, the rigidity of the spindles lies between 7000–10000 $\frac{\text{kg}}{\text{mm}}$. This value again increases by about two times if the spindle is supported on roller bearing at the front end and the spindle is provided with flanged end.

10.8. Dynamic Rigidity of a Machine Tool

Up to now no successful attempt could be made to study the dynamic rigidity of a machine tool.

For each dynamic regime and under each characteristic working condition, the machine will have its own rigidity.

Let us take the case of a forced and damped vibration, considering the tool post of m mass vibrating because of the deformation y in the direction perpendicular to the workpiece, caused by the force of cutting in the aforesaid direction. The equation of motion can be written as

$$m \cdot \ddot{y} + \beta \cdot y + J_{st} \cdot y = P_0 \sin pt \qquad (10.15)$$

where β – damping coefficient proportional to velocity,
J_{st} – statical rigidity,
P_e – amplitude of the external force,
p – frequency of the external excitation force, and
t – time.

From equation 10.15 $\ddot{y} + \frac{\beta}{m} \cdot \dot{y} + \frac{J_{st}}{m} \cdot y = \frac{P_0}{m} \sin pt$

or

$$\ddot{y} + 2\lambda\dot{y} + \omega^2 y = f_0 \sin pt \qquad (10.16)$$

where $2\lambda = \frac{\beta}{m}$ and $\omega = \sqrt{\frac{J_{st}}{m}}$ and is the frequency of natural vibration.

The solution can bi written in the form

$$Y = \frac{P_0 \sin (pt + \varphi)}{m \cdot \sqrt{\omega^2 - p^2) + 4p^2 \lambda^2}} = A \sin (pt + \varphi) \quad (10.17)$$

where phase angle $\varphi = \tan^{-1} \dfrac{2\lambda P}{\omega^2 - P^2}$

and amplitude of vibration $A = \dfrac{P_0}{m \sqrt{(\omega^2 - P^2) + 4p^4 \lambda^2}}$

Dynamic Rigidity

$$J_{dyn} = \frac{dP}{dy} = \frac{dP}{dt} \bigg| \frac{dy}{dt}; \text{ but since } P = P_0 \sin pt$$

$$\frac{dP}{dt} = P_0 p \cos pt. \quad (10.18))$$

Therefore $\quad J_{dym} = m \cdot \sqrt{(\omega^2 - p^2)^2 + 4p^2 \lambda^2} \cdot \dfrac{\cos pt}{\cos (pt + \varphi} \quad (10.19)$

when $pt = \pi$, *i.e.* at the end of the half cycle of vibration

$$J_{dYn} = \frac{m^2 \ (\omega^2 - p^2)^2 + \beta^2 p^2}{m \ (\omega^2 - p^2)} = \frac{m \ [(\omega^2 - p^2) + 4p^2 \lambda^2]}{(\omega^2 - p^2)} \quad (10.20)$$

CHAPTER 11

Controlling Systems in A Machine Tool

In this chapter we have explained only the principles of mechanism used for controlling the shifts in machine tools with particular reference to head stock gear-box. The controlling system and their arrangements play a vital role in the design of machine tools. Functional efficiency, ease in operation, better outward appearance are a few of the essential characteristics of a machine tool. By proper positioning of these controlling systems and proper selection of modes of controlling we can reduce considerably the unnecessary human motions of the operator. Such systems therefore must be designed with due consideration to ease, comfort and speed of working of the man working on the machine. Or in other words, the principles of ergonomics play an important part in designing the proper controlling systems. Principle of motion economy states:

1. Motions of more than one organ should be simultaneous.
2. Motions should be' systematic, rhythmic and as far as possible balanced.
3. Motions involved while an operator work must become habitual.

In the light of the above principles of motion economy we can categorise the considerations for designing the controlling systems :

1. All handles or shifting mechanism must be concentrated in as narrow an area of normal working as possible.
2. All handles must be away from the rotating parts for the purpose of safety and accident prevention.

3. Forces required to operate the various handles must be limited to 6-8 kg.
4. The design of handles should be simplified, number and varieties to be reduced to effect increased production of machine tools.
5. Number of positions should also be limited (8-10 stations or positions).
6. Handle should be so designed that it will indicate the direction of motions.

11.1 Classification

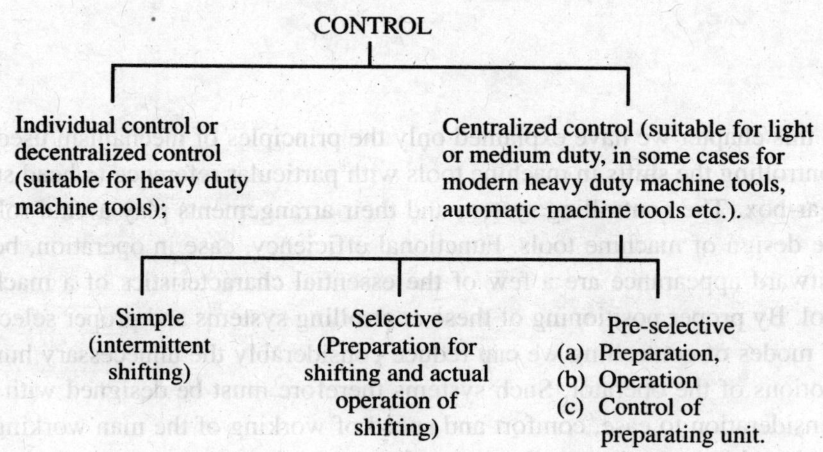

Simple centralized control: In this system we can have either single lever or multilever mechanisms. In the first one with the help of one handle we can have so many manipulations. In Figure 11.1 we have showa,a few simple sketches to illustrate the mechanism of shifting. In Figure 11.2 is shown a simple system where with handle 1 we can rotate a common cylindrical cam to shift three stepped blocks to get 6 different speeds. Shifting is done not necessarily on one shaft. In Figure 11.2(b) (p. 180) is shown the development of the grooves on cam shaft indicating the respective working and neutral position of each shifting lever.

Other examples of centralised controls are shown in

Figure 11.3 Here controlling of two shiftings is done by one rotating disc cam. This is simple and simplicity means low cost and easy maintenance. Main disadvantage in this case is the loss of time (time lag) in manipulation.

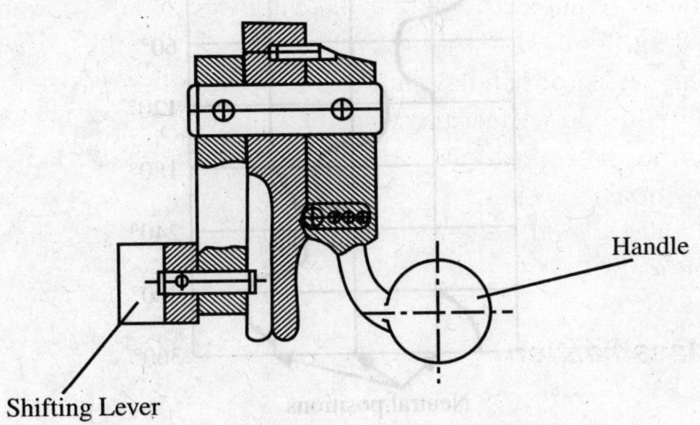

Figure 11.1 (a).

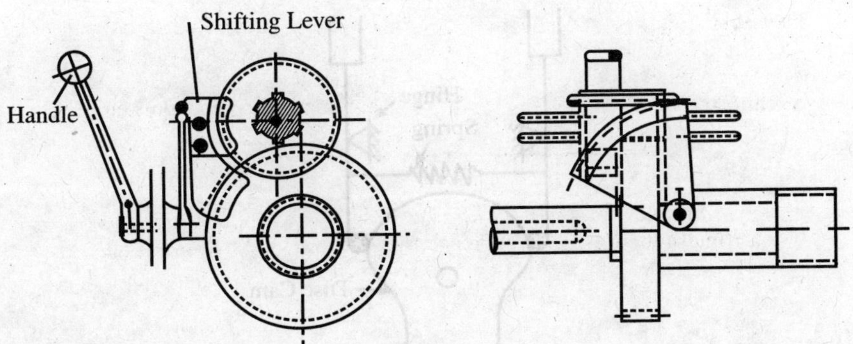

Figure 11.1 (b).

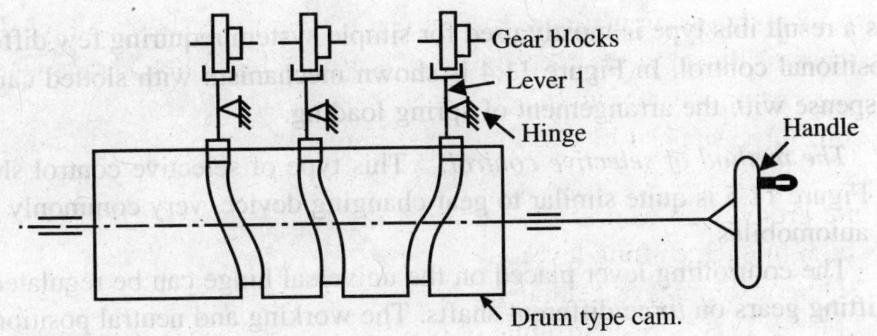

Figure 11.2(a) Simple Centralised Control

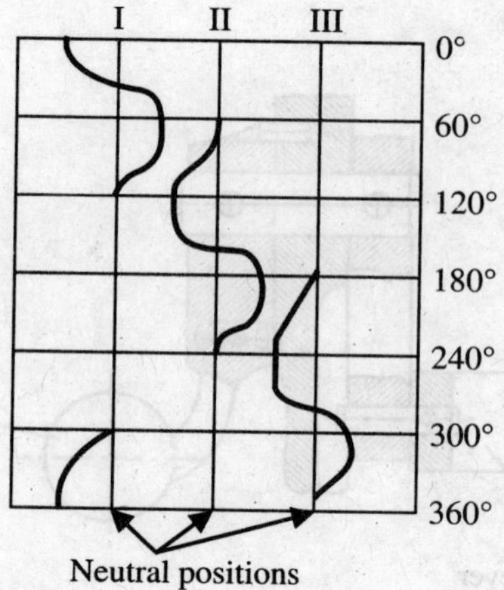

Figure 11.2(b) Development of the Grooves on the Cam-Shaft shown in Figure 11.2(a).

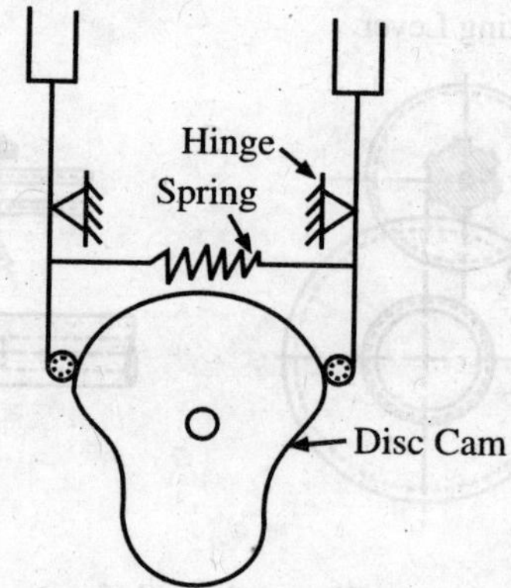

Figure 11.3 Control by Cam.

As a result ibis type is mostly used for simple system requiring few different positional control. In Figure 11.4 is shown mechanism with slotted cam, to dispense with the arrangement of spring loading.

The method of selective control: This type of selective control shown in Figure 11.5 is quite similar to gear changing device, very commonly used in automobiles.

The controlling lever placed on the universal hinge can be regulated for shifting gears on three different shafts. The working and neutral positions of the lever for various shaft positions are shown in the slotted template. Here, from any working position to any other working position the handle is shifted

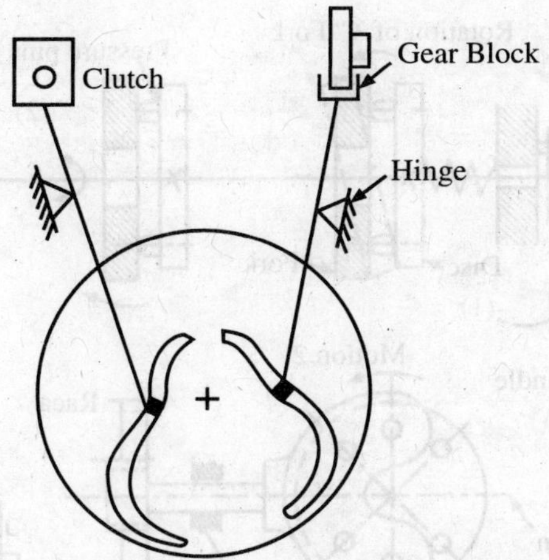

Figure 11.4 Control by Cam.

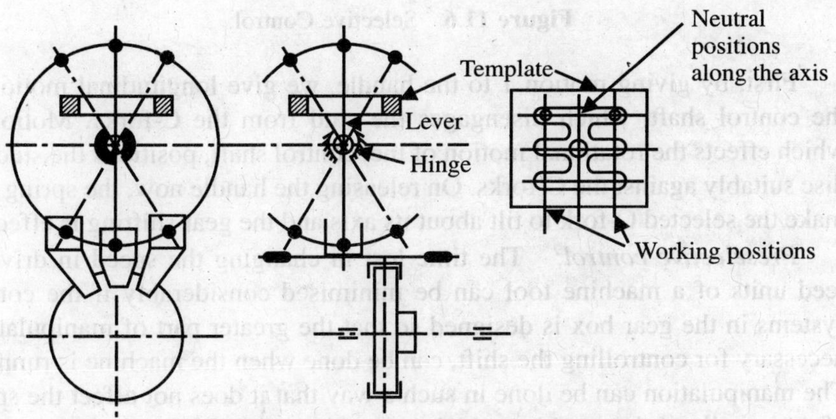

Figure 11.5 Selective Control.

through neutral positions, instead by steps as in simple system, thereby effecting selective control.

In Figure 11.6 another system of selective control has been shown. Three or more discs are fixed on the shaft provided with a handle as shown. On each disc are fixed a pair of studs diagonally opposite. The studded discs are pressing on the C-fork by the action of springs. The C-fork can be tilted, by the proper position of the studs, about its axis. This will cause rotation of the pinion and movement of the rack. The shifting lever is attached to the rack. The principle of working is as follows:

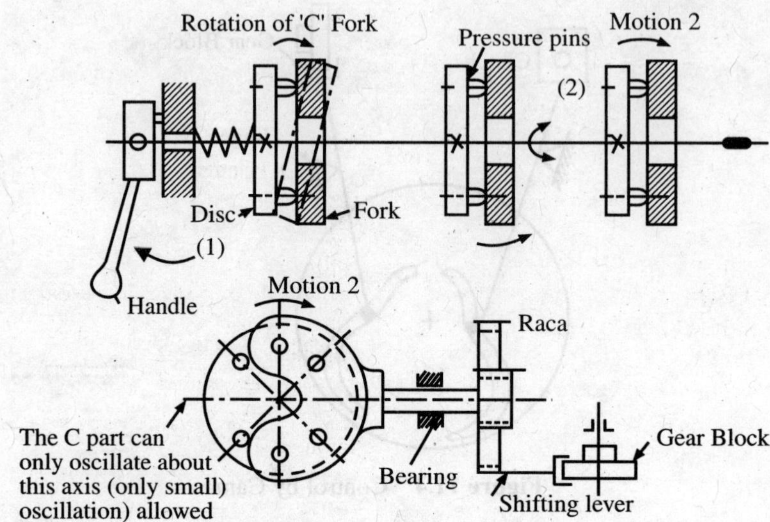

Figure 11.6 Selective Control.

First, by giving motion 1 to the handle, we give longitudinal motion of the control shaft, which disengages the stud from the C-forks. Motion 2, which effects the rotational motion of the, control shaft, positions the studded disc suitably against the C-forks. On releasing the handle now, the spring will make the selected C-fork to tilt about its axis and the gear shifting is effected.

Preselective control: The time lost in changing the speed in drive or feed units of a machine tool can be minimised considerably if the control systems in the gear box is designed so that the greater part of manipulation, necessary for controlling the shift, can be done when the machine is running. The manipulation can be done in such a way that it does not affect the speed of the spindle of the machine which is working. This function can be termed as preparation. After the preparation or greater part of the manipulation is completed the speed of the spindle can be changed very quickly with the help of only one handle or with the help of a press button. Such system of control is popularly know as "preselective control".

A method of preselective control has been depicted in Figure 11.7. Here the angular rotation of the discs to find the definite angular position of the pressure pins means "preparation." There are various types of preparation units. The "operations" mean the actual engagement of the mating pairs. Therefore the preselective control can be effected by two necessary motions in this case : (1) rotation of big disc by handle (preliminary motion) and (2) operation including declutching, shifting the, gear by means of disc and clutching. These three operations are controlled by the handle 2 only by its rotation.

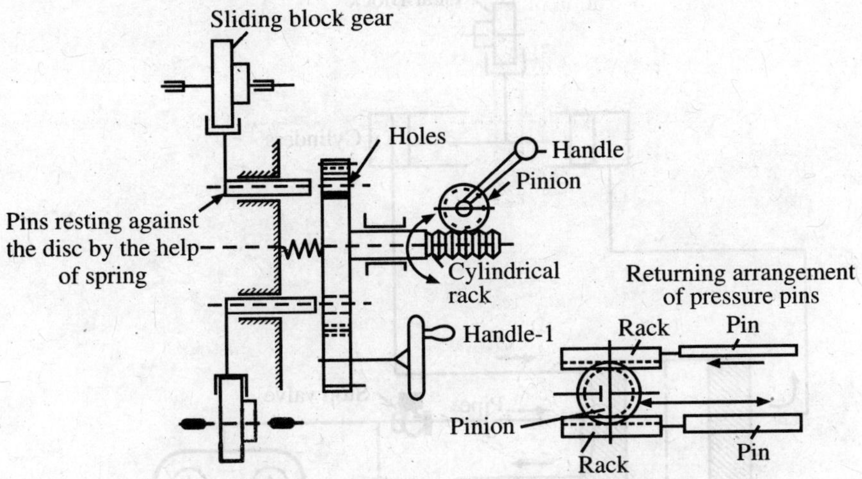

Figure 11.7 System of Preselective Control.

The returning of the shifted gears to the old position is effected by two racks and pinion attached to the gear shifting pins as shown in Figure 11.7. Holes are done in such a way that for a particular position, some bars will be opposite to holes and some will be opposite to the metallic surface of the disc.

In modern machine tools we use more and more hydraulic and pneumatic system of control. One such system is shown in Figure 11.8. This system is used mainly in horizontal boring machine, milling machine etc. This system is anotner example of preselective control.

In system, shown in Figure 11.8, the distributor may be be connected to more than one cylinder. When the pressure in pipe '*a*' increases, the safety valve opens to allow the excess oil to go for lubrication. The system can be used for preselective control also by using a valve on pipe '*a*'. By closing the valve we can rotate the distributor to change the relative pass without changing the block position and then we can open the said valve, so that the block position will be changed now.

Main motions for controlling in this system can be divided into two distinct ones, *viz.*

1. Preselective motion.
2. Motion for shifting.

In Figure 11.9 is shown double-sided centralised hydraulic cylinder for shifting the block in three different positions.

When oil is passed through 1, some pressure is applied on the piston which pushes the right hand side bush and piston will come to extreme right

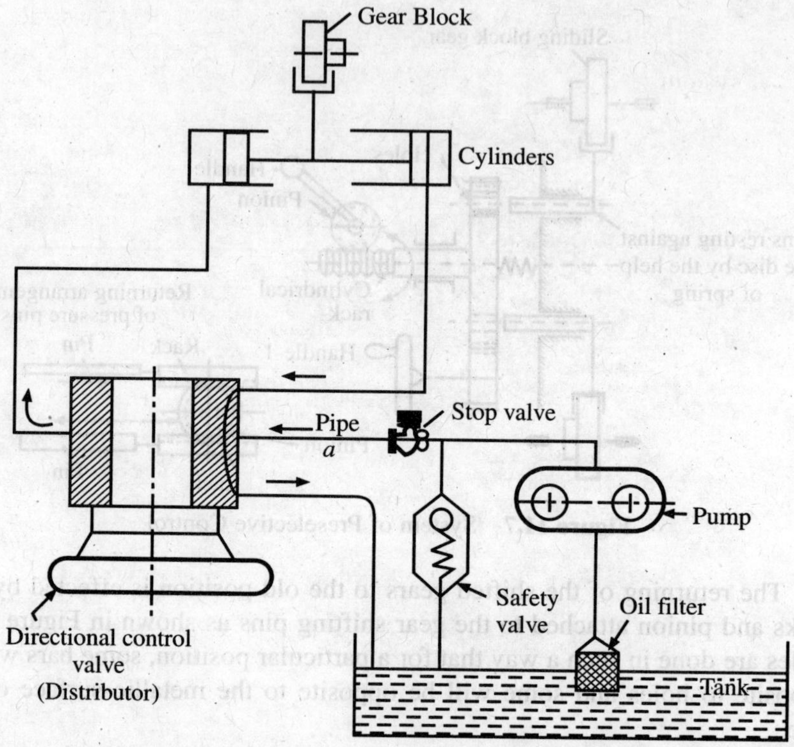

Figure 11.8 Hydraulic Shifting Arrangement.

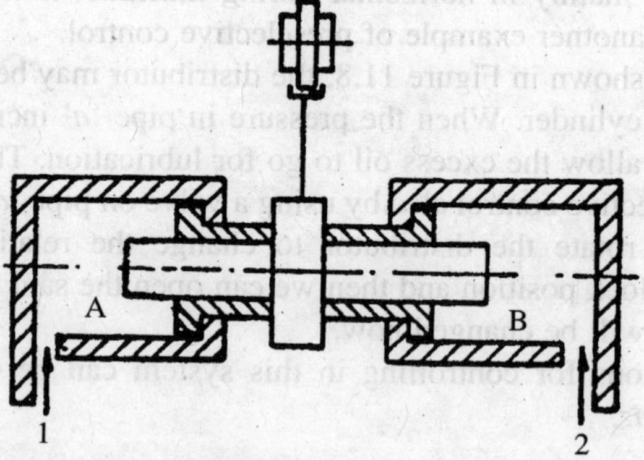

Figure 11.9 Twin Cylinder for Controlling Shifts.

hand position when the fluid passes through 2, the shifting will be to the extreme left. When same pressure is applied through passes 1 and 2, the piston will be in the central position.

In heavy machine tools, control is necessary over a consi-derable distance. Better control in such cases can be obtained by electrical or hydro-electrical system.

In the above case the block shifting system which .is placed in the gear box can be controlled by operator from a distance. The distributor can. be rotated by electric motor controlled from a distance.

Separate motor for each control may be necessary in case of extra heavy duty machine tools.

11.2 Single-Disc Selective Speed Changing System[1,2]

Figure 11.10 shows an elementary scheme of a single-disc selective speed changing mechanism for 16 speeds. It has a number of rack pushers by means of which four double-cluster gears 1,2, 3, 4 are shifted to obtain $4 \times 2 \times 2 = 16$ speeds.

With reference to Figures 11.10, 11.11, and 11.12, it is possible to denote the principal geometric dimensions as under:

Figure 11.10 Diagram for determining the geometric dimensions of elements of the mechanism.

(i) Distance between the axes of the pairs of rack-pushers

$$= C = 2r + d - 2\,m \qquad (11.1)$$

where m denotes the module of the rack pinions Z_1, Z_2, Z_3, Z_4.

(ii) The diameter of the concentric circles on which the centres of holes for the pins of the rack pushers are located is given by

$$D_t = \frac{C}{\sin \dfrac{\beta}{2}} = \frac{C}{\sin \dfrac{K\alpha}{2}} \qquad (11.2)$$

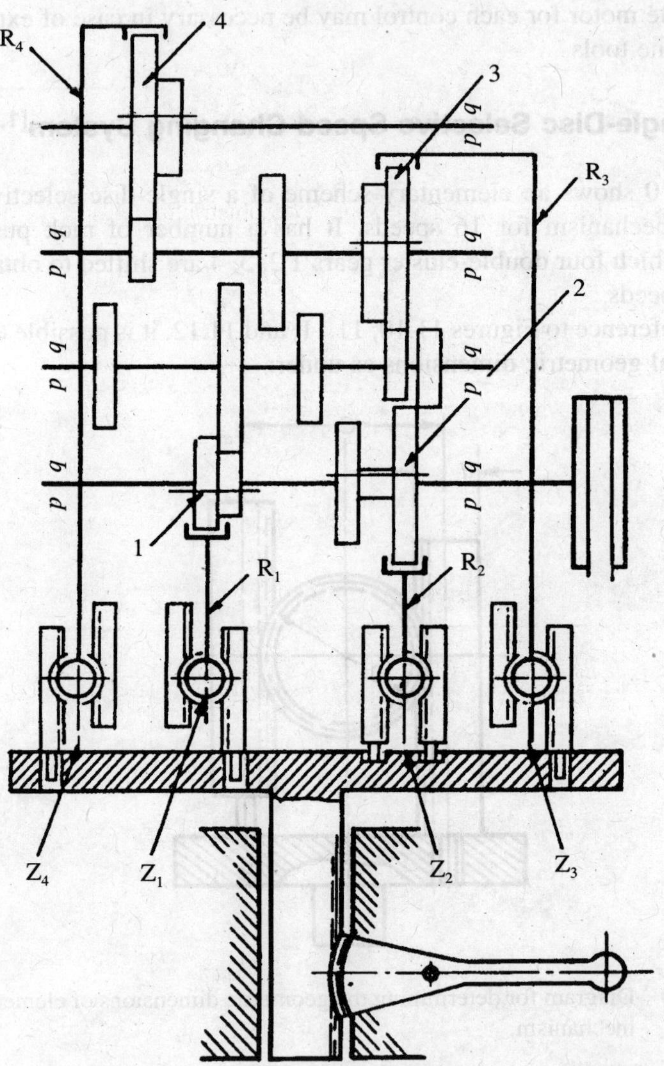

Figure 11.11 Elementary diagram of a single-disk selective speed changing mechanism for 16 speeds

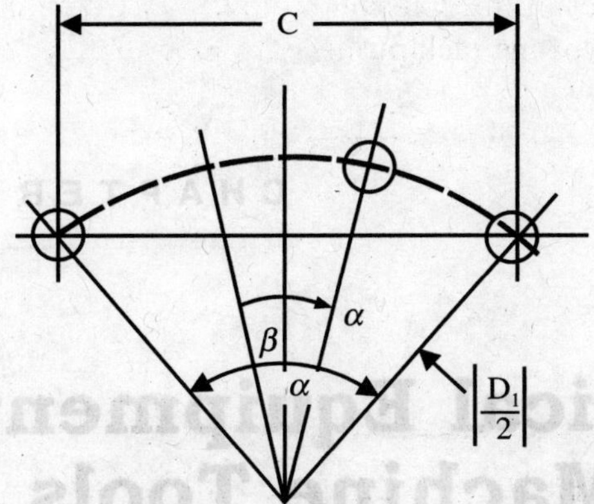

Figure 11.12 Diagram for determining the geometric dimensions of the elements of the mechanism

where K is a whole number

$$\alpha = \frac{360^\circ}{n}$$

n – number of different engagements

Hence $$D_t = \frac{C}{\sin \dfrac{K \cdot 180^\circ}{n}} \quad (11.3)$$

(iii) The length of the travel of the rack pushers depends upon the width 'b' of the toothed rims on the cluster gears and the ratio $\frac{R_t}{r}$, where R_t is the radius of the lever that shifts the corres-ponding cluster gear. As a general value $R_t = (3 \text{ to } 5)r$.

CHAPTER 12

Electrical Equipments in Machine Tools

12.1. Basic Ideas

The cutting force and the force of friction in the mechanisms of a machine tool create a moment of resistance M_c. The turning moment developed by the motor M is equal and opposite to the opposing moment M_c when the motor runs at a constant speed.

If at a given instant $M > M_c$, and the velocity of rotation of the motor increases, then a heavy inertia is set up by the acceleration of the elements of the machine and the motor consequently develops the dynamic moment M_j. The equation of the motion can be written in the form:

$$M = M_c + M_j$$

or in other words
$$M = M_c + J\frac{d\omega}{dt},$$

where ω – angular velocity and t is the time,

J – dynamic moment of inertia of the rotor.

In the absence of acceleration, $\frac{d\omega}{dt} = 0$ and hence $M = M_c$. Such motion is known as stabilised motion.

The time dt for changing the angular velocity by $d\omega$ may be expressed as:

$$dt = \frac{J \cdot d_\omega}{M - M_e} \text{ or, } \int_0 dt = \int_{\omega 1}^{\omega_2} \frac{J \cdot d\omega}{M - M_c}$$

$$t = \frac{J \times 2\pi \cdot (n_2 - n_1)}{60\,(M - M_c)}$$

where n_2 and n_1 denote the rpm corresponding to ω_2 and ω_1 respectively.

By putting $n_1 = 0$ and $n_2 = n$ we can find out the definite time of starting the electrical equipment:

$$t_s = \frac{J \cdot 2\pi \cdot n}{60\,(M - M_c)}$$

By putting $M = 0$; $n_1 = n$ and $n_2 = 0$ we obtain the time for braking the motor

$$t_{br} = \frac{J \cdot 2\pi \cdot n}{60\,M_0}$$

For positive braking purposes the modern machine tools are more and more using electrical braking. In such a case the motor is subjected to a torque in the direction opposite to the usual rotation. Formula for obtaining the time of braking during such token action can be expressed as

$$t_{br} = \frac{J \cdot 2\pi \cdot n}{60\,(M + M_c)}$$

We can combine and say $t_2, br = \dfrac{J \times 2\pi \cdot n}{60\,(M \mp M_c)}$

In most of the machine tools the allowable time of starting

$$t_s \approx 0.25 - 0.4 \text{ sec.}$$

Only in some high speed machines this value is considerably lowered; namely to the extent of

$$t_s \approx 0.1 \text{ sec.}$$

While selecting motor for a particular machine tool, we must look into the characteristic feature of it.

12.2. Selection of Motor for any Executive Organ of a Machine Tool

For the selection of a suitable motor for the drive of a machine tool, it is necessary to find out the following important characteristic. Figure 12.1 shows the displacing organ, namely the sliding table of a particular machine tool, driven by a motor through a lead screw.

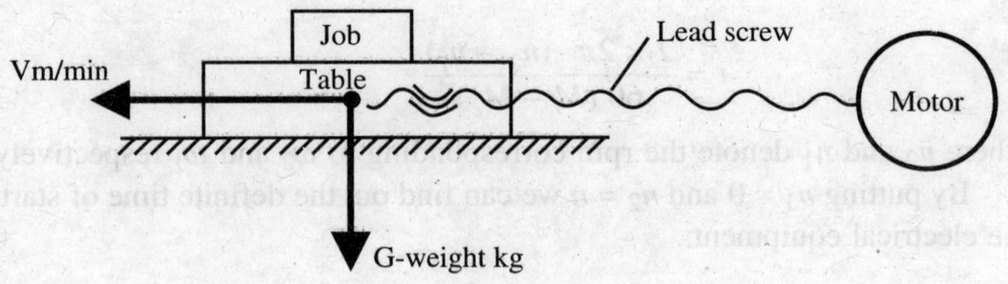

Figure 12.1

The power consumption can be found out from the relation

$$P = \frac{Gf \cdot v}{102 \times 60 \cdot \eta}$$

where P – power in K watts,
η – efficiency of the motor, and
f – coefficient of friction. (Other notations shown in Figure).

The turning moment M of the motor can be expressed by the equation

$$M = 975 \frac{P}{n} = 975 \frac{G \cdot f \cdot v}{102 \times 60 \times \eta \times n}$$

Moment of resistance at the start = Mc_0 = starting torque on the motor:

$$Mc_0 = M \cdot \frac{f_0}{f}$$

where f_0 – coefficient of static friction, i.e. the friction of the guides, when the sliding speed is the smallest possible. For most machine tools $f/f_0 \approx 2$.

12.3. Regulation of Speed in Electrical Control

The speed of an a.c. motor is given by the relation:

$$n = \frac{60\, f' \,(1 - S_{\text{II}})}{p}$$

where n – speed in rpm. of the non-synchronous motor,
f' – frequency of the current,
p – number of pair of poles of the stator coil, and
S_{II} – slip.

The slip 'S_{II}' is equal to $\dfrac{n_0 - n}{n_0}$

where n – speed of motor,

n_0 – speed of rotation of the magnetic poles (synchronous speed).

There are three variable factors in the above equation and as a result, the speed can be effectively regulated by varying either

(a) S_{II}

(b) f'

or (c) p

In machine tools the first two parameters are rarely varied. So the regulation of speed is commonly done by changing the number of pair of poles. To effect this it is necessary to have specially designed electrical switch, that can be operated manually or automatically.

From the above, $n_0 = \dfrac{60\, f'}{p}$

If $p = 2$ and $f' = 50$ cycles/sec,

then $$n_0 = \frac{60 \times 50}{2} = 1500 \text{ rpm.}$$

If $$p = 1, \text{ then } n_0 = \frac{60 \times 50}{1} = 3000 \text{ rpm.}$$

With the winding arranged as shown in the sketch (Figure 12.2) the system behaves like two pairs of poles when the current passes through the direction shown by arrows in full lines. Conversely it acts as one pair of poles when the current is sent in the direction shown by chain dotted arrow. For this it is necessary to have electrical motor of specialised construction Simpler

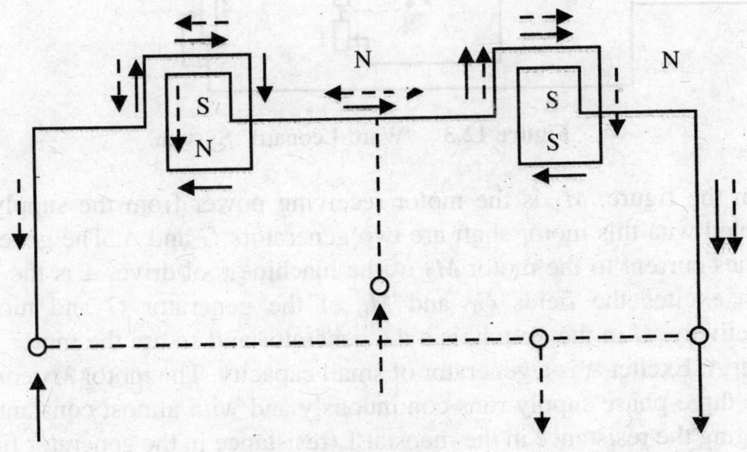

Figure 12.2

method of obtaining two different numbers of pairs of poles is to have the arrangement on the stator of nonsynchronous motor having two independent windings. As is evident the regulation obtained in this case is in very small magnitude.

Ward-Leonard System of Controlling the Speed of the Motor over a Wider Range : This system is also popularly known as motor-generator system. This has wide application in planing machine drives. This type of speed control is very common in machine tool, since it gives a large number of regulation and provides arrangement for self braking. The principle of working is briefly as follows (Figure 12.3: Circuit diagram in case of Ward-Leonard system):

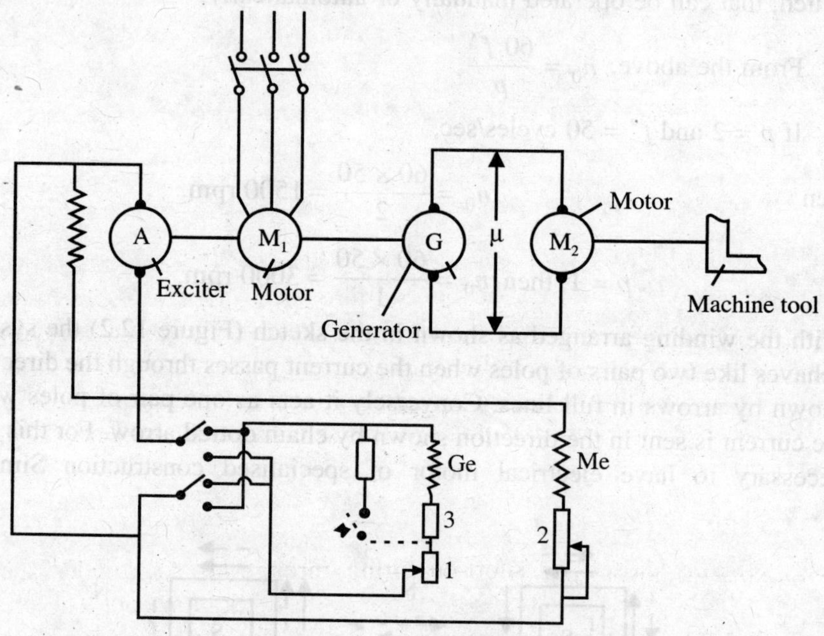

Figure 12.3 "Ward-Leonard" System.

In the figure, M_1 is the motor receiving power from the supply main. 'Coupled wua this motor shaft are two generators G and A. The generator G supplies current to the motor M_2 of the machine tool drive. A is the exciter, which excites the fields G_0 and M_0 of the generator G and motor M_2 respectively. G in the sketch, is a d.c generator and so are the motor M_2 and exciter A. Exciter A is a generator of small capacity. The motor M_1 connected to the three-phase supply runs continuously and with almost constant speed. Changing the resistance in the rheostat 1 (resistance in the generator field) we can change the value of U and consequently, the input to the Motor M_2, directly coupled with the executive organ of the machine tool. Thus in this way it is possible to regulate the speed of the motor. The speed n of the motor M_2 may be written in the form

$$n = \frac{U - I_n\, r_a}{C_0\, \varphi} = \frac{U}{C_0\, \varphi} - \frac{r_n \cdot M}{C_0 - C_m\, \varphi^2}$$

where I_α and r_α are the armature current and resistance respectively, Φ is the quantity of magnetic flux through the motor, C_0. and C_m are two constants and M is the torque produced. For no load speed of the motor $n = n_0$, $M = 0$,

and consequently $n_0 = \dfrac{U}{C_0 \cdot \varphi}$, where n_0 is ideally no-load speed of the motor, and can be changed by changing the value of U. By changing the resistance 2, we can regulate the magnetic flux of the motor. If Φ is constant, the above expression $n_0 = \dfrac{U}{C_0\, \varphi}$ will give rise to straight line mechanical characteristics of the motor. The range of regulation in the motor generator system by varying U does not exceed 2:1 or 3:2. While starting, the rheostat 2 is made completely out, so that the flux through M_2 becomes maximum. Further by regulating the rheostat 1, the excitation of the generator is increased, so as to send more value of U through the armature of M_2. And thus the motor speeds up In many cases during the start the resistance 1 is not at all Used. In the excitation coil of the generator *G* traverses full e.m.f. of the exciting motor A. For forcing the excitation of the generator, the *EMF* of *A* sends 2-3 times more potential than the normal one required for G_0, and so in the scheme, we will find addition of another resistance 3, which while starting, remains closed by short-circuiting through 4. And under such condition a higher potential traverses through the excitation coil. The total range of regulation obtained in motor generator system by varying both the variants, U and Φ, is 10-15. For the purpose of braking, it is necessary that the current through the armature changes in direction Cmoment in such case also changes in sign).

Braking can be obtained by increasing the magnetic flux of the motor by the help of the rheostat 2 or by decreasing the potential of the generator through the rheostat 1. In both Cases $E_{of\ motor} > U$ and the motor works as generator, which produces reverse rotation by the inertia of the dynamic mass. The generator G rotates as the electric motor and starts rotating the machine M_1 with very high speed, which in turn rotates as a generator and gives out energy. For general workshop machine. the "Ward leonard system", with a speed range of 10 : 1 gives a rapid and smooth reversal of table travel and is ideally suited for short stroke work. This system shows only tnree machine set, viz. generator, motor and exciter. But sometimes an additional auxiliary exciter is also provided.

Speed Changes by Controlling Stat or Resistance in A.C. Motor, and Shunt Resistance in AC. Motor. In case of d.c. motor variable speeds can be obtained by controlling the rheostat in the shunt circuit. We have mentioned this in Chapter 3 while discussing the system of electro-mechanical regulation. In case of a.c. motor the stator resistance could be controlled by rheostat to get variable speeds. This regulation could be effectively combined with mechanical regulation to have a large number of speeds.

12.4. Circuit Diagram for Starting the Driving Motor of a Machine Tool

In Figures 12.4 and 12.5 are shown the contactor switch and circuit diagram for starting the motor of a machine tool. As soon as we press the button "start" the current passes through the coil 1 *K* and therefore working contact 1 *K* are closed and the motor starts running. Simultaneously the block contact 1 *K* is also made and this helps in releasing the button "start". As soon as we press the switch "stop" the circuit becomes opened and the machine stops.

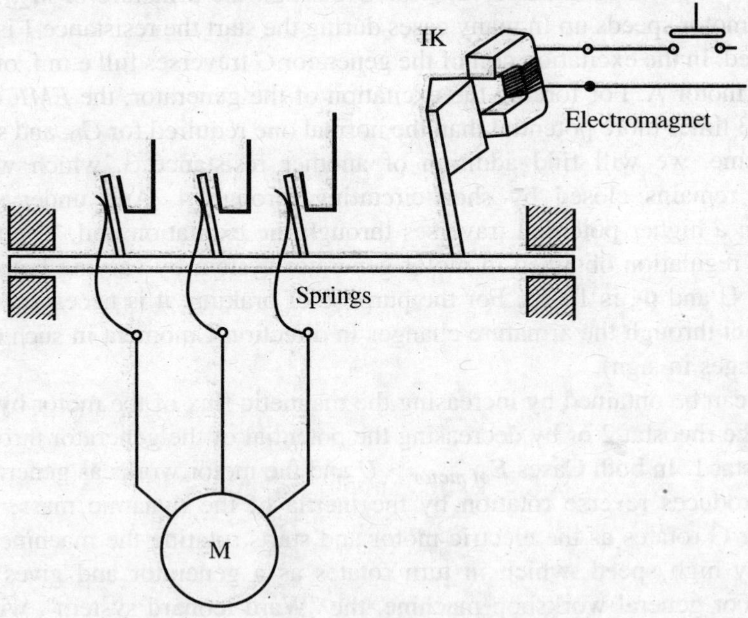

Figure. 12.4

But to stop the machine instantaneously, we can take help of electromagnetic braking, the sketch of which is shown in Figure 12.6. In this case the figure shows how the contacts are actually made against the action of the springs, when the coil 1 *K* is energised. These contacts are known as "working contacts" and are also indicated by position in Figure 12.5.

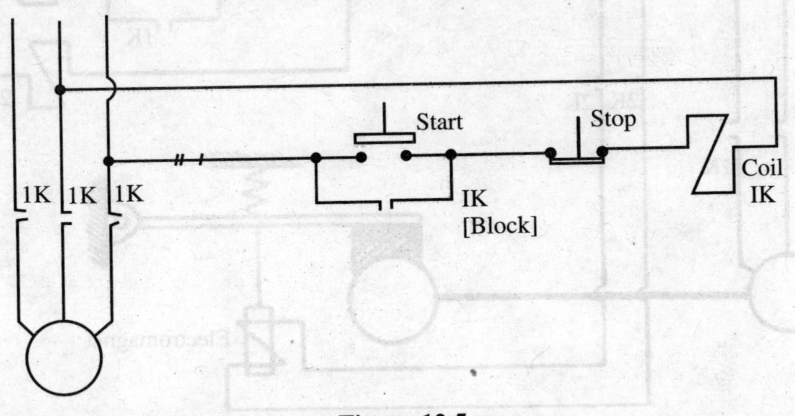

Figure 12.5

12.5. Electrical Brakes

Even after switching off, the motor will run for a certain length of time due to its dynamic moment of inertia. Electrical brakes help to stop the motor instantaneously the switch is made off. The principle of working of such a braking system as shown in Figure 12.6 can be discussed briefly below.

By pressing the button "stop" down, current flows not through the coil 1 *K* any more, but through the auxiliary coil 1 *K*. As a result of which the contacts 2 *K* in the electromagnetic circuit will be closed causing the electromagnet to be energised which, in turn, draws the brake. The machine thereby stops quickly. The time lag is very small in this system. For the purpose of setting a job on the machine or setting the tool or for effecting smaller displacement in machine tools, this is absolutely essential.

12.6. Electromagnets used in Machine Tools Control

Figure 12.7 (p. 195) shows the simple sketch of an electromagnet used in machine tools. As the value of 8 decreases, the induced current through the coil increases, thereby reducing main current in the coil. The nature of variation of the current and the force of attraction can be shown by the graph below (see Figure 12.8).

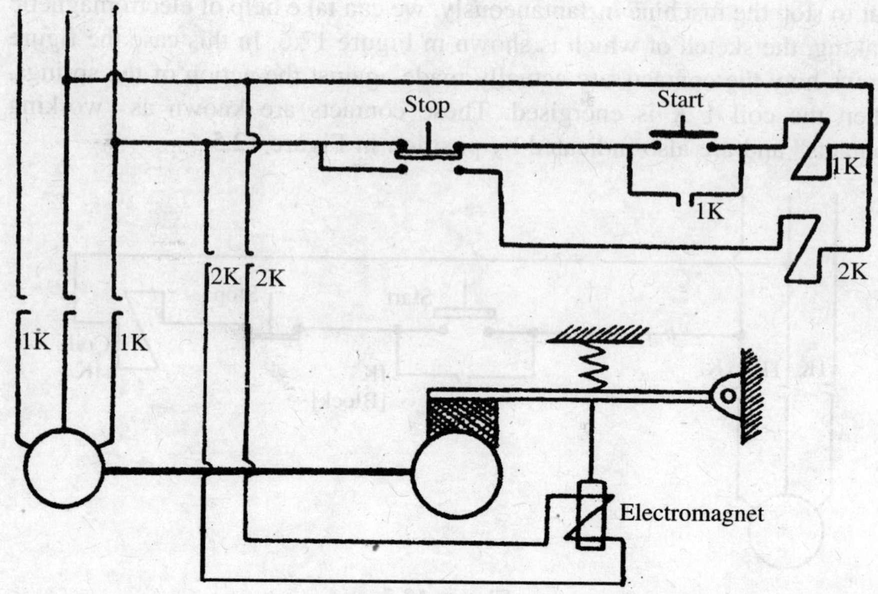

Figure 12.6

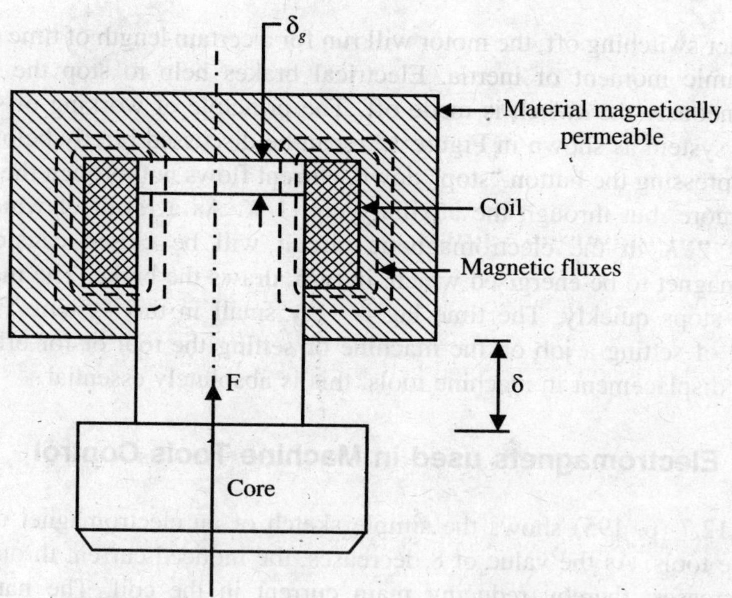

Figure 12.7.

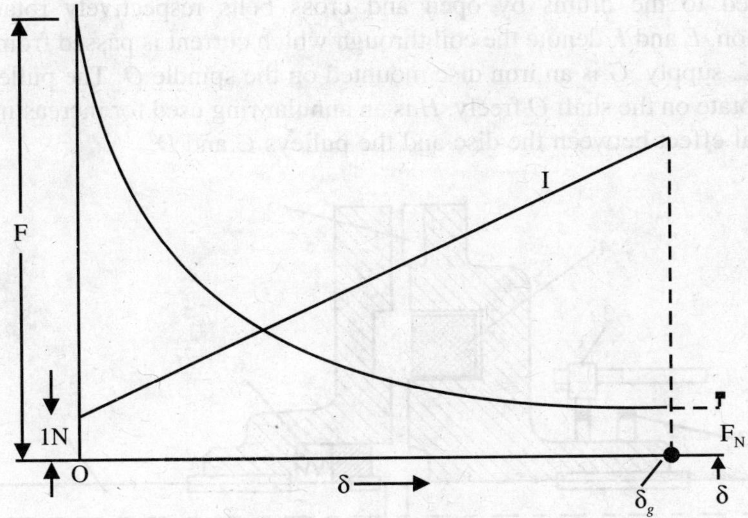

Figure 12.8 Graph showing F & I Against 8.

Induced e.m.f. increases, if δ becomes smaller and hence the inductive resistance. Usually the electromagnets used in machine tool constructions conform to the following characteristic:

F – 1.5 to 25 kg.

Stroke – 25, 30, 50 mm.

Usually a maximum force of attraction near about 15 kg. is obtained when the stroke =15 mm.

12.7. Electromagnetic Clutch

Figure 12.9 shows a simple sketch of a single sided electro-magnetic clutch for the transferring motion from the driver shaft to the driven shaft. Driven disc contact the friction surface and gets the power transmitted from Driver to the driven shaft.

Nomenclature: 1-Driving shaft, 2,3- slip rings for current 4- Electrical coils, 5-Friction (High) surface , 6- Spring, 7- Driven shaft

Now-a-days we are more and more using electromagnetic clutch in modern machine tools. Principles of operation of such electromagnetic clutch can be explained with reference to the sketch shown below.

Figure 12.9(a) double sided electromagnetic clutch. *A* and *B* are two drums (not necessarily of the same size) mounted on sthe motor shaft. These two drums are rotated in unidirec-tion. But the pulleys *C* and *D* which are connected to the drums by open and cross belts respectively rotate in opposition. *E* and *F* denote the coil through which current is passed from 110 volt D.C. supply. *G* is an iron disc mounted on the spindle *O*. The pulleys *C* and *D* rotate on the shaft *O* freely. *H* is an annular ring used for increasing the frictional effect between the disc and the pulleys *C* and *D*.

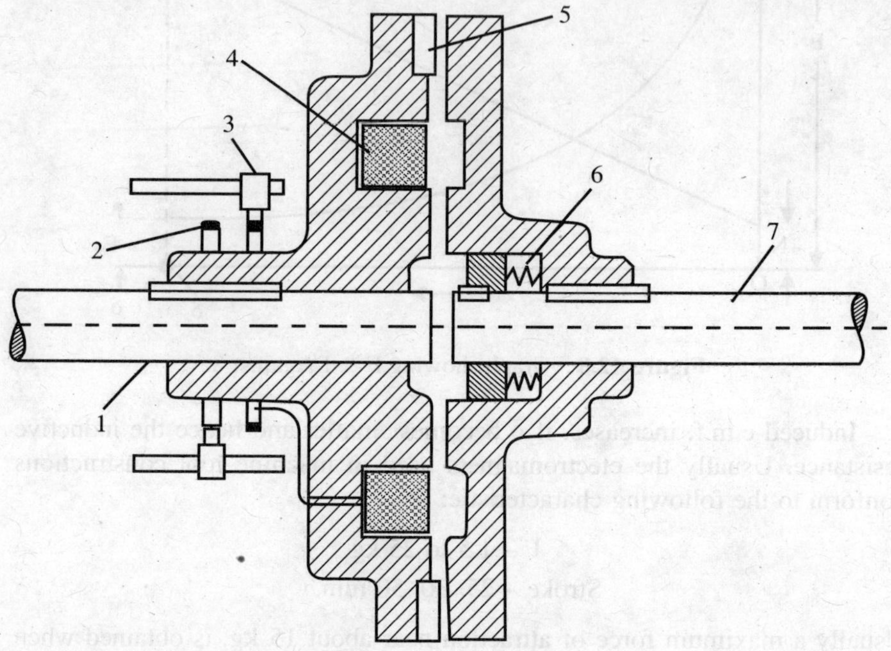

Figure 12.9 Electro-magnetic clutch.

On passing current through the coil *F*, fluxes will be generated in and around the coil and the iron disc which serves as a core will be magnetized. As a resuty *G* and the pulley *C* will come in contact and the spindle *O* will be rotating with the same magnitude and direction of the speed of pulley *C*.

Reverse speed of the spindle will be obtained by passing current through the coil E, the principle of operation being the same. In some modern machines, reversing of the speeds of the pulleys by belting is replaced by gear trains.

These types of clutches are very much in use in Planing machines of various capacities to reverse the table movement.

The above sketch represents only a single disc type twin clutch system. There are multiple disc electromagnetic clutch as well. The gap between G and the friction ring is maintained within 0.2 mm.

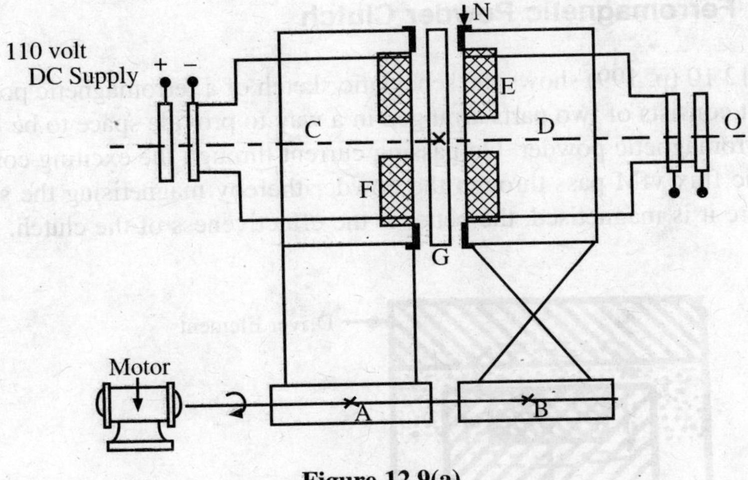

Figure 12.9(a)

Alternating current is not used with this system because of two reasons : (1) magnetic flux generated mil be of pulsating nature; (2) as a result of the above, wear on the disc will be more. Dimensions of the electromagnetic clutch will depend upon the turning moment to be transferred.

For M (turning moment)		outside dia. of the clutch	
	= 0.5 kg m,		D = 200 mm
,,	= 5 kg m,	,,	= 300 mm
,,	= 50 kg m,	,,	= 500 mm

The discs in a multiple disc clutch are made of hardened steel having a thickness of 0.8 mm.

Multiple disc magnetic clutch is much more smaller and compact in size than an ordinary single disc magnetic clutch.[1]

In some types of magnetic clutch, some ferro-magnetic powder is used to attract the core to the disc. In Germany and U.S.S.R. considerable advances have been made in this particular field.

---

[1]Approximate sizes of the multidisc electromagnetic clutches are as follows depending upon the magnitude of the turning moment to be transmitted:

Turning moment	*Outside dla.*
4 kg mm.	78 mm
25 kg mm.	142 mm
160 kg mm.	220 mm

12.8. Ferromagnetic Powder Clutch

Figure 12.10 (p. 199) shows the schematic sketch of a ferromagnetic powder dutch. It consists of two parts arranged in a way to provide space to be filled with ferromagnetic powder. On passing current through the exciting coil the magnetic flux vfM pass through the powder thereby magnetising the same. The more it is magnetised, the better is the effectiveness of the clutch.

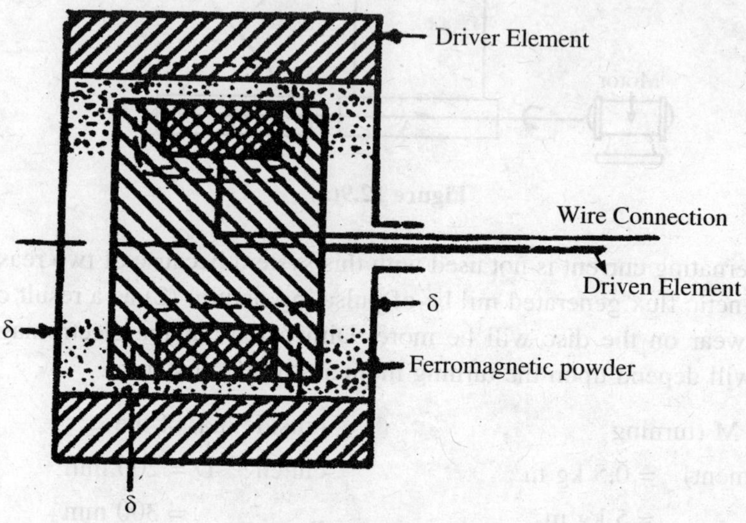

Figure 12.10

This type of clutch is very useful; but it has got a serious disadvantage- The powder after some use has to be taken out and changed periodically. This type of clutch is mostly used in feed mechanisms, mechanism for obtaining precise amount of movement, as well as in dynamometers and braking systems. The powder used is mainly carbonil ferromagnetic powder containing 0.7 to 0.8 per cent C with spherical grains haying diameter 4-8 microns (0.001 mm = 1 micron). This type of powder is very much in use in radiotechnology.

The width of the gap is 0.5-2 mm. The particles are mixed with mixing oil usually in the ratio of 5:1. Coefficient of adhesion obtained in this case is between 0.1 and 0.3. A great portion of this is the contribution of viscous friction of the oil. The air inside such clutch helps to cool down quickly the elements when the clutch is disengaged after a long run ; but it also causes corrosion of the internal part.

12.9. Reversing Mechanism of a Light Duty Planing Machine

With reference to the circuit diagram shown in Figure 12.11, let us denote

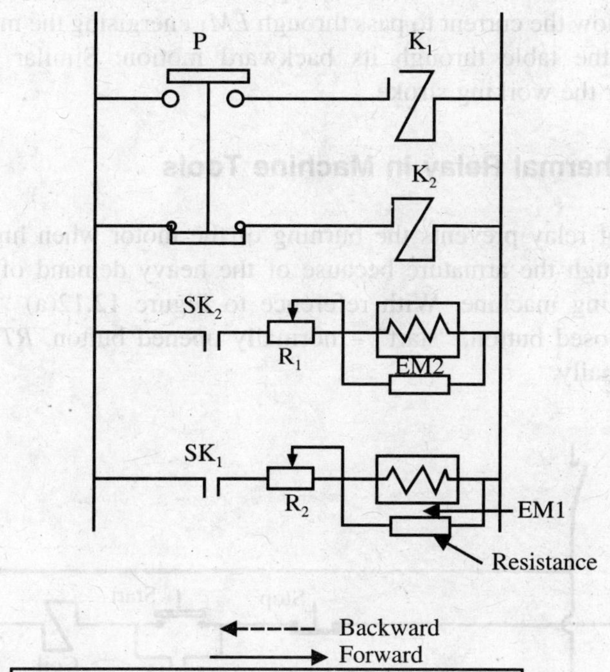

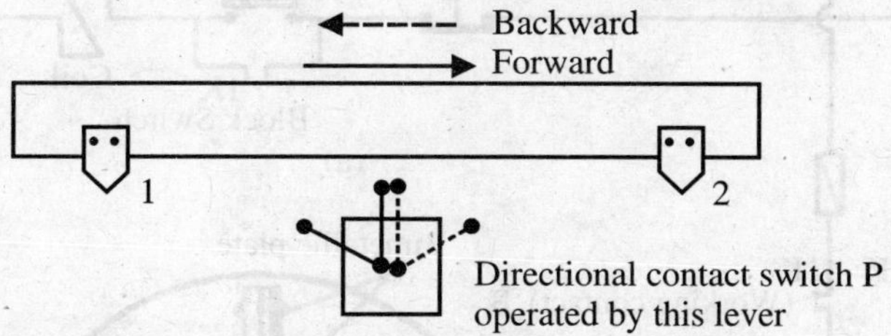

Figure 12.11

K_1 – Coil through which current passes during the backward stroke closing SK_1.

K_2 – Coil through which current passes during the forward stroke, closing the working contact SK_2.

EM_1 – Electromagnet which draws the table mechanism during the backward stroke.

EM_2 – Electromagnet which draws the table mechanism during the forward stroke.

R_1 and R_2 are two rheostats in series. When the projected part 1 knocks the lever of the directional control switch P, the lever attains the position shown by dotted lines. As a result, in the connection diagram, the position of the switch P changes, allowing the current to pass through K_1, as a consequence of which the normally open working contact SK_1 will close. This will allow the current to pass through EM_1 energising the magnet, which will draw the table through its backward motion. Similar operation is repeated for the working stroke.

12.10. Thermal Relay in Machine Tools

This type of relay prevents the burning of the motor when high current is passed through the armature because of the heavy demand of load on the metal working machine. With reference to Figure 12.12(a) "stop" – is a normally closed button, "start" – normally opened button. RT_1 and RT_2 – closed normally.

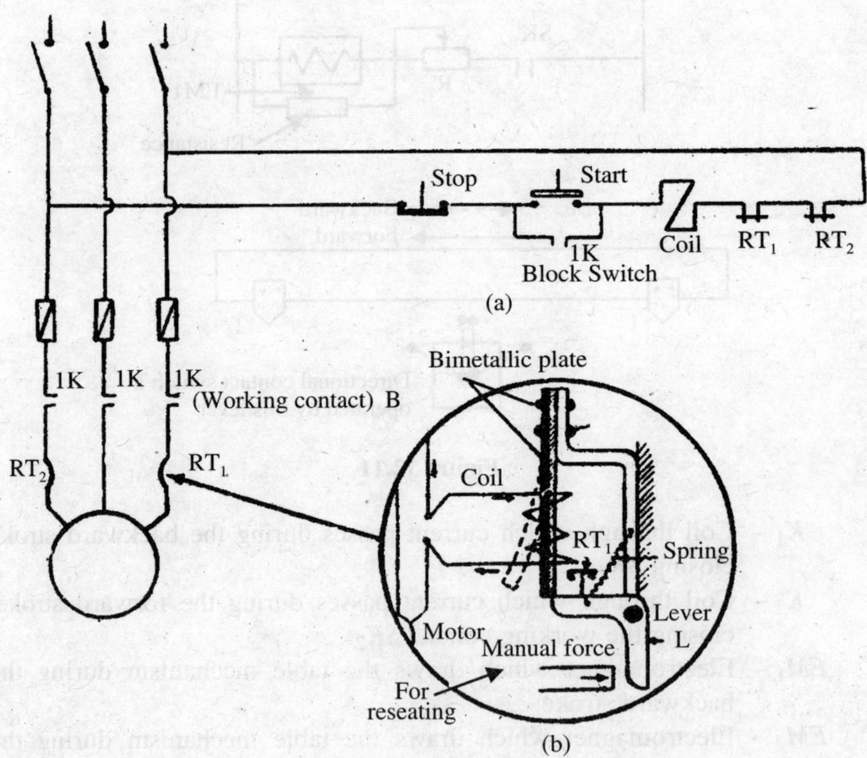

Figure 12.12

When a high current passes through the bimetallic plate *B* in Figure 12.12 (a) it becomes deformed as shown by dotted line because of the heat and, as a result, the lever L under the action of the spring rises and opens the connection *RT* (see Figure 12.12b). From the circuit diagram, as soon as *RT* switch is off, no current can pass through the coil 1 *K* and therefore the working contact, as well as the block contact IK become open and no current passes through the main motor. This is therefore a method of safe-guarding the motor under heavy load.

Once the thermal relay has operated to prevent the motor being burnt due to sudden overloading, it is necessary to reset the lever L manually. After the relay is tipped, it must be allowed to cool before it can be reset and the motor restarted. This time lapse also allows the motor to cool. Such type of relay adequately protects the motor and yet will not take it off the line unnecessarily.

12.11. Electrical Automation in Horizontal Drilling Machine

In Figure 12.13 M_1 and M_2 are the motors for longitudinal feed and spindle rotation respectively, both receiving current from the same supply. The motor Mi can have reverse speed as well.

The illustrated sketch (Figure 12.13) shows the complete electrical control circuit for automatic operation of a horizontal drilling machine. *KF* and *KB* are the magnetic contactors for the forward and backward motion of the motor M_1. *K* are the contact switches for the rotation of the motor M_1, which is used for the rotation of the drill spindle. *PS* 1, *PS* 2, *PS* 3 are the three path directional switches, which operate when the drill point comes to planes 1, 2, and 3 respectively. The slow feeding of the drill during the operation will be executed by the aid of the electromagnet. The principle of operations can be easily explained with respect to the control panel as under: On pressing the button "start" the current passes through the circuit A. It will go through the normally closed switch *PS* 2, *KB* and enter into coil *KF* thereby closing the working, as well as, block contacts *KF*. The motor will run in forward direction. Block contact *KF* of the circuit A enables to release the pressing of the button "start". The motor M_1 will run till the stud *E* presses the directional switch *PS* 2 thereby closing the switch *PS* 2 in circuit *B* and opening the same in circuit *A*. Now no current will flow through the motor M_1 and it will stop. But the current will flow in circuit *B* through normally closed *PS* 3 and coil *KS*, thereby simultaneously exciting the electromagnet. As a result, the working contact *KS* of the motor M_2 will be made, thereby causing the rotation of drill spindle and the electromagnet will draw the carriage against the action of the spring to do the drilling operation.

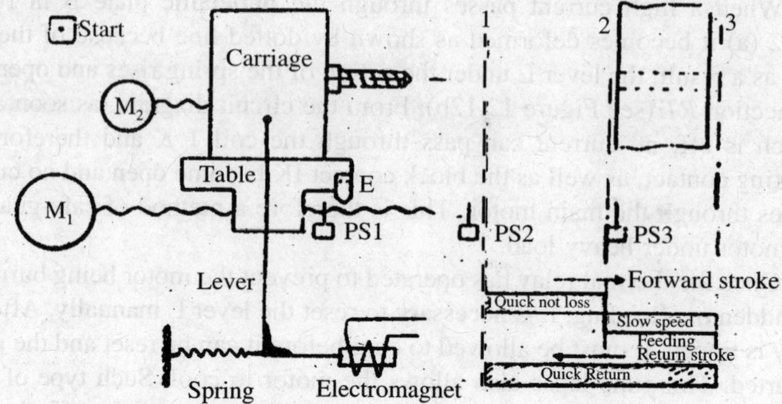

Work carriage can move on the table (at any fixed position of the table) by the help of the electromagnet, to give feed to the drill

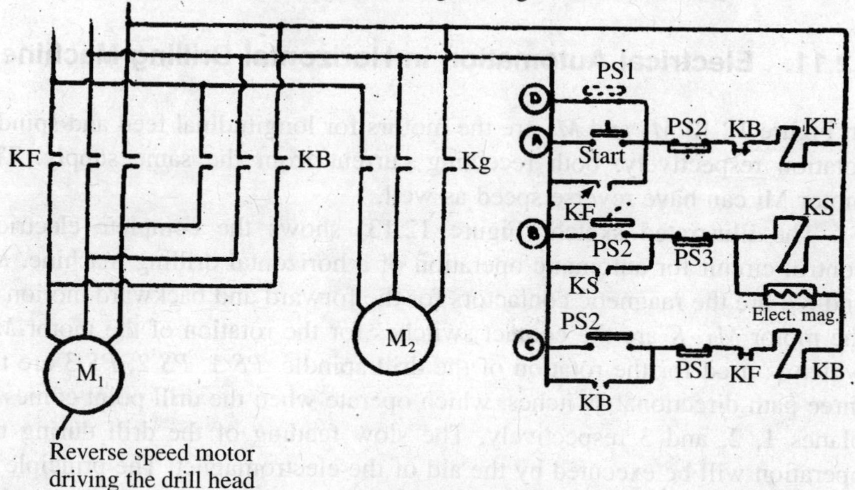

Figure 12.13.

With the carriage is also fitted a stud similar to *E* to make contact with *PS* 3. These two motions will be continued till the stud *E* presses the directional switch *PS* 3. Such pressing will immediately cause opening of contact *PS* 3 in circuit *B* and closing of the same in circuit *C*. As a result the motor M_1 will come to a stand still and the electromagnet will be demagnetized thereby marking the end of the movement of the carriage. Now the current will pass through the circuit *C* and energise the coil *KB* which will mean contacting the working contacts *KB* of motor M_1 and block contact *KB* of circuit *C*. As a result the motor M_1 will be rotated in the reverse direction thus effecting the reverse stroke from the position 3 to 1.

Thus the cycle of operation can be repeated as many times as possible effecting complete automation.

Motor Type Time Relay used in Machine Tool. This type of relay is very much common in machine tools requiring to stop the motion of an executive organ after a predetermined time.

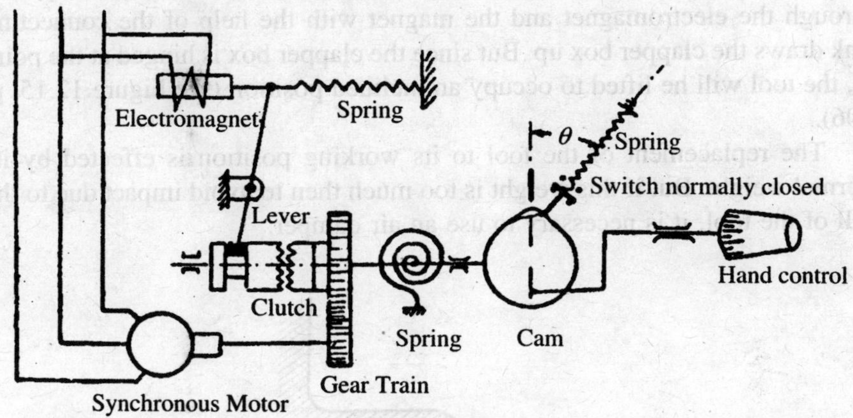

Figure 12.14

The principle of working of this relay shown in Figure 12.14, is as follows.

As soon as the current flows through the synchronous motor, the electromagnet is energised, as a result of which, the lever engages the clutch. The gear on the cam-shaft was rotating freely previously. But as soon as the clutch is engaged, the cam-shaft rotates against the coiled spring and the cam rotates. As soon as the cam is rotated through the pre-set angle (0) the switch is disconnected and no current will flow through the motor which will eventually come to a stop. Arrangement can be made for stopping the motor instantaneously.

$$\text{Time of relay } t = \frac{\theta}{\omega}$$

where $\omega = \frac{2\pi \cdot n}{60}$ radians/sec, n being the r.p.m. of the camshaft. Usually the cam-shaft rotates at very slow speed. The hand controlled graduated wheel is meant for setting the cam through the predetermined angle 0. Similarly there are various other types of relay like magnetic relay, pneumatic relay, speed controlled relay and so on. In the case of pneumatic relay, the relay time $0.4 \angle t \angle 180$ sees. Pneumatic relays are also very much in use in machine tools particularly at the time of starting the motor. The starters resistance could be cut off with the help of these relays.

12.12. Automatic Lifting of Tool During the Return Stroke of a Planing Machine

The lifting of the tool in the return mechanism of a Planing machine is done with the help of electromagnet. Normally each tool post is provided with a separate electromagnet During the return stroke electric current passes through the electromagnet and the magnet with the help of the connecting link draws the clapper box up. But since the clapper box is hinged at the point *O*, the tool will he lifted to occupy an inclined position (see Figure 12.15, p. 206).

The replacement of the tool to its working position is effected by its normal weight. But if this weight is too much then to avoid impact due to the fall of the tool, it is necessary to use an air damper.

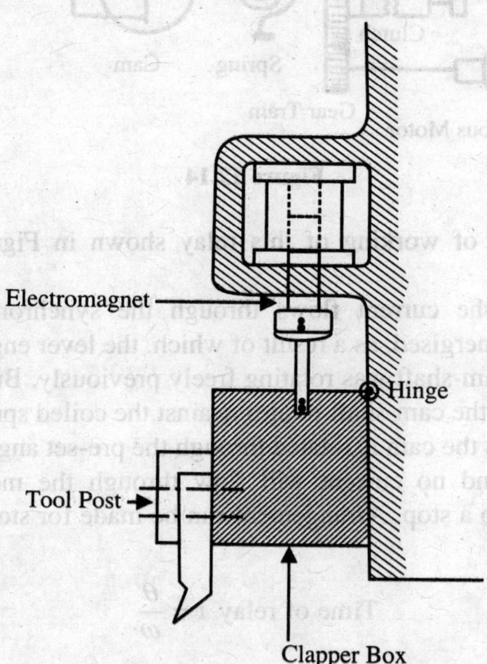

Figure 12.15.

In the case of heavy machine tools the automatic lifting of the tool during the idle stroke can be obtained with, the help of the reversible speed electric motor rotating eccentrically.

12.13. Basic Ideas into Regime of Working of Motors

From the basic equation of motion connecting dynamic torque as given in subpara 12.1, we know that dynamic torque

$$M_J = M - M_c = J\,\frac{dw}{dt} \qquad (12.1)$$

where M_e = Moment of resistance (resisting torque)
w = Angular velocity of motor rotor
J = Dynamic moment of inertia of the rotor

$$\therefore \qquad dt = \frac{J\,dw}{M - M_e} = \frac{J\,dw}{M_J}$$

But M_J is a linear function, normally, with the speed of rotation and hence we may write $M_J = a - bn$, where a and b arc linearity constants. $J\,dw = \frac{GD^2(dn)}{375}$, where G is weight of the rotor in kg and D is the diameter of gyration of the body in metres.

$$\therefore \qquad \int_0^t dt = \frac{GD^2}{375}\,\frac{dn}{(a - bn)}$$

$$t = \left[\frac{GD^2}{375}\left(1 - \frac{1}{b}\right)\ln(a - bn)\right]_{n_1}^{n_2}$$

$$= \frac{GD^2}{375b}\ln\left(\frac{a - bn_1}{a - bn_2}\right)$$

$$= \frac{GD^2}{375b}\ln\frac{M_{j1}}{M_{j2}},$$

where M_{j1} and M_{j2} are torque corresponding to n_1 and n_2 respectively.

Again b is the slope of the curve $M_j = f(n)$

$$\therefore \qquad i = \frac{GD^2}{375}\,\frac{(n_2 - n_1)}{(M_{j1} - M_{j2})}\ln\frac{M_{j1}}{M_{j2}}$$

$$= \frac{GD^2(n_2 - n_1)}{163\,(M_{j1} - M_{j2})}\log\frac{M_{j1}}{M_{j2}} \qquad (12.2)$$

This formula is used for calculation of relay time in prestarting the current in the electric motor. Full load torque, usually known as rated torque, is denoted by M_r

and $$M_r \geq \frac{M_{max}}{\lambda} \tag{12.3}$$

where M_{mam} is maximum torque required by the driven equipment and λ is instantaneous torque overload factor. In d.c. motors this value is restricted by safe commutation, while in a.c. motors, it is determined by the maximum electromagnetic torque available.

Heating and cooling of electric motors

The general equation for heat rise can be written in the form

$$\tau = \frac{Q}{A}(1 - e^{-t/T}) + \tau_0\, e^{-t/T} \tag{12.4}$$

Table 12.1

Values of λ for different types of motor

Type	λ
D.C series and compound wound crane motors	3.5–4.0
General purpose d.c. motor	2.5
General purpose squirrel-cage and wound rotor induction motor	1.7–2.7
Synchronous motor	2.0–2.7

where τ – the temperature at any instant of time t (°C)

T – the time constant (= C/A, where C is the thermal capacity Cal/°C)

τ_0 – the value of τ at $t = 0$ (°C)

$Q\,dt$ – the heat in calories produced in the motor during time dt

A – the emissivity (calories per sec. per °C)

We can rewrite the above equation in the form

$$\tau = \tau_\infty(1 - e^{-t/T}) + \tau_0 e^{-t/T} \tag{12.5}$$

when Q = 0, then $\tau = \tau_0 e^{-t/T}$. This is known as equation for cooling of motor.

τ_∞ in the above equation replaces Q/A and indicates steady state value of this τ. This happens after infinite lapse of time t.

Regime of Working of Drives

(a) Continuous duty

It can be seen from Figure 12.16 that the temperature τ rises till it jecomes τ_∞ and then remains steady.

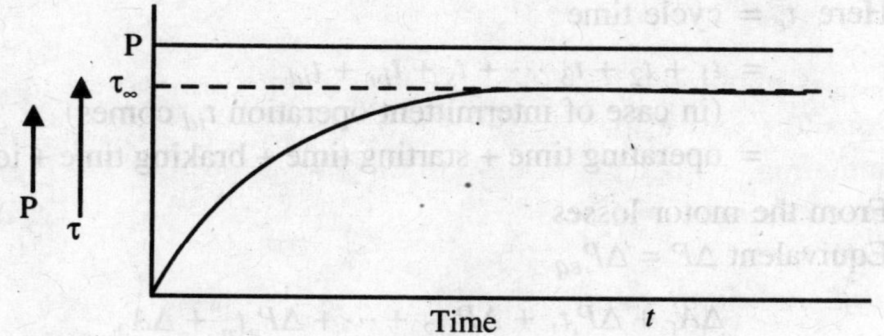

Figure 12.16

This is a requirement for thread milling, heavy duty lathe and boring machines etc.

Nominal power of electric motor should be equal to the power requirement of the machine tool.

$$P_e = \text{Power in cutting} = \frac{F_c V}{102 \times 60}$$

where F_e – cutting force in kg

V – velocity of cutting, m/min

η_e – overall efficiency of machine (considering all mechanical connections from drives to the cutting tool point).

$$\therefore \qquad P = \frac{P_c}{\eta_c} = \text{Power requirement of the motor}$$

$$P_{\max} = \frac{|P_c|_{\max}}{\eta_c} \qquad (12.7)$$

(b) Short-duty operation

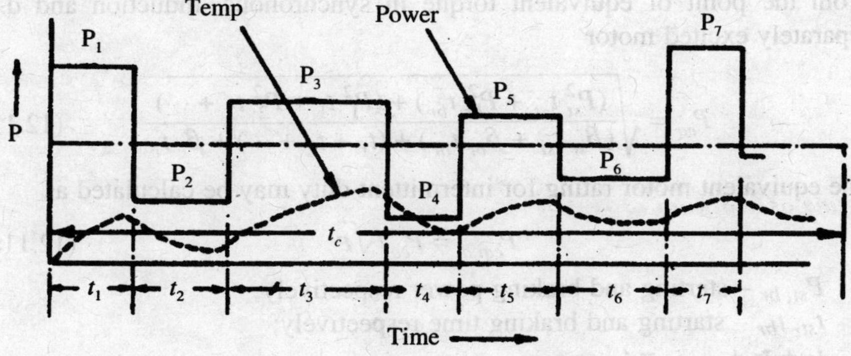

Figure 12.17

Here t_e = cycle time

$= t_1 + t_2 + t_3 \cdots + t_s + t_{br} + t_{id}$

(in case of intermittent operation t_{id} comes)

= operating time + starting time + braking time + idling time

From the motor losses

Equivalent $\Delta P = \Delta P_{eq}$

$$= \frac{\Delta A_1 + \Delta P_1 t_1 + \Delta P_2 t_2 + \cdots + \Delta P_n t_n + \Delta A_2}{t_e}$$

where ΔA_1, and ΔA_2 are power losses in starting and braking.

Neglecting ΔA_1 and ΔA_2 $P_{eq} = \dfrac{\sum_{i=1}^{n} P_1 t_1}{\sum_{i=1}^{n} t_1}$ (12.8)

for motor rating $P_r = KP_{av}$

where $K \simeq 1.1$ to 1.3

$$\therefore \qquad \varepsilon = \text{duty factro} = \frac{\text{operating time}}{\text{total cycle time}} = \frac{t_{OD}}{t_c}$$

$$= \frac{t_1 + t_2 + t_3 + \cdots + t_n}{\sum_{i=1}^{n} t_1 + \sum t_{id} + t_{st} + t_{br}} \tag{12.9}$$

The above method is based on average power loss.

Method of equivalent current

$$I_{eq} = \sqrt{\frac{I_1^2 t_1 + I_2^2 t_2 + \cdots}{\sum t}} \tag{12.10}$$

From the point of equivalent torque in synchronous, induction and d.c. separately excited motor

$$P_{eq} = \sqrt{\frac{(P_{st}^2\, t_{st} + P_{br}^2\, t_{br}) + (P_1^2\, t_1 + P_2^2\, t_2 + \cdots)}{(\beta_{st}\, t_{st} + \beta_{br}\, t_{br}) + (t_1 + t_2 + \cdots) + \beta_c\, t_c}} \tag{12.11}$$

The equivalent motor rating for intermittent duty may be calculated as

$$P_{eqi.d} = P_{eq} / \sqrt{\varepsilon} \tag{12.11a}$$

$P_{st,\, br}$ – starting and braking power respectively;

t_{st}, t_{br} – starting and braking time respectively;

$t_1 + t_2 + \ldots \ldots = t_{\text{operating}}$

t_e = cooling time = idle time

$$\beta_{st} \approx \beta_{br} = \frac{\beta_c + 1}{2} \tag{12.13}$$

β_c = 0.25–0.35 (self cooled protected motor)

β_c = 0.95–0.98 (for enclosed motor)

Note: (i) If power along a certain section of the load curve varies linearly with time, say from P_1 to P_2 then the equivalent power formula assumes the form

$$P_{eq} = \sqrt{\frac{P_1^2 + P_1 P_2 + P_2^2}{3}} \tag{12.13}$$

Time for P_1, time from P_1 to P_2, time for P_2 all assumed same in this case.

(ii) Standard duty motor from the catalogue may have different value of ε_{st} (duty factor). Usually these values are 0.15, 0.25, 0.40, 0.60

$$\varepsilon_{st} \approx 0.15,\ 0.5,\ 0.40,\ 0.60$$

$$P_{eq(st)} = P_{eq(id)} \sqrt{\frac{\varepsilon}{\varepsilon_{st}}} \tag{12.13a}$$

Permissible temperature rise for 'continuous duty' motor

$$\tau = \frac{Q_r}{A} \tag{12.14}$$

For 'short-duration' duty

$$\tau = \frac{Q_k}{A}\left(1 - e^{\frac{-t_k}{T}}\right) \tag{12.15}$$

where Q_k, Q_r – Heat losses in motor under continuous duty and short time service condition respectively

τ – Temperature rise

t_k – Instant time at which temperature rise is considered.

$$\therefore \quad \frac{Q_k}{Q_r} = \frac{1}{(1 - e^{-t_k/T})} = p_T \tag{12.16}$$

pr is known as thermal overload factor.

$$(Q_k - Q_r) = Q \cdot e^{\frac{-t_k}{T}} \tag{12.17}$$

$$\ln\left(\frac{Q_k}{Q_k - Q_r}\right) = \frac{t_k}{T}$$

$$\therefore \qquad t_k = T \ln\left(\frac{p_2}{p_T - 1}\right) \qquad (12.18)$$

and

$$p_T = \frac{Q_k}{Q_r} = \left(\frac{P_k}{P_r}\right)^2$$

$$= \left(\frac{\text{Power in short service condition}}{\text{Power for continuous duty}}\right)^2 \qquad (12.19)$$

$$\therefore \qquad \frac{P_k}{P_r} = \sqrt{\frac{1}{(1 - e^{-t_k/T})}} \qquad (12.20)$$

EXAMPLE

Plot the load curve and select the proper motor for the following intermittent duty.

$$P_1 = 30 \text{ kW for } t_1 = 2 \text{ sec}$$

$$P_2 = 20 \text{ kW for } t_2 = 20 \text{ sec}$$

$$P_3 = 15 \text{ kW for } t_3 = 5 \text{ sec}$$

$$P_4 = 30 \text{ kW for } t_4 = 15 \text{ sec}$$

Between the operations P_3 and P_4 there is a pause ($P = 0$) and duration of the same is 35 sec. At the end of the cycle, there is again a pause of 30 sec.

Solution: Duty factor (Ref. to eqn. 12.9)

$$\in = \frac{t_{op}}{t_e} = \frac{(2 + 20 + 5 + 15)}{(2 + 20 + 5 + 15) + (35 + 30)} = \frac{42}{107}$$

$$\therefore \qquad \in \simeq 0.40$$

The equivalent continuous load power amounts to

$$P_{eq} = \sqrt{\frac{30^2 \times 2 + 20^2 \times 20 + 15^2 \times 5 + 30^2 \times 15}{2 + 20 + 5 + 15}}$$

$$= \sqrt{\frac{1800 + 8000 + 1125 + 13500}{42}}$$

$$= \sqrt{581.5} = 24 \text{ kW}$$

From eqn. 12.11(a)

$$P_{eq(i\cdot d)} = \frac{24}{\sqrt{0.40}} = 37.8 \simeq 38 \text{ kW}$$

Suffix (*id*) standing for intermittent duty.

From eqn. 12.13(a), standard duty motor power

$$P_{eq(st)} = P_{eq(id)} \sqrt{\frac{\in}{\in_{st}}} = 38 \sqrt{\frac{0.40}{0.40}} = 38 \text{ kW}$$

12.14. Classification of Automatic and Semiautomatic Controls

Automatic control system in a machine tool normally provides a precise sequence of engagement, disengagement and manipulation of the speed of travel of various executive units vis-a-vis precise time sequence. The system. normally works on foolproof devices. Any deadlock in the preceding operation or travel of unit should stop subsequent operation of the unit in case the automation is based on strict operational sequence (This is more often a case).

In a machine tool system where units traverse with the help of hydraulic, electric or combined control system, the automatic cycle is controlled along the line of travel of the executive units. In Figure 12.19(a), (b) and (c) are shown the schematic circuit diagrams of 'in-travel', 'time sequence' and 'centralised in-travel' systems respectively. In Figure 12.19(a), where operations are controlled by limit switches, unit II will start only when unit I has travelled a specific distance and engaged the limit switch which disengages unit I but engages unit II and so on. In Figure 12.19(b), the system is on centralised time sequence controlled by rotating drum on which lugs are used for making a specific electrical connection. The positioning of these lugs at angular positions of θ_1, θ_2, θ_3 would mean time sequence of $t_1 = \frac{\theta_1}{w}$, $t_2 = \frac{\theta_2}{w}$ and $t_3 = \frac{\theta_3}{w}$ respectively for engaging units I, II and III respectively. But the centralised in-travel control system has been shown in Figure 12.19(c), where intermittent rotation of commanding drum by ratchet operated by solenoid type relay, controlled by limit switches fixed at pre-specified journey distance, provides 'in-travel' or 'in-path' automation. Since this is resulting from centralised commanding drum, it is called centralised in-travel system.

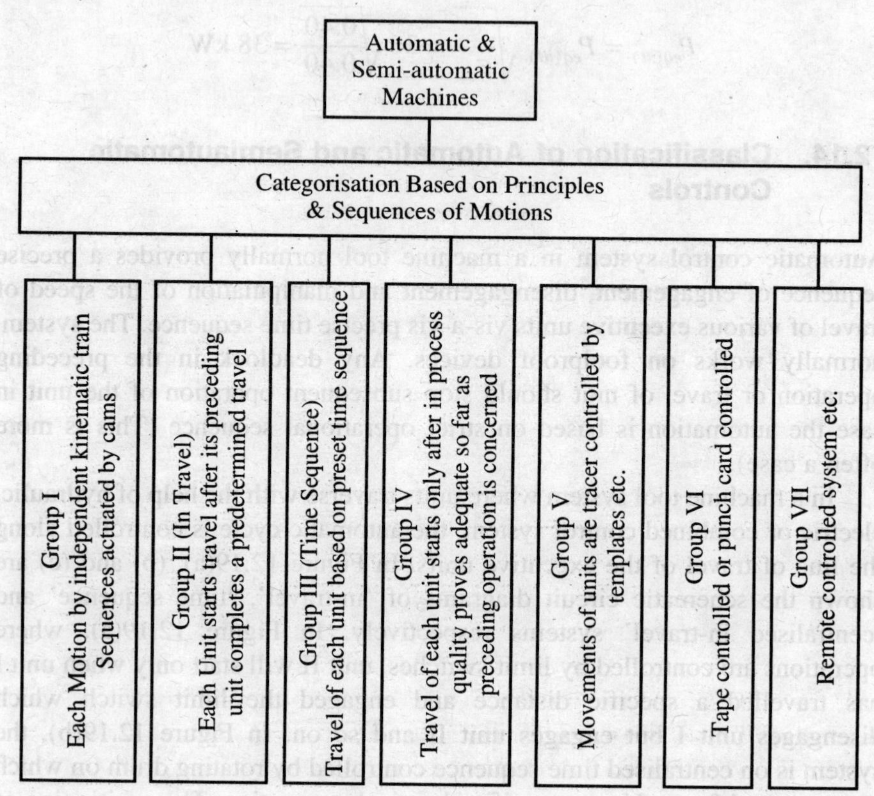

Figure 12.18 Controls in automatic machines.

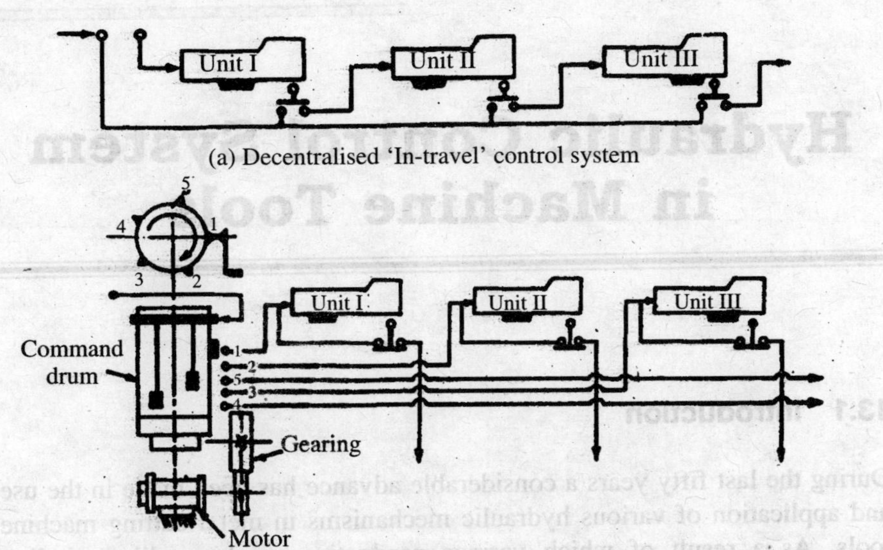

(a) Decentralised 'In-travel' control system

(b) Centralised time sequence control systme

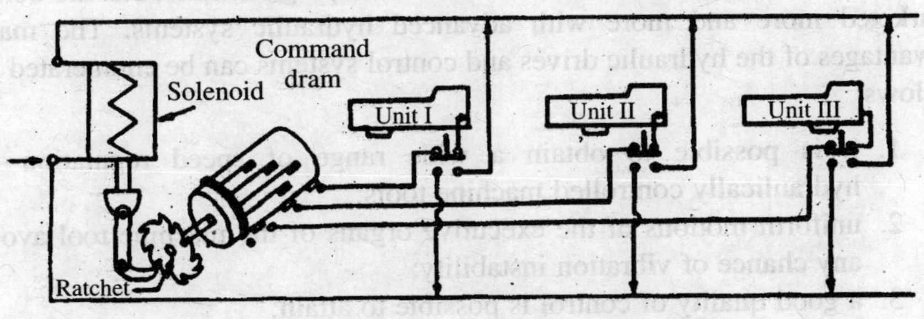

(c) Centralised 'Internal control system'

Figure 12.19 Control systems.

CHAPTER 13

Hydraulic Control System in Machine Tools

13.1 Introduction

During the last fifty years a considerable advance has been made in the use and application of various hydraulic mechanisms in metal cutting machine tools. As a result of which various production machines like grinding machine, broaching machine, lathe machine-shaping machine, ete. are being marketed more and more with advanced hydraulic systems. The main advantages of the hydraulic drives and control systems can be enumerated as follows:

1. it is possible to obtain a wide range of speed regulation in hydraulically controlled machine tools;
2. uniform motions of the executive organs of the machine tool avoid any chance of vibration instability;
3. a good quality of control is possible to attain;
4. automatic lubrication of the motion cylinder parts;
5. less wear;
6. entry of air in the system is protected with the help of the following:
   (a) all executive pipes dispose off fluid below the level of fluid in the tank or reservoir;
   (b) obtaining complete hermetically sealing;
   (c) suction head of the pump should be designed as low as possible ($H_{suc} \leq 50$ cm).

For the above mentioned advantages the hydraulically controlled machine tools are fast superseding the conventional machine tools with manual or mechanical control or drives. The only disadvantages of this system are (1) high initial and servicing cost, and (2) difficulty of access.

Characteristic of the Working Fluid: The working fluid in the hydraulically controlled machine must be of homogeneous chemical composition, high flush point temperature, sufficiently low temperature of freezing and they must not contain any dissolved water, acidity or alkalinity.

Mineral oils used in hydraulic drives of machine tools have under all condition the viscosity in the range of 2-6° E_{50}. This permits of a speed of the motion parts approximately 8 meters/min. or very near to this value. The unit of viscosity $°E_{20}$ $°E_{100}$ is used in U.S.S.R. and Germany.

Absolute coefficient of viscosity μ is related to $°E$ by the following relationship $\mu = 0.00067°E - \frac{0.00058}{°E}$ kg.sec/m^2 the suffixes 20, 50 and 100 of $°E$ denote the temperature at which the viscosity is measured.

The coefficient of viscosity is usually measured at the temperature of 20°C, so that its value at any temperature t°C can be obtained from the relationship

$$\mu_t = \mu_{20} \times \left(\frac{20}{t}\right)^k.$$

For the usual range of oils used in hydraulically controlled machine tools the value of K varies from 1.6 to 2.5.

13.2 Typical Hydraulic Systems in a Machine Tool

Hydraulic drive for the reciprocating motions of the machine tool may be with open or close circulation of the working fluid. In the open circulation system the fluid after doing its useful work in the motion cylinder goes back directly to the tank, where from it is again allowed to circulate by the help of the pump to the motion cylinder.

In the closed circulation the working fluid after doing work in the motion cylinder goes back again to the suction mouth of the pump with the help of a system of returning values, without going through the tank. Such a system gives rise to difficuly of cooling the working fluid, demands careful filtering of it, though, in practice it excludes the formation of cavitation in the internal hydraulic system. In such cases, the amount of fluid consumed is much less than in the open circulation system.

In the hydraulically controlled machine tools, the cooling of the oil can be obtained, if only the tank has enough capacity. However, in such circulation cavitation may be expected at the exit end under high velocity of flow.

The velocity of oil in the pipe line of the hydraulic system of the machine tool varies in wide ranges upto $v = 7$ metres/sec, depending upon thie specification of the pipe line, its form, condition of installation and other variable factors.

If the length of the pipe $l < 100\ d$ where d is the diameter of the pipe, then maximum permissible velocity is 7 metres/sec, but in cases of pipe lines having a greater length of continuity ($l > 100\ d$) the velocity reduces to 3-3.5 metres/sec. In the suction pipe however the velocity $v \leq 1.5$-2 metres/sec. Typical units used in hydraulic circuit of a machine tool are oil tank or reservoir, pump, filter, accumulator, relief valve, throtle valve, directional valve and motion cylinders. In Figure 13.1 is shown a typical hydraulic circuit for a shaping machine. For obtaining quick motion in the return stroke as compared to the forward or the working stroke, two parallel cylinders have been used. The cylinder with bigger diameter is meant for the working stroke or forward stroke, while that with lower diameter helps in obtaining the idle return stroke, when no cutting action is done. From the tank 7 oil is pumped through the gear pump 1 into the bigger motion cylinder 4 through the valve 2. The position of the valve 2 during the working stroke is as shown in the figure. Oil is filtered out before entering the pump and any extra oil after pumping will flow out into the tank through the relief or safety valve 3 which operates in one direction only with the help of spring loaded ball. Any oil trapped in motion cylinder of smaller diameter 5 will flow into the tank 7 through the valve 2 and throtle valve 6.

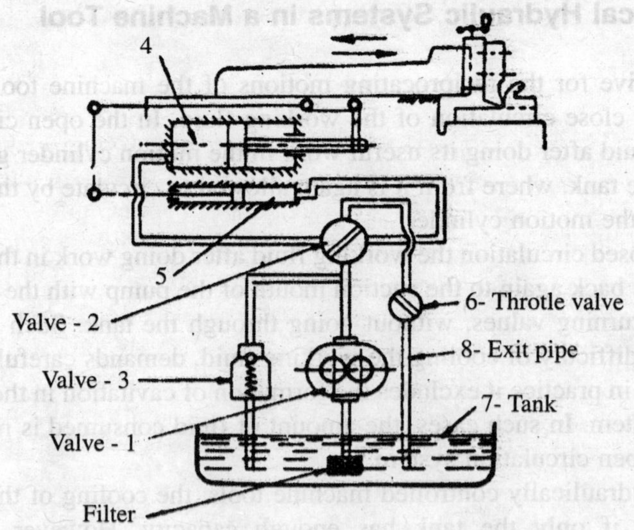

Figure 13.1 Hydraulic circuit for a shaping machine.

During the return stroke, position of the valves 2 and 6 will be changed to the position shown by the dotted line. Oil will now be pumped inside the motion cylinder 5 from the right end causing the ram to retrace its path back with a greater velocity. Oil which in the forward stroke entered into the motion cylinder 4 will go back to the oil tank through the valves 2 and 6.

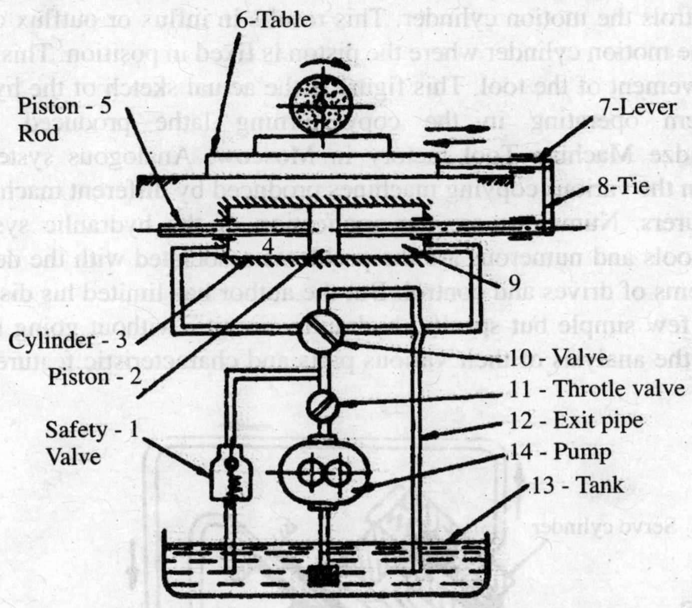

Figure 13.2 Table feed for a grinding machine.

Figure 13.2 shows the hydraulic table feed mechanism of a grinding machine. The system is with throtle valve regulation at the entering end. The principle of operation is very similar to that of the first one. Oil, after being filtered, is pumped by the pump 14 into the left end 4 of the motion cylinder 12 through the throtle valve 11 and directional valve 10. Any excess fluid will flow out into the tank through the one-way relief valve 1. Oil from the back end of the piston namely from the space 9 will go back to the oil tank through the valve 10 as the table gradually moves to the right. During the return stroke the 'position of the valves 10 and 11 will be changed and path of oil can be traced easily. This system does not give any provision for the change in speed between the forward and the backward stroke and hence the table of this type of grinding machine moves forward or backward with the same velocity.

Considerabfe application of the hydraulic systems of drive has been made in the modern broaching machine and in fact it was the broaching in which the hydraulic drive was first applied. But the greatest application of the hydraulic system has been made in the copying machine–*viz.* copy turning lathes, copy milling or diesinking machine, etc.

Figure 13.3 (p. 212) shows the hydraulic servo-mechanism of a copy turning lathe. In such a system the microdisplacement of the tracer point over the profile of the template controls the valve in the servocylinder, which again controls the motion cylinder. This results in influx or outflux of oil to or from the motion cylinder where the piston is fixed in position. This, results in the movement of the tool. This figure is the actual sketch of the hydraulic servosystem operating in the copy turning lathe produced by the Ordzonikidze Machine Tool factory in Moscow. Analogous systems are working in the various copying machines produced by different machine tool manufacturers. Numerous are the application of the hydraulic system in machine tools and numerous are the problems associated with the design of such systems of drives and control. But the author has limited his discussion only to a few simple but specific hydraulic circuits, without going into the details of the analysis of their various parts and characteristic features.

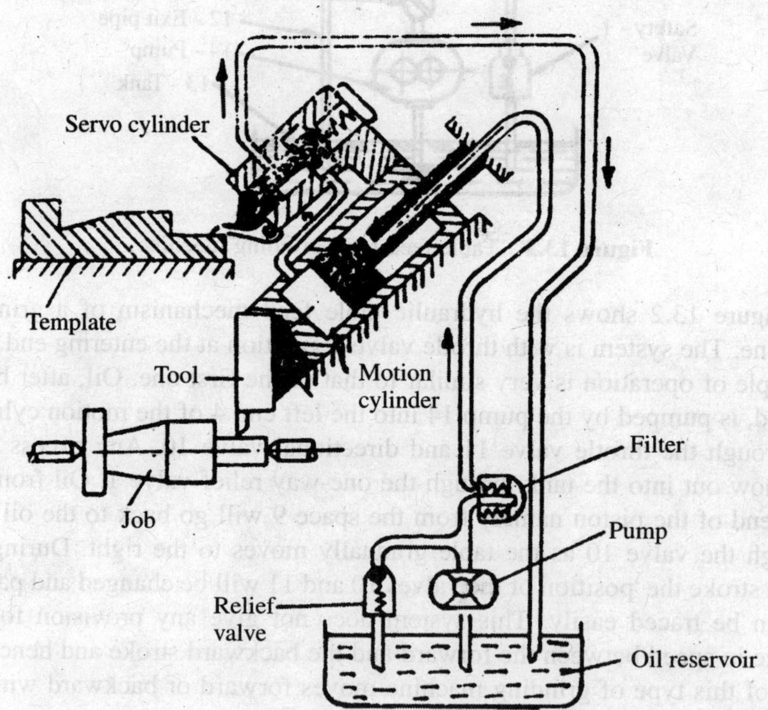

Figure 13.3 Hydraulic and Servo-Mechanism.

13.3 Elements of Hydraulic Systems in Machine Tools

Any hydraulic system consists of two main parts, viz. (a) pump–primary part of the system; and (b) the hydraulic motor–secondary part of the system, deriving its power from the pump. The hydraulic motor can be of reciprocating (hydraulic cylinder) type or rotary type. Hydraulic pump develops the pressure of the working fluid, expending mechanical energy, while the hydraulic motor converts pressure into mechanical work. Thus any hydraulic system may be considered as a power transformer with a closed power cycle.

The control system connecting pump and the hydraulic motor is designed so as to obtain a definite sequence of the various stages of operation in the working cycle. It includes check valves (non-return types), reversing valve (directionally controlled), relays, solenoid operated hydro valves, etc. All such controls form the part of the power pack and may be mounted on the panel.

Energy losses in a hydraulic system are due to (i) volumetric loss, due to leakage of working fluid, (ii) hydraulic loss due to- drop in pressure and (iii) mechanical loss due to friction.

13.4 Resistance Encountered in Flow Through Pipe

Under the condition of laminar flow of working fluid if the Reynold's number $R_e = \dfrac{Vd}{\gamma} \le 2300$, the pressure drop may be calculated from the equation.

$$\Delta p = c\mu V \cdot \frac{1}{d^2} = c\frac{\gamma}{g} \cdot r \cdot V \cdot \frac{1}{d^2} \text{ kg/cm}^2 \qquad (13.1)$$

where γ = density of fluid in kg/m^3;
r = kinematic viscosity of fluid in m^2/sec;
V = average velocity of the fluid in m/sec;
l = straight length of the pipe in m;
d = internal diameter of the pipe in m

The above equation reduces to

$$\Delta p = 10^3 \cdot \frac{c \cdot r}{g} \cdot \gamma \cdot V \cdot \frac{1}{d^2} \text{ kg/cm}^2 \qquad (13.2)$$

considering Δp in kg/cm^2, d in cm, γ in kg/dm^3 (or gm/cm^3) and g in m/sec^2.

$$\text{In other words } \Delta p = \frac{2C}{R_e} \gamma \cdot \frac{1}{d} \cdot \frac{V^2}{2g} \text{ kg/m}^2 \qquad (13.3)$$

Denoting the loss coefficient due to frictional surfaces of the pipe by

$\lambda = \dfrac{2C}{R_e}$, we may write the above equation in the form

$$\Delta p = \lambda \cdot \gamma \cdot \frac{1}{d} \cdot \frac{V^2}{2g} \text{ kg/m}^2 \qquad (3.4)$$

If Q is the quantity flow of oil in m^3/sec, then

$$V = \frac{4Q}{\pi d^2} \qquad (13.5)$$

In the regime of working we are concerned with laminar flow upto $R_e <$ 1600. Assuming the value of d in m, we get

$$\Delta p = \frac{4c}{g} r \cdot \gamma \cdot \frac{l}{\pi d^4} \cdot Q \text{ kg/m}^2 \qquad (13.6)$$

For some materials like light alloy lead etc. for tubes we may have for rough interior surface:

$\lambda = 0.3164\, R_e^{-0.25}$ when $R_e < 100{,}000$

and $\lambda = 0.0032 + 0.221\, R_e^{-0.237}$, when R_e lies between 100,000 and 300,000

For steel tubes

$$\lambda \simeq 0.06 \left(\frac{S}{d}\right)^{0.314}$$

where s is absolute surface roughness lying between $1 \cdot 10^{-5}$ and $8 \cdot 10^{-5}$ m and d is the internal diameter of the tube in m. According to the empirical relationship, advocated by ENIMS[19]

$$\Delta p \simeq 0.072 \frac{V}{d^2} \cdot l \text{ kg/cm}^2 \qquad (13.7)$$

where V – velocity in m/sec;
d – internal diameter in mm;
l – length of the straight portion of the tube, in mm

13.5 Evaluation of Basic Parameters for Design

From equation (13.5)

$$\Delta p = \lambda \cdot \frac{\gamma \cdot l \cdot V^2}{2gd} \text{ and } V = \frac{4Q}{\pi d^2}$$

$$\therefore \qquad \Delta p = \lambda \cdot \gamma \cdot \frac{16 l Q^2}{2g\pi^2 \cdot d^5} \tag{13.8}$$

Since d is constant

$$\Delta p = A\lambda \cdot l \cdot Q^2 \tag{13.9}$$

where $$A = \frac{16\gamma}{2g\pi^2 d^5}$$

($A\lambda l$) is known as hydraulic resistance and hence

$$\Delta p = RQ^2 \tag{13.10}$$

Flowability of the oil is denoted by $S = 1/\sqrt{R}$

and hence $$Q = S\sqrt{\Delta p} \tag{13.11}$$

Modulus of consumption, if denoted by K, then

$$K = Q\sqrt{\frac{\gamma \cdot l}{\Delta a}} = Q\sqrt{i} \tag{13.12}$$

where i is loss in pressure (pressure drop) per unit length of the tube

$$i = \frac{\Delta p}{\gamma \cdot l}$$

From our basic knowledge we know that the consumption of fluid is

$$Q = \mu F \cdot \sqrt{\frac{2g}{\gamma} \cdot (p_1 - p_2)} \tag{13.13}$$

where F is the transverse section of the area

μ is the coefficient of loss, determined experimentally

($p_1 - p_2$) is the pressure difference

$$(p_1 - p_2) = \Delta p = A_2 \left(\frac{1}{\mu^2}\right) Q^2 \tag{13.14}$$

where A_2, specific resistance $= \dfrac{\gamma}{2gF^2}$

$$\text{Denoting } \frac{A_2}{\mu^2} = R_2 \text{ we get } \Delta p = R_2 Q^2 \tag{13.15}$$

Under the circumstances stated above

$$\text{Modulus of consumption } K_2 = \frac{1}{A_2} = \frac{1}{R_2\mu^2} \tag{13.16}$$

$$\text{Hydraulic resistance } R_2 = \frac{A_2}{\mu^2} = \frac{1}{K_2\mu^2} \tag{13 17}$$

Calculation for power required by the pump

Manometric pressure of the pump is found out from the relationship

$$H = \left(\frac{p_1 - p_2}{\gamma}\right) + K_0 Q^2 = H_0 + K_0 Q^2 \tag{13.18}$$

where $$K_0 = \frac{16}{2g\pi^2} \cdot \lambda \cdot \frac{l_n}{d^5} \tag{13.19}$$

Here l_n denotes the summation or total length of the pipes in the system from reservoir to cylinder.

Power consumption by the pump:

$$N = \frac{p \cdot Q1000}{75 \cdot 60 \cdot 1000} = \frac{Qp}{4500} \text{ HP} \tag{13.20}$$

where p is the monometric pressure kg/cm^2

Q delivery capacity of the pump in litres/min

If Q_m – theoretical capacity of the pump at $p_0 = 0$ atmos.

Q – actual capacity of the pump under given pressure of p atmos.

$Q_{i(p)}$ – loss (capacity) at pressure p atmos

then $Q = (Q_m - Q_{i(p)})$; and the efficiency η of the pump is expressed as

$$\eta = \frac{Q}{Q_m} = \left(1 - \frac{Q_{i(p)}}{Q_m}\right) \tag{13.21}$$

If the power of the electrical motor supplying power to the pump is N_E then $\Delta_{NE} = N_E - N$

$$N_E = N_w \times \eta_m$$

where N_w is wattmeter reading of power supplied to electric motor and η_m is the efficiency of the motor

$$\therefore \quad N_E = 173 \cdot 10^{-5}, \text{E.I} \cos \phi \eta_m \text{ kW} \tag{13.22}$$

where E – voltage
I – current in amperes
$\cos\phi$ – power factor, characterising the motor.

13.7 Energy Losses

In a hydraulic system energy losses are made up of (a) volumetric losses, due to leakage of working fluid, (ii) hydraulic losses due to drop in pressure, and (iii) mechanical losses due to friction.

The speed of the hydraulic motor (reciprocating or rotary) is varied by regulating fluid flow into or out of the motor in unit time. Hydraulic drives with power rating $N \geq 0.45$ kW are economically sound; electrical drives are normally used for $N < 0.45$ kW. The speed response of a system and its dynamic properties are evaluated for practical purposes by ratio of maximum torque T_{max} (or maximum pulling force P) to the moment of inertia I of the rotor, by the value $T_{max}/I = a$.

13.8 Compressibility Factor

Compressibility factor of oil is characterised by bulk elasticity

$$\beta_T = \frac{1}{\rho T} \cdot \left(\frac{d\rho}{dp}\right)_T \text{ sqcm/kg} \tag{13.23}$$

where ρ-density kg/cm^2 p–pressure, kg/cm^2 T–temperature, °C

$$\frac{d\rho}{\rho} = \frac{d\left(\frac{M}{V}\right)}{\frac{M}{V}} = -\frac{dV}{V} \tag{13.24}$$

where M = mass in kg and V = volume in cm^3

$$\beta_T = -\frac{1}{V} \cdot \frac{dV}{dp} \text{ sqcm/kg} \tag{13.25}$$

Compressibility factor is reduced with the increase in pressure. Oil density is dependent on pressure

$$\rho_T = \rho_0(1 + Cp + Dp^2)_T \tag{13.26}$$

ρ_0, ρ_T – density at atmospheric pressure and temperature and that at temperature T

C, D – certain coefficients depending iipon temperature.

$$d\rho = \rho_0(Cp + Dp^2)$$

$$\frac{d\rho}{dp} = \rho_0\,(C + 2Dp) \tag{13.27}$$

$$\beta_T = \frac{C + 2Dp}{1 + C_p + Dp^2} \tag{13.28}$$

Hydraulic drives require normally a pressure ranging from 70-200 kg/sqcm. in machine tools and presses.

13.9 Efficiency of Hydraulic Pump Considering Losses

Total power of the pump should be such that it will cover up the volumetric losses as well as frictional losses

$$N = N_{\text{eff}} + N_{\text{vol}} + N_{\text{fr}} \tag{13.29}$$

***N* denoting power and suffixes denoting effective, volumetric and frictional respectively**

$$N_{\text{eff}} = \frac{p \cdot Q}{A} \tag{13.30}$$

where p – **pressure in hydraulic cylinder (motor)**
Q – **volume of oil flow rate**
A – **conversion factor**

$$\eta_0 = \eta_v \cdot \eta_m \cdot \eta_h \tag{13.31}$$

where η_0 – **overall efficiency**
η_v – **volumetric efficiency (leakage loss counted)**
η_m – **mechanical efficiency (friction counted)**
η_h – **hydraulic efficiency** $\left(= 1 - \frac{p_1}{p_0}\right)$

where p, p_0**–pressure of oil leaving the channel and at entry to the pump respectively.**

CHAPTER 14

Programme Control in Machine Tools

14.1 Introduction

Considerable Advance has been made during the recent years in the field of development of programme controlled machine tool, offering tremendous scope for achieving higher rates of production.

An attempt is made in this chapter to discuss the various systems of automatic control of machine tools with particular reference to programming by magnetic tape and numerical means.

Numerous are the techniques of control of machine tools, but the author prefers to limit his discussion only to a few specific cases of centralised programme control and essential functioning of feed back systems.

14.2 Automation in Machine Tools

Figure 14.1 shows classified systems of control resulting in automation in machine tools. Automatic control of feed mechanism and displacements of the executive organ of the machine tools is done by the method of programming. By programming we mean the scheduling of the various motions of the executive organs of the machine tool with the help of some transferring document like magnetic tape, steel tape, cinefilm, perforated paper etc. Various other methods of copying a particular profile can be done by some electrical or hydraulic copying systems working on "feed back". Those machines may also be included under broad classification of the systems of programme control.

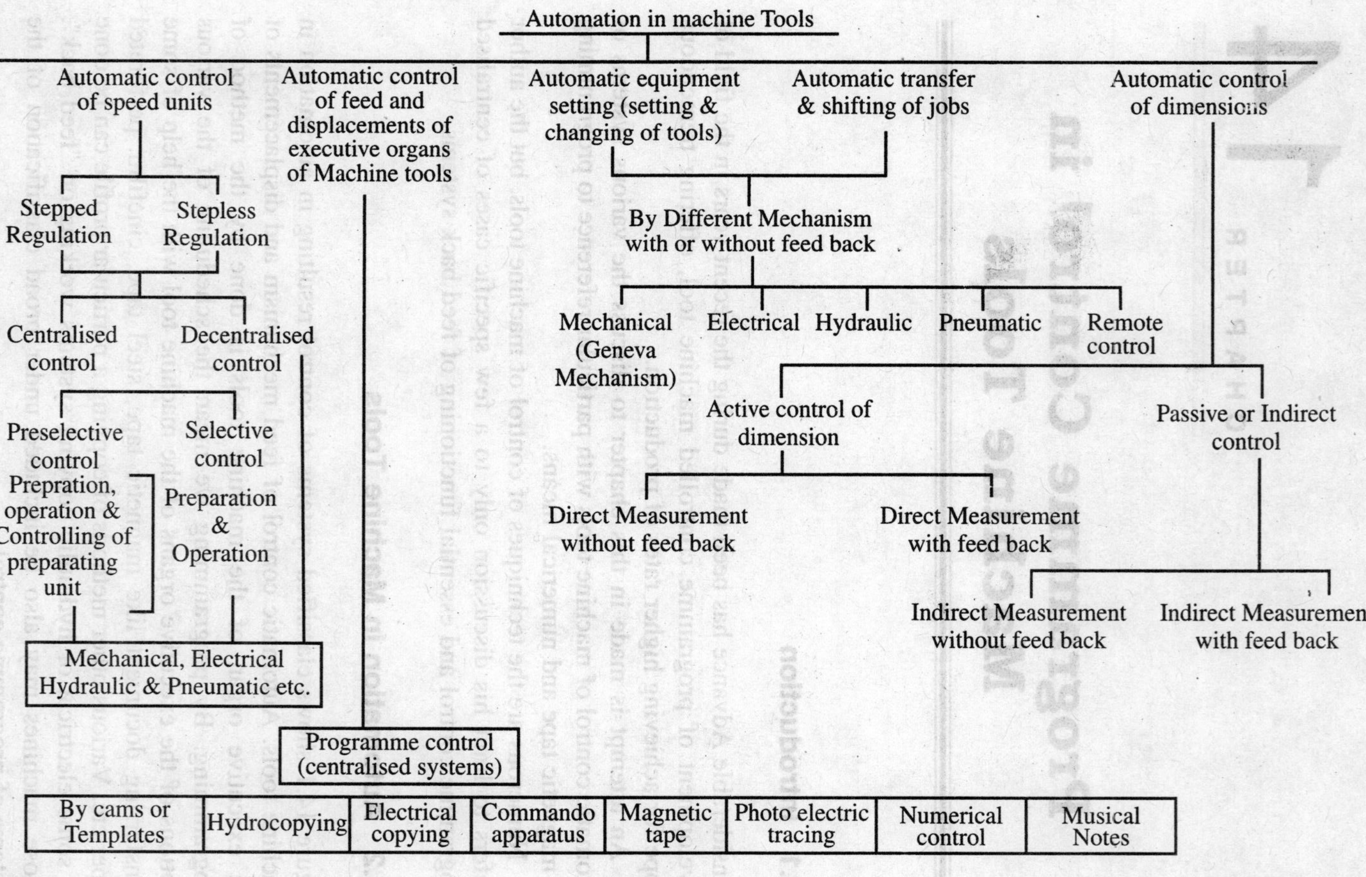
Automation in machine Tools
Automatic control of speed units
Automatic control of feed and displacements of executive organs of Machine tools
Automatic equipment setting (setting & changing of tools)
Automatic transfer & shifting of jobs
Automatic control of dimensions
Stepped Regulation
Stepless Regulation
Centralised control
Decentralised control
Preselective control
Selective control
Prepration, operation & Controlling of preparating unit
Preparation & Operation
Mechanical, Electrical Hydraulic & Pneumatic etc.
Programme control (centralised systems)
By cams or Templates
Hydrocopying
Electrical copying
Commando apparatus
Magnetic tape
Photo electric tracing
Numerical control
Musical Notes
By Different Mechanism with or without feed back
Mechanical (Geneva Mechanism)
Electrical
Hydraulic
Pneumatic
Remote control
Active control of dimension
Passive or Indirect control
Direct Measurement without feed back
Direct Measurement with feed back
Indirect Measurement without feed back
Indirect Measurement with feed back

Figure 14.1

14.3 Magnetic tape Controlled Machine Tool

Figure 14.2 shows the line diagram of a magnetic tape control system having single as well as multi-chanel programming. The principle of working of such system is simple. The recording of the impressions on the magnetic tape is done by working on the first part manually. As soon as the switch 1 is pressed the generator 2 sends current through the recording head 3 and the motor 9 through the contactor switch 8. So long the motor which controls the executive organ moves, the current will flow through the recording head and impressions will be made. The different motors in this particular system will be run by different frequency currents.

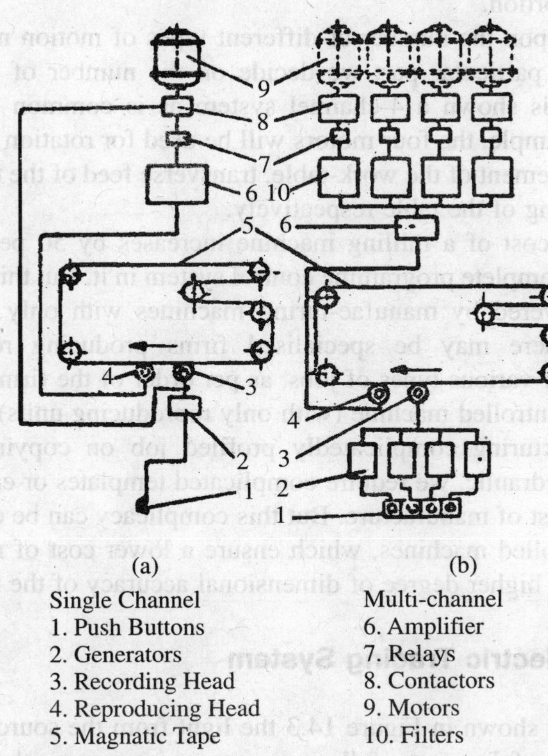

Figure 14.2 Magnetic Type Control System.

When the recorded tape moves under the reproducing head 4, in the coil is generated the same amount of current, as was passed through the recording coil at the time of recording. The reproducing head takes up the current which is amplified and then sent through the filter. The-filter sorts out the currents of different frequencies. These currents pass through the time relays and switches to start the motors of the different executive organs of the machine tools.

Instead of magnetic tape we can conveniently use drum with ferromagnetic coating. The system considered here is based on the principle devised by the Institute of Physics of the Academy of Sciences of Ukrainian Republic in U.S.S.R.

Recently a programme controlled milling machine working on such a system was exhibited by ENIMS (Machine, tools Research Institute of U.S.S.R.) in the Soviet Industrial Exhibition held in New York.

This system has got certain obvious advantages; for example, any error in recording impression on the tape can be easily rectified by erasing or by cutting off the portion.

Depending upon the number cf different types of motion necessary for manufacturing a particular part we decide on the number of channels. In Figure 14.2 (b) is shown a 4-channel system as is common in a milling machine. For example, the four motors will be used for rotation of the outer, longitudinal movement of the work-table, transverse feed of the table and the raising or lowering of the table respectively.

Usually the cost of a milling machine increases by 50 per cent if we incorporate the complete programme control system in it. But this cost can be considerably lowered by manufac-turing machines with only reproducing arrangement. There may be specialised firms producing recording on magnetic tape for various types of jobs, as per order of the firms using such magnetic tape controlled machine (with only reproducing units).

For manufacturing complicatedly profiled job on copying maehine-mechanical or hydraulic, we require complicated templates or cams, thereby increasing the cost of manufacture. But this complicacy can be elminated by using tape controlled machines, which ensure a lower cost of manufacture, uniformity and a higher degree of dimensional accuracy of the parts made.

14.4 Photoelectric Tracing System

In this system as shown in Figure 14.3 the light from the source is directed through a series of lenses to fall as a convergent ray on the line of the drawing which is of a considerable thickness. The light ray is of the dimension 0.01-0.5 mm. The diameter of the ray or beam is 2/3 times less than the thickness of the lines drawn with deep black ink. The intensity of light after being reflected by the mirror falls on the photoelement which generates current to be amplified and then fed to the motor of the executive organ through proper relays.

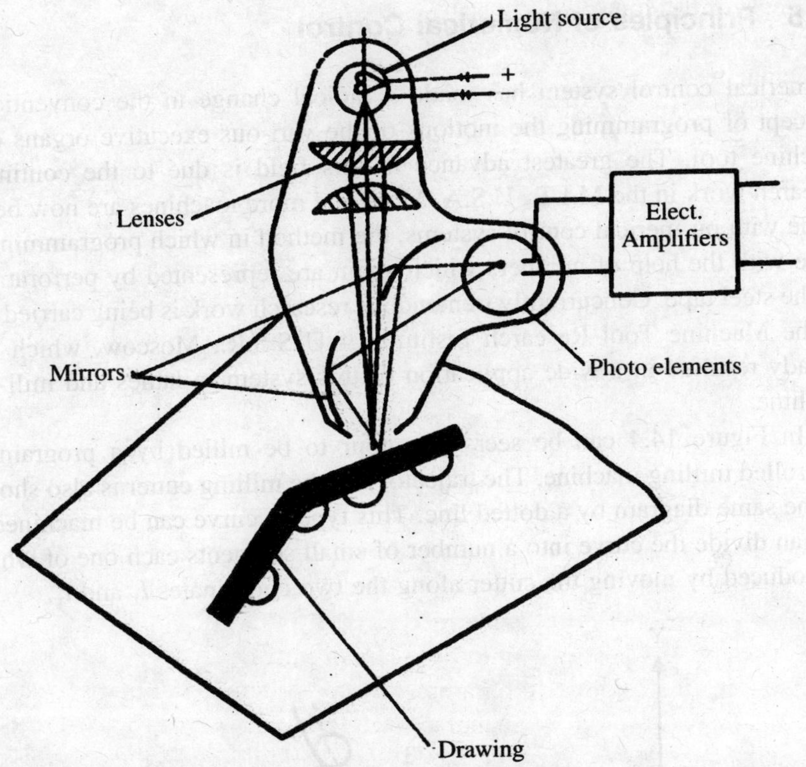

Figure 14.3 Photoelectric Tracing.

This type of recording head is rigidly connected to the supports or the spindle on which the milling cutter is fixed. The ray of light is allowed to fall on the drawing in such a way that half of the diameter of the beam of ray will be on the lines when moving on straight line of the drawing and the other half will be reflected from the paper. At the time of taking a bend the intensity of light reflected will change, since the light is reflected from a greater portion of the circular beam and this will change the current through the photoelement and the equalizing relay will operate to make the feed mechanism work in appropriate direction till the equilibrium is established.

The principle is quite simple though the method is very seldom in use now-a-days. This is because of the magnified scale drawing of the part which is needed in this case. Of course it is needless to say that the magnified line drawing only replaces the templates used in any copying machine. This method has its limited application in copy milling machine.

14.5 Principles of Numerical Control

Numerical control system has made a radical change in the conventional concept of programming the motions of the vari-ous executive organs of a machine tool. The greatest advance in this field is due to the continued research work in the M.I.T., U.S.A. More and more machines are now being made with nu-merical control systems, the method in which programming is done with the help of numbers which again are represented by perforations on the steel tape. Concurrently tremendous research work is being carried out in the Machine Tool Research Institute of U S.S.R., Moscow, which has already resulted in a wide application of this system in lathes and mill-ing machine.

In Figure 14.4 can be seen a contour to be milled by a programme controlled milling machine. The trajectory of the milling cutter is also shown on the same diagram by a dotted line. This type of curve can be machined if we can divide the curve into a number of small segments each one of which is produced by moving the cutter along the two coordinates l_x and l_y.

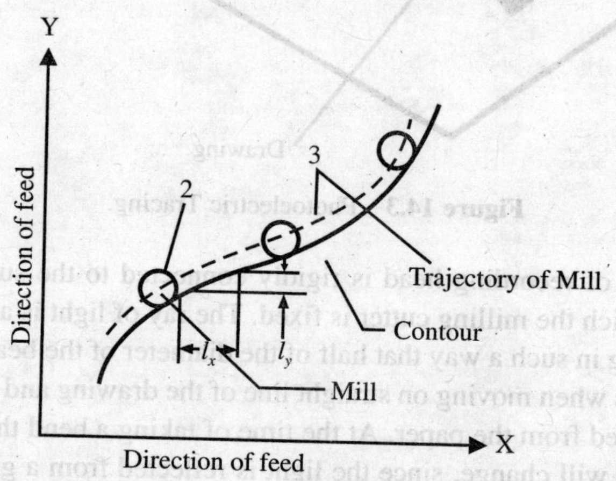

Figure 14.4 Basis of numerical control (M.I.T. System).

If we consider the feed per one electrical impulse, which is known as elementary pace to be equal to 0.0125 mm, the l_x and l_y can be representes by numbers n_x and n_y respectively, where

$$n_x = \frac{l_x}{t_e} \quad n_y = \frac{l_y}{t_e}$$

Here t_e represents elementary pace or feed of the tool per one electrical impulse and as per recommendation of M.I.T. is approximately equal to

0.0125 mm. These numbers n_x and n_y can be represented by two separate systems with the purpose of systematizing the methods of making perforations on the tape. The system are: (1) Binary system and (2) Decimal system. In Binary system:

$$n_{x,y} = \sum_{i=0}^{i=N-1} a_i \cdot 2^i \tag{14.1}$$

where i = exponent = 0, 1, 2, 3 . . . $N - 1$

a_i = 0 or 1

N in the above is total number of coefficients.

If n_x = 312, then it can be represented by this system as follows :

$$312 = 1.2^8 + 0.2^7 + 0.2^6 + 1.2^5 + 1.2^4 +. 1.2^3 + 0.2^2 + 0.2^1 + 0.2^0$$

or n_x = 100111000. 1 – denotes thevposition of holes on the document.

In the decimal system :

$$n_{x,y} = \sum_{i=0}^{i=N-1} a_i \cdot 10^i \tag{14.2}$$

where a_i = 0, 1, 2,3 . . . 9

and N = total number of digits in n_x or n_y.

Digits	312	*codes*
Hundreth	3	3.10^2
Tenths	1	1.10
Units	2	2.10°

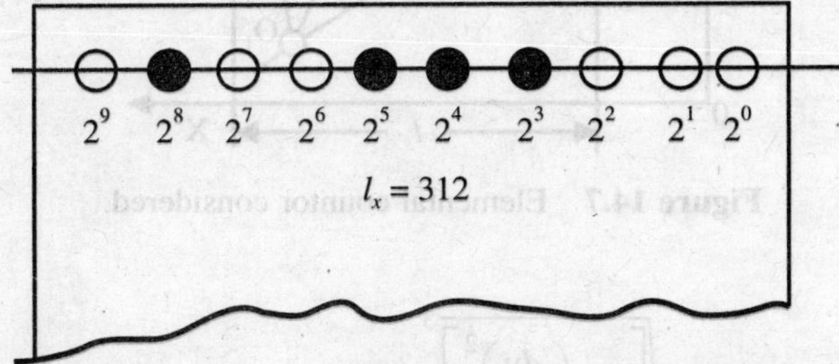

Figure 14.5 Disposition of holes in binary system. Numerical control.

The error by moving the cutter along the coordinates can be estimated as under. Suppose an elemental portion of the above contour which has equation $Y = f(x)$ is shown by ac;

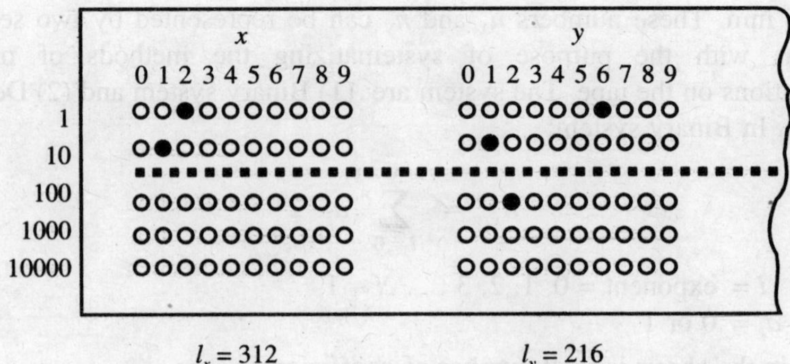

Figure 14.6 Disposition of holes in decimal system.

$x_0 y_0$ – co-ordinates of the point a;
$x_1 y_1$ – co-ordinates of the point c;

By following the method of moving by co-ordinates we actually go through straight line ac thereby making a theoretical error of δ.

From Figure (14.7)

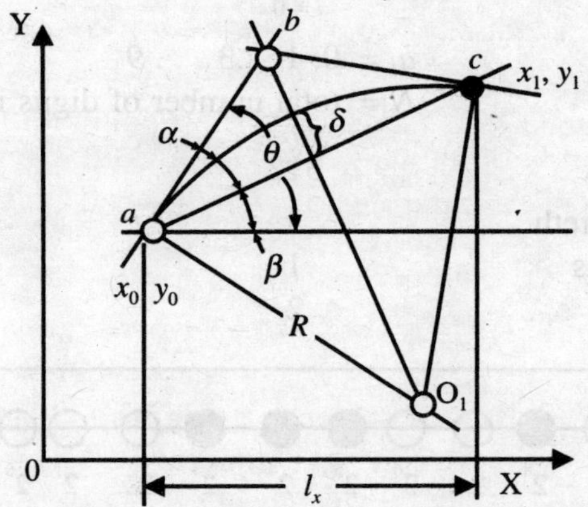

Figure 14.7 Elemental countor considered.

$$R = \frac{\sqrt{\left[1 + \left(\frac{dy}{dx}\right)^2\right]^3}}{\frac{d^2 y}{dx^2}}$$

$$\beta = \varphi - \alpha$$

where $\varphi = \text{arc lg}\left(\frac{dy}{dx}\right)_{x_0 y_0}$; $\cos\alpha = \frac{R-\delta}{R}$

and consequently $\alpha = \text{arc cos}\left(1 - \frac{\delta}{R}\right)$; $l_x = l\cos\beta$,

where $ac = 2\sqrt{R^2 - (R-\delta)^2} = 2\sqrt{2R\delta}$.

Therfore $lx = 2\sqrt{R^2 - (R-\delta)^2}$, $\cos\beta = 2\sqrt{2R\delta}.\ \cos\beta$

Wherefrom $x_1 = x_0 + lx$

$y_1 = f(x_1)$.

Thus we find out the co-ordinates of the point c if we presume certain permissible value of the theoretical error. This theoretical error

$$\delta \leq 0.05 \text{ mm.} \tag{14.3}$$

In the similar way the timing of an operation can also be represented by number of perforation.

The control of the motions of the executive organs is done by the flow of electrical impulses. Light passing through these holes and falling on the photoelements or electronic system gives signals and relays the timing, velocity and other necessary informations regarding feeding of the tool. For copying automatically a complicated profile, this method is the most suitable one. As a result of which numerical methods are being used in modern milling mcahines.

In the desperate bid to replace the brawn power by more of brain power and brain power by more of control power and concurrently to eliminate the errors and limitations imposed by human fallibility and fatigue, the machine tool designers all over the world have concentrated their efforts to design machine tools with practically all types of controls on it. As a result the machine tool today is more complicated than what it was even ten years ago. Machines with automatic inspection unit for automatically ensuring the dimensional accuracy of the job are now being increasingly manufactured. Most of those built-in inspection units are again working on feed back systems.

The successful application of the programme control systems in a machine tool requires considerable researches in the field of micro displacement of machine tool, transmission systems used in feed mechanism, systems of lubrication, elimination of stick-slip, etc. since the fundamental objective of producing programme-controlled machine is to have complete automation of manufacturing processes.

Though programme controlled machine tools are not much in, it will not be long before we will turn to these machines for increasing the output of our productive units. Programme controlled milling machines, jig boring machines, lathes etc. have been commercially made in many parts of the world, notably U.S.A..and U.S.S.R.

14.6 Method of Disposition of Punched Holes in Cases of Binary and Decimal

Supposing we have a number 1500, in binary system. This could be analysed simply as

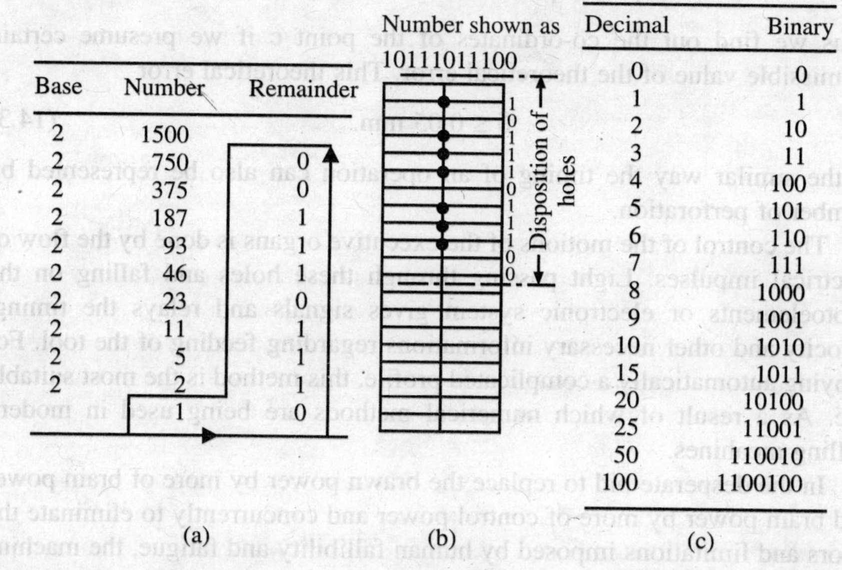

Base	Number	Remainder
2	1500	
2	750	0
2	375	0
2	187	1
2	93	1
2	46	1
2	23	0
2	11	1
2	5	1
2	2	1
	1	0

Decimal	Binary
0	0
1	1
2	10
3	11
4	100
5	101
6	110
7	111
8	1000
9	1001
10	1010
15	1011
20	10100
25	11001
50	110010
100	1100100

(c)

Figure 14.8

With reference to Figure 14.9 we have shown a typical punched tape used in numerically controlled machine.

(i) Block signal shows that one block is over and the other is starting.
(ii) Feed is in number of pulses.
(iii) In the channel, shown as parity check, the holes are punched, so as to make the total number of holes across the tape in any row an odd number.

Thus if a reader misses a hole, even number of contacts will be made. There are relays arranged, such that only when odd number of contacts are made, the connection to the commander (or commanding apparatus) is complete. A special circuit is used to check whether the information is being properly recorded and read. This reduces the possibility of errors occuring, but does not of course completely eliminate them.

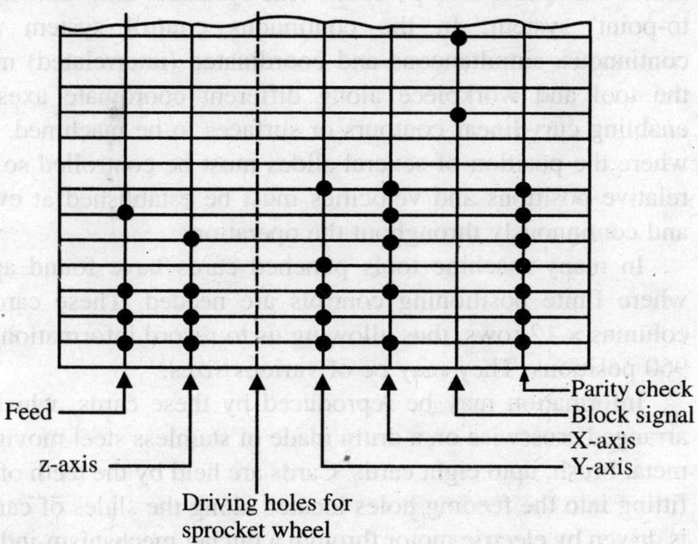

Figure 14.9 Typical punched tape.

Binary system addition and multiplication

Any numbers on binary system may be added or multiplied for quick coding as shown

$$4 = 100$$
$$+$$
$$\frac{5 = 101}{1001 = 9}$$

$$4 \times 5 = 100 \times 101 = \quad \begin{array}{r} 100 \\ \times 101 \\ \hline 101001 \end{array} = 20$$

14.7 Principle of Operation of NC Systems

NC systems can be divided into two, in accordance with the methods of processing viz. (i) finite positioning control and (ii) continuous controlor contour control.

(i) Finite positioning control

In the first system it is the coordinate dimensions which are used without definite linear motion between the points machined.

Processing operation is performed when the operative unit has reached its coordinate position. This system is also known as 'point-to-point' system. In the continuous control system there are continuous, simultaneous and coordinated (interrelated) motions of the tool and workpiece along different coordinate axes, thereby enabling curvilinear contours or surfaces to be machined. It is used where the position of several slides must be controlled so that their relative positions and velocities must be established at every point and continuously throughout the operation.

In many machine tools punched cards have found application where finite positioning controls are needed. These cards are 80 columns × 12 rows, thus allowing us to record information in any of 960 positions. They may be of various sizes.

Information may be reproduced by these cards, which may be arranged crosswise on a drum made of stainless steel moving against metal brush, upto eight cards. Cards are held by the teeth of the drum fitting into the feeding holes located along the slides of cards. Drum is driven by electric motor through a ratchet mechanism-indexing one pitch equal to distance between rows on a punched card. Photoelectric method using photocell placed below the bed picking up light through the holes on the punched card and thereby transforming it into voltage pulse conforming to the punchings. Conforming to the particular coordinate dimension.

(ii) Continuous control

Let us take the case of a contour milling where cutter has to travel 10 mm along x-axis and 5 mm along. y-axis. If it is possible to infeed 10 signals, distributed uniformly to the control system for longitudinal travel of the workpiece simultaneously with five such signals for the cross travel.

Each signal of the tally code corresponds to one elementary displacement of the operative unit. The magnitude of the displacement of the operative unit from one pulse is sometimes known as pulse value, For accurate contour machining pulse value should be less (of the order of 0.01 mm).

While contour machining, error in approximation should be well within 15-25% of the total tolerance on the machining error of the work.

Let us assume pulse value as 25 microns, then to traverse a system coordinate $l = 2$ mm, we require 2/0.025 = 80 pulses. If the feed of the cutter is S mm/min then time required will be

$$t = \frac{60l}{S} = \frac{60\sqrt{\Delta x^2 + \Delta y^2}}{S} \text{ sec}$$

where the increments of the coordinates are expressed in millimetres.

If we are machining a contour on radius R we take n angular segments for the purpose of dividing the contour into large number of elementary coordinates. If the increment angle is $\Delta\phi$, $\Delta\phi$ should be dependent on magnitude of R and the error allowed, i.e., δ.

$\Delta\phi - R$ is usually a exponential curve diminishing with radius R. For radius $R = 100$ mm $\Delta\phi < 2\text{-}3°$. Such experimental curves are available in Ref. 89.

CHAPTER 15

Built-in-inspection Units in Machine Tools

15.1. Introduction

With the growth of newer and better techniques of manufacture and demand for a higher degree of quality, it becomes necessary to modernise a machine tool. But-the type and character of modernisation varies from machine to machine. Gone are the days when a machine tool designer based his individual decision on intuition and experience, rather than on scientific principles. A comparative study of a particular machine built ten years ago and the same built today will justify the statement.

Availability of a variety of control systems and servo-mechanisms has enabled us to automatise the important operations like feeding and clamping of jobs, feeding of tools, automatic inspection of in-process parts. Automatic transfer machine, the logical development arising out of the need of flow production, has created possibilities of great savings. Shifting and positioning of in-process parts from station to station, automatic removal of processed parts, transfer and control mechanisms-have led to full mechanisation of production lines. Brawn power has been augmented by mechanical power and brainpower by more of control power.

In this chapter the author explains briefly some of the automatic built-in inspection units used in production machines with particular reference to grinding machines.

15.2. Systems of automatic Inspections

In Figure 15.1 is shown the classification of the various methods available for securing the dimenstional accuracy of a product while it is still on the machine.

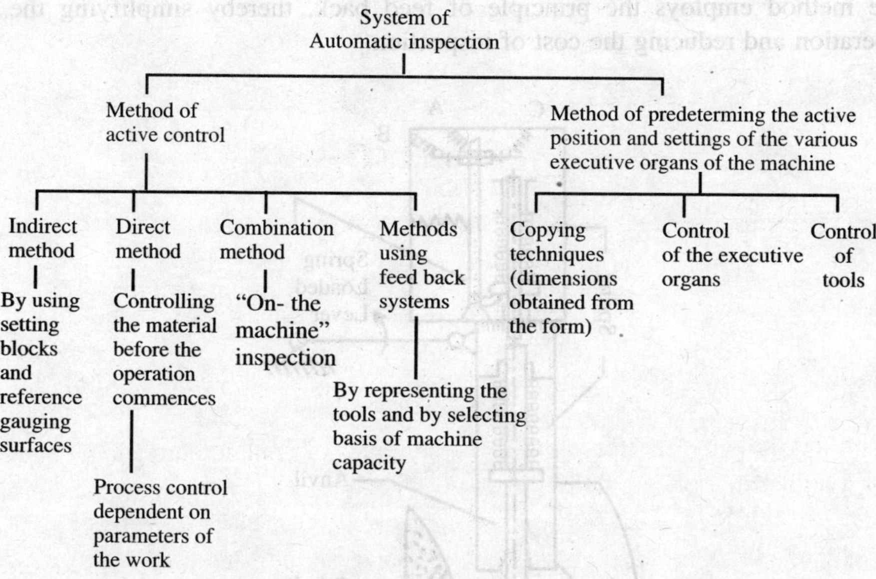

Figure 15.1 Automatic Inspections.

By this method given dimension can be achieved with the help of (tool settings and equipment) arrangements limiting the displacement of the working parts of the machine or by coordinating their various motions. Methods used under this heading are usually those having feed back arrangement. Automatic tool changing, automatic displacement etc can be effectively controlled. It is always necessary to install the built-in-inspection unit at such a convenient point on the machine that it can be readily used whenever needed and that its sensibility should not be affected by the thermal deformations that the machine may undergo while under dynamic condition.

15.3. Some Typical Built-in-inspection Equipments

In Figure 15.2 has been shown the schematic arrange-ment of one such inspection unit used in cylindrical grinding machine. This equipment can be used as a permanent accessory of the machine, which can be fitted on the wheel head and swivelled round whenever measurement is needed.

If the job is oversize or undersize the electrical contact between A & B or A & C will be made. Such electrical contact will start the feed mator in forward or reverse direction respectively to feed in or feed out the grinding wheel in transverse direction. Or in otherwards, the feeding of the grinding wheel in or out will be controlled by the size of the on-the machine job. Feeding is stopped when the job size gets within the prescribed limits. Thus the method employs the principle of feed back, thereby simplifying the operation and reducing the cost of inspection.

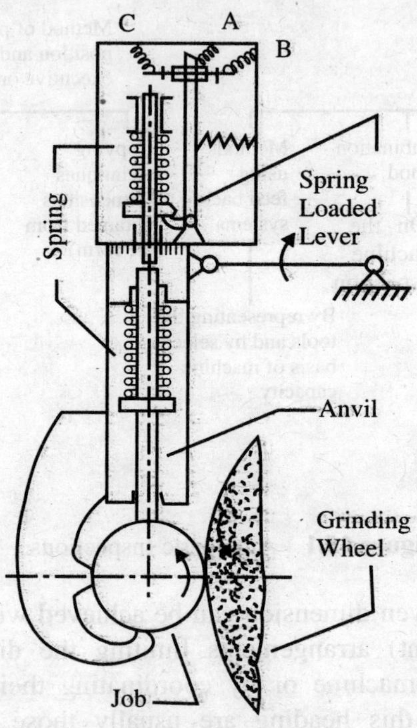

Figure 15.2

In Figure 15.3. we find an inspection unit for controlling the size of in-process parts in Internal grinding machine. This is a two-point contact system and therefore more stable than single contact units. This unit enables us to measure the size directly. Oversize or undersize of the part dimension is indicated by the movement of the pointer on the dial towards the right or left respectively. But this method is not able to feed back the information to the appropriate executive organ of the machine tool with a view to rectifying before it is too late. The divisions on the dial can be painted in different colours to indicate oversize or undersize of parts.

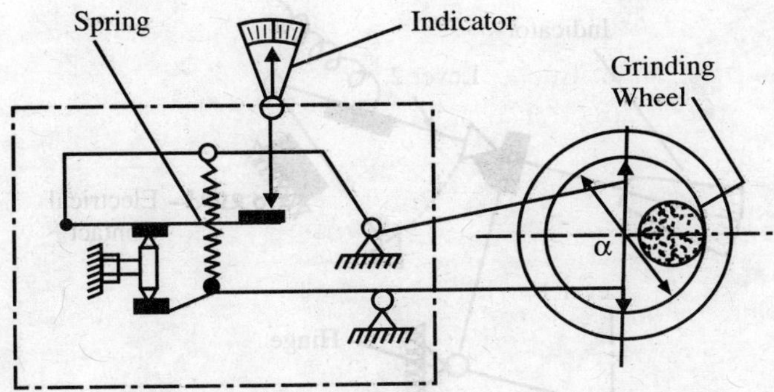

Figure 15.3.

The inspection equipment shown in Figure 15.4 and used in cylindrical grinding operations relays only one way information. For example, as soon as the correct size is attained, the contact lever will swing to the position shown by the dotted line, thereby making electrical contact which will cause the motor (used for giving transverse feed to the grinding wheel) to run in the reverse direction to take the wheel away from the job. There are preset directional switches to limit the outward feed of the wheel.

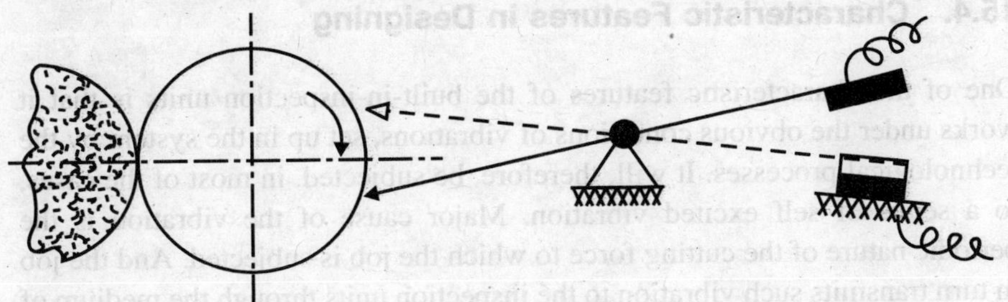

Figure 15.4.

Figure 15.5 shows equipment for measuring outside diameter and is analogous to the system described in Figure 15.4. When the size is attained lever 1 will allow the lever 2 to fall on the contact switch to outfeed the wheel.

The unit shown in Figure 15.5 is similar to that, in Figure 15.2, only one is for measuring internal diameter, while the other is for measuring the outside diameter. The system in Figure 15.6 unlike that in Figure 15.3 has only one point contact. Obviously this equipment is much less rigid in comparison to that shown in Figure 15.3.

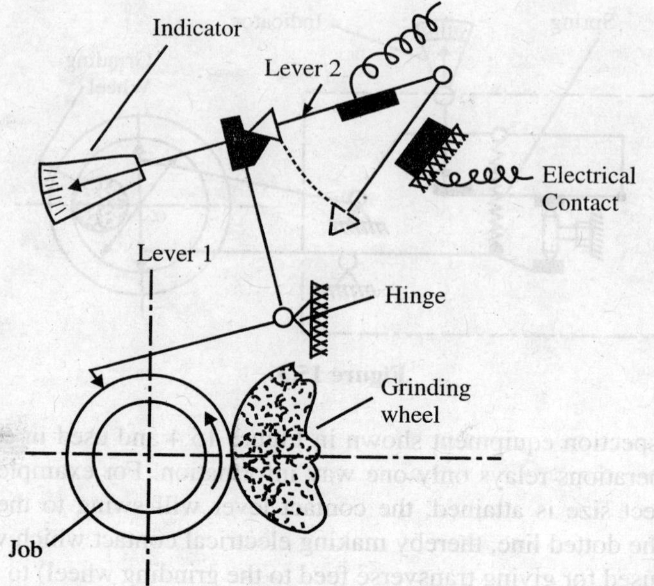

Figure 15.5.

15.4. Characteristic Features in Designing

One of the characteristic features of the built-in-inspection units is that it works under the obvious conditions of vibrations, set up in the system by the technological processes. It will, therefore, be subjected, in most of the cases, to a series of self excited vibration. Major cause of the vibration is the periodic nature of the cutting force to which the job is subjected. And the job in turn transmits such vibration to the inspection units through the medium of contact points. Self excited vibration of the measuring equipment may also be caused by the bad quality of the surface of the workpiece.

Effect of such vibration on the accuracy of measurement by such built-in-inspection units, is dependent on the type of construction of the inspection equipment itself. Its effect will be maximum in the case of inspection equipments having single point contact. To increase the accuracy and sensibility of such equipment, it is therefore necessary that the natural frequency of vibration of the measuring anvil of the equipment should be higher than the frequency of the self-excited vibration. This can be achieved by either increasing the stiffness of the spring with which the anvil is loaded or decreasing the mass of the anvil.

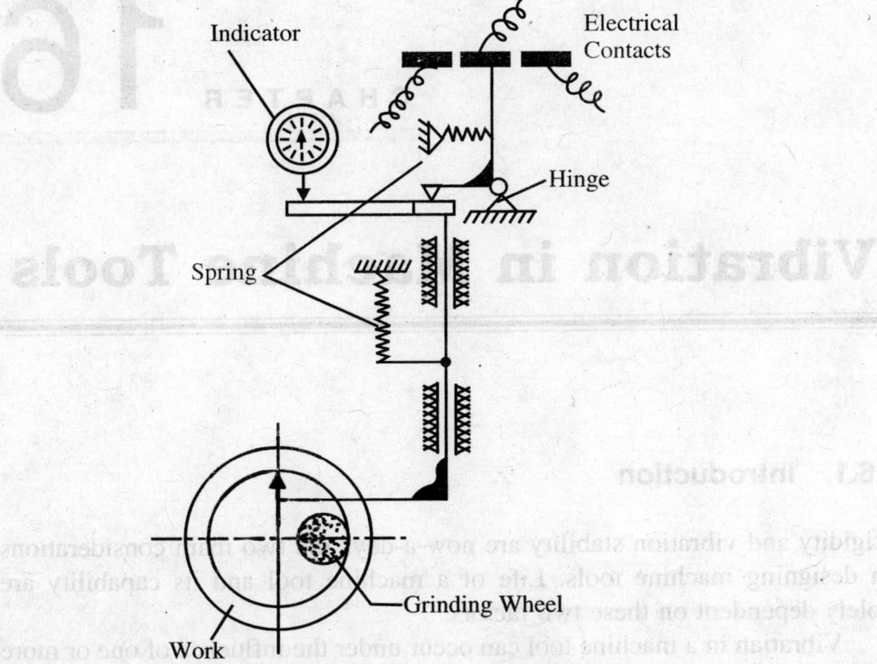

Figure 15.6 Single Contact Equipment.

15.5. Conclusion

The author confined his discussion only to a few specific inspection equipments, used in grinding machines as permanent accessories. Analogous systems are working in other machines also.

But all such inspection equipments, discussed in this chapter, are incapable of incorporating automatic compensation of errors, which may be caused because of wear of the tools or instruments and the thermal deformation of the parts of the executive organs of the machine tool.

It is not impossible to design built-in-inspection units which will compensate for the above errors; but such designs are bound to be complicated and unstable under the dynamic condition of the machine tool.

CHAPTER 16

Vibration in Machine Tools

16.1. Introduction

Rigidity and vibration stability are now-a-days the two main considerations in designing machine tools. Life of a machine tool and its capability are solely dependent on these two factors.

Vibratian in a machine tool can occur under the influence of one or more of the following factors, *viz*:

(a) Unbalanced rotating force in the drives of the Machine tools;
(b) Transmission of the vibratory forces to jobs and to tools;
(c) Transmission of the vibratory motions of the other machines in the vicinity through the foundations of the machine.
(d) Cutting processes, rigidity of head-stock and/or tailstock centres, variable chip thickness etc. causing "Regenerative" or "Self excited" vibration. This type of vibratian has been explained later in this chapter. (Vibration caused by the motion of the tool does not occur necessarily at high speeds; but tool chatter does show much complicated pattern of chatter bands at high speeds).
(e) Frictional relaxation vibration known as stickslip.

Effect of vibration on a machine tool can be much too disastrous leading to the failure of tools and certain executive organs. Quality of the job will suffer, thereby lowering the productive capacity.

16.2. Forced vibration

Let us analyse one of the simplest occurences of vibration in a machine tool, viz. the forced vibration (shown in Figure 16.1.)

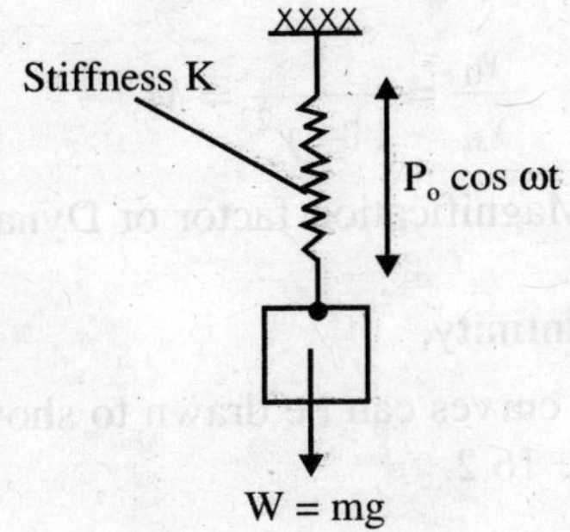

Figure 16.1.

The basic equation of forced vibration can be written in the form :

$$\frac{W}{g} \cdot \frac{d^2 y}{dt^2} + K \cdot y = P_0 \cdot \cos \omega t. \tag{16.1}$$

where W = weight of the vibrating mass, y - displacement from its mean position, K – stiffness of the connecting mechanism which can be likened to an elastic member – and $P_0 \cos \omega t$ – the periodic force.

$$\frac{d^2 y}{dt^2} + \frac{K \cdot}{W} g \cdot y = \frac{P_o}{W} g \cdot \cos \omega t.$$

The general solution of the above differential equation can be expressed as under:

$$y = A \cos \omega t + B \sin \omega t = y_0 \cdot \cos \omega t \tag{16.2}$$

$$\frac{d^2 y}{dt^2} = -\omega^2 \cdot y_0 \cdot \cos \omega t \tag{16.3}$$

Therefore the equation (16.1) can be written in the terms of equations (16.2) & (16.3).

From which
$$y_0 = \frac{P_0 / K}{1 - \left(\dfrac{\omega}{\omega_n}\right)^2} \tag{16.4}$$

where ω_n – natural frequency of vibration and hence $\dfrac{\omega}{\omega_n} = \gamma$ = ratio of frequencies.

$$y_0 = \frac{y_{st}}{1 - \gamma^2}, \text{ where } y_{st.} = \text{static deflection} = \frac{P_0}{K}$$

$$\frac{y_0}{y_{st}} = \frac{1}{1 - \gamma^2} = M, \tag{16.5}$$

where M is described as Magnification factor or Dynamic Magnifier.
where $\gamma = 0$, $M = 1$
$\gamma = 1$, M tends to infinity.

Various characteristic curves can be drawn to show the magnitude of M dependent on γ, see Figure 16.2.

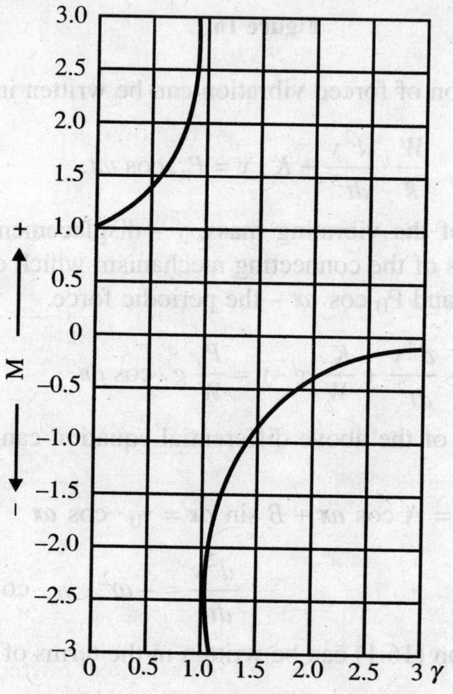

Figure 16.2

$\gamma = \frac{1}{18}$ or $\frac{1}{20}$ is quite a good figure to be used by machinc tool designers.

Usually the machines are separated from the foundation by means of layer of rubber, cork, felt, nylon, neoprene (impregnated fibres), fibre glass etc. Machine tools may be mounted on concrete floor having a layer of 1" sand or felt. This type of material is usually known as "Isolators". Isolators can be of two types, *viz.* (a) active isolators and (b) passive isolators. Active isolators prevent the outside transmission of the vibratory forces, while the Passive isolators are used to prevent vibration of the neighbouring sources being transmitted to the machine under consideration. Sometimes these isolators are known as "shock mounts". The advantages of shock mounts in a machine tool can be enumerated as follows :

(1) the machine becomes portable, no connection between the isolation and the shop floor;
(2) cuts down labour cost;
(3) reduces wastes and minimises rejections, thereby improving the quality of the products made:
(4) shocks are eliminated;
(5) foundation costs are minimised;
(6) life of the machine tool is increased;
(7) impact machines may be grouped together; and
(8) Excellent damping.

Example: A machine has to be isolated by placing 1" thick felt on the four legs of the machine in the form of discs. If the total weight on the isolators = 100 lbs and speed of the machine in the highest dynamic condition is 3600 rpm, calculate the area of the each felt disc, assuming that the machine gives a vibration transmissibility of 30%.

Solution : Dynamic magnifier $M = \dfrac{1}{1-\left(\dfrac{\omega}{\omega_n}\right)^2} = \dfrac{1}{1-\left(\dfrac{f}{f_n}\right)^2}$

where f = r.p.s. of impressed frequency
and f_n = natural frequency in r.p.s.

Natural frequency f_n is usually less than impressed frequency f.

From the data given $-0.30 = \dfrac{1}{1-\left(\dfrac{f}{f_n}\right)^2}$ or $\left(\dfrac{f}{f_n}\right)^2 = 4.03$

Since f_n – 60 r.p.s. $\quad f_n = \dfrac{60}{\sqrt{4.03}} \approx 30$ r.p.s.

Permissible pressure of felt for this frequency f_n can be obtained from the characteristic curves supplied by the manufacturer. If the permissible pressure corresponding to this frequency f_n = 10 p.s.i,

$$\text{then area of the felt} = \frac{100}{10} = 10\,\text{in}^2$$

If the weight of the machine is taken up equally by the four legs, then the area of each felt disc will be equal to $\frac{10}{4} = 2.5\,\text{in}^2$. It should be noted that the permissible pressure on the felt disc plotted against f gives a falling characteristic which becomes approximately asymptodic at ends.

16.4. Self-excited Vibration in Machine Tools

Self induced vibration which often occurs in machine tool operations is known as regenerative vibration or chatter, since it originates by the process itself. The nature of the force under the dynamic condition acting on the tool point is again in turn dependent on the amplitude of the vibration. Slight disturbance in the process caused by the varying chipthickness, varying rate of penetration of the tool into the job, or variation in the angular speed of the job may cause such a type of vibration. Sometimes even a sudden impact load on the tool may result in chatter. The generation of such vibration may again sometimes be caused depending upon the surface geometry or slenderness of the workpiece.

Much work have been done in this field by R.N. Arnold(5), S. Doi(17) S.A. Tobias and W. Fishwick(49), S.G. Bhattacharya(16), J. Tulsty(50) and others.

Such type of self induced vibration produces detrimental effects, such as, high deformation of the various members of the machine, reduction of the tool life, bad quality of the finished product and considerable noise in the vicinity.

Self excited vibration in a machine tool may result from the dynamic instability of the machine parts or of the tool used in cutting. Sometimes both may be the causes of self excited vibration. Under the dynamic condition, that is, under the working condition, the frequency of vibration of the tool is much more than that of the various parts of the machine tools.

The general equation of motion under such a condition can be written in the form :

$$m \cdot \frac{d^2y}{dt^2} + \beta \cdot \frac{dy}{dt} + K \cdot y = \Delta_1 \cdot P_y + \Delta_2 \cdot P_y + \Delta_3 \cdot P_y \tag{16.6}$$

where $m \cdot \frac{d^2y}{dt^2}$ – interia force ot the viorating mass in question (say, tool).

$\beta \cdot \frac{dy}{dt}$ – Resisting force due to damping, where β is the damping coefficient expressed in terms of force per unit velocity;

$K \cdot y$ – elastic force;

$\Delta_1 \cdot P_y$ – variable force generated in the machining system with the falling characteristic of the force of friction;

$\Delta_2 \cdot P_y$ – variable force generated in the system, with the change in the cutting angle of the instrument; and

$\Delta_3 \cdot P_y$ – variable force generated in the oystem with the change in the shear phase.

Self excited vibration exists so long as the exciting process exists. It ceases immediately as the source is taken off. Let us consider now a much simplified equation

$$m \cdot \frac{d^2y}{dt^2} + \beta \cdot \frac{dy}{dt} + K \cdot y = F \cdot \left(\frac{dy}{dt}\right) \tag{16.7}$$

where $F \cdot \left(\frac{dy}{dt}\right)$ – the exciting force is a function of the velocity and proportional to it.

Such a case of self excited vibration can have three distinct consequences.

$$m \cdot \frac{d^2y}{dt^2} + (\beta - F)\frac{dy}{dt} + K \cdot y = 0 \tag{16.8}$$

Various possibilities are $\beta - F > 0$; $\beta - F = 0$; $\beta - F < 0$.

The vibration patterns for all such cases are as shown below:

$\beta - F > 0$	Dying amplitude
$\beta - F = 0$	Uniform amplitude
$\beta - F < 0$	Increasing amplitude

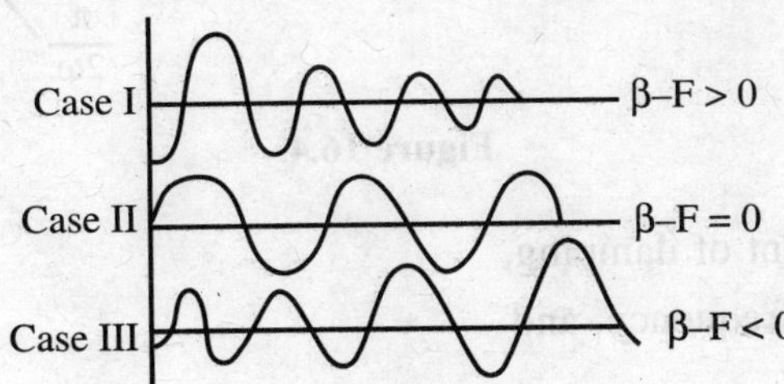

Figure 16.3.

The third case is very interesting and important too, since the increasing amplitudes of vibration will lead to dangerous consequences.

The exciting force in a self excited vibration may vary with velocity or with tha chip thickness while cutting. In the case when $\beta - F > 0$, the vibration dies out. From all these analyses the following conclusions can be made with the purpose of preventing regenerative vibration:

(1) To have high damping coefficient of the system;
(2) to have large stiffness of the tool; and
(3) damping material in the tool holder also is a necessity.

Theory of tool chatter has been developed taking all these factors into consideration.[5]

16.5. Other Types of Forced and Damped Vibration as shown in Figure 16.4.[25]

Let us now consider a case when a material is being turned in a centre lathe between the centres. Also let us assume that the tailstock has a very small rigidity, in comparison to headstock.

Under such condition the law of vibration set up can be expressed as under:

$$y = A_1 \cdot e^{-\beta t} \cdot \cos \omega t, \tag{16.9}$$

where A_1 – initial amplitude as shown in 16.4,

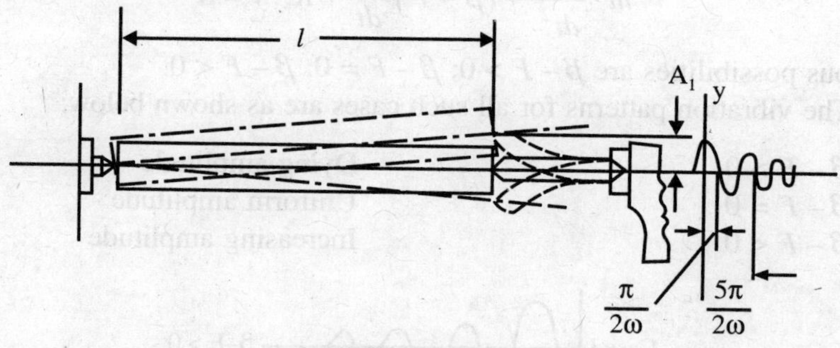

Figure 16.4.

β – coefficient of damping,

ω – angular frequency, and

t – time.

To find out the exact amouat of work done by the resisting force, it is absolutely necessary to know the value of the coefficient of friction μ'. The damping decrement $\theta = \frac{A_1}{A_2}$ where A_1 and A_2 are two successive amplitudes of damped vibration. The value of these amplitudes can be obtained from oscillograph.

The force of resistance of the system is proportional to the velocity $\frac{dy}{dt}$ of the Oscillating motion.

$$\therefore \qquad F = \mu' \cdot \frac{dy}{dt} \qquad (16.10)$$

Elemental work done by the force of resistance = dR

$$= \mu' \cdot \frac{dy}{dt} \cdot dy.$$

$$\therefore \qquad dR = \mu' \cdot \left(\frac{dy}{dt}\right)^2 \cdot dt, \qquad (16.11)$$

$$dR = \mu' (A_1^2\, \omega^2 \cdot e^{-2\beta t} \sin^2 \omega t + A_1^2 \cdot \beta\omega \cdot te^{-2\beta t} \sin 2\,\omega t$$

$$+ A_1^2 \cdot \beta^2 \cdot e^{-2\beta t} \cdot \cos^2 \omega t) \cdot dt \qquad (16.12)$$

(The damping coefficient $\beta = \theta \cdot f = \frac{\theta \cdot \omega}{2\pi}$ where f – frequency of oscillations)

If we solve this equation for one period of vibration starting from the moment when $y = 0$, then we can find out the work done by the resisting force during one period.

$$R = \mu' \cdot A_1^2 \cdot \omega^2 \cdot \int_{\frac{\pi}{2\omega}}^{\frac{5\pi}{2\omega}} e^{-2\beta t} \cdot \sin^2 \omega t \cdot dt + \mu' \cdot A_1^2 \cdot \beta \cdot \omega \qquad (16.13)$$

$$\int_{\frac{\pi}{2\omega}}^{\frac{5\pi}{2\omega}} e^{-2\beta t} \cdot \sin 2\omega t \cdot dt + A_1^2 \beta \int_{\frac{\pi}{2\omega}}^{\frac{5\pi}{2\omega}} e^{-2\beta t} \cos^2 \omega t\; dt \qquad (16.14)$$

By simplifying we obtain

$$R = \frac{1}{4} \cdot \mu' \cdot A_1^2 \cdot \omega^2 \cdot \frac{1}{\beta} \cdot e^{\frac{-\pi \cdot \beta}{\omega}} \left(1 - e^{\frac{-4\pi \cdot \beta}{\omega}}\right), \tag{16.14}$$

For the obtaining of the dimped vibration it is neccessary that this work should be equal to the loss in kinetic energy during this period. Loss in kinetic energy

$$dt = \frac{1}{2} \cdot (\rho \cdot dx) \left(\frac{dy_1}{dt}\right)^2. \tag{16.15}$$

where p – mass per element length of the work piece.

$\frac{dy_1}{dt}$ – Velocity of vibrating motion of the element

$$\frac{dy_1}{dt} = \frac{dy}{dt} \quad \frac{x}{1},$$

$$\frac{dy_1}{dt} = -A_1 \beta e^{-\beta t} \cdot \cos \omega t - A_1 \cdot \omega \cdot e^{-\beta t} \cdot \sin \omega t$$

When $t = \frac{\pi}{2\omega}, \frac{dy}{dt} = -A_1 \cdot \omega e^{\frac{-\pi}{2} \cdot \frac{\beta}{\omega}}$

$$\frac{dy_1}{dt} = -A_1 \cdot \omega \cdot e^{\frac{-\pi\beta}{2\omega}} \cdot \frac{x}{l}.$$

But $dT = \frac{1}{2} \cdot \rho \cdot A_1^2 \, \omega^2 \cdot e^{\frac{-\pi \cdot \beta}{\omega}} \cdot \frac{1}{l^2} \cdot x^2 \cdot dx.$

When the total deflecticn of the part is zero, that is $t = \frac{\pi}{2\omega}$

$$T_1 = \frac{1}{2} \cdot A_1^2 \cdot \omega^2 \cdot e^{\frac{-\pi\beta}{\omega}} \cdot \frac{1}{l^2} \cdot \rho \cdot \int_0^l x^2 \cdot dx$$

or
$$T_1 = \frac{1}{6} \cdot A_1^2 \cdot \omega^2 \cdot e^{-\pi\beta/\omega} \cdot m \tag{16.16}$$

Where m – total mass of the workpiece.

when $t = \frac{5\pi}{2\omega} \qquad \frac{dy}{dt} = -A_1 \omega e^{\frac{-5\pi\beta}{\omega}}$

Analogically:

$$T_2 = \frac{1}{6} \cdot A_1^2\ \omega^2 \cdot e^{\frac{-5\pi\beta}{\omega}} \cdot m$$

$$\Delta T = T_1 - T_2 = \frac{1}{6} \cdot A_1^2\ \omega^2 \cdot e^{\frac{-\pi \cdot \beta}{\omega}} \left(1 - e^{\frac{-4\pi \cdot \beta}{\omega}}\right). \qquad (16.17)$$

Since $R = \Delta T$,
we have:

$$\frac{1}{4} \cdot \mu'\ A_1^2 \cdot \omega^2 \cdot \frac{1}{\beta} e^{\frac{-\pi \cdot \beta}{\omega}} \left(1 - e^{\frac{-4\pi\beta}{\omega}}\right)$$

$$= \frac{1}{6} \cdot A_1^2 \cdot \omega^2 \cdot m \cdot e^{\frac{-\pi}{\omega}\beta} \left(1 - e^{\frac{-4\pi}{\omega}\beta}\right)$$

$$\therefore \qquad \beta = \frac{3}{2} \cdot \frac{\mu'}{m}. \qquad (16.18)$$

Therefore the coefficient of damping should be equal to $\frac{3}{2} \cdot \frac{\mu'}{m}$, where μ' is the coefficient of friction and m is the mass of the workpiece.

16.6. Stick-Slip Vibration in Machine tools [43]

General concepts: The phenomenon of stick-slip sliding is commonly observed in the moving parts of certain machine tools, apparatus and devices. In some heavy machine tools, for example, it is found that tables (weighing several tons) moving in guides execute long jumps extending to several millimetres. Such intermittent motion of moving parts, obviously, has harmful effects both on the accuracy and sensitivity of the precision tools and appliances.

On the basis of experimental data on the sliding of different metallic and nonmetallic bodies, Papenhuysen,[36] in 1938, discovered that every flexible system has its own critical velocity, and (*i*) if the average velocity of sliding of such a system is equal to or less than this critical value, motion occurs with alternate stick and slip; and (*ii*) if the average velocity is greater than the critical, stick-slip motion is rendered impossible.

Bowden and his co-workers,[15] after extensive experiments, suggested that frictional resistance between unlubricated metals is due, primarily, to the shearing of small metallic junctions or welds formed at high points of contact between the moving surfaces where very high temperatures and pressures may occur. The intermittent joining and breaking of the rubbing surfaces cause large fluctuations in the frictional force and surface temperatures.

Block explained the dependsnce of the stick-slip phenomenon on the basis of the physical and dynamical characteristics of the sliding system and the rubbing surfaces. His analysis, however, was confined to the sliding part of the motion system only, wherein it was assumed that the friction remained constant during the stick period.

Ishlinsky and Kragelsky, on the contrary, determined the critical velocity on the basis of the stick period, and found that static frictional force changed as a function of the period of stopping, as shown in Figure 16.5. However, their analysis was not proved by experimental observations.

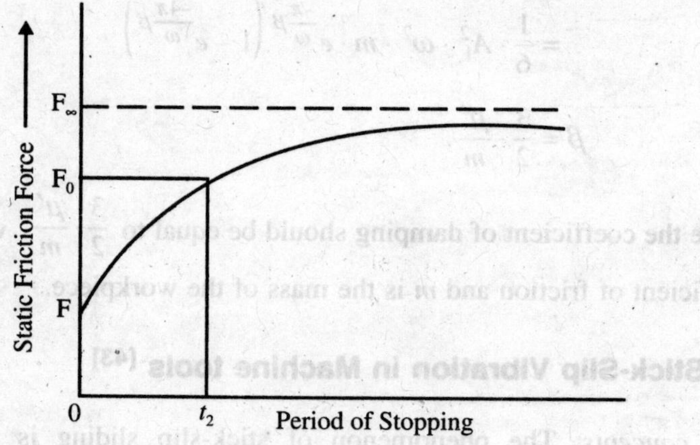

Figure 16.5

Deryagin, Push and Tolstoif391 analyzed the concept of critical velocity on the basis of combined stick and slip periods of motion, taking into Consideration the vibration of friction force both during the period of motion and of stopping.

Dynamical Analysis

Figure 16.6 represents an elastic system having a mass, m coupled by a spring of stiffness, k_y to a drive having a constant velocity v.

In the equations given below, wherever x' and x'' occur, they should be taken to signify $\dot{x}$ and $\ddot{x}$, that is to say, the velocity and acceleration respectively.

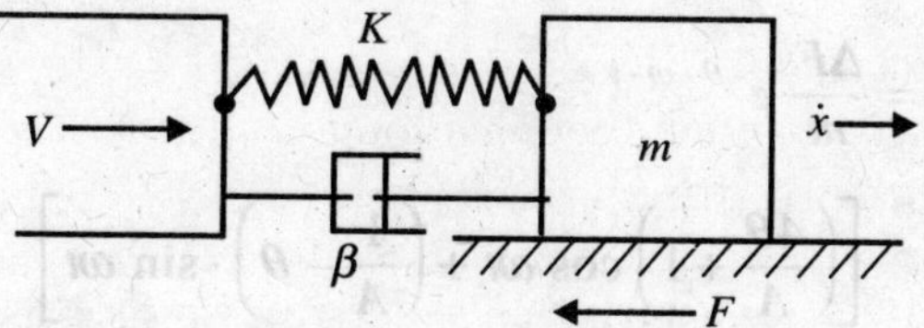

Figure 16.6

The motion of the mass or the 'slip' begins when the compressive force in the spring just overcomes the static frictional force. The differential equation of motion for the mass whicr holds for this period of slip is given by:

$$mx'' + \beta(x' - v) - k\left(x_0 - \frac{\beta v}{k} - vt - x\right) + F = 0, \qquad (16.19)$$

where β = coefficient of equivalent viscous damping of system, $x_0 = \frac{F_0}{k}$, F_0 being the static frictional force at the end of the stick period, and F = kinetic frictional force.

The solution of equation (16.19) for small values of damping usually obtained in practice ($\theta^2 \angle\angle 1$) is

$$x = \left(x_0 - \frac{\beta v}{k} - \frac{F}{k}\right) + vt + e^{-\omega t} \times (c_1 \sin \omega t + c_2 \cos \omega t) \quad (16.20)$$

where $\theta = \frac{\beta}{2\sqrt{mK}}$, $\omega = \sqrt{\frac{K}{m}}$ and c_1 and c_2 are constants to be determined from initial conditions *viz.* $x = x' = 0$ at $t = 0$.

Hence $$C_1 = -\left[\frac{\upsilon}{\omega} + \theta\frac{(F_0 - F)}{K}\right]; \text{ and}$$

$$C_2 = -\frac{(F_0 - F)}{K} - \frac{\beta\upsilon}{K}.$$

Expressions for displacement, velocity and acceleration of the mass may now be written from equation (16.20) as

$$x = \frac{\beta \cdot \upsilon}{K} + \frac{\Delta F}{K} + \upsilon t - \frac{\upsilon}{\omega} \cdot e^{-\theta\omega t}$$

$$[(20 + A) \cos \omega t + (1 + A\theta) \sin \omega t] \qquad (16.21)$$

Therefore, $x' = u\ \{1 - e^{-\theta\omega t}\ [\cos \omega t - (A + 3\theta) \cdot \sin \omega t]\}$, (16.22) and $$x^{v} = \frac{\Delta F}{m} e^{-\theta \cdot \omega \cdot t}$$

$$\left[\left(\frac{4\theta}{A}+1\right)\cos \omega t + \left(\frac{1}{A}-\theta\right)\cdot \sin \omega t\right]. \qquad (16.23)$$

where $\Delta F = F_0 - F$ and $A = \dfrac{\Delta F}{v\sqrt{K \cdot m}}$.

Critical Velocity

For stick-slip motion to occur, under limiting conditions, both $x' = 0$ and $x'' = 0$ at the end of the slip period $t = t_1$; when this occurs, the velocity of the drive has its critical value $v = v_e$ and the corresponding value, of

$$A = A_e = \frac{\Delta F}{v_e\sqrt{K \cdot m}}$$

Thus, for this critical situation, equations (16.22) and (16.23) give

$$1 = e^{-\theta \cdot \omega t_1}\,[\cos \omega t_1 - (A_e + 3\theta)\sin \omega t_1] \qquad (16.24)$$

and

$$(A_e + 40)\cos \omega t_1 + (1 - A_e\,\theta)\sin \omega t_1 = 0 \qquad (16.25)$$

From this,

$$20 = \tan^{-1}\left(\frac{A_e + 4\theta}{A_e \cdot \theta - 1}\right) = \log\,(1 + A_0^2 + 6\,A_e \cdot \theta). \qquad (16.26)$$

So the desired critical velocity, v_e is given by

$$v_c = \frac{\Delta F}{\sqrt{K \cdot m}} \cdot \frac{1}{A_c} = \frac{\Delta F}{\sqrt{K \cdot m}} \cdot \frac{1}{f(\theta)}, \qquad (16.27)$$

where $A = f(\theta)$ is given by equation (16.26). It is found that A_e may be approximately represented by a curve $f(\theta) = \sqrt{4 \cdot \pi\,\theta}$, so that the expression for critical velocity may be written as

$$v_c = \frac{\Delta F}{\sqrt{4 \cdot \pi \cdot \theta\,K \cdot m}} \qquad (16.28)$$

Considering the variation of frictional force during the period of stick, this variation may be represented by (see Figure 16.5)

$$F_0 - F = (F_\infty - F)\,(1 - e^{-\delta t_2}), \qquad (16.29)$$

Where F_∞ = value of static frictional force attained after infinite time lapse, and δ = a constant indicating the rate at which kinetic friction changes to its static value.

Equation (16.29) may be approximately written as

$$F_e - F = (F_\infty - F) \cdot \frac{\delta t_2}{(1 + \delta t_2)} \tag{16.30}$$

Also writing $k \times v_0 \times t_2 = F_0 - F$ and $\omega t_2 = A_e$, the critical velocity is approximately given by

$$v_e = \frac{(F_\infty - F)}{\sqrt{K \cdot m}} \times \frac{1}{\dfrac{\omega}{\delta} + \sqrt{4\pi \cdot \theta}} \tag{16.31}$$

This equation is shown plotted in (Figure 16.7).

The foregoing analysis is true when the velocity, v, of the drive is constant. However, in all practical cases, thus velocity fluctuates and may be represented by

$$v_f = f_0\,(1 + \alpha \cdot \cos pt), \tag{16.32}$$

where v_f = fluctuating velocity, V_e = mean velocity,

α = ratio of amplitude of fluctution to mean value of velocity, and

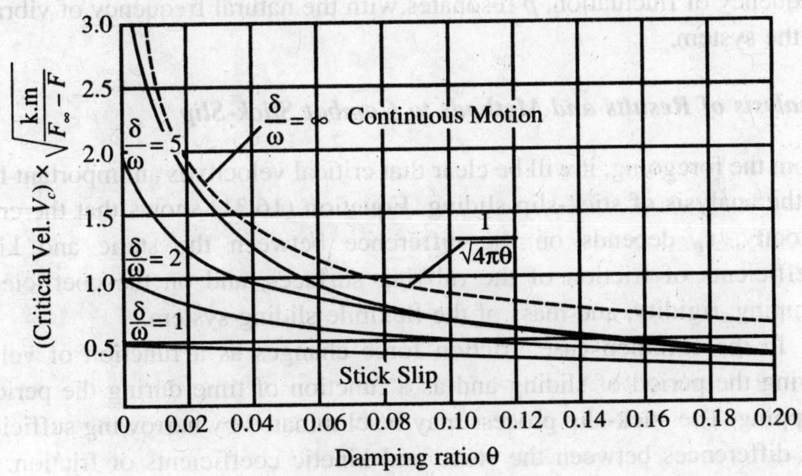

Figure 16.7

p = frequency of fluctuation (see Figure 16.8). In such a case, the critical velocity, $v_c f$, was found by Push, Singh and Verma [38] as

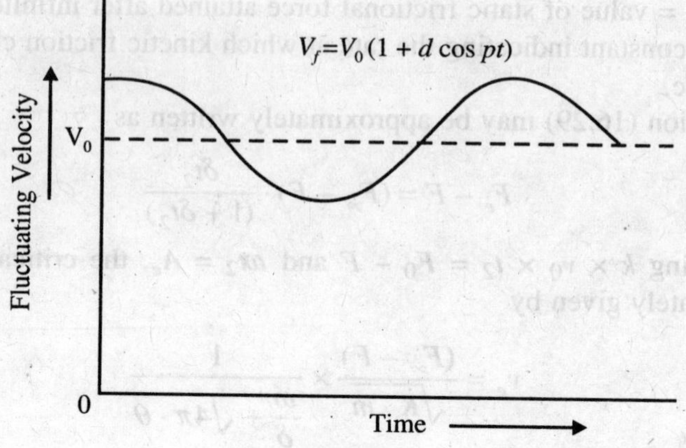

Figure 16.8

$$\frac{v_e f}{v_e} = f\left(\theta, \alpha, \frac{p}{\omega}\right) \tag{16.33}$$

$$= \text{a mathematical functions of } \theta, \alpha \text{ and } \frac{p}{\omega}.$$

It was observed that the critical velocity increases considerably when the frequency of fluctuation, p resonates with the natural frequency of vibration, of the system.

Analysis of Results and Methods to Combat Stick-Slip

From the foregoing, it will be clear that critical velocity is an important factor in the analysis of stick-slip sliding. Equation (16.31) shows that the critical velocity, v_e depends on the difference between the static and kinetic coefficients of friction of the rubbing surfaces, and on the coefficient of damping, rigidity, and mass of the flexibile sliding system.

In the common case, friction force changes as a function of velocity during the period of sliding and as a function of time during the period of stopping. The stick-slip process may be eliminated by narrowing sufficiently the differences between the static and kinetic coefficients of friction. This may be done by selecting suitable materials for the rubbing surfaces and by using certain special lubricants with additives. Figure 16.9 shows the stick-slip characteristics, as determined by Merchant,[35] of such lubricant as paraffin oil, to which a polar additive, oleic acid, is added.

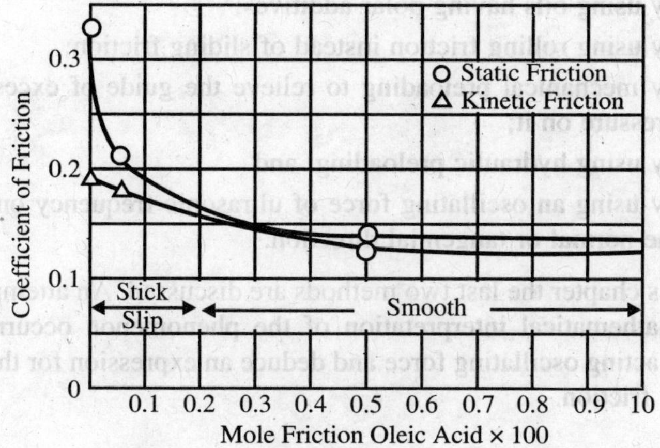

Figure 16.9

The difference between the static and kinetic coefficients of friction may also be reduced by other methods such as by the application of forced vibration on the sliding system, or by the use of rolling friction instead of sliding friction. Roller guides are successfully employed in machine tools with this object in view.

Increased damping in the sliding system may help to reduce stick-slip considerably. Damping can be augmented externally by the use of damper, or by employing grease lubrication in place of fluid lubricants. On the other hand, damping may be increased by the application of special materials with high internal friction, as for example, plastic in place of metals.

Increasing the rigidity of the drive is an important measure in combating stick-slip sliding. This can be done, for example, by the use of a unidirectional drive which must be located towards the end of kinematic scheme, as near to the driven body as possible.

16.7. Minimization of Stick-Slip Vibration in Machine Tools

Usual methods available. At low speeds of sliding, the coefficient of friction is high, but as the speed increases, the value of kinetic friction comes down. The difference between static and kinetic friction is of great importance in the design of machine tool slides. If this difference is high, then 'stick-slip* vibrations of the slide take place. Such stick-slip vibration, which usually occurs at low sliding speeds, affects the accuracy of the guided motion and thereby the quality of work done on the machine.

Various methods can be adopted to prevent the occurrence of stick-slip vibration in modern machine tools. Some of these are enumerated below.

(i) by using oils having polar additives;

(ii) by using rolling friction instead of sliding friction;

(iii) by mechanical preloading to relieve the guide of excessive normal pressure on it;

(iv) by using hydraulic preloading; and

(v) by using an oscillating force of ultrasonic frequency on the slide in the normal or tangential direction.

In this chapter the last two methods are discussed. An attempt is made to give a mathematical interpretation of the phenomenon occurring under a normally acting oscillating force and deduce an expression for the coefficient of kinetic friction.

Influence of Oscillating force on Stick-Slip Sliding.

Influence of oscillation on the characteristic of kinetic friction of the slide was investigated by the author.[8]

The driving mechanism of the table of a machine tool may be represented by the schematic diagram shown in Figure 16.10. In the figure, *A* represents the drive having a constant velocity υ, *B* the sliding carriage having a mass *m*, and *C* the mechanical connection between the drive and the carriage having a stiffness *k*.

Let us consider a particular case where an oscillating force $N = N_0 (1 + a \sin \omega t)$, as shown in Figure 16.10, acts on the moving carriage. Here, $a = \dfrac{\Delta N}{N_0}$ is the amplitude of vibration of the normal force, ω is equal to $2\pi f$ where f = frequency of oscillation.

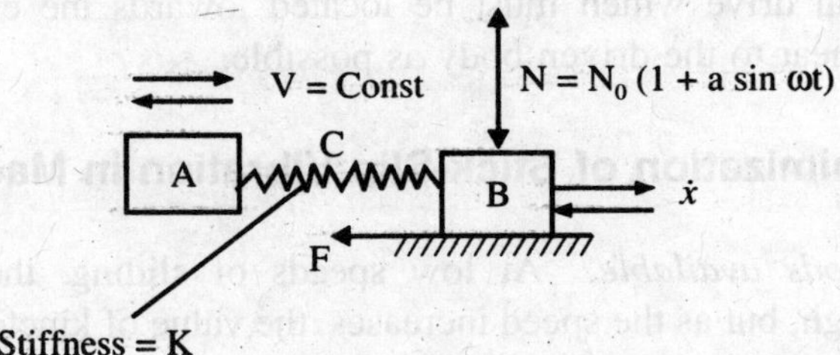

Figure 16.10.

The force of friction, which is equal to $\mu' N$, is represented by a sinusoidal curve, as shown in Figure 16.11. μ' here denotes the coefficient of friction.

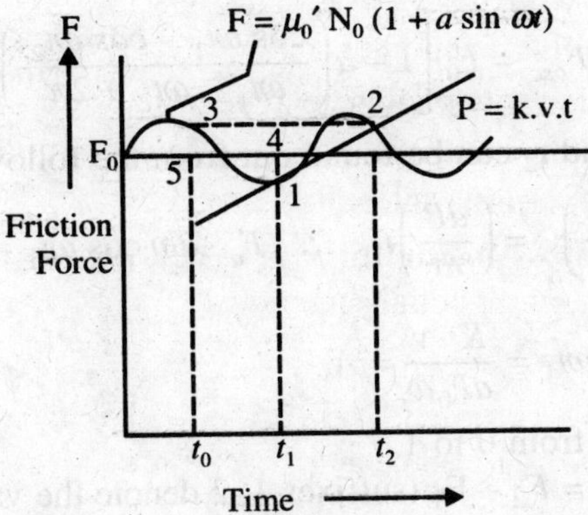

Figure 16.11.

It is assumed that *F* in Figure 16.11 is not dependent on the amplitude of stick-slip vibration.

The resisting force in the mechanical connection *C* is given by $P = kvt$, where k = stiffness of the part *C*, v = velocity of the drive, and t = time. When the force of friction is greater than the resisting force *P*, the motion of the table cannot take place. For movement of the table or for sliding, it is very necessary that $F = P$. If the curve $P = k\,v\,t$ which touches the sinusoidal curve at point 1 also intersects it at point 2, then the points 2 and 3 can be regarded as the points where jumping motion (slip in stick-slip motion) of the carriage starts. But point 1 denotes the position where' from the period of haltage or sticking starts. Consequently, the area of the curve bounded by the points 3 t_0 t_1 1 represents the work done by friction during the period of sliding.

Force of friction during, the period of sliding

$$= \int_{t_0}^{t_1} F_0\,(1 + a \sin \omega t)\,dt. \tag{16.34}$$

Average work done by force of friction F_0 = area $54t_1\ t_0$, i.e.

$$F_{osc} = \frac{\int_{t_0}^{t_1} F_0\,(1 + a \sin \omega t)\,dt.}{\int_{t_0}^{t_1} dt}$$

Since $t_0 = t_2 - \dfrac{2\pi}{\omega}$ we can write F_{osc} in the following form

$$F_{osc} = F_0\left[1 - a\left(\frac{\cos \omega t_1 - \cos \omega t_2}{\omega t_1 - \omega t_2 + 2\pi}\right)\right]. \tag{16.35}$$

The values of t_1 and t_2 can be found out from the following equations:

$$\left(\frac{dF}{dt}\right)_{t_1} = \left(\frac{dP}{dt}\right)t_1, \quad \therefore \quad F_o \cdot a\omega \cos \omega t_1 = K \cdot v$$

$$\cos \omega t_1 = \frac{K \cdot v}{aF_0\omega} = A. \tag{16.36}$$

A can have values from 0 to 1.

Again $P_2 - P_1 = F_2 - F_1$ (suffixes 1, 2 denote the values at points 1 and 2 respectively). Therefore

$$Kv\ (t_2 - t_1) = F_0 \cdot a\ (\sin \omega t_1 - \sin \omega t_1)$$

$$A\ (\omega t_2 - \omega t_1) = \sin \omega t_2 - \sin \omega t_1 \tag{16.37}$$

Equation (16.35) may be written in the form

$$F_{oso} = F_0\ [1 - a\ \psi\ (A)] \tag{16.38}$$

where ψ(A) is a mathematical function of A of trigonometric nature and makes this equation a transcendental equation.

The relation between ψ(A) and A is shown in Figure 16.12. For the sake of simplicity, this curve may be approximated to a straight line joining the two extreme points.

Under this condition, the equation of the curve will be ψ(A) ≈ 1 – A. Putting the value in equation (16.38), we get the following important relation.

$$F_{osc} \approx F_0\ [1 - a(1 - A)] \tag{16.39}$$

$$\mu' \approx \mu'_0\ [1 - a(1 - A)] \tag{16.40}$$

where $0 \le A \le 1$, and μ′ is the coefficient of kinetic friction when an oscillating force is applied.

From the above relation, it is quite evident that μ' will have its minimum value when $A = 0$, but this value increases as A or v increases.

Equation (16.40) can be written in the form

$$\mu' \approx \mu'_0\ (1 - a) + \frac{K \cdot v}{N_e \cdot 2\pi \cdot f} \tag{16.41}$$

Putting definite values in equation (16.41), a fcmily of curves could be drawn which shows the nature of kinetic coefficient of friction in relation to the velocity of sliding. Figure 16.13 shows such curves for values of $a = 0.2, 0.3, 0.4$ and 0.5. For plotting the curves, it is assumed that

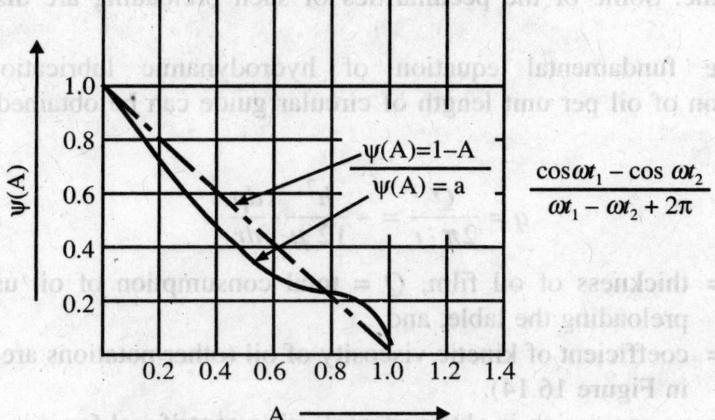

Figure 16.12.

$$\frac{K \cdot v}{N_e \cdot 2\pi \cdot f} = 0.2 \quad \text{and} \quad \mu' = 0.15$$

The characteristic of the coefficient of kinetic friction is very favourable (as can be seen from Figure 16.13) for elimination of stick-slip sliding. The possibility of slick-slip will be less as the frequency of oscillation is increased. Usually, to use this principle- in machine tools, it is necessary to have a high frequency motor for the oscillatory motlion.

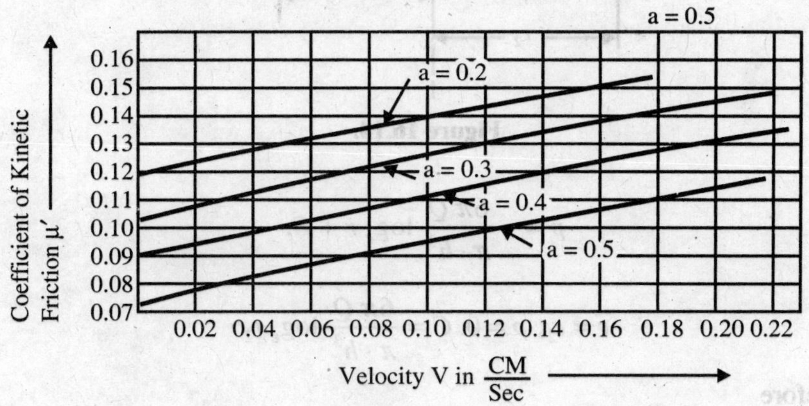

Figure 16.13.

Hydraulic Preloading

The effect of hydraulic preloading to relieve the guide of excessive normal pressure was investigated by Push[38] on the circular table mechanism of a vertical lathe. Some of the peculiarities of such preloading are discussed here.

In the fundamental equation of hydrodynamic lubrication, the consumption of oil per unit length of circular guide can be obtained in the form

$$q = \frac{Q}{2\pi \cdot r} = -\frac{h^2}{12\,\mu} \cdot \frac{dp}{dr} \tag{16.42}$$

where h = thickness of oil film, Q = total consumption of oil used for preloading the table, and

μ = coefficient of kinetic viscosity of oil (other notations are shown in Figure 16.14).

The above expression is obtained neglecting centrifugal force in the oil.

$$\frac{dp}{dr} = -\frac{6\,\pi\,Q}{\pi \cdot rh^3}$$

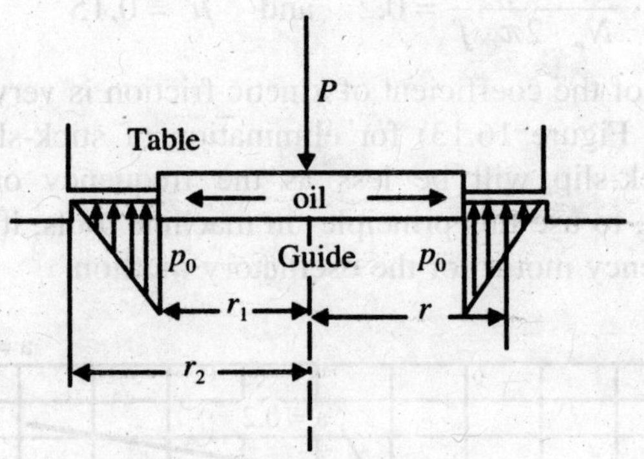

Figure 16.14.

$$p = -\frac{6\pi\,Q}{\pi \cdot h^3}\log_e r + C.$$

when $\quad r = r_2,\ p = 0.\ C = \dfrac{6\pi\,Q}{\pi \cdot h^3}\log_e r_2$

Therefore

$$p = \frac{6\pi\,Q}{\pi\,h^3}\log_e \frac{r_2}{r}$$

When $\quad r = r_1,\ p = p_0 \ or\ p_0 = \dfrac{6\mu Q}{\pi h^2}\log \dfrac{r_2}{r_1}.$

Therefore

$$p_0 = p_0 \frac{\log_e \frac{r_2}{r}}{\log_e \frac{r_2}{r_1}} \tag{16.43}$$

If $r = r_1 + x$ and $r_2 = r_1 + b$, where b = width of the circular guide, then the equation can be written in the form

$$p = p_0 \left\{ \frac{\log_e \frac{r_1 + b}{r_1 + x}}{\log_e \frac{r_1 + b}{r_1}} \right\},$$

or

$$p = p_0 \left\{ 1 - \frac{\log_e \left(1 + \frac{x}{r_1} \right)}{\log_e \left(1 + \frac{b}{r_1} \right)} \right\},$$

Again since log. $(1 + \alpha)$ can be expressed in the form

$$\alpha - \frac{\alpha^2}{2} + \frac{\alpha^2}{3}$$

the above equation can be written in the form

$$p = p_0 \left\{ 1 - \frac{\frac{x}{r_1} - \frac{1}{2}\left(\frac{x}{r_1}\right)^2 + \frac{1}{3}\left(\frac{x}{r_1}\right)^3}{\frac{b}{r_1} - \frac{1}{2}\left(\frac{b}{r_1}\right)^2 + \frac{1}{3}\left(\frac{b}{r_1}\right)^3} \right\} \tag{16.44}$$

Since in the case of heavy machines (as in the case of heavy duty vertical boring machines), the ratio $\frac{b}{r_1} < 1$ and r, is much greater than b, the above equation can profitably be reduced to

$$p = p_0 \left(1 - \frac{x}{b} \right) \tag{16.45}$$

Equation (16.45) gives the straight line characteristics of the pressure distribution.

Under the action of oil, the table will be tilted as shown in Figure 16.15. Assuming that maximum thickness of oil film under this tilting of the table is h_{max}, at any angle φ, the thickness of the film h_o, is obtained by

$$h_0 = \frac{h_{\max}}{2}(1 + \cos\varphi) = h_{\max} \cdot \cos^2 \frac{\varphi}{2}. \tag{16.46}$$

Condition for Uniform Thickness of Oil

Let Q_0 = original consumption, when thickness of oil is h_0,

Q_φ = consumption of oil when thickness is $h\varphi$

$$Q_\varphi = \int_0^{2x} \frac{h^3{}_{\max}}{96\,\mu} \frac{d\rho}{dr} \cdot r\,(1 + .\cos\varphi)^3\, d\varphi. \tag{16.47}$$

But

$$Q_\varphi = -\frac{h^2{}_0}{12\mu} \frac{dp}{dr} \times 2\,\mu \cdot r. \tag{16.48}$$

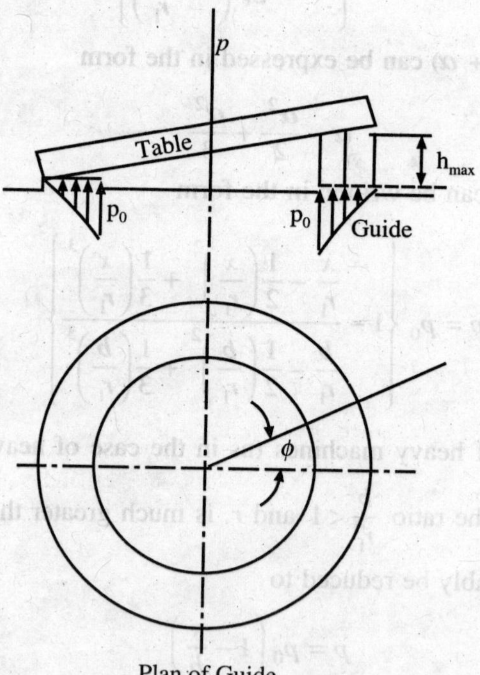

Figure 16.15

Assuming that Q_φ = Q_0 and that the law of distribution of pressure in each radial section is obtained by the relation

$$p = p_0\left(1 - \frac{x}{b}\right), \text{ we get } h_{max} = 1.47\ h_e$$

This gives the optimum thickness of oil which will allow preloading of the table without the table being tilled. In machine tools where hydlaulic preloading is used, particularly in heavy vertical boring machines, oil which is fed through the centre, as shown in Figure 16.14 flows under pressure outwards in radial directions.

This method reduces the normal pressure on the guide, and thereby the wear of the guideways. The friction characteristics obtained in this arrangement, particularly at tow sliding speeds, are favourable for the minimization of stick-slip vibration to a great extent.

By Using Oils With Polar Additives

By giving certain additives to the lubrication oil and thereby reducing its kinetic viscosity, we reduce considerably the difference between static and kinetic coefficient of friction.[9]

Positional error due to stick-slip

$$\Delta S \propto \frac{f_s - f_{kin}}{K}$$

where f_s, and f_{kin} are static and kinetic coefficient of friction respectively, and K–rigidity of the mechanism, connecting the sliding unit to the driving unit.

In the graph shown below (Figure 16.16) the author has shown the characteristic curves of f_{kin} dependent on V_s the velocity of sliding of the table, when the table and the guides are lubricated by three oils *A*, *B* and *C* having coefficients of kinetic visc sity 0.075, 0.021, 0.007 respectively, in Kg sec/m^2 at t = 20°C.

In the case *C* which is known as anti-stick-slip oil the difference between the static and kinetic coefficient of friction is the minimum possible, thereby eliminating the possibility of stick-slip sliding.

16.8 Vibration Isolated Tool Holders

One of the various msthods of combating vibration, so as to reduce its disastrous effects on the tools under dynamic condition, is to use specially designed tool holders.

The author has shown below three such designs which serve useful purpose for eliminating or isolating tool vibration. The first one

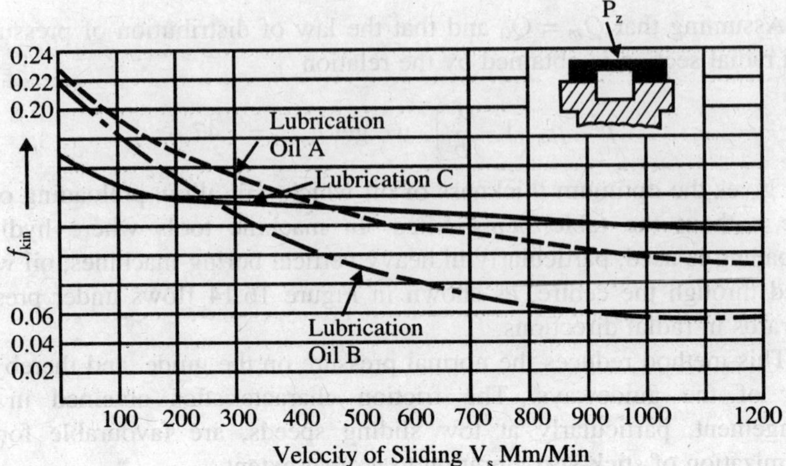

Figure 16.16

(Figure 16.17) may be used in precision boring machine. The boring bar is designed hollow with arrangement for fitting it with oil damper and spring loaded plunger. This type is known as Hydrodamper. The other two sketches shown in Figures 16 18 and 16.19 arc used for isolating the vibration of the lathe tools, For precision lathe work, the use of such tools are found to be very much useful.

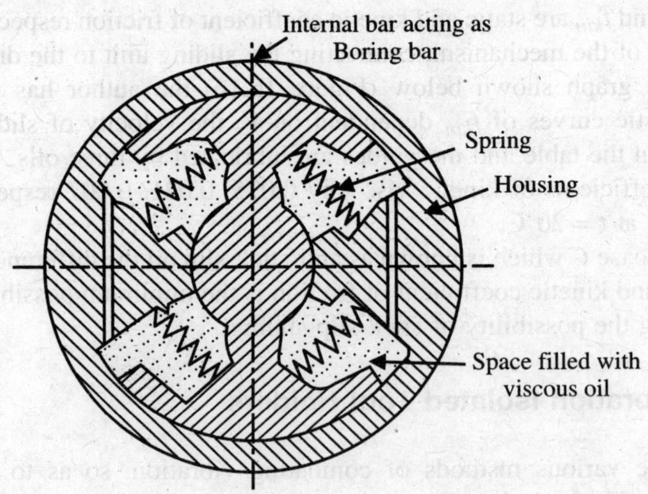

Figure 16.17 Shock Absorbing Boring Bar.

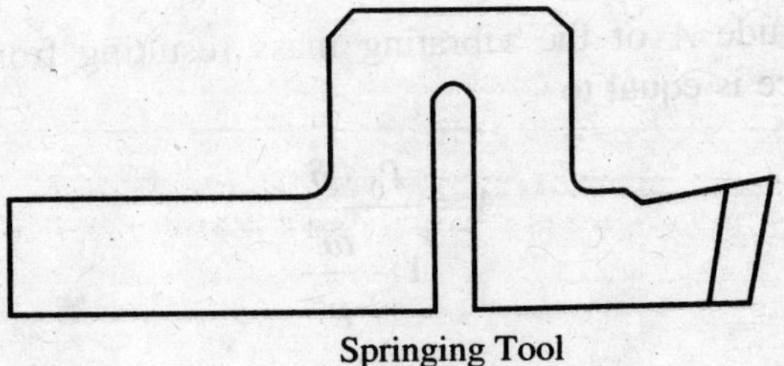

Springing Tool

Figure 16.18

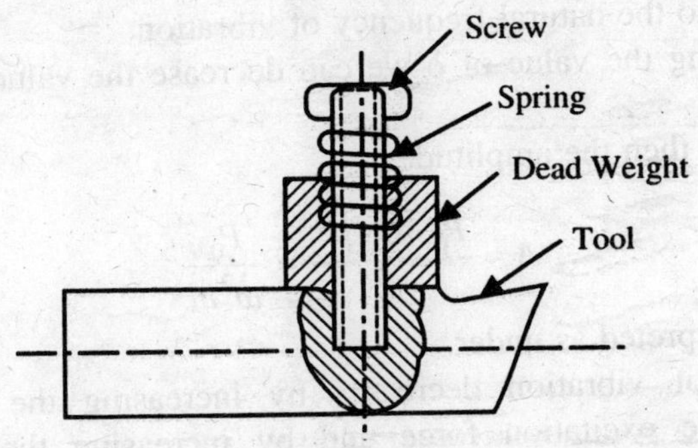

Figure 16.19

Let us represent a simple vibrating system as shown by skeleton diagram (Figure 16.20). Excitation force is denoted by

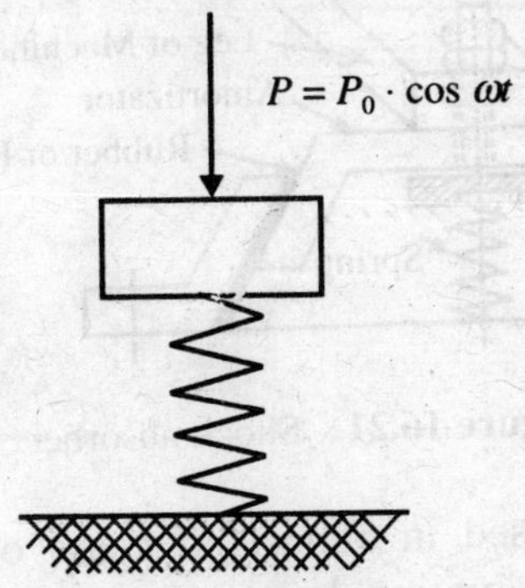

Figure 16.20

P and is an harmonic excitation force represented by P_σ coivsr where P_0 – maximum magnitude of the excitation force

$$\omega = 2\pi f,$$

where *f* is the frequency of oscillation of the exciting force.

The amplitude A of the vibrating mass resulting from the harmonic excitation source is equal to

$$A = \frac{P_0 \cdot \delta}{1 - \frac{\omega^2}{p^2}},$$

where
$$p = \frac{1}{\sqrt{m\delta}},$$

This p is equal to the natural frequency of vibration.

By increasing the value of δ we can decrease the value of the natural frequency p.

If $p \angle \angle \omega$, then the amplitude

$$A \approx \frac{P_0 \cdot \delta \cdot p^2}{\omega^2} = \frac{P_0}{\omega^2 m}$$

This can be interpreted as under.

Amplitude of vibration decreases by increasing the value of the frequency of the excitation force and by increasing the mass of the amortizator "m".

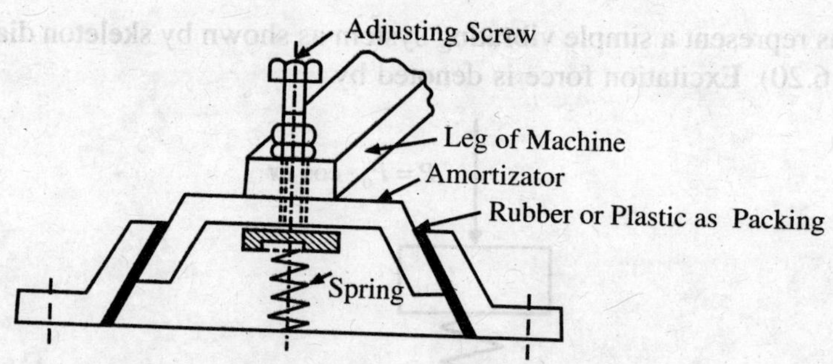

Figure 16.21 Shock absorber.

This principle is utilised in almost all cases of amortizaton used in machine installation, tool vibration dampers etc.

CHAPTER 17

Microdisplacements in Machine Tools

17.1 Introduction

Microdisplacements in machine tools are needed for precision movement of the tools related to the work. Various techniques have now been in use to obtain the precision movement of tool holder or any traversing executive organ of the machine tool.

The authors prefer to present a few important methods which could have an effective utilisation in light duty or heavy duty precision machine tools.

17.2 Magnetostrictive Drive

With reference to Figure 17.1, there are some materials which when subjected to electromagnetic potential undergo finite change in

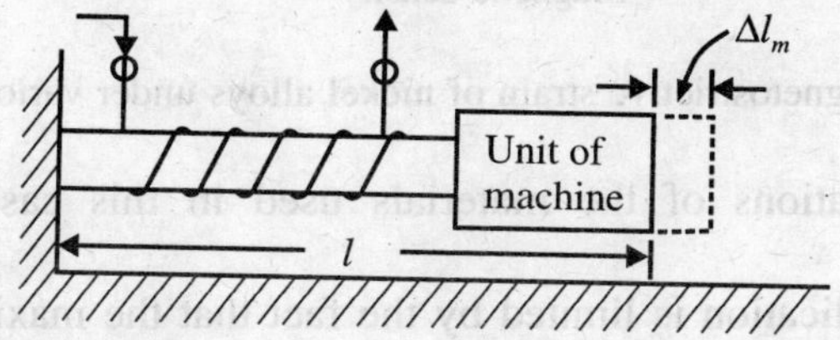

Figure 17.1 Magnetostrictive drive.

length. In some cases this Δl_m i.e. the change in length is positive and in some cases this may be negative. Accordingly we get positive magnetostriction or negative magnetostriction

$$\Delta l_m = \lambda \cdot l$$

where λ is known as related magnetostrictive strain. This of course depends upon, apart from the property of the material, the actual number of ampereturns. Figure 17.2 shows various curves showing $\lambda - H$, where H is the ampere turns/cm for various magnetostrictive alloys.

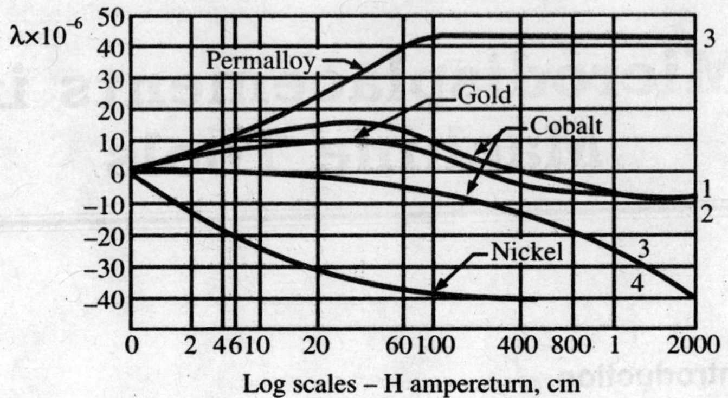

Figure 17.2 λ-H curves for various alloys.

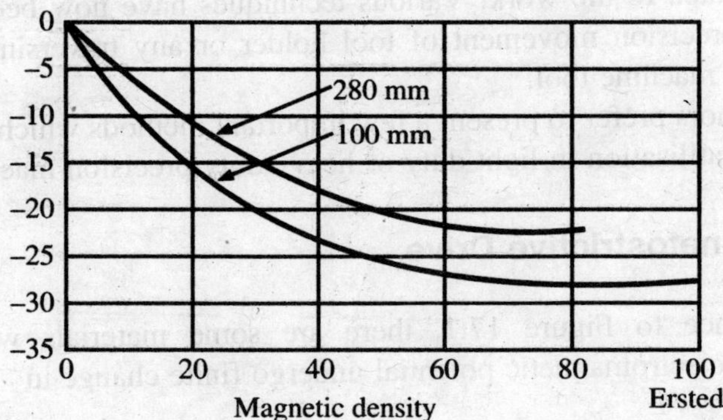

Figure 17.3 Magnetostrictivc strain of nickel alloys under various base lengths.

The specifications of the materials used in this case are given in Table 17.1.

Practical application is limited by the fact that the maximum extension over a length of 100 mm, using magnetostriction property lies between 6 and 7 microns. Magnetostrictive property also varies with temperatures widely even within a narrow range.

Considering the traversing mechanism with the help of hydraulic cylinder and further microdisplacement by magnetostriction, the model can be simply represented as above. Here the portion of the length under the magnetic field being l_1 between the poles and the body has a stiffness K (elastic stiffness). The microdisplacement of the unit can be expressed as:

$$a = \lambda_s \cdot l_1 - \frac{F}{ES} \cdot l_2 \tag{17.1}$$

Table 17.1 Specifications for materials commonly used

Material	*Composition*	*Density gm/cm³*	*Magneto-strictive strain* λ_s	*Young's modulus kg/cm²*	*Ultimate tensile strength kg/cm²*
Nickel	–	8.9	-35×10^{-6}	2.1×10^6	47
Cobalt-Vanadium alloy (Permalloy)	49% Co, 1.5 – 1.8% Va Rest Fe	8.08	70×10^{-6}	2.2×10^6	45
"	60% Co Rest Fe	8.25	90×10^{-6}	2.2×10^6	67

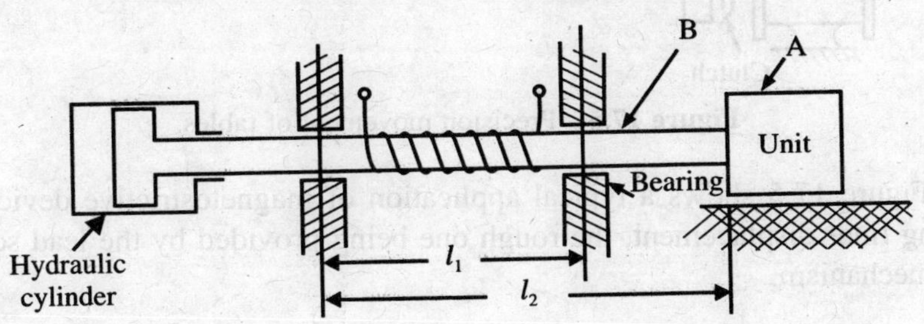

Figure 17.4 A block diagram for practical case.

where F is kinetic force of friction between the bed and the unit, sliding over the bed, acting on the body as axial force; and S is traverse area of cross section of the connecting unit B. Also

$$a_{\min} > \frac{F_0}{K}; \text{ and } a_{\max} \leq \left(\lambda_s \cdot l_1 - \frac{F}{K}\right) \tag{17.2}$$

where F_0 and F are static and kinetic force of friction respectively.

From Eqn (17.2): If $\quad a = 0, \lambda_s \cdot l_1, = \dfrac{F}{ES} \cdot l_2 \quad$ (17.3)

$$\sigma_{per} = \frac{F}{S} = \lambda_s \cdot \left(\frac{l_1}{l_2}\right) \cdot E \tag{17.4}$$

where σ_{per} is the maximum permissible value of stress for design purposes. In practice the exact amount of microdisplacement of the unit may be expressed as:

$$a = \lambda l_1 - \frac{F}{ES} l_2 \pm \frac{F}{ES} \cdot l_2 \tag{17.5}$$

Considering the possibility of dispersion of the dimension and hence the maximum positional error due to this micro movement may be of the order of

$$\delta_{max} = 2 \cdot \left(\frac{F}{ES}\right) \tag{17.6}$$

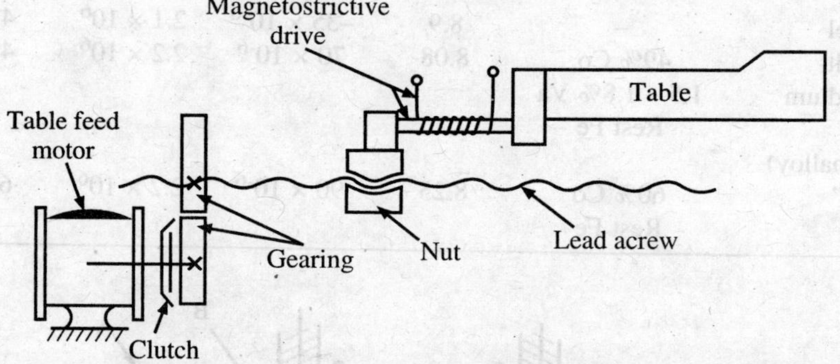

Figure 17.5 Precision movement of tables.

Figure 17.5 shows a typical application of magnetostrictive device for giving finer displacement, the rough one being provided by the lead screw-nut mechanism.

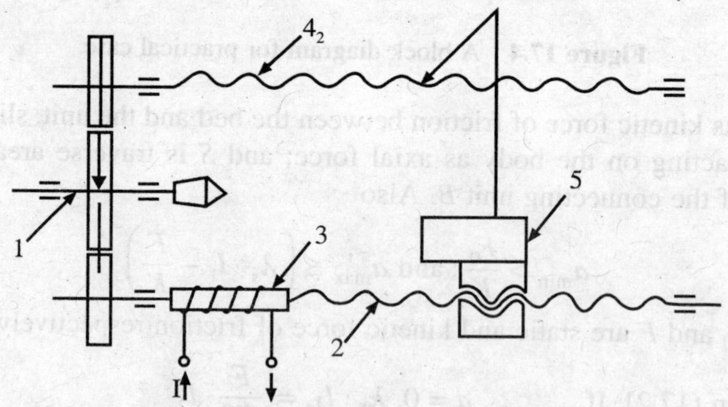

Figure 17.6 Lead screw error compensation [73]

In Figure 17.6 has been shown a typical schematic diagram for compensation of lead screw error commonly known as 'back lash' by using magnetostrictive device.

The tool carriage 5 requiring precise movement is provided with lead screw-nut mechanism as shown in 2. It. has a portion of length (magnetostrictive) 3 made of material having good magnetostriction properties. By passing different ampere-turns (dependent on current 1) the precision movement of carriage 5 is possible upto a fraction of one micron. The spindle unit with gear trains is shown as 1 in the above figure. The screw 4 is a master screw. The lead screw for feed movement 2, and the accurate (standard) master screw with minimum possible error 4, move synchronously to provide micro-traverse of the unit 5. Hence any change due to error of the lead screw at a particular part is sensed in relation to master screw and this feeds back to pass requisite amount of current (i.e. ampere-turns-since number of turns is constant) through 3, which compensates the error of the lead screw. This system may be conveniently used in cases where table requires accurate coordinate movements, such as in NC machine tools or in jigboring machines.

17.3 Thermodynamic Drive[13]

Sometimes the carriage has its support (connecting link) fitted with electrical coil .inside through which current is sent. Depending upon the temperature attained the support B in Figure 17.7 undergoes change in length causing precision positional displacement of the carriage.

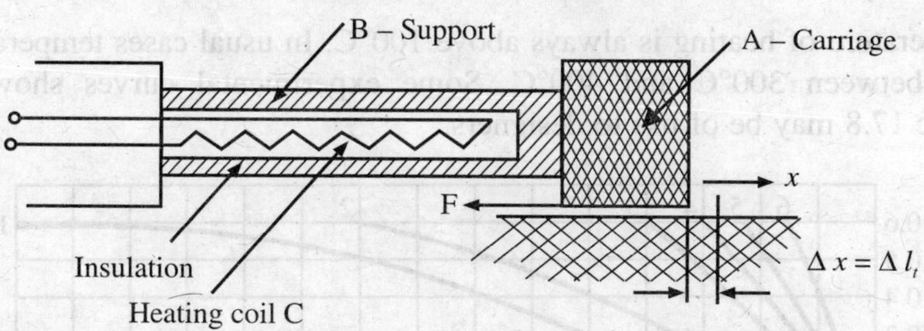

Figure 17.7 Block diagram showing thermodynamic drive of feed unit.

With reference to Figure 17.7, we know

$$\Delta l_t = \alpha \cdot l \cdot \Delta t \tag{17.7}$$

where α – Coefficient of linear expansion

l – Length of unit, containing the coil for expansion.

Δt – temperature rise.

If C_1 is a coefficient defined by the characteristics of the force of friction dependent on the velocity of movement and damping in the system etc., then

$$a = \Delta l_t \pm \frac{C_1}{K} \tag{17.8}$$

'a' denoting resulting positional displacements

$$a - \Delta l_t = \pm \frac{C_1}{K} \tag{17.9}$$

Dividing both sides by Δl_t and using equation 17.7 we get:

$$\frac{a - \Delta l_t}{\Delta l_t} = \frac{\pm C_1}{\alpha \cdot l \cdot \Delta t \cdot K} \tag{17.10}$$

Assuming the axial force = F and the area of transverse cross section of the connecting link as S we get

$$\frac{F}{SE} = \frac{\Delta l_t}{l} \tag{17.11}$$

where E is the Young's modulus of the material.

Thus $$K = \frac{F}{\Delta l_t} = \frac{SE}{l} \tag{17.12}$$

or in other words $$K \cdot l = S \cdot E \tag{17.13}$$

Substituting eqn (17.13) in eqn (17.10) we get

$$\frac{a - \Delta l_t}{\Delta l_t} = \pm \frac{C_1}{\alpha \cdot \Delta t \cdot SE} \tag{17.14}$$

Temperature of heating is always above 100°C. In usual cases temperatures vary between 300°C and 400°C. Some experimental curves shown in Figure 17.8 may be of use to designers.

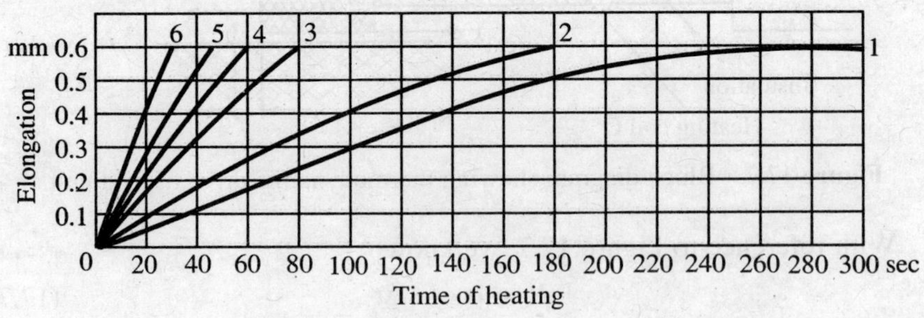

Figure 17.8 Displacement characteristic based on various feed velocity 1–0.1 mm/min; 2–0.2 mm/min; 3–0.4 mm/min; 4–0.55 mm/min; 5–0.7 mm/min; 6–1.4 mm/min;

17.4 Minimisation of Positional Error by the Use of Oscillating Normal Force

Earlier in Chapter 16, we have seen how stick slip vibration in machine tools and hence the positional error, caused thereof, may be minimised to a great extent by using oscillating force of ultrasonic frequency (Ref. to para 16.7 and Figure 16.10). With the notations same as in Chapter 16, we may write the oscillating force of friction F_{osc} as in eqn. (16.39)

$$F_{osc} = F[1 - a(1 - A)] \tag{17.15}$$

where $A = \dfrac{KV}{aF_0\omega}$ and lies between 0 and 1, F_0 being the value of the force of friction while the sliding velocity starts from 0.

Frictional work done $= F_{osc} \cdot V$

$$= [F_0V - F_0aV\,(1 - A)] \tag{17.16}$$

Since $A < 1$, we get $V < \dfrac{aF_0\omega}{K}$ (17.17)

For minimum frictional energy in the drive under such a case

$$\phi(Z) = [F_0aV(1 - A)] \text{ should be minimum} \tag{17.18}$$

Since $A = \dfrac{K \cdot V}{aF_0\omega}$ our objective, in effect, will be to maximise

$$|\,\phi(z)\,|_{max} = F_0 \cdot a \cdot \omega - \frac{KV^2}{\omega} \tag{17.19}$$

This is satisfied when $F_0a = \dfrac{2KV}{\omega}$ (17.20)

$$V = \frac{F_0a\omega}{2k} \tag{17.21}$$

$$A = \frac{KV}{aF_0\omega} = \frac{1}{2} \tag{17.22}$$

This gives therefore the optimum velocity of the drive of the carriage to avoid stick slip in the form

$$V_{opt} = \frac{F_0a\omega}{2K} \tag{17.23}$$

Substituting the above values in eqn. (17.15) we get

$$F_{osc} = F_0[1 - 0.5a] \tag{17.24}$$

or coefficient of friction $f_{osc} = f_0[1 - 0.5a]$. (17.25)

The values of 'a' under the operating condition should be as low as possible. The 'a' normally lies between 0 and 1. It is not difficult to show that, under a predetermined value of $(f_0)_{v \to 0}$, Δf is the least when 'a' lies between 0.8 and 1.

17.5 Recirculating Ball Screws

In the Chapter 8, we have already discussed the increasing uses of recirculating ball screws in numerically controlled machine tools to avoid positional error due to stick slip. Minimum variation in Δf, with high contact stiffness and adequate means of preloading have resulted in the extensive use of such units to reduce positional error or error in microdisplacement. For optimisation of the design parameters of such a ball screw nut pair, to be used in a machine, we may proceed as follows:

Let us denote j as rigidity under an axial load Q, i as number of threads in which balls recirculate, R_0 as the effective radius of the pitch line, R_1 as the radius of the balls and δ the total contact deformation.

We know (from Chapter 8):

$$J \propto Q^{1/3} \cdot i^{2/3} \cdot R_0^{2/3} \cdot R_1^{-1/3} \qquad (17.26)$$

From tne designers' point of view such a pair should work with high rigidity so that error in microdisplacements becomes minimum. This fact becomes, therefore, a guided objective, while designing precision traversing mechanism.

From the technical constraints and prevalent practices we encounter a few other restrictions, such as

$$\left.\begin{aligned} i &\leq 4 \\ \frac{R_0}{R_1} &\leq 12 \end{aligned}\right\} \qquad (17.27)$$

From the loading consideration also we have restrictions, such as maximum load and maximum contact deformation not to exceed a certain amount. For example, we may like to select, depending upon the requirement, a value of load given by

$$Q \leq 2000 \text{ kg} \qquad (17.28)$$

And under this load, the total contact deformation not to exceed 20 microns

$$\delta \leq 20 \text{ microns} \qquad (17.29)$$

Now the case may be solved by simple linear programming problem, taking logarithmic transformation. Or the problem may be solved as it is by geometric programming, to determine the optimum values of the design parameters to obtain maximum rigidity under given conditions.

17.6 Surface Topography and Contact Stiffness

Elastoplastic deformations of the surface asperities in contact, particularly at moveable joints, cause a change in the overall stiffness of the structure and hence contribute to error in microdisplacement. Frictional behaviours of the sliding units are dependent on the nature of the contacting surfaces, their asperities deformation and the actual area of contact. Normal, elastic deformation of the contact surface[74][75] is given by

$$\delta_e = \frac{1}{m} \cdot \ln\left(\frac{p}{p_m}\right) \tag{17.30}$$

Here m is a constant for particular pair of contacting surfaces, but dependent on the surface irregularities of the joint faces, mean pressure on the interface and flow stress of the material.

p and p_m are pressure and mean pressure at the interface respectively.

The total elastoplastic deformation of the surface asperities, as advocated by Basu and Chikate[75], takes the form

$$\delta_p = \left(\frac{h_1 \cdot h_2}{h_1 + h_2}\right) \times \frac{2p}{\sigma_f} \times \left(\frac{1-p}{2\sigma_f}\right) \tag{17.31}$$

where h_1, h_2 are the C.L.A. values of the interfacing surfaces and σ_f is the flow stress of the material. In their work, the authors[75] estimated the value of 'm' theoretically as :

$$m = \left(\frac{h_1 + h_2}{2h_1 h_2}\right) \times \left[\frac{\sigma_f^2}{p_m\,(\sigma_f - p_m)}\right] \tag{17.32}$$

and verified the.validity of the same by experimenting with various mating surfaces having variable working parameters.

But theoretically to define and predict the behaviour of the complete surface, analysis of infinite number of surface profile records are necessary. The exact topography of a mating surface is a function of numerous parameters, such as material, cutting conditions, tool geometry, rigidity of machine tool workpiecc system etc., and contains both periodic and random character. Random content which is a probabilistic time invariant function can only be explained by statistical methods. One can get the idea of ergodicity of a surface by analysing the ensemble of surface profile records for autocorrelation function and superimposing the autocorrelograms, so obtained. If the decay rate or correlation length of all autocorrelograms is the same, the surface is ergodic. Experimental work carried out by Agarwal, Patki and Basu[76][77] shows that all machined surfaces are rot ergodic and hence single profile record is not enough for the complete characterisation of the surface. However, non-ergodic surfaces can be analysed after filtering off low-frequency waves of varying amplitudes and by introducing mean and mean surface correction. According to Chikatc[75] the asperities undergo elastoplastic deformation under load and with the increase in the load the deformations tend to become plastic. But according to Agarwal[76][77] theoretical calculation shows that all engineering surfaces undergo plastic deformations even under small load and even when the bulk material is within elastic limit.

All these recent findings have changed the basic concept ot contact compliance and opened up a vast field for machine tool designer to investigate into. Uptill now the behaviours of contact surfaces have been studied under static load. It is indeed interesting to us to know the same under dynamic loading conditions, which will enable us to predict the parametric influence on moveable joints in machine tools.

17.7 Error Due to Stick Slip Motion[73]

Referring to Figure 17.9, if the driving unit moves by an amount *a* and the carriage (mass *m*) connected with elastic stiffness *K* moves

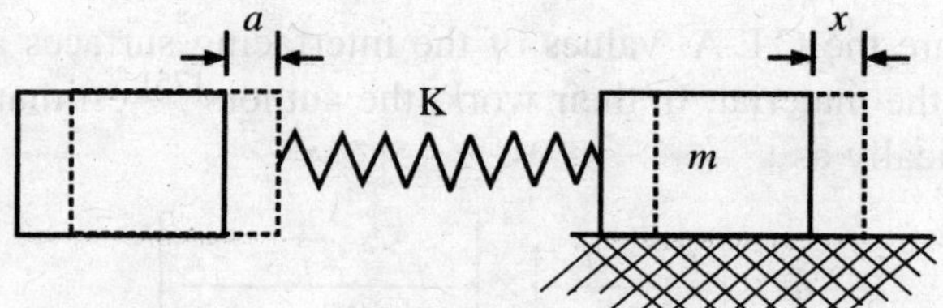

Figure 17.9 Displacement of executive organ.

by an amount *x*, normally less than *a*. Accuracy in the positional displacement is denoted by

$$\Delta = a - x \tag{17.33}$$

It is needless to say that

$$a < \frac{F_0}{K} \tag{17.34}$$

where F_0 is force of static friction (limiting force of friction). Under such condition, the movement of the carriage cannot take place and $x = 0$. Under such environment, if we assume that the initial deformation is absent, then the positional error is given by

$$\Delta_1 = \frac{F_0}{K} \tag{17.35}$$

In overall motion system given in Figure 17.9, in regard to combination motion of the carriage, we may consider a few cases for obtaining displacement under actual condition.

If the drive has a velocity V less than the critical velocity, shown in Chapter 16 (eqn. 16.28), the carriage unit will move with stick slip as shown in Fig. 17.10.

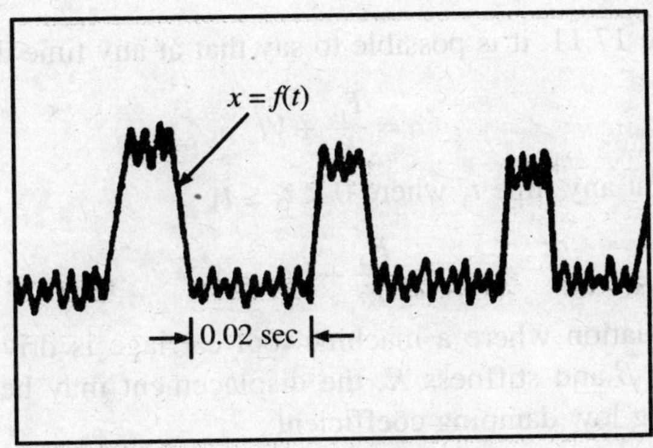

Figure 17.10 Oscillogram trace of velocity less than its critical value.

The displacement in such case is given by[73]

$$S = V(t_1 + t_2) = \frac{\Delta F}{K} \cdot f_0 \tag{17.36}$$

where t_1 and t_2 are time for slipping and sticking respectively

ΔF = Difference between the static and dynamic force of friction at velocity V

$f_0 = f\left(\theta, \dfrac{V}{V_c}\right)$ that is, it is function of θ and ratio of velocity to critical velocity (see chapter 16). θ depends on damping characteristics and stiffness of the drive

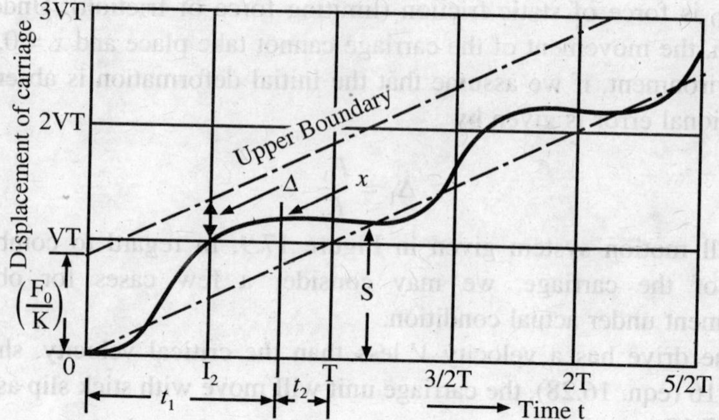

Figure. 17.11 Displacement when the velocity is small.

From Figure 17.11, it is possible to say that at any time displacement

$$a = \frac{F_0}{K} + Vt \tag{17.37}$$

Positional error at any time t_x where $0 < t_x \leq t_1$

$$\Delta_x = \frac{F_0}{K} + Vt_x - S \tag{17.38}$$

In a realistic situation where a machine tool carriage is driven by a drive having damping β and stiffness K, the displacement may be expressed as under considering low damping coefficient

$$x = \frac{\Delta F}{K} - \frac{\Delta F}{K} e^{-\theta \omega t} (\cos \omega t + \theta \sin \omega t) \tag{17.39}$$

During the period of stoppage (i.e. $\omega t = \pi$) we get

$$x = \frac{\Delta F}{K} (1 + e^{-\pi\theta}) \tag{17.40}$$

Under such a case, the positional error is given by

$$\Delta_x = \frac{F}{K} - \frac{\Delta F}{K} e^{-\pi\theta} \tag{17.41}$$

Considering the entire cycle of stick and slip

$$b = \Delta_1 - \Delta_2$$

$$= \frac{F_0}{K} - \frac{F}{K} + \frac{\Delta F}{K} e^{-\pi\theta} \tag{17.42}$$

$$= \frac{\Delta F}{K} (1 + e^{-\pi\theta}) \tag{17.43}$$

In the absence of damping ($\theta = 0$) and hence the resulting positional error

$$b = \frac{\Delta F}{K}(1+1) = \frac{2.\Delta F}{K} \tag{17.44}$$

In fact at any instant resultant positional error will vary between 1 and 2. We may say

$$b = C_b \cdot \frac{\Delta F}{K} \tag{17.45}$$

where $C_b \doteq$ any value between 1 and 2.

Equation (17.43) is very important to machine tool designer since it indicates the maximum magnitude of positional error.

CHAPTER **18**

New Concepts in Machine Tools Design

18.1 Introduction

With the increasing need of precision movement, easy operation and long life and durability, the production engineers have been faced with a tremendous challenge in regard to conceptual as well as realistic transformation of fruits of technological innovation into the manufacturing units to make them technoeconomically viable. Earlier a designer of machine tool was mainly concerned with the functional design based on strength calculation, and basic requirement of dimensional accuracy. Even the simple rules of value analysis and lower cost of production were not taken very seriously. Faced with inroads of competition resulting from innovations in electronics and materials science, each manufacturer of machine tool today has to be extra cautious to pass on to the customer some extra benefit in regard to (i) process capability, (ii) reliability even after long use, (iii) minimisation of error in microdisplacement, and (iv) longer technoeconomic life. It is more so needed for the machine tool manufacturer to get into race for its survival, if not for its advancement. Sophistication has taken the shape of essentiality today, when one thinks of multi-axes machining of complex profiles using NC or CNC machine. These are certainly the machines, which are going to have wide applicability in few years to come, and a knowledge in regard to their technology of manufacture and exploitation is needed to absorb the technology already developed as well as to do further work on the same suited to our specific needs.

18.2 Probability Concept in Design

The basic concept of process variability could be denoted by $PV = 6\ \sigma$, σ being the standard deviation of the process as shown in Figure 18.1(a). We also know that the process tolerance, viz. the maximum range of variation permissible for the size of dimension in a particular process $P_{tx} = p_i \cdot T_x$, where *pt*, is known as tolerance factor and T_x design tolerance for the dimension *X*. The p_i values depending upon the process capability of machine, often vary between 0.6 and 1.0. Designer's specification in regard to tolerancing of a particular part for a functional requirement must be based on the machine available and its process capability known, this, therefore,

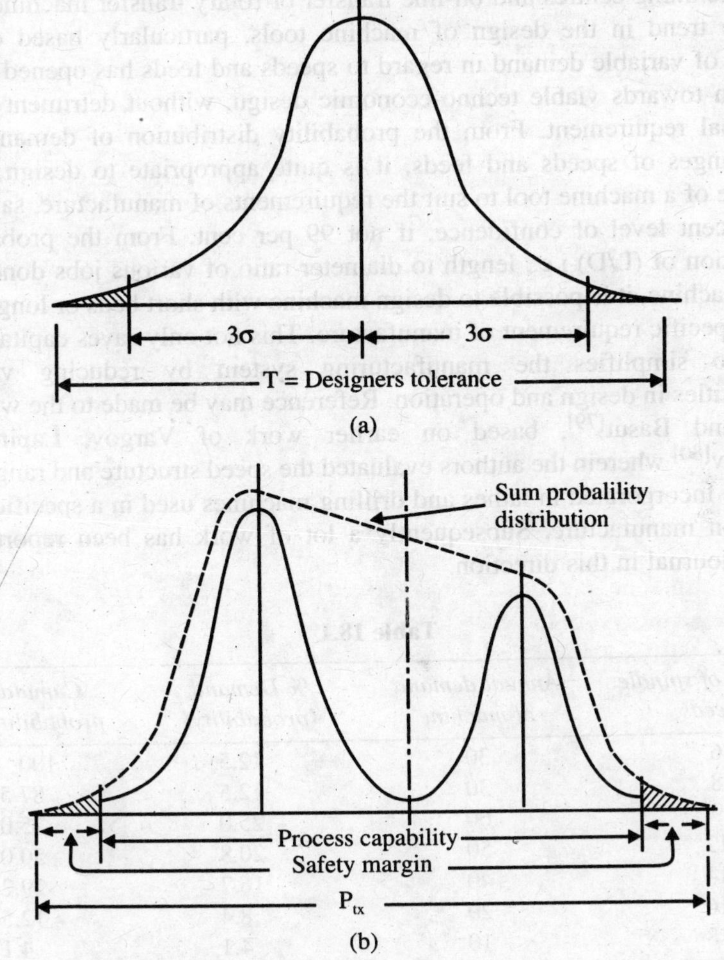

Figure 18.1 Process capability and design tolerance.

constitutes an important parameter, almost dictating the type of the manufacturing machine or its initial design specification. But each machine tool today is a complex one consisting of units involving time dependent variable parameters, each one of which has a specific probability distribution independent of the other, and unit having multiple tool posts carrying tools having different wear rates, causing different temperature variations apart from other systematic time dependent deviations. This may result in probability sum distribution, thereby giving different magnitudes of process tolerancing and safety margins at the higher and lower side respectively. Such type of time dependent variable factors normally cause deviation in the process base as well (Figure 18.1(b)).

This has unlimited application, while designing automated machines with machining centres and on-line transfer or rotary transfer machines.

The trend in the design of machine tools, particularly based on the concept of variable demand in regard to speeds and feeds has opened a new direction towards viable techno-economic design, without detriment to the functional requirement. From the probability distribution of demands for some ranges of speeds and feeds, it is quite appropriate to design speed structure of a machine tool to suit the requirements of manufacture, say with 95 per cent level of confidence, if not 99 per cent. From the probability distribution of (L/D) i.e.. length to diameter ratio of various jobs done on a lathe -machine, it is possible to design machine with short beds or long beds, to suit specific requirement of manufacture. This not only saves capital cost, but also simplifies the manufacturing system by reducing various complexities in design and operation. Reference may be made to the work of Mitra and Basut[79], based on earlier work of Vargov, Lapin and Nafedeev[80] wherein the authors evaluated the speed structure and range that could be incorporated in lathes and drilling machines used in a specific field of wagon manufacture. Subsequently a lot of work has been reported in various journal in this direction.

Table 18.1

Number of spindle speeds	*Annual demand of machine*	*% Demand (probability)*	*Cumulative probability (%)*
6	30	12.5	100
8	30	12.5	87.5
9	60	25.0	75.0
12	50	20.8	50.0
14	40	16.7	29.2
16	20	8.4	12.5
18	10	4.1	4.1

The example quoted below will enumerate right type of machines based on cumulative probability of demand. The table indicates the the demands, of various drilling machines, with specific reference to a particular manufacturing programme of the customers, as faced by machine tool manufacturer.

Following the technique of decision making[81]

If marginal profit in producing one unit above the stipulated production $= \phi_p$

Marginal loss if this unit is not sold $= \phi_L$

Probability that the marginal product be sold $= p$

Probability that the marginal product will not be sold $= (1 - p)$

$$p \times \phi_p = (1 - p)\, \phi_L \quad \text{or} \quad p = \frac{\phi_L}{\phi_p + \phi_L}$$

In case ϕ_L = Rs. 500 and ϕ_p = Rs. 1000, $p = 33.33\%$

From the table we thus select cumulative probability 50 per cent or more and hence machines having speed range from 12 upwards.

The above example shows a typical method of taking strategic management decision based on probabilistic pattern of demand and is certainly valuable to machine tool manufacturer.

18.3 Unified Systems Approach to Machine Tools Problems[82]

System concept may be regarded as "organised technological common sense". System analysis is usually done to find out the process of transformation of input into output within a set of restraints and environment. 'Systems Approach' requires following systematic steps to be followed:

(i) Identification of the system with proper location, commonly known as 'systems envelope' or 'control surface'.

(ii) Description of the system. Here we classify the (a) systems function and (b) systems structures.

(iii) Apply (ii) in subsystems as well, in proper relationship to the system.

By systems functions we mean, firstly, separation of the system from the environment by proper choice of systems envelope; secondly compilation of all inputs and outputs and their functional relationship.

By systems structures, we mainly identify elements of the system/ subsystem-their properties relevant to the problem and find out their characteristic interdependence and interactions.

A machine tool may be considered as a system having large'num-ber of sub-systems interconnected in series and parallel. H. Czichos has done considerable work in analysing various tribo-mechanical units through systems approach.

While analysing a machine tool, we may divide the system into several sub-systems, viz. Gear Box, Slideways and Carriage, Tool-post, etc. Identify the system or sub-system through structure. Let us consider a gear box consisting of gears shafts, bearings, etc. each one having certain properties and interrelationship amongst them propertywise. Thus a structure may be denoted as a set.

$$S = \{[A], [B], [R]\} = \text{set} \tag{18.2}$$

$[A]$ = Subset of elements $[a_1, a_2..a_n]$
$[B]$ = Subset of relevant properties $[p_1, p_2...p_n]$
$[R]$ = Subset of relations $[r_1, r_2...r_k]$

Systems envelope gives a constrained enclosure for converting inputs into outputs.

Let us assume that the two elements of a A are (i) carriage and (ii) slideway, over which the carriage moves. Properties are (i) type of material of either element, (ii) surface roughness at the interface of each element, (iii) velocity loss in gears .and (iv) wear due to pitting etc. Inter-relationship may be (i) pattern of variation of fiction force depending upon contact pressure, velocity of sliding, type of lubrication, etc. (ii) wear rate based on same parameters, such as velocity of slide, contact pressure, surface roughness, lubrication etc. These are mostly time dependent values. In Figure 18.2(a) has been shown typical wear values of such unit in simulated test set up. Curve 1 is the mean wear pattern curve showing running in wear, steady wear and disastrous wear, while curves 1 and 2 are the same for, say, two extreme conditions obtainable on a machine tool. The shaded portion shows the zone of progressive wear based on time. Assuming a suitable value of permissible wear, as shown in Figure 18.2(a), we get functional life span. Experimentally determined failure density function and cumulative probability of failures are not very difficult to conceive.

The overall life of machine is nothing but a system having many such constituent sub-systems, each one of which could be analysed to know its separate, cumulative probability curve of failure based on life, as shown in Figure 18.2(b). From this it is possible to know the functional life of the system, i.e. the machine, by using random number simulation approach (Monte-Carlo method).

In analysing the micro movement of slides very often we come across errors due to (i) manufacturing error of transmission screws–pitch and

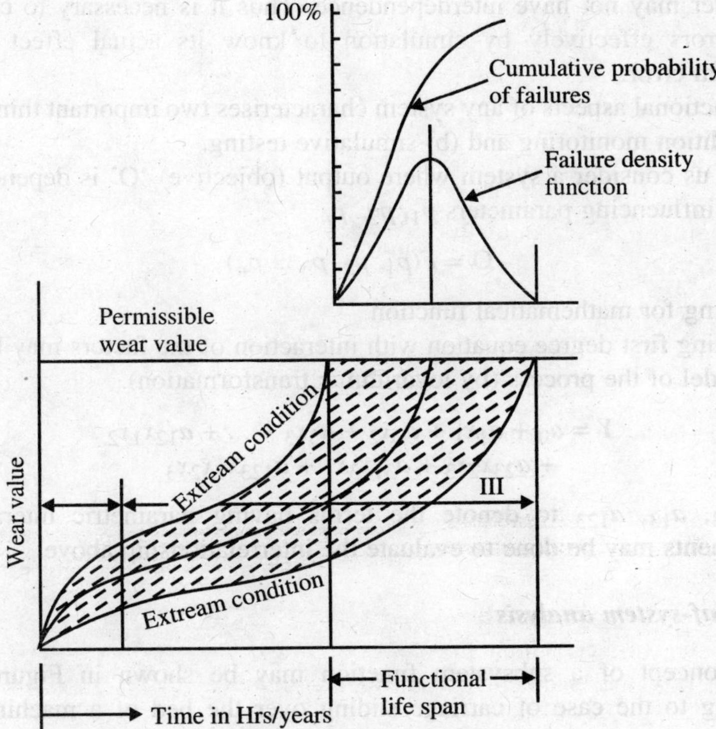

Figure 18.2(a) Wear and failure function

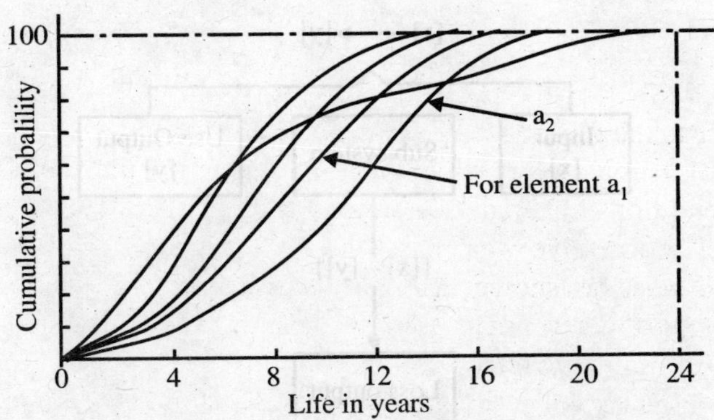

Figure. 18.2(b) Cumulative probability of failure.

dimensional error in effective diameter, (ii) frictional behaviour and relaxation, (iii) super-imposed vibration; sometimes of a self-excited type, (iv) errors in surface topography and due to contact compliance. In many cases such errors may be systematic, or random or both. Each contributing parameter may not have interdependence. Thus it is necessary to combine such errors effectively by simulation to know its actual effect on the positional error.

Functional aspects of any system characterises two important things, viz. (a) condition monitoring and (b) simulative testing.

Let us consider a system where output (objective) 'O' is dependent on various influencing parameters $p_1, p_2 ... p_n$.

$$O = F(p_1, p_2, p_3 \ldots p_n) \tag{18.3}$$

F standing for mathematical function

Taking first degree equation with interaction of the factors may betaken as a model of the process (by logarithmic transformation).

$$\begin{aligned} Y = a_0 + a_1x_1 + a_2x_2 + a_3x_3 + \ldots + a_{12}x_1x_2 \\ + a_{23}x_2x_3 + a_{31}x_3x_1 + a_{123}x_1x_2x_3 \end{aligned} \tag{18.4}$$

a_{12}, a_{23}, a_{13}, a_{123} to denote the terms having parametric interactions. Experiments may be done to evaluate the interrelationship above.

Basic suif-system analysis

Basic concept of a subsystem function may be shown in Figure 18.3. Referring to the case of carriage sliding over the bed of a machine tool, functional characteristics of the system may be dilated as shown in Figure 18.4, which basically represents an 'Input-Output' analysis.

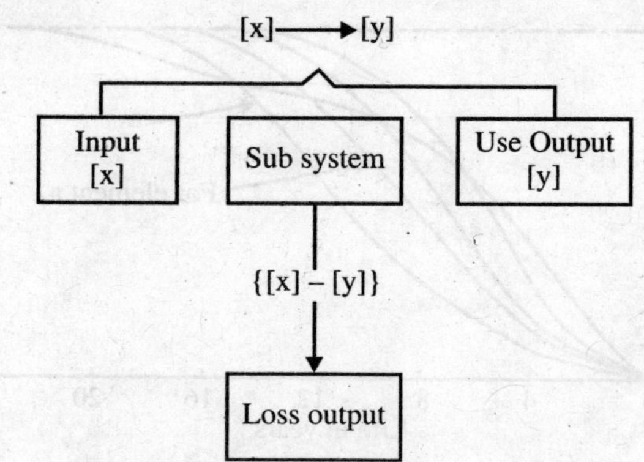

Figure 18.3 Basic subsystem analysis.

Loss output is mainly due to loss in friction and wear resulting from temperature, material, surface roughness, energy absorption by the interfaces etc.

Some of the inputs are often disturbing, viz. vibration from external source, radiation etc. Sometimes again external vibrations (particularly of ultrasonic frequency) may produce 'jerk-free' motion at the 'use-output' end. Such case though rare is beneficial.

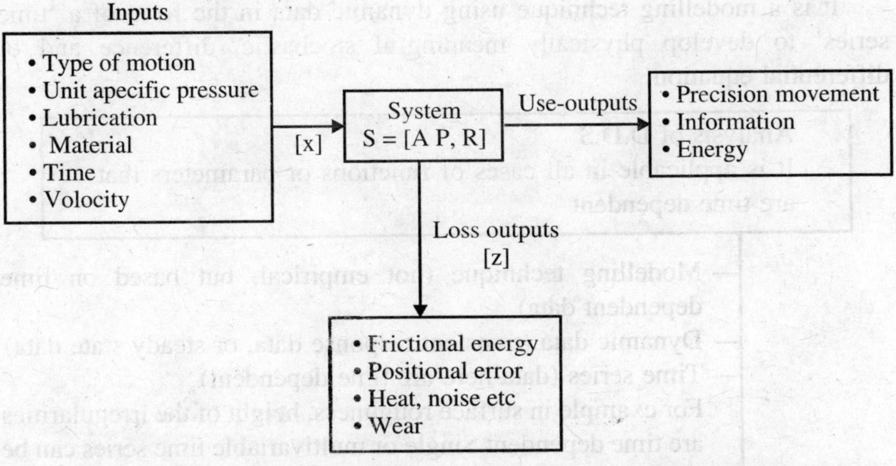

Figure 18.4 Functional system elaborated.

When the input and output are both time dependent, the system represents a dynamic state. In studying dynamic behaviour of systems, we use Laplace transform method by following the systematic steps below :

(i) use Laplace transform to form equations of motion
(ii) solve to get transfer and response functions
(iii) inverse transform to obtain time response

If system is the dynamic equilibrium, we call it a steady state, where the equations could be written in matrix form

$$\begin{pmatrix} Y_1 \\ Y_2 \\ \vdots \\ Y_q \end{pmatrix} = \begin{pmatrix} t_{11} & t_{12}\cdots\cdots & t_{1p} \\ t_{21} & t_{22} & t_{2p} \\ \vdots & \vdots & \vdots \\ t_{q1} & t_{q2} & t_{ap} \end{pmatrix} \begin{pmatrix} X_1 \\ X_2 \\ \vdots \\ X_q \end{pmatrix} \tag{18.5}$$

$$[Y] = [T] \cdot [X]$$

where use-output is related to input through transform matrix $[T]$.

In cases where the system is in stochastic condition we attempt to evaluate the regime of proper functional behaviour by applying theory of probability.

Dynamic Data System[83 – 86]

The latest in the field of condition monitoring comes modelling system known as Dynamic Data System (DDS). It can be used for universal application in all fields of engineering.

It is a modelling technique using dynamic data in the form of a 'time series' to develop physically meaningful stochastic, difference and or differential equation.

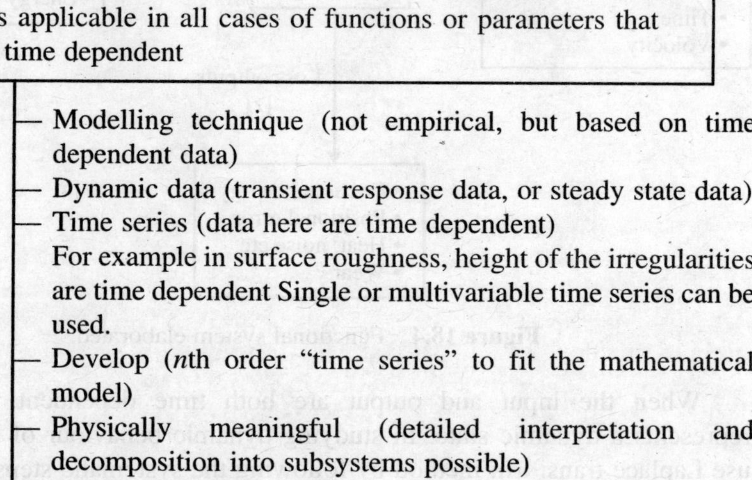

Though DDS uses time series techniques, sometimes function of variables other than time is included in the idea of a 'time series'. Surface profile heights of a machined surface is based on distance and could be considered as a time dependent series. In DDS either a single or several series can be used. The latter involves "multivariate time series analysis", which is especially useful for control purposes.

In DDS data is obtained first, without any preconceived notion of the model, followed by stepwise procedure to build up a model. After the model is satisfactorily built up, it is then decomposed to show what is happening in the system. A high order model may be decomposed into different combination of lower order subsystems. This is done to prove that the model built up is really a physically meaningful model.

The final model form developed by DDS methodology is the Auto Regressive Moving Average model (ARMA). A typical model that may occur is the following ARMA (2, 1) model, which takes the form

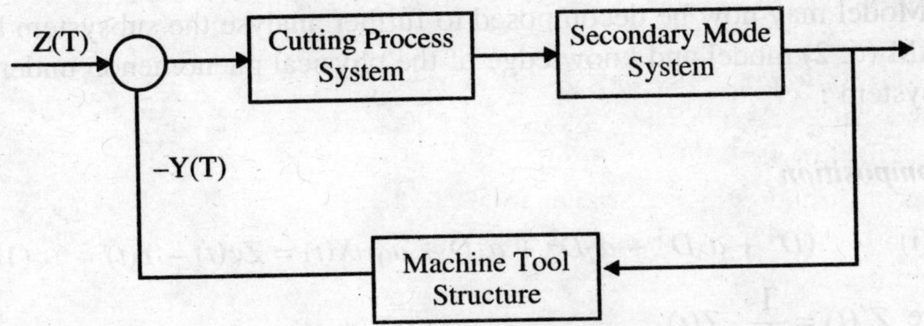

Figure 18.5 Metal cutting process.

$$X_t = \phi_1 X_{t-1} + \phi_2 X_{t-2} + \alpha_t - \theta_1 \alpha_{t-1} \tag{18.7}$$

However for the physical interpretation of a system, a continuous model in the form of stochastic differential equation is more meaningful. A semi-equivalent ARMA (2, 1) model, if converted into a stochastic differential equation, is given by

$$\ddot{x} + \xi w \dot{x} + \omega_n^2 x = Z(t) \tag{18.8}$$

where ξ – damping factor, ω_n –natural frequency of vibration and $Z(t)$ is the forcing function. This dynamic model is valid for any metal cutting process in a machine tool.

The estimation of the stochastic differential equation is unique for ARMA (2, 1) model by recognizing the fact that the 0 parameter is functionally related to the parameters ϕ_1 and ϕ_2. Furthermore: the spectrum analyses can be achieved by converting either the discrete or the continuous model into spectrum form.

With reference to the cutting process shown in Figure 18.5 by block diagram, if we consider

X(*T*)–cutter workpiece relative displacement

Y(*T*)– feedback

Z(*T*)–random influence, including cutting force

DDS modelling strategy is used to identify the dynamics and reliability of the cutting process. The characteristic roots for the AM (6,2) model refers to the autoregressive polynomial

$$(D^6 + a_5 D^5 + a_4 D^4 + a_3 D^3 + a_2 D^2 + a_1 D + a_0)\, X(t)$$
$$= (b_2 \mathrm{D}^2 + b_1 D + 1)\,(Z(T)) \tag{18.9}$$

where $D = d/dt$ = differential operator

a_0, a_1, a_2, a_3 are autoregressive parameters

b_1, b_2........ are the moving average parameters.

Model may now be decomposed to further analyse the subsystem based on AM (6, 2) model and knowledge of the physical phenomenon underlying the system :

Decomposition

(i) $$(D^4 + a_3D^3 + a_2D^2 + a_1D + a_0)\ X(t) = Z¢(t) - Y(t) \tag{18.10}$$

where $Z'(t) = \dfrac{1}{g_0} \cdot Z(t)$

(ii) $$(D^2 + g_1D + g_0)\ Y(t) = (MD + K)\ X(t) \tag{18.11}$$

Combining the above equations, the values of the coefficients g, g_0, a_0, a_1, a_2, a_3; and M and K can be calculated from the parameters of the AM (6, 2) and comparing the actual data. While g_1 and g_2 are the coefficients characterising the machine tool structure dynamics, a_0, ... a_3 lead into dynamics or other two systems, and K and M indicate effects of vibration on the force.

DDS has been successfully used uptill now in various problems of metal cutting, machine tools, transportation, printability of paper, malfunctioning of power plant etc.

CHAPTER 19

Industrial Robots and Their Applications

19.1 Introduction

Industrial robots may be classified as machines with manipulators that can be easily programmed to do variety of manual tasks automatically. A robot consists of one or more manipulators (arms), end-effectors (hands), a controller and power supply. Sensors are now being used increasingly to provide informations about environment and feedback of performance in tasks accomplishment. According to Japanese Industrial Standard JIS-BO 134-1979, 'A robot is defined as a mechanical system which has flexible motion functions analogous to the motion functions of the living organisms or combines such motion functions with intelligent functions and which acts in response to human will. In this context, intelligent functions mean the ability to perform at least one of the following: judgement, recognition, adaptation or learning'.

Contrary to the popular notion 'the robotics have an enormous amount to offer the manufacturing and assembly fields, but are not confined to choosing the machine to do a job, currently done by people'. Uses of industrial robots should be effective primarily in two areas, viz.,

(i) labour intensive areas, such as, packing, warehousing, plant maintenance, etc.; and
(ii) adverse environmental situations, such as, accident-prone, hazardous as well as monotonous areas of work.

In the area of steel making or further processing of steel, industrial robots, therefore, should find place in accident-prone handling and hazardous environment resulting from operations at hostile temperature conditions, constrained by human inaccessibility, with a view to reducing the operational time, and increasing productivity. Figure 19.1 shows a Puma Handling Robot used commercially.

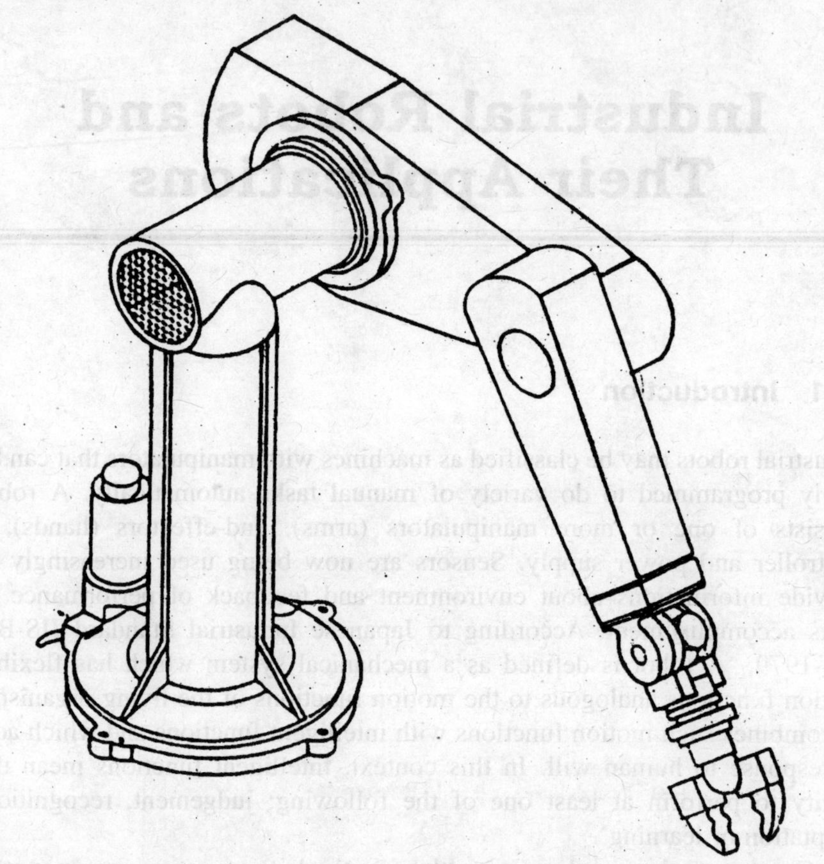

Figure 19.1. Puma robot in commercial use.

19.2 Basic functions of robotic elements

Apart from power source and central processing unit (or controller) the major elements and their functions are classified as shown in Figure 19.1(a). A rational sub-groupings of the various types of industrial robots, now in use, can be shown as in Figure 19.1(b). Some classification based on functional aspects of robots has been shown in Figure 19.1(b).

'Non-servo' robots can provide sequence of point-to-point (PTP) motions, wherein manipulator-actuated members move till the limits of travels are reached. This requires adjustment by mechanical limit switches, end stops etc., for each axis movement. Robots known as 'Bang-Bang', 'Pick and place', 'limited sequence' are examples of this. Positional accuracy in all cases is typically reliable and repetible and adequately high, better than ± 0.5 mm. In 'servo-controlled' robots, we get sequence of PTP motions with controlled intermediate velocities to follow any programmed trajectory.

Normally six degrees of freedom are needed to locate a point and its orientation by the end-effector. But for 'pick and place' only 2-3 degrees of freedom may be sufficient. Such independent coordinate motions are arranged within the manipulator through pneumatic, hydraulic or electric transmission systems.

Figure 19.2 shows an 'industrial robot's arm movement while Figure 19.3 shows basic manipulator's geometries.

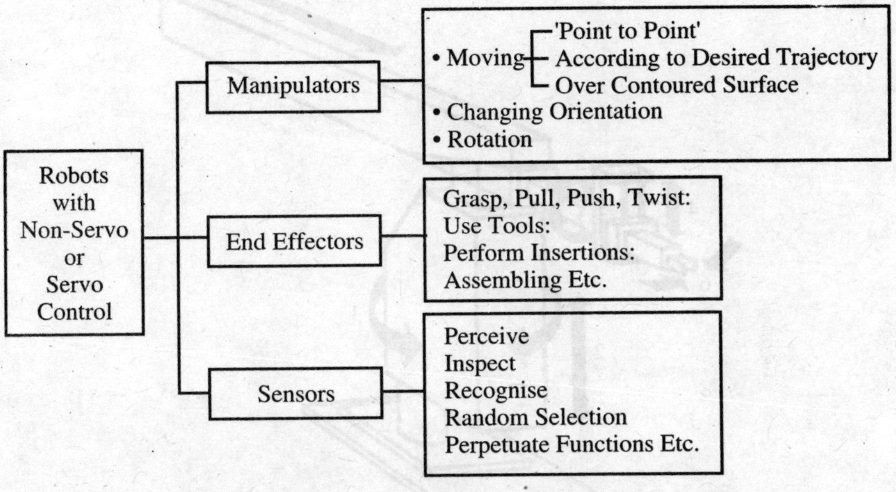

Figure 19.1(a) Elemental functions (This should not be considered as a system diagram).

Degrees of freedom can be further augmented by using angled arms and offsets. Mobility of the industrial robots is further increased by built-in mechanical linkages Sometimes arms may by mounted on rails, hung from ceilings etc. This multiply the range of installed configurations, as shown in Figures 19.4(a), and (b).

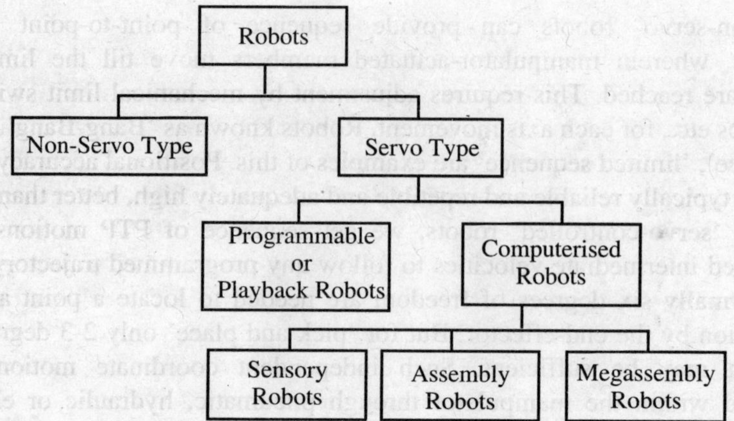

Figure 19.1(b) Various types of industrial robots.

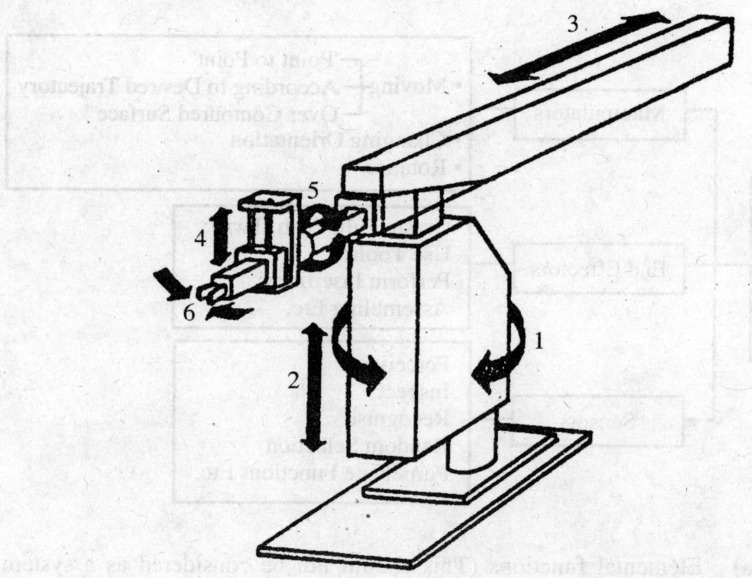

Figure 19.2 An industrial robot's arm movements.

19.3 Mobility of Robots

In the field of handling materials now-a-days we have robot carts and robot vehicles Crawling robots are now under experimentations and so are the walking machines, which were first developed a decade ago are largely under research in places like Ohio State Uuiversity, Carnegie Mellon University, Tokyo Institute of Technology and others. According to Engelberger[96] (1980) 'Practical robot devices with legs could have advantages, in such applications as arctic transport, mining, agriculture, fire-fighting, explosive orodance disposal, unmanned ocean floor surveys and planetory exploration'.

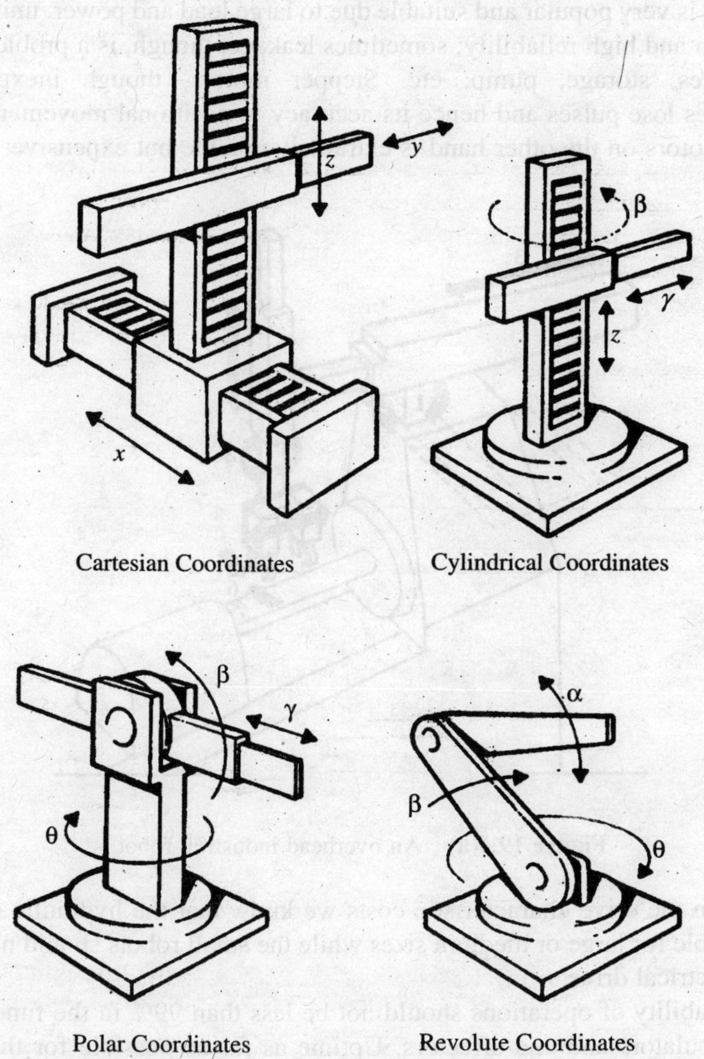

Figure 19.3 Basic manipulator geometries.

19.4 Reliability in Operation

Actuation of the various movements of arm/wrist, joints/ manipulators is normally by pneumatic, mechanical, electro-mechanical, electro-hydraulic or by electrical means. While pneumatic is probably less expensive and simple, it is not suitable for all types of robots. It is mostly for 'pick and place' robots, whose trajectory motions are controlled by mechanical stops. Hydraulic actuation is very popular and suitable due to large load and power, uniformity of motion and high reliability; sometimes leakage, though, is a problem. But it requires, storage, pump, etc. Stepper motors, though inexpensive, sometimes lose pulses and hence its accuracy in positional movement. D.C. torque motors on the other hand is extremely reliable but expensive.

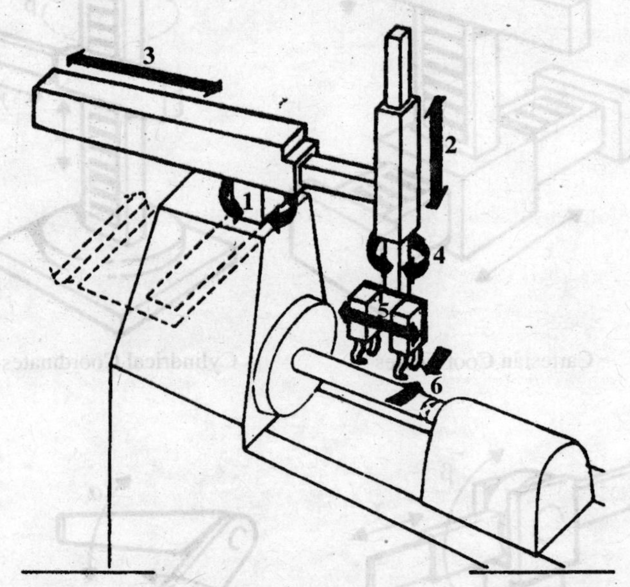

Figure 19.4(a) An overhead industrial robot.

From the drive characteristic costs we know that the hydraulic systems are suitable for large or medium sizes while the small robots should normally have electrical drive.

Reliability of operations should not be less than 99% in the functioning of manipulators and end-effectors. Uptime as far as possible for the entire system should be as high as 98 %. Apportioning of the reliability of the units should be optimised for the adequate permissible reliability of the system.

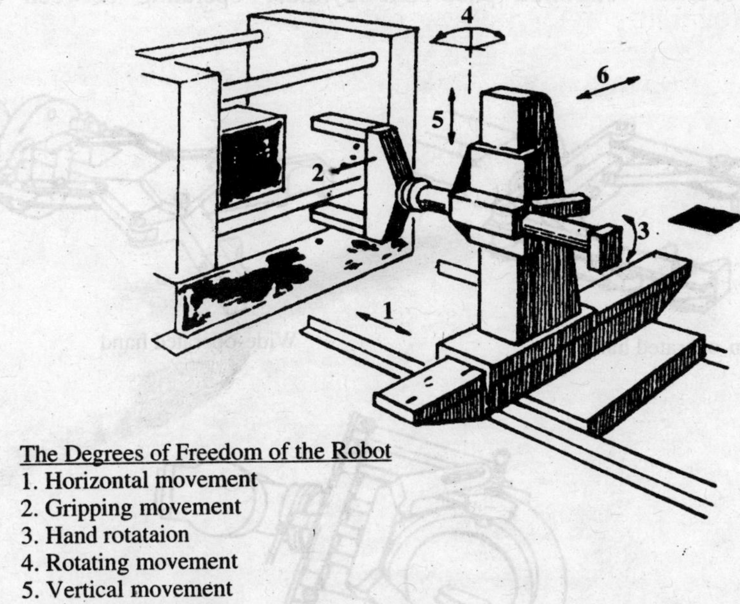

The Degrees of Freedom of the Robot
1. Horizontal movement
2. Gripping movement
3. Hand rotataion
4. Rotating movement
5. Vertical movement
6. Lateral hand travel

Figure 19.4(b) A floor type rail mounted industrial robot.

Considerable research is being done to analyse the limiting factors in the manipulation of universal robots. Inhibition of end-effectors is due to lack of dexterity and programmability of the hands. Positional errors, repeatability and conformance to configurations are some of the important areas of research.

19.5 Control

Figure 19.5 shows various types of grippers commonly used. All robots require the use of suitable drive and control system for rotary as well as translatory motion of the robot body and arm. Similarly the movements of the hand and gripper have to be controlled. While the drive system mainly consist of hydraulic, pneumatic or electric, the control in almost all the cases consists of microprocessor-based systems. For large robots sometimes mini-computers are also employed for control purposes. In case of pneumatic system robots, simple sequence controllers (for very limited operations) or microprocessor-based systems are used. The microprocessor in this case is used with a stepping motor drive screw for moving stops to new positions. The microprocessor is thus able to control the travel of the pneumatic cylinder. Figure 19.6(a) illustrates a schematic block diagram of microprocessor controlled pneumatic cylinder operating between fixed points.

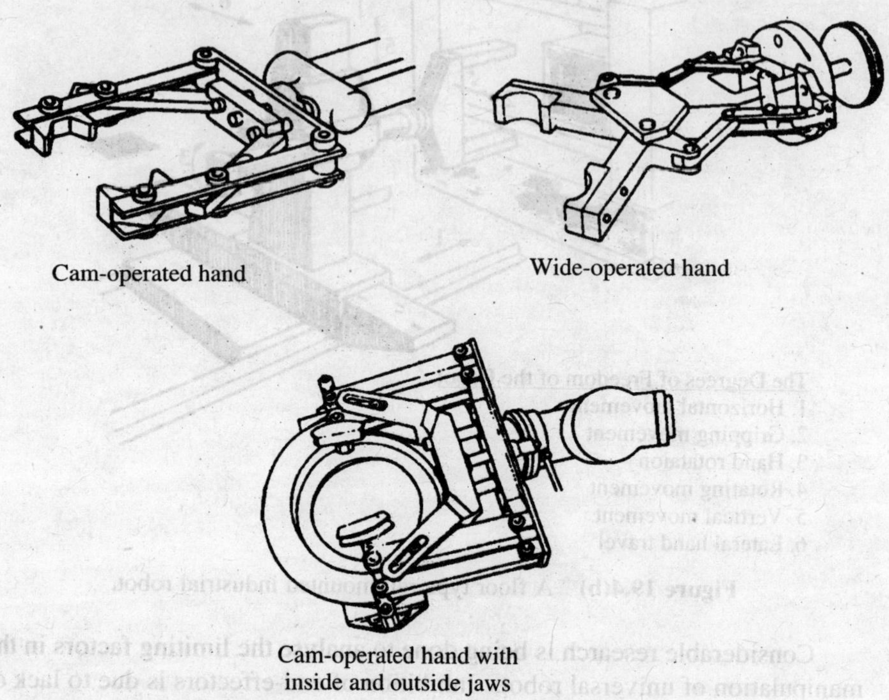

Figure 19.5 Examples of end-effectors or grippers.

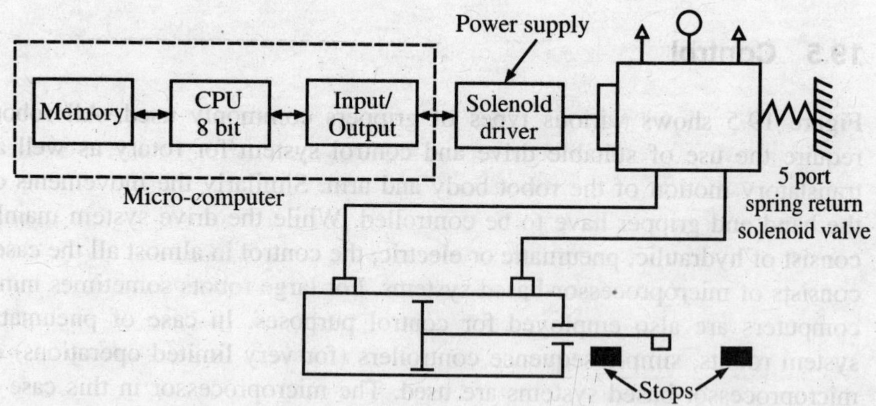

Figure 19.6(a) A microprocessor-based pneumatic system controller.

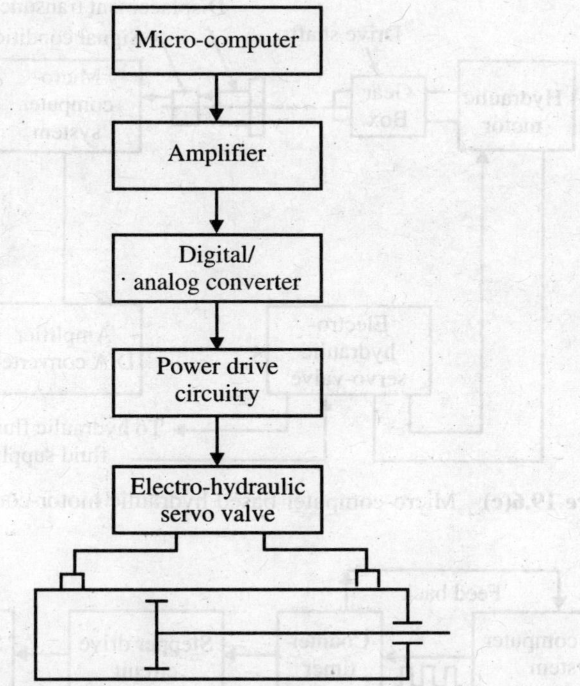

Figure 19.6(b) Micro-computer control of hydraulic cylinders.

Microprocessor control is particularly advantageous in the case of robots based on the application of hydraulic system. While such robots are more expensive than pneumatically actuated robots, the hydraulic systems (Figure 19.6) are useful for producing a piecewise linear motion which can be stopped at pre-set programme points as shown in Figures 19.6(a) and 19.6(b).

For electric drive system D.C. motor and stepper motors are widely used, both of which are controllable by microprocessors. The schematic arrangements are shown in Figures 19. 6(d) and 19.6(e).

Apart from the differences in the drive system the robotic devices and control systems used vary according to the functions to be carried out by them. For example, the devices used for unloading die casting or injection moulding machines require at least 3 axes control in the body of the robot and usually a wrist rotation, a yaw motion and a rolling motion. Sensory feedback of the pressure applied by the grippers is particularly useful. Control of all the six degrees of freedom to obtain a precise position generally requires the use of a minicomputer, a powerful 16-bit microcomputer or individual micro-processors for each of the motion axis. This latter configuration involving the use of a number of microprocessors, require a central micro-computer which can coordinate the control functions

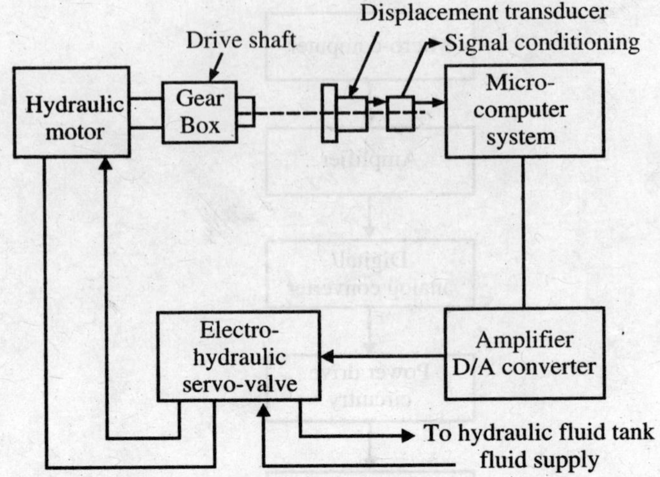

Figure 19.6(c) Micro-computer based hydraulic motor-control system.

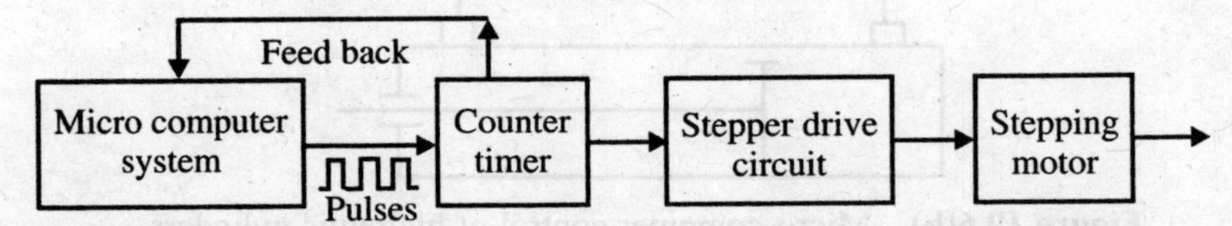

Figure 19.6(d) A micro-computer controlled stepping motor (upto 500 gms payload)

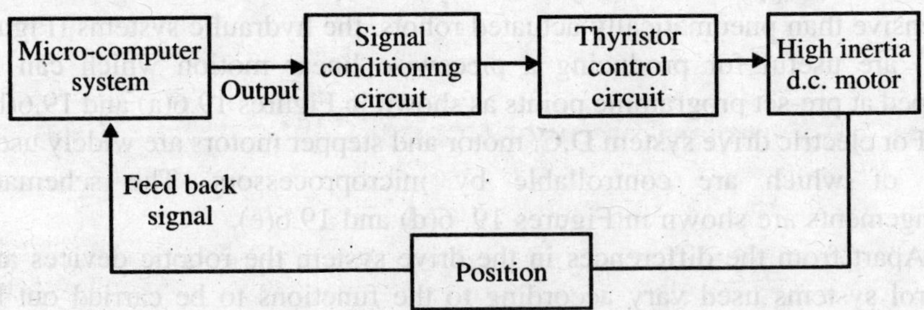

Figure 19.6(e) A micro-computer based d.c. motor control systems (upto 5 kg. pay load).

of individual micro processors. Considerable memory and high speed is required to calculate and control the movements of all the six axes. The availability of 16 bit microcomputer and multi-bus system reduce the cost of robot control system. Alternately, teach system is used whereby the operator defines the robot body, arm and gripper movements. It is thus not required for the computer to calculate the movements. The schematic diagram of such a system is shown in Figure 19.7.

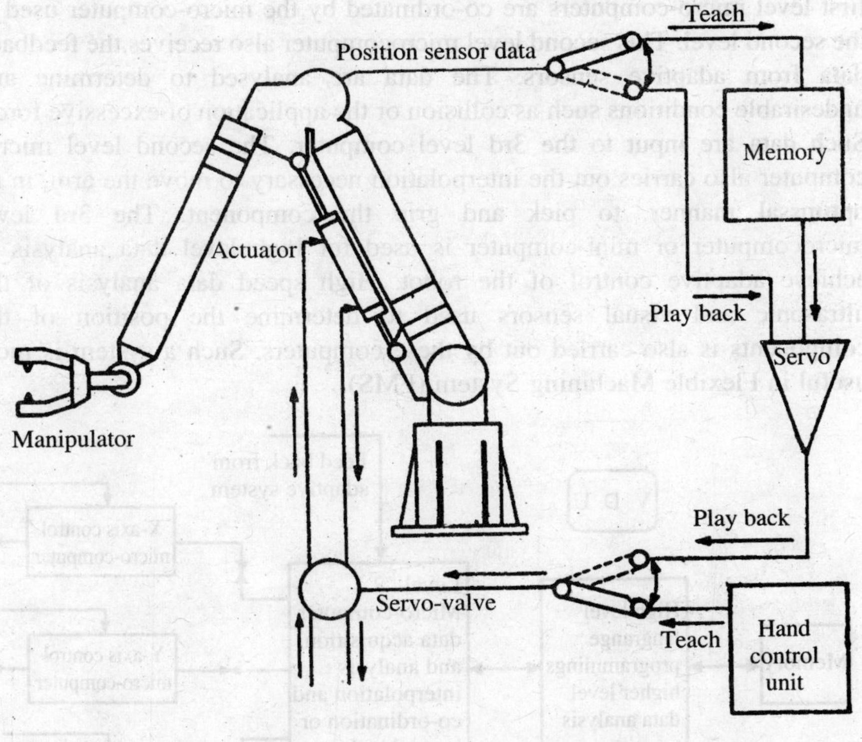

Figure 19.7 Programming an industrial robot by teach method.

19.6 Hierarchial Computer Control Configuration Robotic System

Computer control of robots often involves the use of number of distributed microprocessor/computers for data acquisition, analysis and control. Such an arrangement is necessary to reduce the cost of the control system configuration and process the data at required high speed and also for a number of tasks such as the co-ordinated motion of the arms and grippers, taking with the data acquired by means of suitable sensors mounted on the robot body, arm and gripper. The data acquired by visual sensors also have to be processed and analysed to move the robot arm. A typical hierarchically distributed system configuration might make use of the micro-processors/ computers at three levels as shown in Figure 19.8.

In such computer control system configuration shown, the first level micro-computers control the translational and rotary movement of the robot body and the movement of the gripper. The control actions carrid out by the first level micro-computers are co-ordinated by the micro-computer used at the second level. This second level microcomputer also receives the feedback data from adaptive sensors. The data are analysed to determine any undesirable conditions such as collision or the application of excessive force. Such data are input to the 3rd level computer. The second level micro-computer also carries out the interpolation necessary to move the arm, in an optimssal manner, to pick and grip the component. The 3rd level microcomputer or mini-computer is used for high level data analysis to achieve adaptive control of the robot. High speed data analysis of the ultrasonic and visual sensors used to determine the position of the components is also carried out by these computers. Such a system is most useful in Flexible Machining System (FMS).

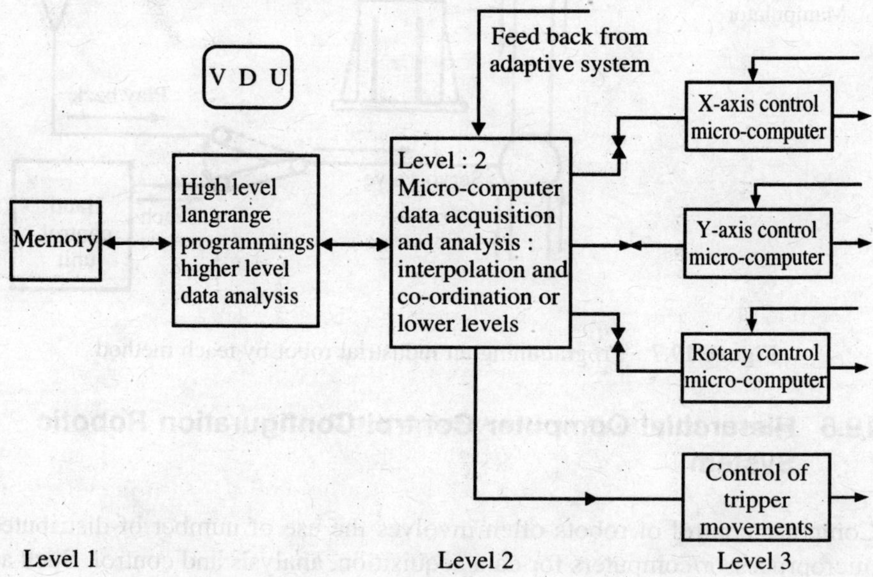

Figure 19.8. Hiemrchial computer control for robots.

19.7 Sensing

Sensing is an important function of robots uses for industrial work of varying natures. The concept of sensing, parameters of sensing and the methods available in practice require careful consideration. This will be clear when looked at the block diagram below:

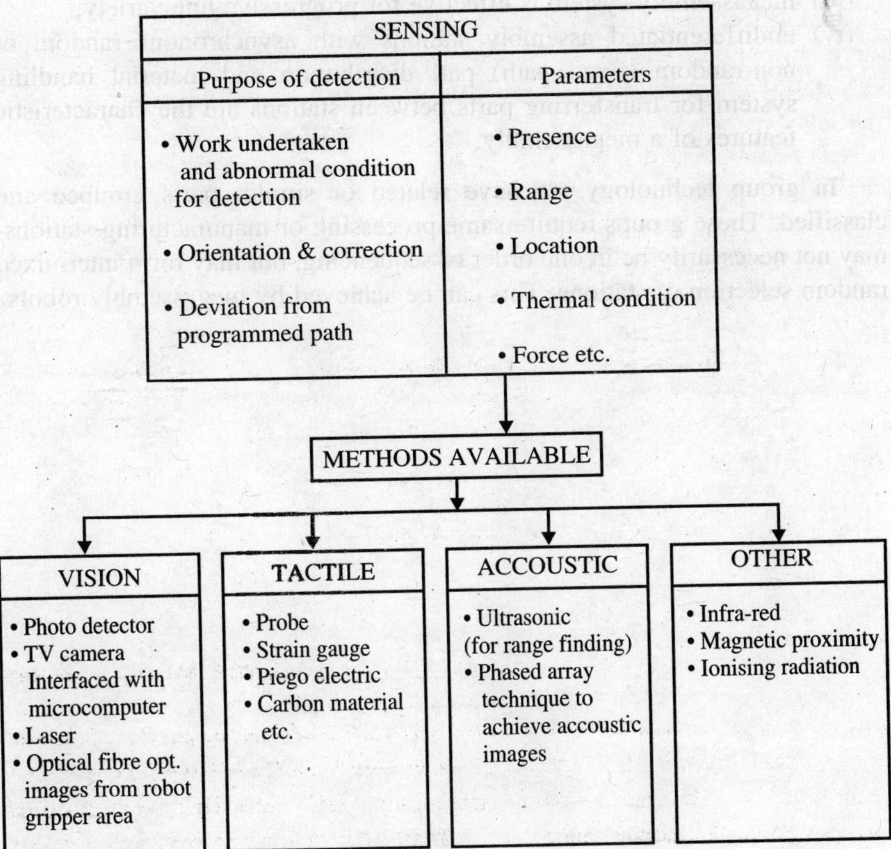

Figure 19.9. Sensing in sensory robot.

19.8 Assembly and Megassembly Robots

Normally assembly systems use 1-3 robot arms and perform a portion of completed process or sub-assemblies.

Megassembly is one where 10 or even more robot arms are used to complete a total assembly tasks in progressive steps. Such megassembly robot known as a 'Sleeping Giant' has certain basic advantages. To name a few–

(i) it allows simpler and faster operation;

(ii) computing system, particularly in a network structure, involves substantial cost for the first unit, while cost per station, in case of multistations operations, drops inversely as the number of stations increases;

(iii) megassembly system is effective for progressive line variety;

(iv) undifferentiated assembly stations with asynchronous random or non-random (fixed path) part distribution and material handling system for transferring parts between stations are the characteristic features of a megassembly.

In group technology we have related or similar parts grouped and classified. These groups require same processing or manufacturing stations-may not necessarily be in one order of sequencing, but may form intermixed random selection of stations. This can be achieved by megassembly robots.

CHAPTER 20

NC-CNC-DNC Machines

20.1 Introduction

Numerical Controlled Machine tools have been defined as "a system in which actions are controlled by the direct insertion of numerical data at some point. The system must automatically interpret at least portion of the data". Numerical data referred to above means physical data converted to digits.

The development of N.C. machines refers back to 1940 when J.T. Parson of Parson's Corporation tried to generate a curve automatically by milling cutters by providing coordinate motions. In 1949, the U.S. Airforce Commissioned Servo-mechanism Laboratory of Massachusetts Institute of Technology for developing a workable N.C. system based on Parson's. This was the first developmental work on two-axis point-to-point control system using perforated tape. 1954 was the year of completion of such a project and symbolic languages for communication with computer was under way for N.C. programmes. First installation of such a machine commercially was done in 1957. In fact between 1955 and 1960 many experimental machines were demonstrated in various exhibitions and trade shows in U.S.A. Then came the APT, computer Based Part Programming language APT which means, 'Automatically Programmed Tool' was released for commercial uses in 1962.

First generation of inflexibly programmed control systems, based on electronic valves and relays were in use right from 1954 onwards. With the gradual uses of transistors and printed circuit boards, first and second generation N.C. machines were produced. The progress in microelectronics from Low degree of integration S.S.I. (Small Scale Integrated Circuit), to Medium degree integration (MSI), and large scale integrated circuit (LSI) and even very largo scale integrated circuit (VLSI) resulted in production of third and fourth generation N C. Machines-paving the path towards the development of 'Inbuilt' dedicated Computer Controlled Machine tools.

20.2 Principles of a CNC Machine

Fundamentally a (C.N.C.), Computer Numerical Controlled Machine utilises a 'dedicated stored programme computer' within the machine. It performs all the basic Machine Control Unit (MCU) functions as per software stored in the computer in-built in machine. Operating programmes in the present CNC systems offer enough scope for editing and modifications, as per requirement. Like a Computer, C.N.C. control units are based on 'stored programme' held in the Computer memory. Also the executive programme, resident in the memory of the Computer, makes such a system flexible in that a new executive can be input (through paper tape) and the system can operate on a different type of machine tool. Hence the term 'soft wired system' because of the fact that the control logic for the machine is generated by software programme, rather than by wired logic circuits. Some of the salient advantages are as under:

Because of the memory availability we can store part programmes also in the computer.

Possible to edit part programmes at ease.

Possible to have user-oriented sub-routine programmes.

High reliability and repeatability of operation.

High degree of precision in the closed loop control system and possibility of maintaining the precision.

In-built diagnostic programmes make the trouble shooting and maintenance functions easier.

Figure 20.1 shows the system layout for a Computer numerical controlled machine tool. From the block diagram it is easier to understand the main constituent units of the machine and the feedback control circuits used for each executive organ of the machine tool.

20.3 Classification of CNC or NC Machines Through Control Axes

Basic classification of a CNC machine tool is done by identifying the following characteristics: (vide Figure 20.2).

Positional, denoted by P

Straight line, denoted by L

Contouring system, denoted by C

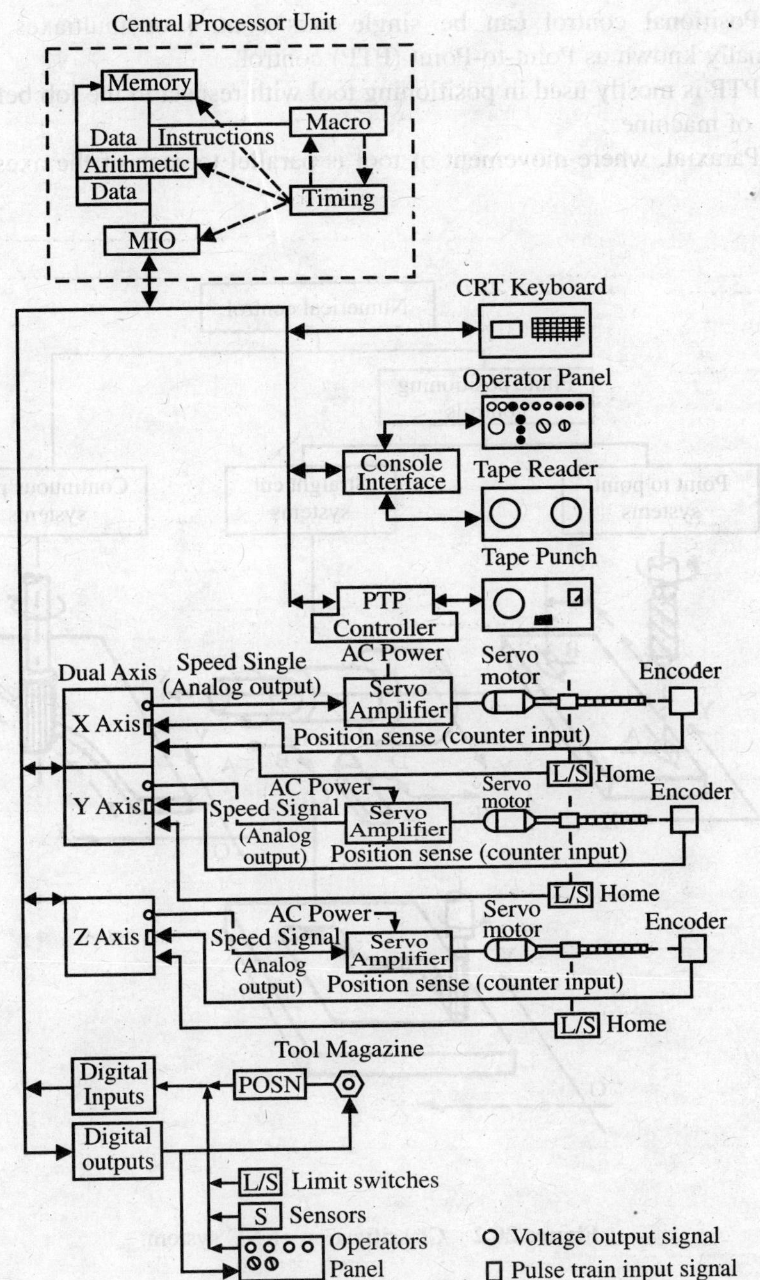

Figure 20.1 C.N.C system.

Positional control can be single axis control or multiaxes control normally known as Point-to-Point (PTP) control.

PTP is mostly used in positioning tool with respect to the job before the start of machine.

Paraxial, where movement of tool is parallel to each of the axes of the work.

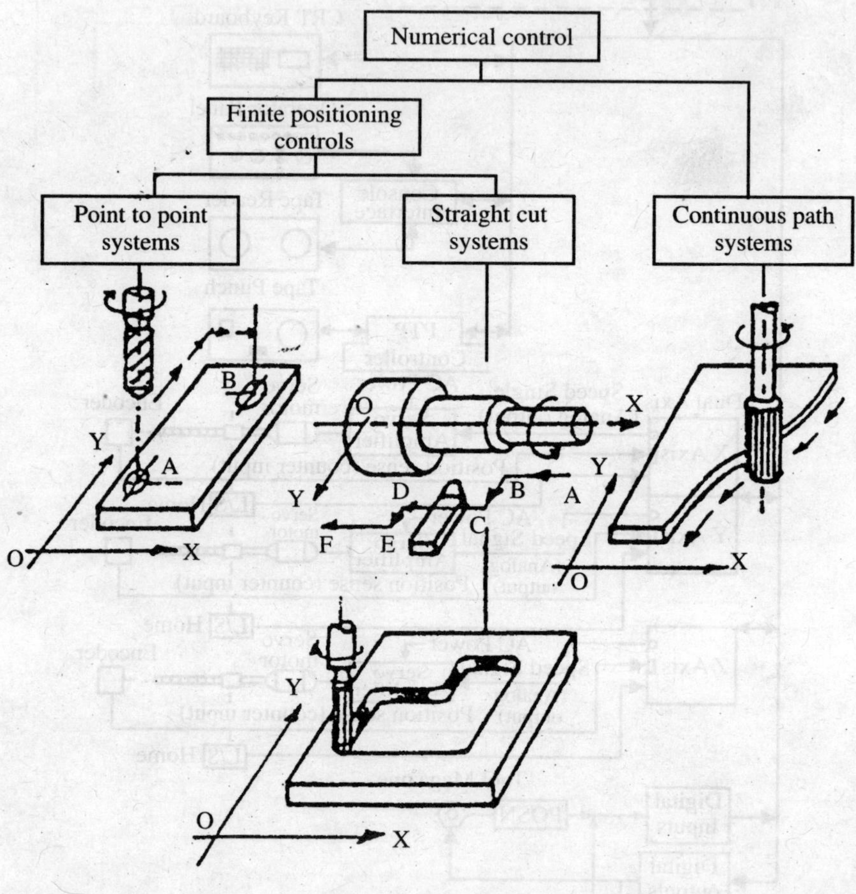

Figure 20.2 Classification of NC system.

When we indicate a machine by identification 2P, *L*.... we mean 2-axes positional control; and 1-axis Linear Control. Examples of such machines are PTP Drills.

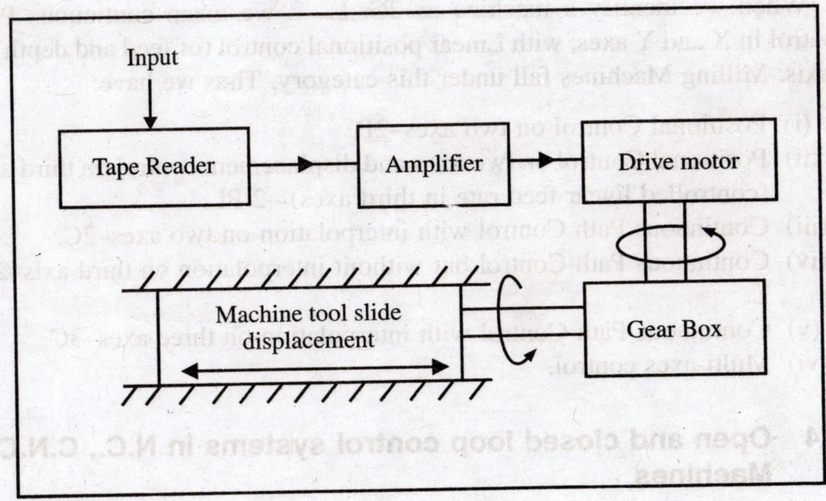

Figure 20.2(a).

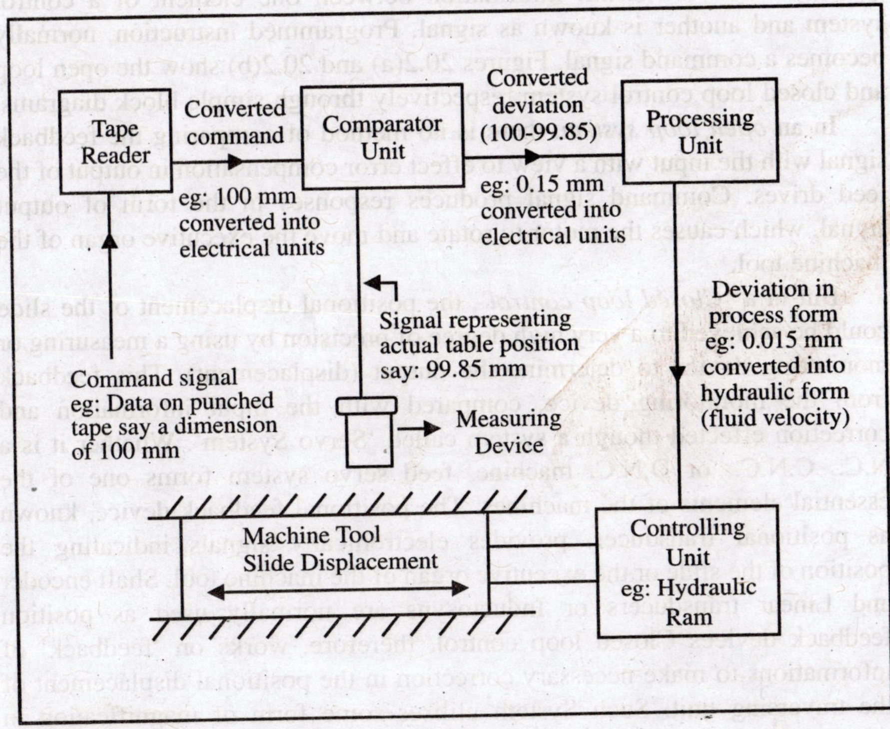

Figure 20.2(b).

When we identify a machine as 2C, L...... we mean continuous Path Control in X and Y axes, with Linear positional control (of feed and depth) in Z axis. Milling Machines fall under this category. Thus we have:

(i) Positional Control on two axes–2P.
(ii) Positional Control on two axes and displacement control on third axis (controlled linear feed rate in third axes) –2 PL.
(iii) Continuous Path Control with interpolation on two axes–2C.
(iv) Continuous Path Control but without interpolation on third axis–2C, L.
(v) Continuous Path Control with interpolation on three axes–3C.
(vi) Multi-axes control.

20.4 Open and closed loop control systems in N.C., C.N.C. Machines

Basic programmed instruction to any control system is a signal. Any physical quantity used to transmit information between one element of a control system and another is kuown as signal. Programmed instruction, normally becomes a command signal. Figures 20.2(a) and 20.2(b) show the open loop and closed loop control system respectively through simple block diagrams.

In an *open loop system*, there is no method of comparing the feedback signal with the input with a view to effect error compensation in output of the feed drives. Command signal produces responses in the form of output signal, which causes the motor to rotate and move the executive organ of the machine tool.

But in a '*Closed loop control*', the positional displacement of the slide could be achieved to a very high degree of precision by using a measuring or monitoring device to determine the output (displacement). This feedback from the monitoring device, compared with the input information and correction effected though a system called 'Servo System'. Whether it is a N.C., C.N.C., or D.N.C. machine, feed servo system forms one of the essential elements of the machines. The positional feedback device, known as positional transducer, provides electronically signals indicating the position of the slide or the executive organ of the machine tool. Shaft encoder and Linear transducers or Inductosyns are normally used as position feedback devices Closed loop control, therefore, works on 'feedback' of informations to make necessary correction in the positional displacement of the traversing unit. Such System utilises some form of magnification in which mechanical position is the controlled quantity. Such an integrated system is called 'Servo mechanism'.

20.5 Working of N.C. Machine Tool

Normally any N.C. Machine will consist of (i) Drive unit, (ii) Elements of machine tool system, (iii) Control System or Director, (iv) Feedback Servo system, (v) Electrically operated control equipment or 'Magnetic Box', and (vi) Manual Control (operator's Control). The Director in (iii) is sometimes called N.C. console.

In the Control System, tape instructions are read by a tape reader. These instructions undergo electronic processing and the system sends electrical commands to the drive units, and to the electrical control cabinet called magnetic box. Command signals sent to the drive unit determine the lengths of the movement and appropriate feed rates.

Commands directed to the magnetic box call for several other operations of the machine tools e.g. spindle motor starting and stopping, selection of spindle speeds, actuation of tool change, coolant supply etc.

The positional transducer feeds back the information regarding actual position achieved to the control system. In the control cabinet this is compared with the input command and until the difference between the two signals, known as 'error' is brought to zero, the drive unit is actuated by suitable amplifiers from the error signal. The block diagram of a common N.C Machine has been shown in Figure 20.3.

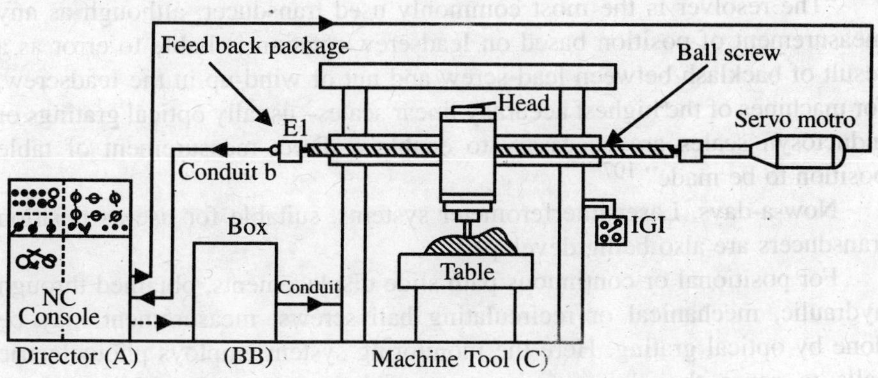

Figure 20.3 Schematic diagram of N.C. machine.

20.6 Transducers and Monitoring of Displacement

Displacement of the machine tool elements in each axis requires to be monitored for error compensation and for this purpose the basic elements required are:[122]

(i) "in feedback system, transducers on each controlled axis to. measure the position or displacement of the cutting tool or workpiece;

(ii) Servo motors on each controlled axis to produce the required relative movement of the cutting tool and workpiece;

(iii) Logic circuits to enable the instructions contained in the control programme to be carried out.

"Transducers are essentially of two basic types– those which measure the position or displacement of the cutting tool or workpiece in terms of the position or displacement of those parts of the machine tool carrying the tool or workpiece and those which measure the rotation of a leadscrew driving the executive element of the machine tool. The first type consists of a scale on the fixed part of the machine extending, parallel to the axis along which measurement is to be made and the reading head carried on the moving part, –magnetic, electro-magnetic and optical scales are used. Transducers of second type are resolvers, encoders or digitisers and it is usually necessary to have several geared together in order to remove ambiguity due to the fact that the full traverse of the machine involves many rotations of the leadscrew.

It is also necessary to distinguish between analogue and digital transducers according to the form of output signal. Resolvers and inductosyns are analogue devices, whereas optical gratings, encoders and digitisers are digital transducers.

"The resolver is the most commonly used transducer, although as any measurement of position based on Ieadscrew rotation is liable to error as a result of backlash between lead-screw and nut or wind-up in the Ieadscrew, for machines of the highest accuracy linear scales– usually optical gratings or inductosyn scales–are necessary to enable a direct measurement of table position to be made".[107]

Now-a-days, Laser-interferometer systems, suitable for use as position transducers are also being developed.

For positional or continuous path slide displacements, obtained through hydraulic, mechanical or recirculating ball screws, measurement may be done by optical grating. Here the monitoring system employs photoelectric cells to sense the changing intensity of light caused by Moire Fringe phenomenon associated with optical grating. The system consists of highly reflective metal strip having scale grating along tbe lengths of axis measurement. Similar grating, having same lines per mm is made on the short length fixed on reading head. This is known as Index grating. Gratings on the moveable element are parallel lines. While that an index grating, even though made of the same pitch contain lines inclined at one pitch distance. Relative movement of the slide therefore causes change in light intensity (fringes) falling on the photoelements. The output signal follows a sinusoidal form. If grating is 60 lines per mm, pitch of the line is 0.016 mm. If now the index grating lines are inclined at one pitch distance and there are n number of photocells spaced across the width of grating, then we get n pulses per cycle and the displacement corresponding to each pulse is $\frac{0.016}{n}$ mm. Assuming $n = 4$ photocells. We get displacement $\frac{0.016}{4} = 0.004$ mm.

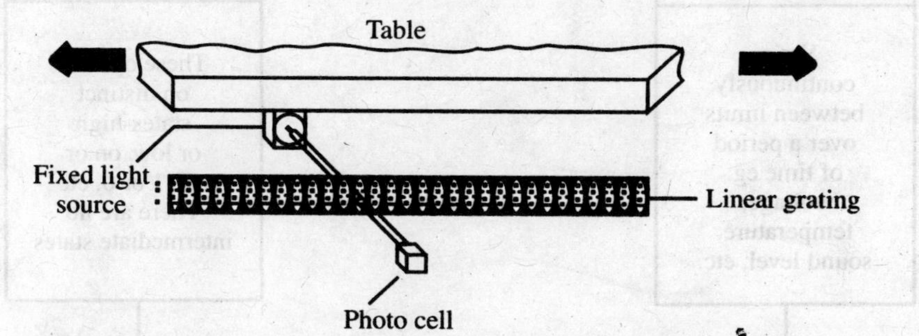

Figure 20.4. Position measuring transducer using lioear gratings.

20.6.1 *Open and closed loop*

In a closed loop control system, we always need *feedback* and this feedback are normally of two types (i) *velocity feedback* and (ii) *positional feedback.*

Velocity feedback can be achieved by providing within the servomotor a *tachogenerator*, which gives voltage proportional to the angular speed. It may, therefore, be possible to call a tachogenerator as a transducer.

For *positional feedback* it is necessary to measure the distance moved by the axis slides. For such purpose we normally use transducers. But while measuring position it must measure with respect to a prescribed datum. From such base point the distance moved can be expressed either in incremental form or absolute form, as shown below in Tables 20.1(a) and 20.1(b):

Table 20.1(a)

Point	*X coordinate*	*Y coordinate*
	Absolute System	
1	35	45
2	60	70
3	70	30
4	80	70

Table 20.(b)

Point	*X coordinate*	*Y coordinate*
	Incremental System	
1	35	45
2	25	25
3	10	–40
4	10	40

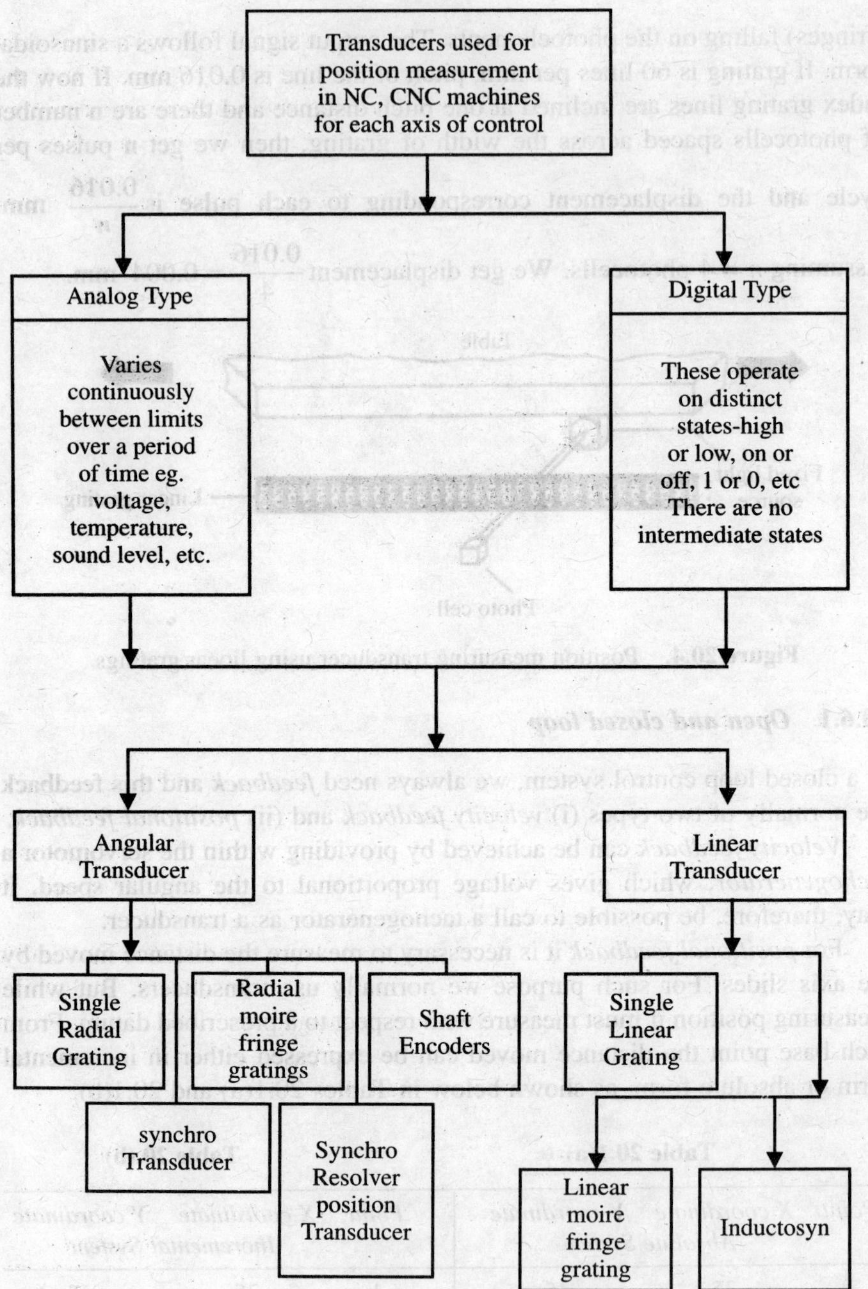

Figure 20.5 Block diagram showing classification and functioning of transdicers.

20.6.2. *Encoders*

In a shaft Encoder a series of photocells will transmit binary codes depending upon the light intensity passing through the transparent region and blocked by opaque regions, etched on the rotating disc known as Encoder, fixed on the rotor shaft attached to the axis Ieadscrew (see Figure 20.6). Accuracy depends on the number of photocells, sometimes a dozen cells are used to increase the accuracy.

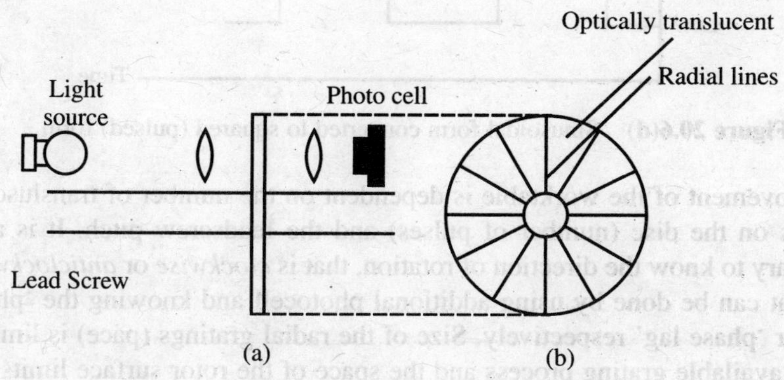

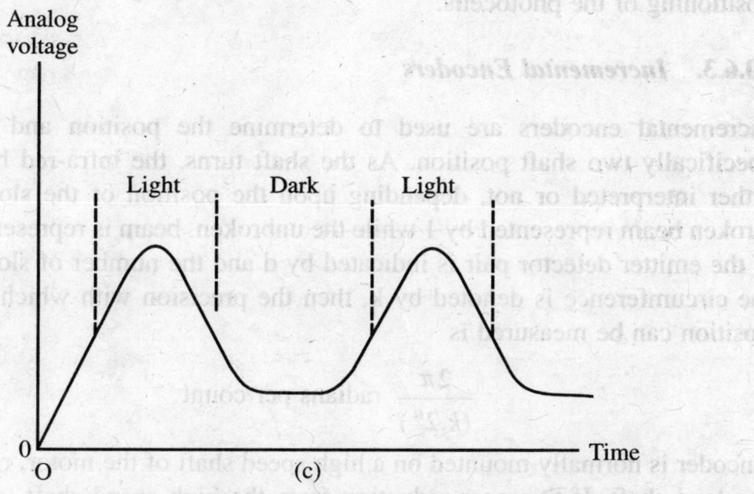

Photo cell voltage as the radial grating disc rotates

Figure 20.6(a), (b), (c).

This gives a sinusoidal pattern, which could be converted into square shaped pattern of pulse output, with the use of a electrical device as under.

Each of the discrete pulses corresponds to the transluscent region of the disc and indicates an angular movement of the Ieadscrew.

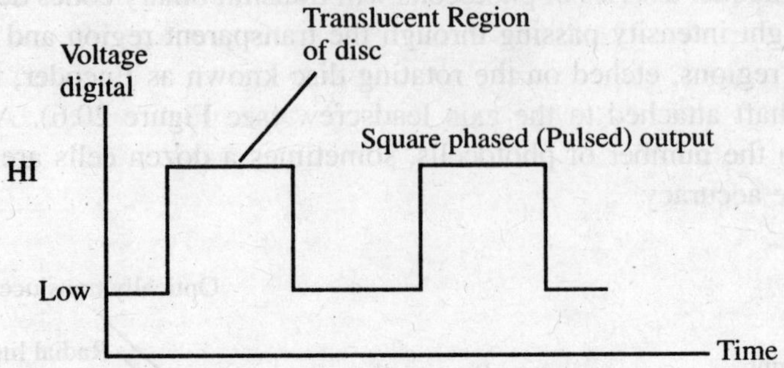

Figure 20.6(d) Sinusoidal form converted to squared (pulsed) form.

Movement of the worktable is dependent on the number of transluscent regions on the disc (number of pulses) and the Ieadscrew pitch. It is also necessary to know the direction of rotation, that is *clockwise* or *anticlockwise* and that can be done by using additional photocell and knowing the 'phase lead' ar 'phase lag' respectively. Size of the radial gratings (pace) is limited by the available grating process and the space of the rotor surface limits the positioning of the photocells.

20.6.3. *Incremental Encoders*

Incremental encoders are used to determine the position and are used specifically two shaft position. As the shaft turns, the infra-red beams are either interpreted or not, depending upon the position of the slotted disc. Broken beam represented by 1 while the unbroken. beam is represented by 0. If the emitter detector pair is indicated by d and the number of slots around the circumference is denoted by k, then the precision with which the shaft position can be measured is

$$\frac{2\pi}{(k.2^d)} \text{ radians per count}$$

Encoder is normally mounted on a high speed shaft of the motor, rather than the load shaft. If this gear reduction from the high speed shaft to the load shaft is m : 1, then overall incremental load shaft precision is given by

$$\Delta\varphi_{m\cdot} = \frac{\pi}{k.m.2^{d-1}} \text{ radians per count}$$

Slotted disc encoders

This type of encoders are having circumferential concentric slots. A slotted disc with 3 bit gray codes and using emitter detector. There are three light sources as emitters on one side and correspondingly three detectors of light on the other side. For measuring their pulses through large significant bit, medium significant bit and small significant bit, higher source passing through concentric circular slots fall on the photo element to produce pulses. In encoders it is necessary to use gray codes instead of Binary codes.

In Figure 20.6 (e) circumferential slots have been arranged using binary codes and Figure 20.6 (f) shows circumferential slots for Gray codes where 0 is black and 1 is white. The code is increasing count in the CW direction. The disc is coded from outside ring (Lest Significant Bit) to inside ring (Most Significant Bit).

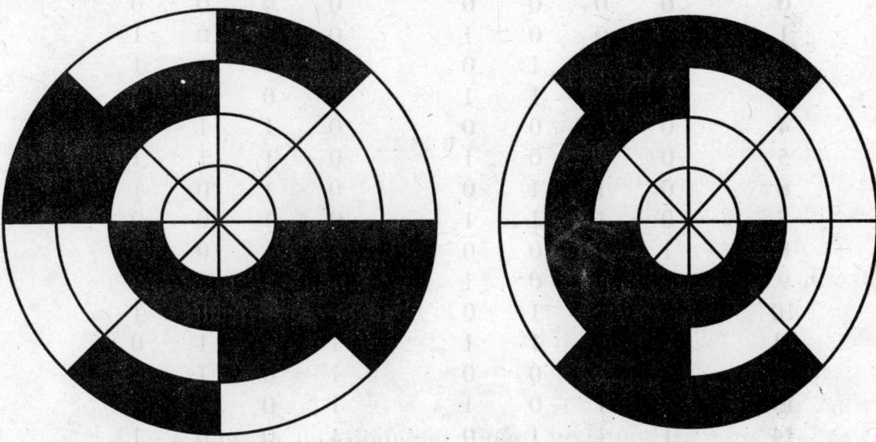

Figure 20.6(e) Binary coded disc. **Figure 20.6(f)** Gray coded disc.

For understanding the codes, it is necessary to see the table of numbers, as shown below:

Table 20.2(a) shows Gray code compared with Binary codes upto eight digits, since $2^3 = 8$ as there are three tracks whereas Table 20.2(b) shows Gray code compared with Binary codes upto 16 digits as there are 4 tracks.

If 'n' is the number of tracks, the absolute encoder will have a resolution of 2^n, then angular width of each increment is given by formula as given below.

$$\text{Angular Width} = \frac{360°}{(2^n)}$$

In case the resolution wanted is 0.3515° per revolution then

Table 20.2(a) Gray code compared with Binary codes upto eight digits

Number	*Binary Codes*			*Gray Codes*		
0	0	0	0	0	0	0
1	0	0	1	0	0	1
2	0	1	0	0	1	1
3	0	1	1	0	1	0
4	1	0	0	1	1	0
5	1	0	1	1	1	1
6	1	1	0	1	0	1
7	1	1	1	1	0	0

Table 20.2(b) Gray code compared with Binary codes upto 16 digits

Number	*Binary Codes*				*Gray Codes*			
0	0	0	0	0	0	0	0	0
1	0	0	0	1	0	0	0	1
2	0	0	1	0	0	0	1	1
3	0	0	1	1	0	0	1	0
4	0	1	0	0	0	1	1	0
5	0	1	0	1	0	1	1	1
6	0	1	1	0	0	1	0	1
7	0	1	1	1	0	1	0	0
8	1	0	0	0	1	1	0	0
9	1	0	0	1	1	1	0	1
10	1	0	1	0	1	1	1	1
11	1	0	1	1	1	1	1	0
12	1	1	0	0	1	0	1	0
13	1	1	0	1	1	0	1	1
14	1	1	1	0	1	0	0	1
15	1	1	1	1	1	0	0	0

$$\frac{360°}{(2^n)} = 0.3515°$$

Or in other words $n = 10$.

A four track 16 sector pure Binary Coded disc used as an Absolute Encoder disc, for tracks to indicate 2^0, 2^1, 2^2, 2^3 respectively.

Moire Fringe Pattern

In the case of 'Moire Fringe' radial grating, we have fixed annular disc and a rotating disc. Rotating disc is fixed on the rotor shaft attached to the axis leadscrew, while the stator grating disc is fixed to the stationary part. There is a distinct air gap between the two. By the previous method, as discussed above, the Moire Fringe pattern generated will rotate i number of times during each full rotation of the shaft, assuming i number of lines are engraved on the discs. Clockwise or anticlockwise-the direction are known by phase sensing (lead or lag) with the help of additional photocell. A typical Moire fringe pattern generated by the radial grating can be shown as under.

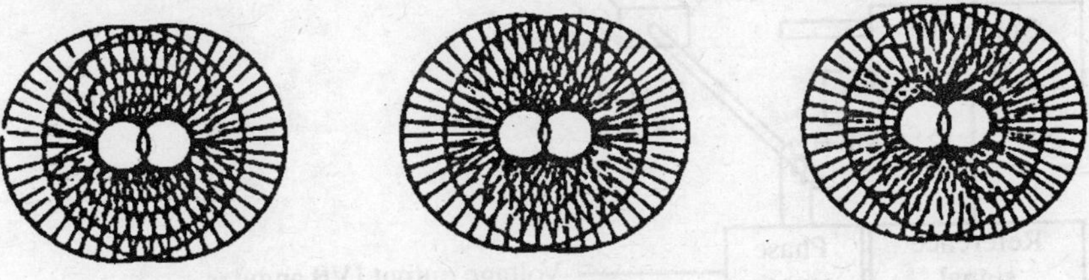

Figure 20.6(g) Moire fringe.

20.6.3 *Synchro and synchro resolvers and inductosyn*

Synchro position transducer is shown in Figure 20.7. The central spindle carries a winding attached to the leadscrew and rotates within the stator. By stator we mean the series of stationary windings around the periphery of the 'synchro'. Both synchro position transducer and synchro-resolver position transducers are based on the principles of magnetic induction. A synchro is an electro magnetic position transducer cornprising of a rotor and a stator with a number of winding. The level of the output voltage depends upon the angular position of the rotor. The peak value of the voltage induced varies

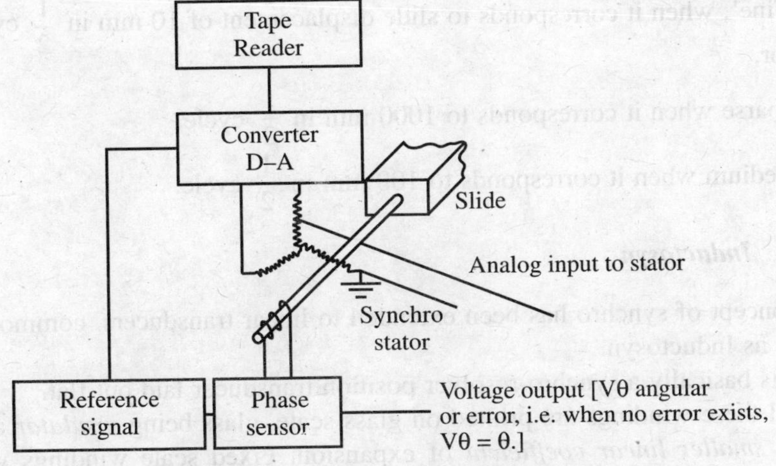

Figure 20.7 Synchro position transducer.

sinusoidally with the angle of rotation of the rotor. The voltage reduces to zero, when the axis of the motor coil coincides with the field vector of the stator.

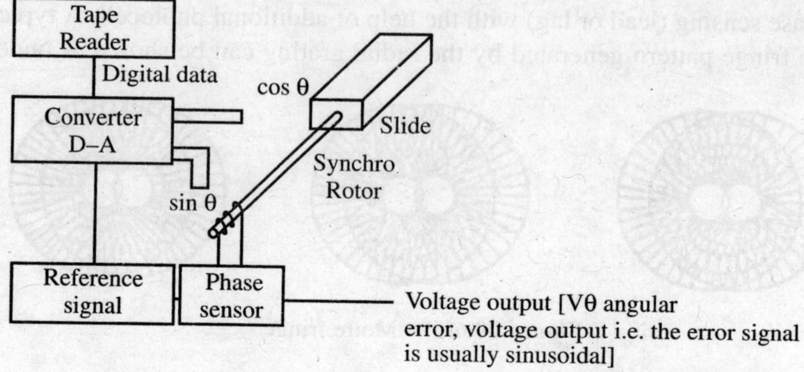

Figure 20.8. Synchro resolver position transducer.

The synchro-resolvcr has two windings at exactly 90° to each other and resolves the voltage into the components $V \sin \phi$ and V $\cos \phi$, where ϕ is the angular position of the rotor shaft.

If the rotor is coupled directly with the leadscrew then one revolution will correspond to one pitch of leadscrew as linear displacement of slide.

A number of synchros may be coupled to get 10:1 or 100:1 reduction/ amplification of feed traverse.

Normally synchros are:

'Fine', when it corresponds to slide displacement of 10 mm in $\frac{1}{2}$ cycle of rotor,

Coarse when it corresponds to 1000 mm in $\frac{1}{2}$ cycle.

Medium when it corresponds to 100 mm in $\frac{1}{2}$ cycle.

20.6.4 *Inductosyn*

The concept of synchro has been extended to linear transducers, commonly known as Inductosyn.

It is basically a synchro-resolvcr position transducer laid out flat.

All these windings are printed on glass scale, glass being *insulator* and having *smaller linear coefficient* of expansion. Fixed scale windings will have a voltage induced in it.

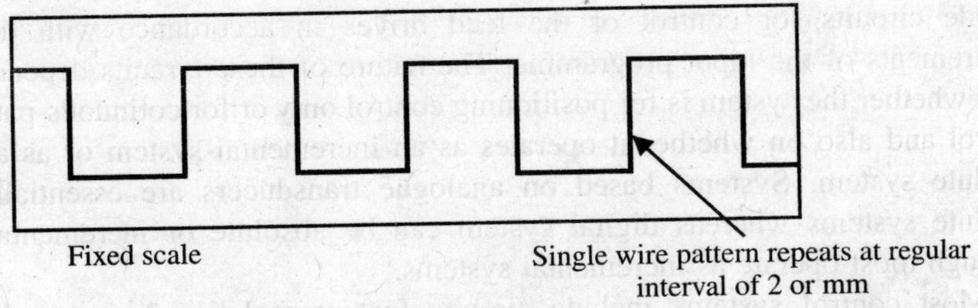

Figure 20.9(a).

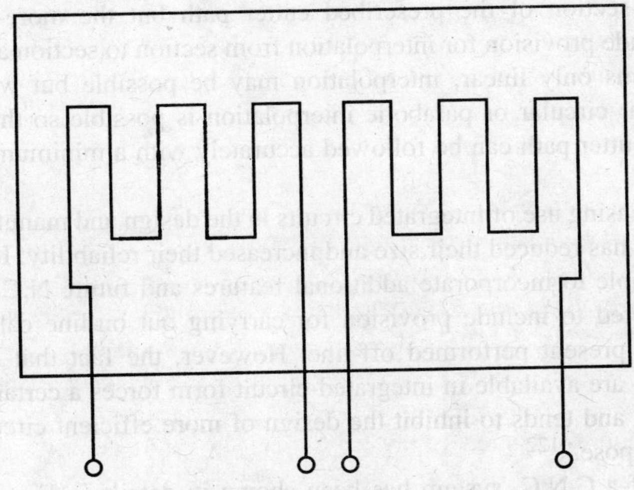

Figure 20.9(b).

Difference in the phase between the induced and the supply voltage will give positional movement of the slide.

20.6.5 *Control in C.N.C machine tool*

We have "feedback" or "no feedback" depending upon whether it is a closed loop or open loop control of the axis traverse. It is possible to depict the control system of a C.N.C. machine as shown under:

"The control system must also include circuits for the selection of spindle speeds, for the control of auxiliary functions such as the supply of coolant, and for operating the automatic tool-changing mechanism if fitted. Essentially these are simple iogic circuits but the control system must also include circuits for control of the feed drives in accordance with the requirements of the input programme. The nature of these circuits depends upon whether the system is for positioning control only cr for cotinuous-path control and also on whether it operates as an incremental system or as an absolute system. Systems based on analogue transducers are essentially absolute systems whereas digital system can be absolute or incremental, although most operate as incremental systems.

Most control systems include circuits for interpolation between the discrete points on the cutter path that are described in the input programme. With the simpler types of system provision is made for interpolation only within each section of the prescribed cutter path but the more-elaborate systems include provision for interpolation from section to section also. With simple systems only linear, interpolation may be possible but with many systems linear circular or parabolic interpolation is possible so that almost any desired cutter path can be followed accurately with a minimum of input information.

The increasing use of integrated circuits in the design and manufacture of N.C. systems has reduced their size and increased their reliability. It has also made it possible to incorporate additional features and future N.C. systems can be expected to include provision for carrying out on-line calculations which are at present performed off-line. However, the fact that complete block circuits are available in integrated-circuit form forces a certain degree of uniformity and tends to inhibit the design of more efficient circuits for a particular purpose."[122]

C.P.U. of a C.N.C. system has been shown in details in Figure 20.11. This consists of essentially (i) control unit, (ii) Arithmetic unit, and (iii) Intermediate access or internal memory unit. Speed of accession of informations in the internal memory is very high and in a matter few nanoseconds. Control unit is responsible for all coordinating functions carried out by the computer. External storage and backing store is suitable for using a C.N.C. machine with an external programme also as and when needed.

20.7 N.C. Retrofitting

This is a method of converting any manually controlled machine tool into a computer controlled machine tool.

It involves the following:

Fitting D.C. servo motors on each axis of the machine. Replacement of conventional leadscrews with recirculating ball screws.

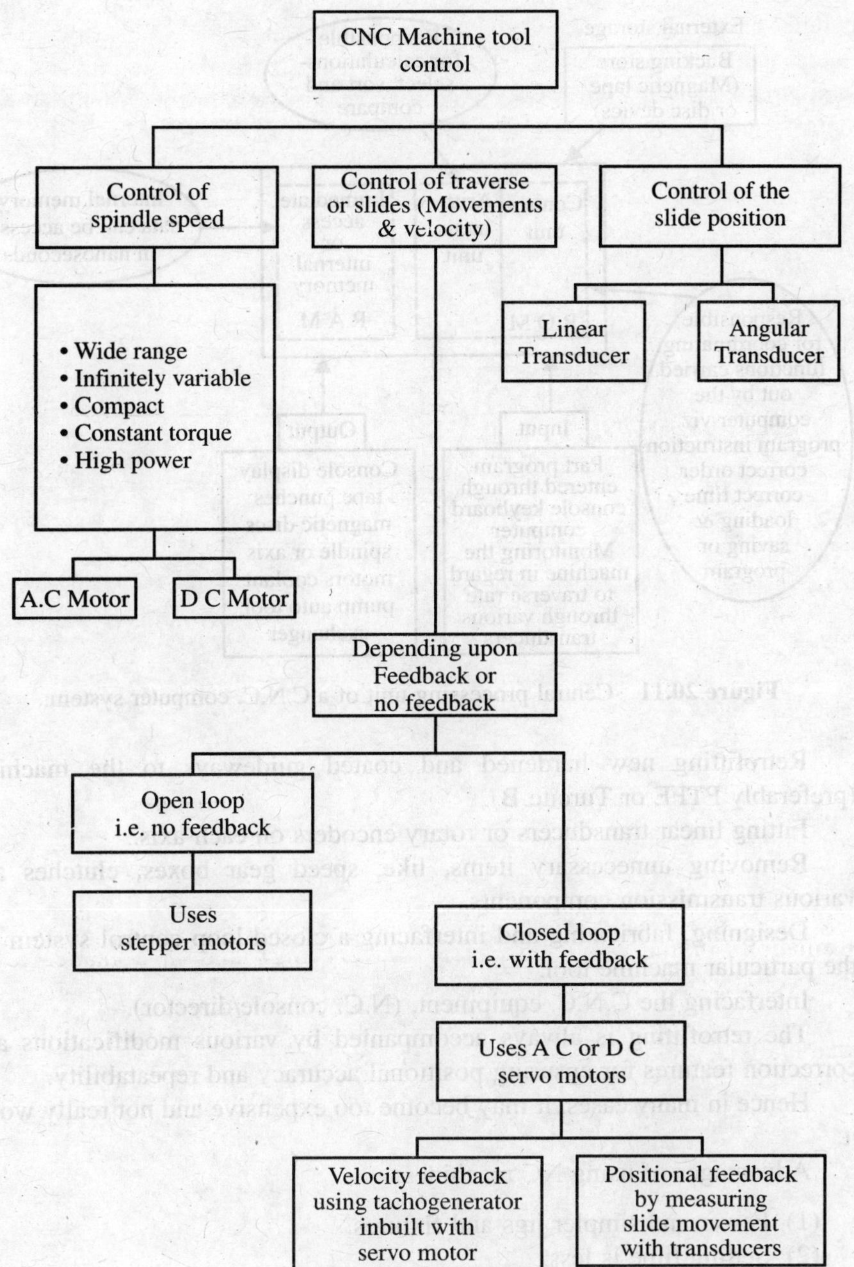

Figure 20.10 Control applied to C.N.C. machines.

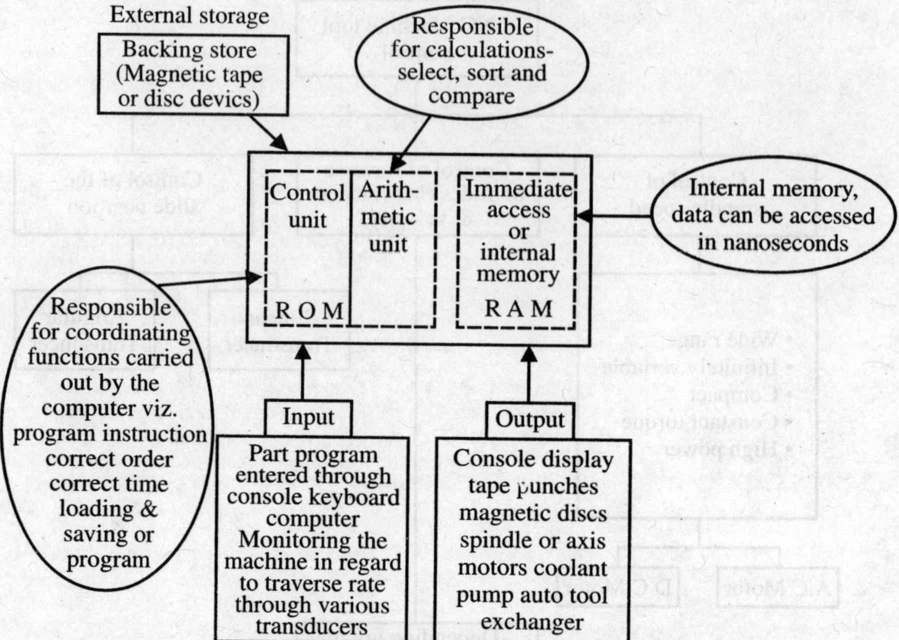

Figure 20.11 Central processing unit of a C.N.C. computer system.

Retrofitting new hardened and coated guideways to the machine, (preferably PTFE or Turcite B).

Fitting linear transducers or rotary encoders on each axis.

Removing unnecessary items, like, speed gear boxes, clutches and various transmission components.

Designing, fabricating and interfacing a closed loop control system for the particular machine tool.

Interfacing the C.N.C. equipment, (N.C. console/director).

The retrofitting is always accompanied by various modifications and correction features for ensuring positional accuracy and repeatability.

Hence in many cases, it may become too expensive and not really worth it.

Advantages of using NC machines:

(1) Fewer and simpler jigs and fixtures,
(2) Setting time is less,
(3) Reduction of the skill requirement of the operator,
(4) Increased flexibility. Modifications can be introduced by merely modifying the control tape instead of modifying jigs and fixtures (By changing the software),

(5) Fewer machine tools are needed,
(6) Rationalisation of production Control System.

20.8 Programme and Computer-man Interaction

Computer programmes, solving routine tasks and display units, for graphical representation of workpieces, tools, tool paths or manufacturing processes, are easy to use and effective aids for rationalisation of production planning.

Significant advantages of this graphical interactive method are:

(i) The interactive mode yields a high flexibility of planning and specific requirements can be met easily.
(ii) Planning process is minimised and easy to control.
(iii) For N.C Programming no test of the punched tape is necessary.
(iv) The dialogue leads through the programme and therefore the user does not have to learn any computer language.
(v) Clamping, the selection of tool and tool movements plans can be generated automatically.

For input in dialogue *Alpha numeric* keyboard is used. An additional feature is the possibility to define coordinate data with an adjustable cross wire on the screen. Intermediate results are displayed on the terminal while the plotter, line printer, paper tape punch are used to issue machining plans.

Programme system consists of 4 processors connected by a monitor routine.

Geometry processor transforms the workpiece description into an internal data structure.

Production planning processor is used for processing the geometrical problems and for graphical display of results.

Cutting processor determines the values of cutting variables.

Output processor compiles from computer internal data for manufacturing.

20.9 Present status

Though India has started in big way to go into areas of high technology related to manufacturing sciences, our effort is insignificant in comparison to other countries of the world. There exists a tremendous technology gap in our indigenous effort to make sophisticated NC, CNC machines. Also there is not much adequate response from the user industries. Absorption of technology, to some extent requires uptodate knowledge in the existing technology know-how and a vision to appreciate the countries, need in future.

20.10 Electronics Revolution and Computer Growth

The development of NC system from the gas filled electronic valve construction, through solid state transistors, large scale integration (LSI) circuits and the current VLSI and microprocessor have contributed to increased reliability and lower cost of the system.

Here it would be pertinent to take a look at the way miniaturisation and micro miniaturisation in electronics has been advancing. FANUC, one of the reputed manufacturers of NC systems have made the following transition of fundamental hardware technology in NC systems.

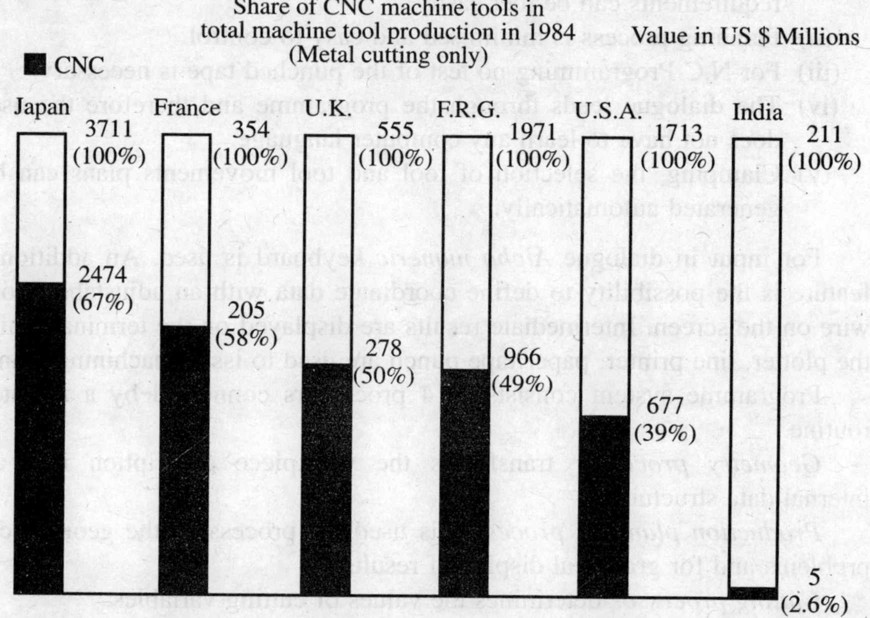

Source: American Machinist & Alternated Manufacturing February 1986

Figure 20.12(a) Shows the dismal picture of CNC machine tools production in India vis-a-vis advanced countries of the world.

One of their first generation systems had transistor elements and the NC system consisted of 280 printed circuit boards (PCB) and extensive wiring.

With the advent of integrated circuits (IC) only 40 PCBs were used with reduced wirings for the system.

Further ibtegration called MSI replaced nearly 10 ICs and the system involved only 5 PCBs.

With the introduction of LSI techniques the system employs only 2 PCBs using LSI, MSI & ICs in a combination. The trend with micro electronics had been that while costs lowered, reliability of systems increased.

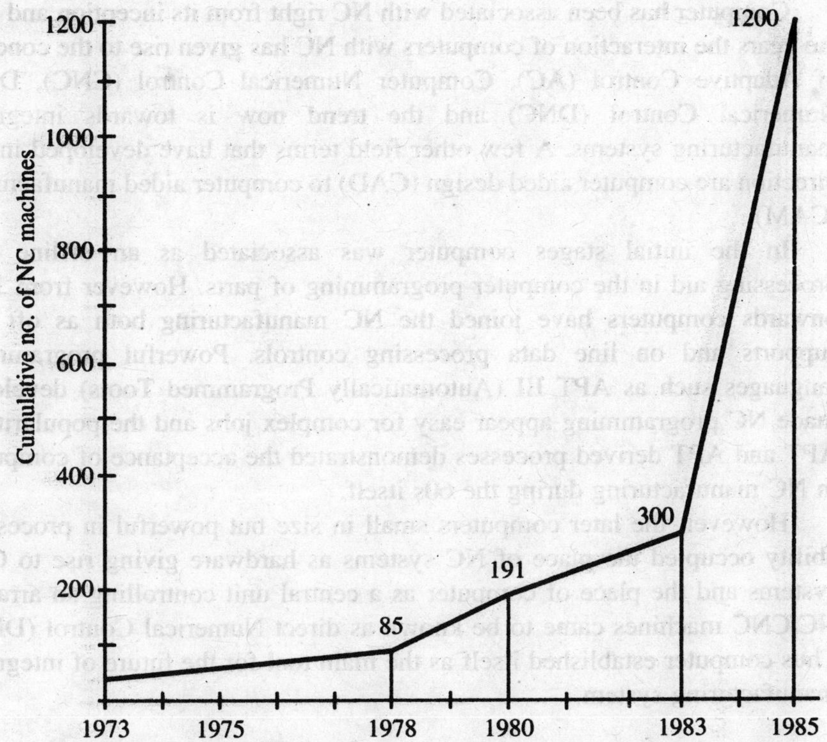

Source: Perspective Plan for machine tools and Indian Machine Tools census 1966

Figure 20.12(b) Shows a definite trend in the increased uses of N.C. machine tools in India during the period 1983-85. Though data for 87-88 not known, it is expected to have a high trend.

It is in this context that the dedicated mini-computers came to the market as computer, Numerical Control (CNC) systems. The inherent feature of memory in the Computer Numerical Control system gave the new technology–'Software Oriented systems'. Here was a system which was flexible unlike the rigid hardwired circuit NC systems used hitherto.

The appearance of the microprocessor on the scene brought a new range of computer numerical control (CNC) systems which are highly cost effective. These with the semiconductor memory devices have reduced the size of the system as compared to the core memories used in the earlier mini-computers. The distinguishing features that marked the mini-computers, as capable of higher processing ability a few years back, are gradually disappearing as micro-computers are steadily entering the arena overcoming this disadvantage.

Computer has been associated with NC right from its inception and over the years the interaction of computers with NC has given rise to the concepts of Adaptive Control (AC), Computer Numerical Control (CNC), Direct Numerical Control (DNC) and the trend now is towards integrated manufacturing systems. A few other field terms that have developed in this direction are computer aided design (CAD) to computer aided manufacturing (CAM).

In the initial stages computer was associated as an offline data processing aid in the computer programming of parts. However from 1970 onwards computers have joined the NC manufacturing both as off line supports and on line data processing controls. Powerful programming languages such as APT III (Automatically Programmed Tools) developed made NC programming appear easy for complex jobs and the popularity of APT and APT derived processes demonstrated the acceptance of computers in NC manufacturing during the 60s itself.

However, the later computers small in size but powerful in processing ability occupied the place of NC systems as hardware giving rise to CNC systems and the place of computer as a central unit controlling an array of NC/CNC machines came to be known as direct Numerical Control (DNC). Thus computer established itself as the main tool for the future of integrated manufacturing system.

20.11 Direct Numerical Control (DNC)

The association of computer with NC right from its inception stage is quite noteworthy. Inclusion of a mini-computer within the N.C. system came at a much later date than the involvement of Digital Computers with NC manufacturing.

As early as 1967 Fanuc developed a computer operated machining group control system called system K. Under this group control system, seven NC lathes were controlled by a central computer. This computer kept control of the part programme loading, batch quantity regulation and delivery schedules. As more and more computers came to be used in such a concept the term 'Direct Numerical Control' became popular. By 1974 there were 70 such DNC installations in Japan. In the U.S.A. also the DNC was catching up fast, pioneered by NC machine tool builders, system builders and computer manufacturers. However, the DNC installation being expensive and requiring highly skilled software knowledge could be installed in large corporations.

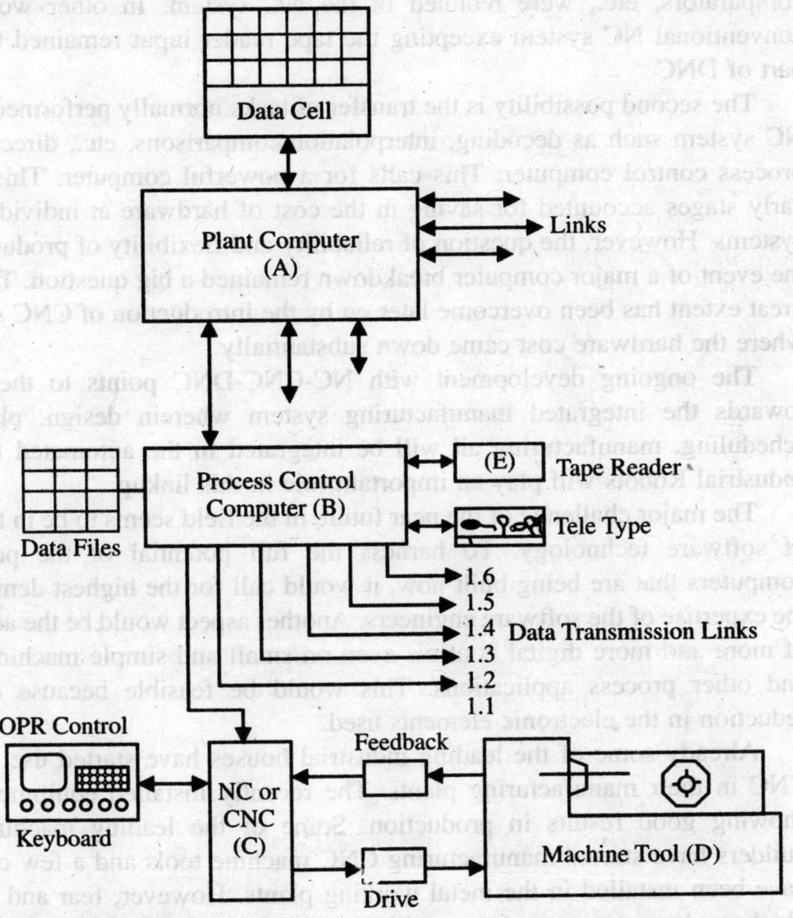

Figure 20.13 Shows the typical block diagram of a DNC installation.

The main plant computer (A) is the one used for part programming and for other data processing operations. With this is interlinked the process control computer (B) which is assigned to control a group of NC systems and machine tools (C and D) through data transmission links (1_1, 1_2, 1_3, etc). The process control computer (B) with its own data files can store active programmes and supply to the NC systems as required. New programmes can be introduced through the tape reader (E) or will be provided by the main plant computer (A).

Two distinct approaches were noticed in the DNC field. The first, in which the process control computer fed the programmes and functions such as character check, decoding, buffer store, internal interpolations, comparators, etc., were retained in the NC system. In other words the conventional NC system excepting the tape reader input remained there as part of DNC

The second possibility is the transfer of tasks normally performed by the NC system such as decoding, interpolation/comparisons, etc., direct to the process control computer. This calls for a powerful computer. This in the early stages accounted for saving in the cost of hardware at individual NC systems. However, the question of reliability and flexibility of production in the event of a major computer breakdown remained a big question. This to a great extent has been overcome later on by the introduction of CNC systems where the hardware cost came down substantially.

The ongoing development with NC-CNC-DNC points to the future towards the integrated manufacturing system wherein design, planning, scheduling, manufacturing all will be integrated in the automated factory. Industrial Robots will play an important role in this linkup.

The major challenge of the near future in the field seems to be in the area of software technology. To harness the full potential of the powerful computers that are being built now, it would call for the highest demand on the expertise of the software engineers. Another aspect would be the adoption of more and more digital controls even on small and simple machine tools and other process applications. This would be feasible because of cost reduction in the electronic elements used.

Already some of the leading industrial houses have started use of NC/CNC in their manufacturing plants. The recently installed equipments are showing good results in production. Some of the leading machine tool builders have started manufacturing CNC machine tools and a few of these have been installed in the metal working plants. However, fear and mental blocks and resistance to change will take some time before NC can be made popular in India. Signs of growth for this technology are already evident. The imports and indigenous NC machine tools in the past 4 to 5 years have accounted for nearly 75 to 100 million rupees in purchase value. With the growth of electronics in this country NC holds a great promise for the future of manufacturing in India.

20.12 Part programming

Various part progamming languages, available commercially, are given below in a tabular form:

APT (Automatically Programmed Tool)	Originated at MIT 1952
APT II	MIT in sponsorship with Aerospace Industries Association, 1961
ALRP (APT Long Range Programme)	Aerospace Industries Association & Illinois Institute of Technology Research Institute at Chicago (1961)
ALRP. Changed to CAM-I Computer-aided Manufacture-International	" Since 1969
NELAPT (A subset of APT) (NEL stands for National Engineering Laboratory)	National Engineering Laboratory, UK, vocabulary same .s APT. But some words are condensed to carry same meaning. Example: C-for 'Circle' V–for 'Vector' GB–'Go Back'
Ex APT (Extendedy subset of APT) ExApt–I ExApt–II ExApt–III	Developed by Prof H. Opitz et al., of Aachen. Maintained by Exapt Association of Industries.
Autospot	Developed in U.S.A. (Simultaneously with APT), for point-to-point, circular-arc machining on machining centres with ATC. But this is not fully compatible with APT.

20.13 Part Programming

A set of instructions needed for machining a component (known as a part), by using NC, CNC machine tool can be written as

NGXYZ ABFSTMEOB

N— Sequence Number
G— Preparatory Function
XYZ AB— Co-ordinate dimensional data
F, S— Feed and Speed function respectively
T— Tool Function
M— Miscellaneous Function
EOB— End of Block

Commands which instruct the machine (various coordinate axes and other executive units) for different movements, like, straight positioning, contouring etc. are called Preparatory Function and denoted by G. Usual modes of function G are:

G00 PTP Positioning
G01 Linear contouring
G02 Spindle CW, circular interpolation
G03 Spindle CCW
G04 Dwell
G28 Thread cutting macro block I
G29 Thread cutting macro block II
G33 Thread cutting mode
G70 Inch dimension (coordinate)
G71 Metric dimension (coordinate)
G90 Absolute mode, w.r.t. orginal datum
G91 Incremental mode, w.r.t. current datum
G94 I.P.M., MMPM mode
G95 IPR, MMPR mode
G98 Servo drive adjust cycle
G99 Servo drive calibration cycle

In a similar fashion we ha"ve the classification of Miscellaneous Function M codes and Tool Function T codes. Some of the M codes are being reproduced here:

M00 Programme stop
M01 Optional stop
M02 E.O. program
M03 Spindle CW
M04 Spindle CCW
M05 Spindle 'off'
M08 Coolant on
M09 Coolant off
M17 Chip conveyor on
M18 Chip conveyor 'off'
M30 End of tape, tape rewind
M39 Auto recycle.

Tool functions are normally 4 digit T word.

In APT programming the coordinates are normally put as 5 digit number. Though from machine to machine, there is some variation. Program format used by a particular machine is included in the manual for user's benefit.

In some cases a dimension of 162.43 in X axis may be denoted by X 16243, 18.43 in Y axis may be denoted by Y 01843, +ve or –ve signs are also provided to denote the direction of coordinate movement. Since the sequences can be large in number, N code may start from N001.

Linear and circular are the two distinct types of interpolation generally used in a programming.

Coordinates of the destination point are provided in the Block and relevant machining parameters. So far as the slope of the straight line or otherwise extent of intercept are concerned, the calculations are done in the data processing unit. For example if the destination point has coordinates (45, 58) along a-straight line we write the address format as

N010 G01 X 45000 Y 58000 F 120.........

For circular interpolation to proceed along an arc having a centre (0, 0) to a point (80, 0) in the programme, we write the format

N011 G02 X 80000 Y 00000 I 25000 J 35000

where I, J corresponds to coordinates of the centre with an origin at (25, 35).

Normally the programme is written in the three different methods:

(i) Word address format
(ii) Fixed sequential format
(iii) Tab sequential format

(i) **Word address format:**

N015 G81 X 30000 Y 43000 Z 72000 F 100 T01 M03.

(ii) **Fixed sequential format:**

Same instruction may be written in fixed sequential format as:

015 81 30000 43000 72000 100 01 03.

Here there can be errors in the programme writing as well as in its consequent interpretations. In word address format, there is no chance of such errors creeping in and hence that is supposed to be more advantageous.

(iii) **Tab sequential format:**

In this system each element of information is separated by TAB. In case of repetition, the programme can be repeated by TAB TAB TAB.

Figure 20.14 shows a typical component being machined on an HMT NH CNC machine. The part programming for this component is shown in Figure 20.15.

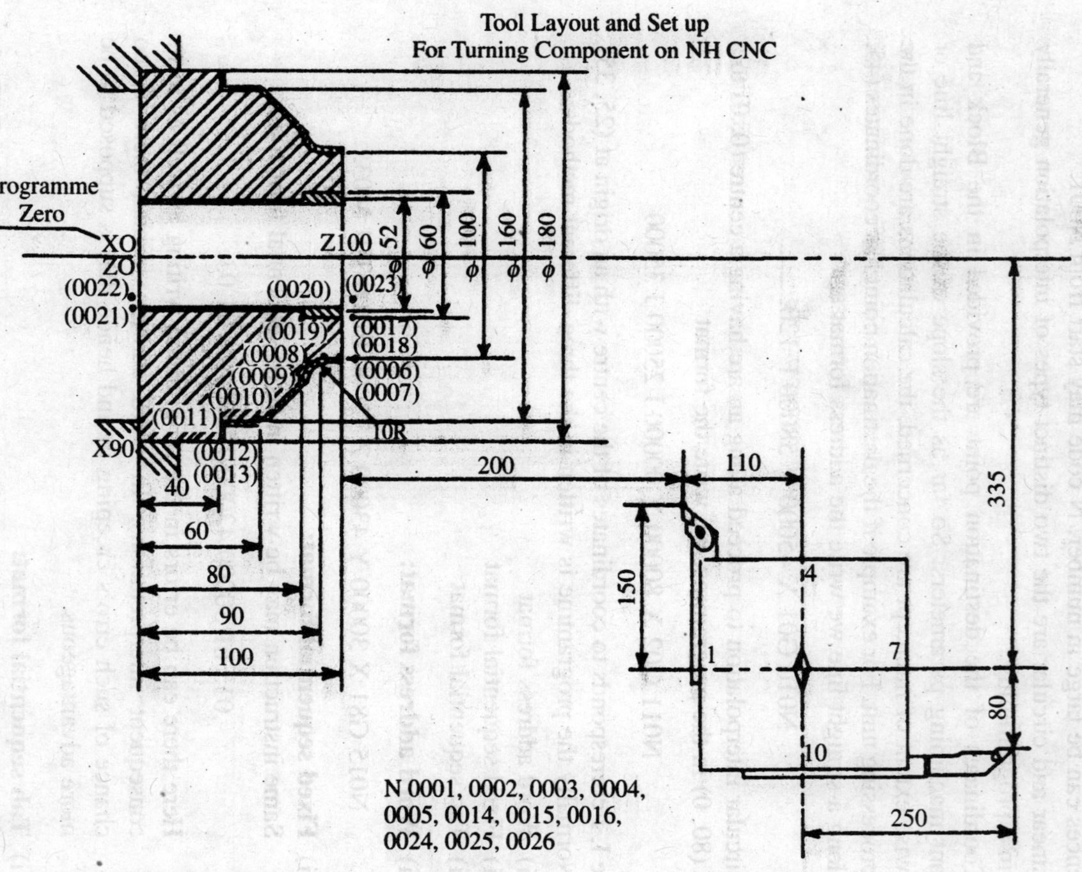

Figure 20.14 Components with dimensions and tool station for a program.

NC PROGRAMME

(Component (or NH CNC Lathe)

N	G	X	Z	I	K	F	M	S	T
N0001	G90								
N0002	G71								
N0003	G00	X0	Z0						
N0004							M03	505	T0101
N0005	G92	X + 185000	Z + 300000						
N0006	G00	X + 50000	Z + 103000				M08		
N0007	G95								
N0008	G01		Z + 90000			F200	M17		
N0009	G03	X + 60000	Z + 80000	I10000	K0				
N0010	G01	X + 80000	Z + 60000						
N0011			Z + 40000						
N0012		X + 92000							
N0013	G94								
N0014	G00	X + 185000	Z + 300000				M09		T0100
N0015								S06	T0702
N0016	G92	X + 255000	Z + 160000						
N0017	G00	X + 30000	Z + 103000				M08		
N0018	G95								
N0019	G01		Z + 80000			F150			
N0020		X+26000							
N0021			Z – 2000						
N0022	G94	X + 20000				F2000			
N0023			Z+103000				M18		
N0024	G00	X+255000	Z + 160000				M09	S00	T0700
N0025	G92	X0	Z0						
N0026							M30		

Figure 20.15 Progromme in NH CNC lathe of HMT.

20.14 CNC Lathe

Figure 20.16 shows the outside dimensions and capacity chart of HMT CNC Lathe. Figure 20.17 and Figure 20.18 show the component with full dimensions for NC. programming on this lathe and the two alternative layout of tools for machining the component respectively.

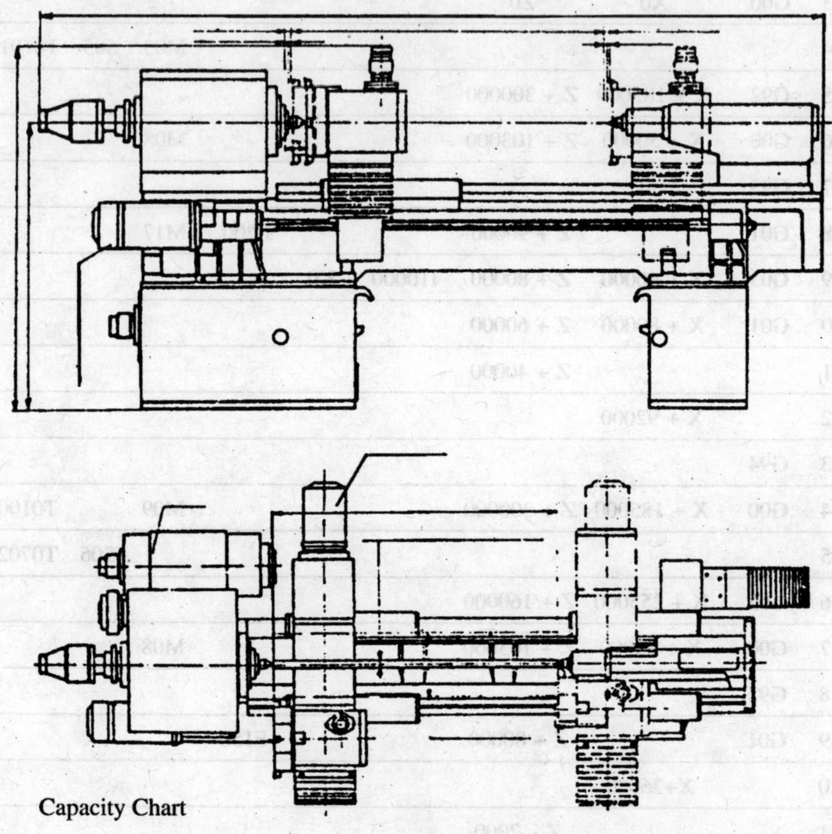

Figure 20.16 Capacity chart for HMT NH CNC.

20.15 Machining Centre HMT-KTM (Horizontal Machining Centre)

HMT-KTM Horizontal Machining Centres adopt a unique modular approach. The major sub-assemblies, such as the headstock, tool changer, drives, power distribution panel, controls and pallet shuttle, are standardised.

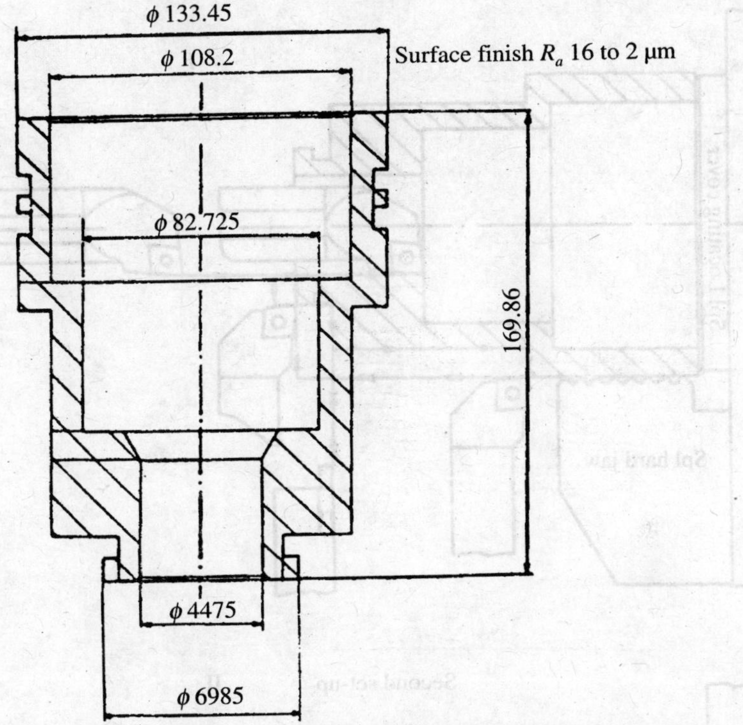

Figure 20.17 A component for machining.

The basic structural elements can be altered to obtain different traverses and configurations. Since these structures are weldments, changing is easy and inexpensive. This ensures flexibility. The degree of commonality reduces the learning period. Programmes, operators and maintenance personnel wno are familiar with one model will have no difficulty in switching to another. Spares and tooling are common to all the models.

HMT-KTM Horizontal Machining Centres have been designed to serve as an ideal starting point for the FMS, the flexible manufacturing system.

Tool magazines of higher capacity, pallet changers and extensive part programme storage-essentials for the crucial first step towards the FMS-are offered on these machines.

Even so, HMT-KTM Horizontal Machining Centres are simple stand-alone machines that can meet your present requirements.

Slideways

The hardened and ground steel ways are standard. They are rectangular and bolted to the supporting member. The moving members are fully supported

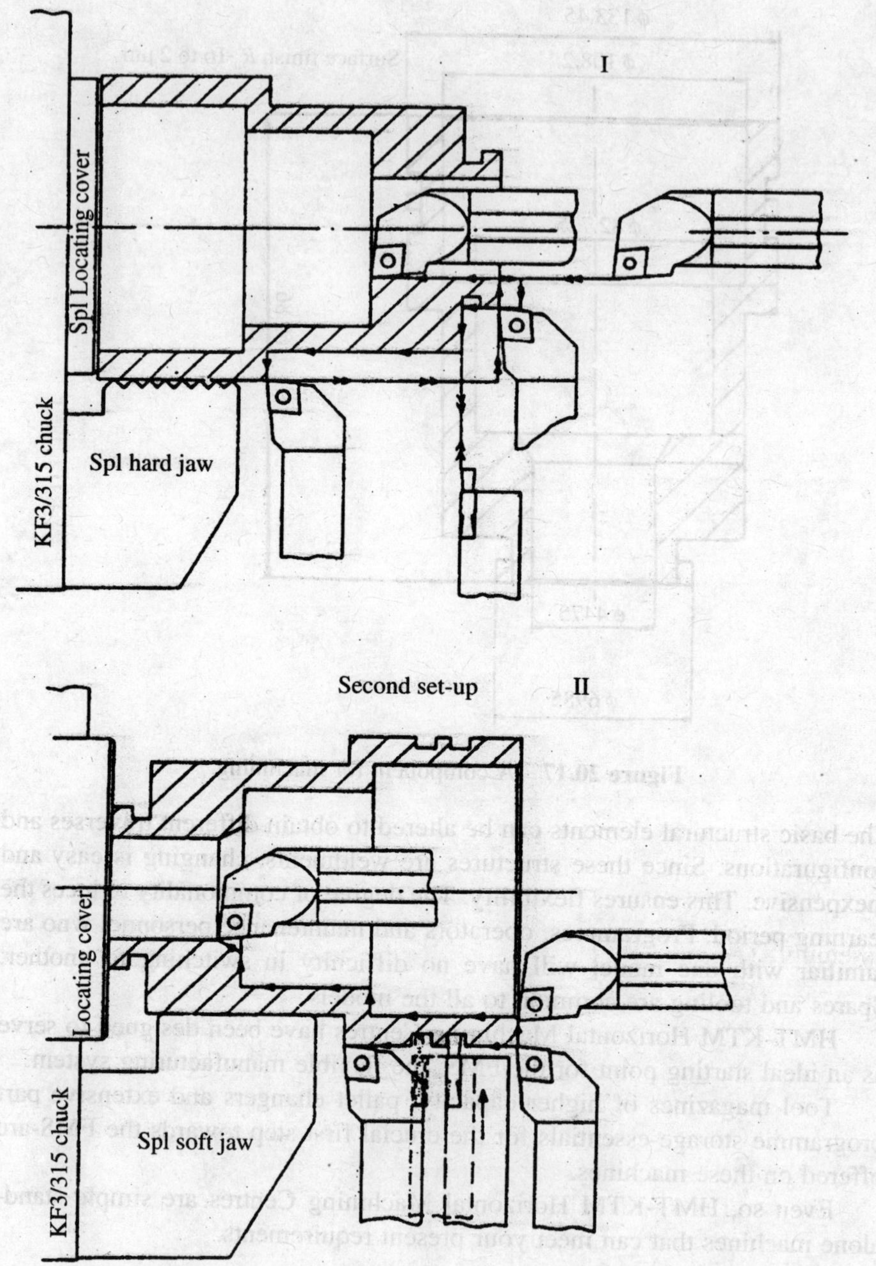

Figure 20.18 Two alternative tool layout for a component shown in Figure 20.17.

over full travel. Tychoways and turcite bearings are used to guarantee stability, rigidity and accuracy. Tychoways provide vertical and lateral support. Turcite-lined keeper plates provide reverse support and damping. An automatic lubrication system ensures the adequate supply of oil to each sideway. The Y axis covers are flexible roller type covers and the X axis covers are of the overlap sliding type. (See Figure 20.19)

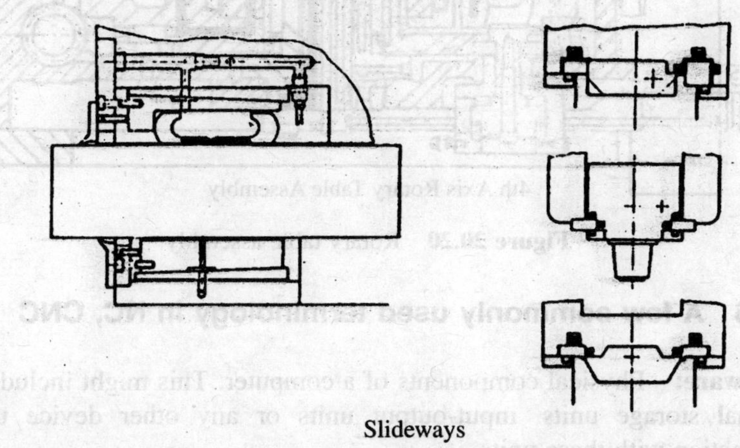

Slideways

Figure 20.19 Slideways used in HMT-KTM Horizontal Machine Centre.

Rotary Table (See Figure 20.20)

The HMT-KTM 760 and HMT-KTM 1000 have a full rotary 4th axis table as a standard feature. This can either be used for positioning or rotated while machining. The table is driven through a dual lead worm and worm wheel assembly. The main load is taken up by the bearing ring. Radial support is provided by taper roller bearings on the central post. The pallet is accurately located by four precision cones on the saddle of rotary table and clamped hydraulically. The rotary table is clamped in position by disc-spring-actuated and hydraulically released clamps. The table is programmed directly in degree with feed rates in mm/min. Feedback is through an encoder. The drive for the rotary table is contained within the X axis table.

Auto tool changer

The auto tool changer is a standard module 40 and 60 tool magazines are available. The tool magazine is mounted on the left hand side of the column, outside the machining area, away from the swarf and the coolant.

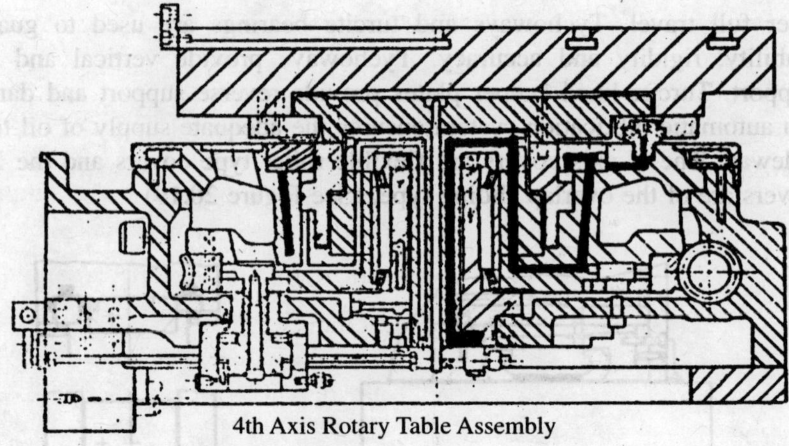

Figure 20.20 Rotary table assembly.

20.16 A few commonly used terminology in NC, CNC

Hardware: Physical components of a computer. This might include CPU, external storage units, input-output units or any other device used in conjunction with these units.

Software: A set of computer programmes.

Program: A set of instructions written in a programming language, either machine code or high-level language for a computer to solve some given problem.

CPU: Central processing unit, a part of the computer consisting of *arithmetic logic unit, control unit and internal storage*.

Disk: A magnetic storage device upon which data and instructions can be written and stored.

Floppy Disk: A small storage disk 8" or 51" diameter enclosed in a square jacket. It is coated with magnetic material and can store information.

Character: A symbol, a digit or letter or special sign which is acceptable to a computer as the smallest unit of information.

Microprocessor: A processor built on an integrated circuit (silicon chip) that can function as CPU.

Mini Computer: A medium to small-scale computer.

Mainframe: A computer having a wide range of facilities and capabilities.

Signal: Any physical quantity used to transmit information between one element of a control system and another.

Feedback: Transmission of signal from a later to an earlier stage.

Openloop: Without monitoring feedback.

Servomechanism: An automatic monitored kinetic control system which includes a power amplifier in the forward path.

Gain: Amplification of an element or system in which input and output signals are of the same physical kind.

Hunting: A prolonged self-sustained oscillation of undesirable amplitude.

Magnetic Tape: A strip of plastic material e.g. polyester or polyvinyl chloride usually 6 or 25 mm wide, to one face of which is bonded a thin layer of (λ-Fe_2O_3) particles. Oxide particles are of 0.001 mm in size and are distinguished from non-magnetic (α-Fe_2O_3) by crystal structure. It usually has single, double, four or multiple tracks.

CHAPTER 21

Robot Language-state of the Art

21.1 Introduction

Today industrial robots are considered to be key elements in FA (factory automation) system as well as in many other applications related to production engineering. A general survey reveals that most of the industrial robots meant for repetitive and fixed sequence operation (like those used in welding painting, etc.) are controlled by means of teaching playback system. This system is very much user-friendly from the operation point of view. However, they lack the capability of altering the sequence, add, delete, or change the sequence of motions and positional data. For tasks like assembly and inspection, the main operation of the robot is done by point-to-point control and in this case it is necessary to have a function that alters the sequence of motion easily, one that interlocks with the peripheral units and various sensors to improve adaptability. This, therefore, necessitates computation, decision taking capability of the robots, through microprocessor based system and programming facility. However, the complexity of operations on the part of robots makes it difficult to develop programs, and an effective method of programming is one of the keys to success of automation. This is the main reason for which all leading manufacturers of robots are engaged for the exploration of suitable efficient robot language.

This chapter discusses the programmability of robots with suitable classifications and reviews the available programming languages with a few examples.[106]

21.2 Robot Language Outline

To operate a robot, it is necessary to give instructions by certain method. There are four primary methods for programming a robot:

Physical Set-up

Here the operator sets up programmes by physically fixing stops, setting switches, arranging wires, etc. This is a characteristic of the simpler robots.

Lead Through

Here the operator leads the robot through the desired positions and locations by means of a remote teach box. These points are recorded and used to generate the robot trajectory during operation.

Walk Through

Here the robot arm is physically manipulated through the desired motions (which are recorded then played back by the robot control during operation).

Writing A Software Program

A software program is written and executed when desired. The emphasis in programming research today is on software programming of computer controlled robots. Work on sensor controlled manipulation is extending the scope for program-mability. Interacting with the robot by means of software provides more flexibility than the other programming methods, and allows for conditional actions of flexible adaptations.

The basic components of a robot language are:

(i) Instructions which determine the actions of the robot;
(ii) Tools which construct a sequence of procedures for a task;
(iii) Media to transfer a sequence of codes into other departments and which accumulate this data.

Although these demands are met in other high level languages, a person using a robot language is not always a computer engineer, and this calls for the following conditions to be met, viz.,

(i) instruction must be simple, and (ii) media must be readable and unified.

21.3 General Description of Programming Langnage

All processing languages are characterized by their capacity to define data structures as well as by the algorithmic structure. For the level of programming, closest to the machine, there is-only one type of data: The binary unit, which has only two values (0 and 1).

The programs written in machine code are:

Difficult to write correctly;
Difficult to correct, if errors are found; and
Difficult to reread and therefore to maintain.

Since the earliest days of programming, other languages, known as high level languages have been developed to make program writing easier and more reliable. To be used by a computer, a program written in a high level language always requires another program which is executed (in machine code) by the same computer or possibly a separate one. Two solutions are possible:

- The compiler translates the program written in high level language into a machine code, which can then be executed by the computers; and
- The interpreter directly executes the program written in high level language.

The latter solution is most commonly chosen for BASIC and for robotic languages because, despite the lower performance, it is more flexible to apply.

Robot languages are generally derived from the real time versions of the major programming languages (essentially FORTRAN, BASIC, PASCAL, ADA) for their general programming facilities, that is, on the level of the data structures and associated operations and on the algorithmic level.

21.4 Real Time

With a few exceptions, all standard programming languages have been developed to resolve problems, in which the concept of time synchronization (real time) is not involved. The difficulty arises essentially from the sequential nature of a program, whereas a control problem is parallel and combinational. This problem is solved by the rapidity of computers in relation to the process to be controlled. This allows the process to loop very rapidly on a set of relatively independent programs giving the impression that each program is a permanent active function. Programming of these functions is generally achieved by declaring them independent tasks with rules for activation and deactivation, frequency of call and priority. The rules are defined in the main program and are activated by a monitor. It is this monitor function, absent in general programming languages, which produces the specificity of real time languages.

21.5 Geometric Modelling

Because of the nature of the tasks to be undertaken, robot programming languages must exercise perfect control over the movements of robot segments in the work space. This approach, which is valid for programming by traiping is extremely difficult in symbolic programming, when the positions acquired are not as a result of training but calculation. In high level robotic languages, the programmer has access to variables which define the positions and orientations of various segments in a cartesian set of coordinate axes. The approach generally is to choose a set of coordinate axes associated with the segment and to describe the segment contained in it (Figure 21.1).

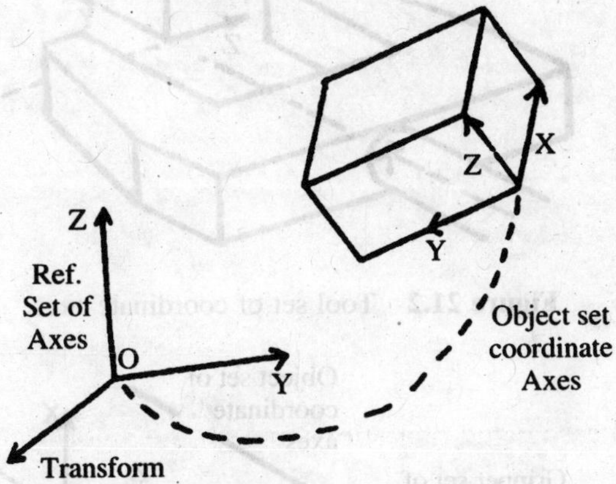

Figure 21.1 REF set of axes transform.

In theory, three variables are sufficient to define the rotation. The most commonly used method in robotics is the 3 × 3 matrix, in which each row represents the coordinates of the projections of the new coordinate unit vectors in the old one.

Although the level of redundancy is high (nine variables instead of three variables), this method of representation offers a number of advantages:

The matrix variables are often used for latter calculations;

The method avoids the problems of singularity;

The representation is relatively easy to visualize.

21.6 Tool and object sets of coordinate axes

In manipulation work, a preferential set of coordinate axes is one which describes the location of the end effector (e.g. gripper or tool). This set, known as the tool set of coordinate axes, generally has the tip of the tool or the central gripping point of the gripper as its origin and its axes are oriented as shown in (Figure 21.1). In the case of variable tools or grippers, the tool set of coordinate axes is deduced from an intermediate set, called the wrist set of coordinate axes, using transformation which depends on the tool used and can be modified by the user. Robot programming is essentially to make the tool set of coordinate axes coincide with the others linked to the task to be executed. If an object is to be manipulated the user must define one or more gripping sets of coordinate axes for each object, bearing in mind the characteristics of gripper, so as to make its set of coordinate axes coincide with those calculated for gripping motions (Figure 21.3).

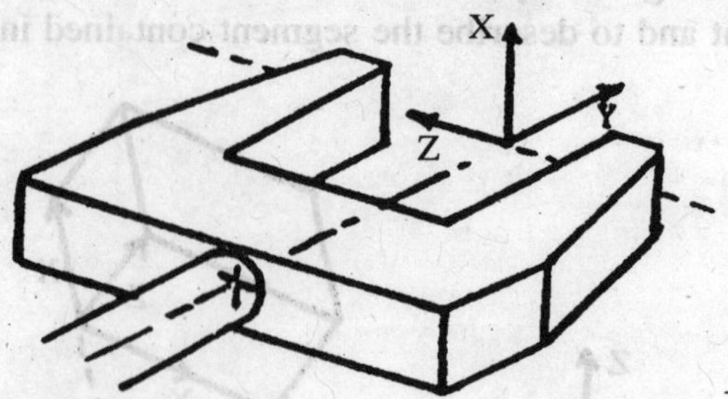

Figure 21.2 Tool set of coordinate axes.

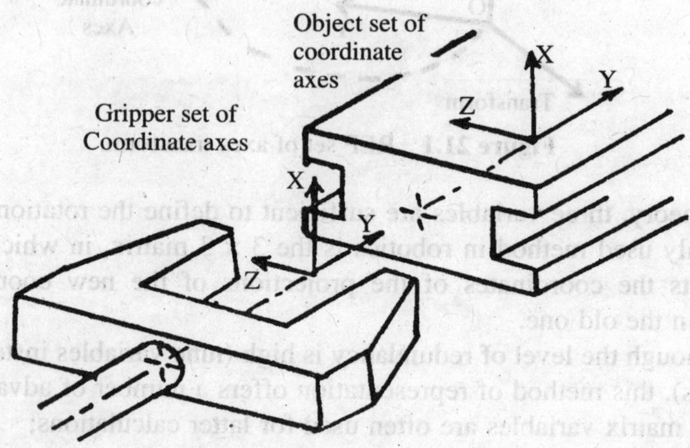

Figure 21.3 Gripping on object.

Some robot languages (e.g. AL, LM) allow automatic updating of the set of coordinate axes when the objects are moved. Some sets of coordinate axes can be linked to a single object. Using a linking instructions (like ATTACH IN LM), several sets of coordinate axes can be moved simultaneously. The inverse instruction (like DETACH) allows the sets of coordinate axes to be made independent of another.

21.7 Movements

Generally speaking, programming of robot motions is carried out in the point-to-point mode with the destination of the tool set of coordinate axes specified as:

21.7.1 *Move gripper to object*

Or by specifying relative movements in relation to the current positions:

MOVE GRIPPER BY ΔX, ΔY, ΔZ, $\Delta\alpha$, $\Delta\beta$, $\Delta\gamma$

Depending on the language, these two basic instructions may or may not accept the variants described hereafter.

21.7.2 *Structure of the trajectory*

The trajectory may be executed in the free mode, the coordinated actuator mode or linear interpolation in the Cartesian mode. Depending on the language and the robot, the user either has free choice of these three modes, to choose between two or in some cases, no choice at all.

Example: MOVE GRIPPER TO OBJECT IN CARTESIAN MODE

21.7.3 *Speed or time of execution*

In many applications, it is desirable to perform the trajectory either at a given speed or in a given time. It is often possible to provide a speed factor relative to the nominal speed.

Example: MOVE GRIPPER TO OBJECT AT SPEED = 0.25

21.7.4 *Intermediate points*

In complex trajectories, or if it becomes necessary to avoid obstacles, it is useful to be able to specify a trajectory made up of several segments without the robot stopping at each intermediate point.

Example: MOVE GRIPPER VIA R_1, R_2, R_3 TO OBJECT several movement instructions can be linked-
CONTINUOUS
MOVE GRIPPER TO R_1
MOVE GRIPPFR TO R_2
MOVE GRIPPER TO R_3
MOVE GRIPPER TO OBJECT
END

21.7.5 *Approach and departure points*

In robotics it is frequently useful to approach the final position with a final phase which is not necessarily in the direct trajectory arising from the preceding point. It is, therefore, generally necessary to specify an intermediate point close to the end point. A simplified version of this approach is to specify only an approach distance. The intermediate point is generated automatically as a set of coordinate axes with the same orientation as the destination set of coordinate axes and translated along the tool axis by a specified distance.

Example: MOVE GRIPPER TO OBJECT WITH APPROACH = 50

A similar specification can be made for the start of the trajectory with a starting distance.

21.8 Sensors

Programmable robots usually can accommodate sensors to make their actions take account of their environments. Software can make use of continuous or binary variables to initiate branching sequences in a program or to terminate an action. The detection of a contact with an object or an unexpected force could be such a case.

Example: MOVE TOOL TO DEST UNTIL FZ > 10

To modify a movement once it has started, using servo control of external variabtes, it is necessary to have access'to the movements of the articulations in the reference set of coordinate axes both for position and velocity.

Example: MOVE TOOL WITH VX = 50 and
VY = (A–B) * 5

where A and B would be two differential values supplied by a proximity sensor.

Vision sensors employ an external process to examine a scene and send, on demand from the main program, a certain number of variables which depend on the vision system.

21.9 Tools

In addition to carrying out displacements, a robot must be able to operate tools. These may be suited at the end of the arm or at a fixed post. Generally speaking, tools can be classified into (t) those controlled in the binary mode, and (ii) those controlled in the continuous path mode. Depending on the languages and the tools, standard output control commands specific to the tool can be used.

Example: ACTUAL TOOL I (A, B)

where A, B are operating parameters of Tool I.

One specific tool to be considered is the proportional opening gripper. It is not only possible to give a closing or opening command with a required gap but it is also possible to check that the correct gap has been obtained.

Example: CLOSE GRIPPER TO 12.5 IF NOT GO TO WITHOUT OBJECT.

21.10 Example of programming

Machine loading and unloading READ, GRASP, MACHINE, MACHINE-APPRO, DEPOSIT, WAIT all these variables are for the destination or transit point coordinate sets.

MOVE GRIPPER TO WAIT-OPEN GRIPPER
WHITE GO = TRUE GO
Go is an input variable describing the state of the station control switch.
MOVE GRIPPER TO GRASP WITH APPROACH = 10
IF PART-OK THEN CLOSE GRIPPER
ELSE INCIDENT

PART-OK is a switch which indicates the presence of a part at the grasping station, INCIDENT is a function which calls the operator.

MOVE GRIPPER TO MACHINE VIA MACHINE-APPRO

MACHINE-APPRO is an approach set of co-ordinate axes for loading the machine.

SIGNAL CLOSE-JAWS

This sends an output command to close the jaws of the lathe.

```
OPEN GRIPPER
MOVE GRIPPER TO WAIT VIA MACHINE-APPRO
SIGNAL START MACHINE
                    (activa.ion of the lathe cycle)
WAIT END-MACHINE
                    (wait for command to end cycle)
MOVE GRIPPER TO MACHINE VIA MACHINE-APPRO
CLOSE GRIPPER
WAIT 2 SIGNAL OPEN-JAWS
MOVE GRIPPER TO DEPOSIT VIA MACHINE-APPRO
    WITH APPROACH = 20
                    (finished piece deposited)
OPEN GRIPPER
MOVE GRIPPER TO WAIT WITH DEPART = 20
(return to intermediate position and recommencement of cycle)
END
```

21.11 Some Commercial Languages

Some of the major programming languages available on the market are presented here, though this list is not exhaustive. The selection criterion has mainly been the availability of reference manuals. The basic principles of programming languages are same as described earlier, only the type of format, type of instruction, number of instructions are different for difierent programming languages.

21.11.1 *ARL (Assembly Robot Language)*

For complicated tasks in assembly and inspection, there has been an increased demand for high level robot control synchronized with sensors or peripheral units. The progress of microcomputers has made it possible to execute these kinds of high level controls and in order to control tasks, such as assembly and inspection by language, an Assembly Robot Language (ARL) was developed.

21.11.2 *HARL (Hitachi Assembly Robot Language)*

It is a commercial version of ARL especially suited for point-to-point assembly and handling tasks. By this method, motion control data is programmed in the same way as conventional computer language and after programming, location data used in the program is taught in order to make locational information of motion statements. There are 18 typical HARL instructions and programming is possible in the robot language by using these instructions.

21.11.3 *AL (Arm Language)*

It is currently the language which has undergone the most important development. It first appeared in the course of original research at the Stanford University into programming robots and the use of WAVE language, and has been the object of constant improvements since 1974.

21.11.4 *VAL (Vic Arm Language)*

An interactive version of AL, known as VAL and its associated hardware have been adapted to all Unimate robots which previously used an old form of point control. A new version of VAL, called VAL II, is at present being tested for commer-cial use.

21.11.5 *AML (A Manufacturing Language)*

It is a language represented by IBM with its robots, and was introduced into the market in 1982. The Cartesian hydraulic robot RS-1 (or 7565) is produced by IBM and the 7535 electric robot from the Scara range is produced in Japan by Sankyo. IBM Automated Parts Assembly System (AUTO PASS) language attempts to eliminate the need for issuing detailed instructions to the robot. The program automatically determines the grip points and motion paths from geometric data base.

21.11.6 *IRL (Intuitive Robot Language)*

This is a language developed by the Swiss Company Microbo, affiliated to the large watchmaking companies, for its range of high precision assembly robots.

21.11.7 *LM (Language de Manipulation)*

This was developed in the 1MAG robotics laboratory at the University of Gernoble in 1979. It adopts most of the main concepts of AL, but it is used on a microcomputer (68000).

21.11.8 *MCL*

MCL was developed within the framework of ICAM projects to resolve in a unified way all the problems associated with robot programming. MCL is an extension of APT and its aim is the programming of flexible units, that is, of a set of machines served by one or more robots. At present, MCL can control Cincinnati T3 and Westinghouse Allegro robots.

21.11.9 *PLAW (Programming Language for Arc Welding)*

This is particularly well suited to 'intelligent' welding, that is, welding which involves the use of sensors.

21.11.10 *Other languages*

RAIL was developed by Automatix and this is the first robot language that can be applied to problems of manipulation as well as problems of vision.

ROL (Robot Language) is based on the new approach to the robotics market. ROL was also developed from research with the aim of designing a complete commercial system for computerized control (including hardware and software) adaptable to any robot.

SIGLA (Sigma Language) was the first commercial language available for use with an industrial robot. It was developed by Olivetti for its Cartesian Sigma robots, and was considerably influenced by numerical control languages, but nonetheless allows control of several arms, with loops and tests on the sensors.

Some languages are intended to operate with artificial sensory input that enables the robot to act more independently. For example, the standard Research Institute Robot Programming Language (RPL) includes capabilities for interpreting video signals, enabling the robot to visually identify parts. And Draper Industrial Assembly Language (DIAL) developed at Charles Stark Draper Laboratory uses electronic force feedback to duplicate human sense of touch in assembling components.

21.12 Conclusions

Once a robot is controlled by robot ianguage, the facility of the robot is dependent on the function of the robot language. The facility of input and that of modification of robot language have also become important factors.

The intelligence of programmable production machines is considered to be steadily increasing, in relation to the development of microelectronics. It is also necessary to continue research with regard to various other FA devices. In this situation, not only robot language but also a unified language applicable to all the other FA devices is crucial.

Using these languages, it is expected that a high level of automation for manufacturing various types of products with sophisticated interaction between many devices will be realized.

CHAPTER 22

Flexible Manufacturing System

22.1 Introduction

A flexible manufacturing system is developed with the objective of having centrally supervised, but loosely connected production machines or a group of production equipment interconnected by an automated handling system. This requires uses of Computer Numerical Controlled machine tools and or Machining Centres using manipulating robots for handling— control being executed from distributed computer or microprocessor interfaces.

Though NC machine is the base for FMS, CNC offers more flexibility in operation of any manufacturing system. Computer aided design followed by computer aided manufacture is the major step forward towards FMS.

22.2 FMS–its Meaning, Objectives and Significance

Objectives of FMS may be limited to:

(i) medium size production run, using NC, CNC or DNC machine;
(ii) facility for manufacturing product components falling into certain families–for each one of which a programmable manufacturing schedule or operations layout could be drawn;
(iii) self-contained manufacturing facility that can do planning, machine scheduling and quality control.

An FMS consists of:

1. The host manufacturing control computer.

2. Work centres, i.e., machine tools with their programmable logic controllers (PLC) and or numerical controllers (NC), processing centres with their PLCs, manual work stations with PLCs or data acquisition terminals (DATs).
3. A transport system with its controller.

"For each component the host module organises the work centres to produce components either on an interchangeable (operation can be carried out on any machine in machine– mix X) or complementary (as before, but a further operation must be carried out on any machine of machine–mix Y) basis. Machines need not be mutually exclusive to different mixes and the work stations can operate autonomously or under direction of the host.

The layout of an FMS will depend on the machines' capabilities, capacities and component-throughput required. In 'Stand-alone' mode machine controllers communicate with other controllers in the communication control hierarchy (see diagram). This decentralised design between levels enables a higher system uptime, if one continues either unaffected apart from a drop in system capacity or with all machines, in a lower level, working in 'Stand-alone' mode".

As the lot size increases the production cost per piece comes down, similarly flexibility in manufacturing increases as we move up from the automatic transfer lines to FMS and further to NC, CNC control. Figure 22.1 shows the flexible manufacturing concept based on workpiece spectrum and lot sizes-keeping both productivity and flexibility into consideration.

Present day design methodology aims at identifying the components based on similarity in size and geometry and specific nature of complexity in manufacture, forming into groups to undergo same process (on basis of GT) and then designing eacb group to evolve a suitable computer aided software. Thus right from CAD it can come to CAM (Computer Aided Manufacture) through computer aided drafting. CAD-CAM is of course the pre-requisite of achieving FMS.

FMS system normally presents a hierarchical control spread over two or three tiers. The 1st tier consists of administrative level of control, and the 2nd tier deals with supervisory control. Production scheduling, control and maintenance of the management data base etc., form the part of administrative control, while supervisory control consists of scheduling machine tools, handling equipment–and their control and operator guidance. Both are controlled through computer using various languages.

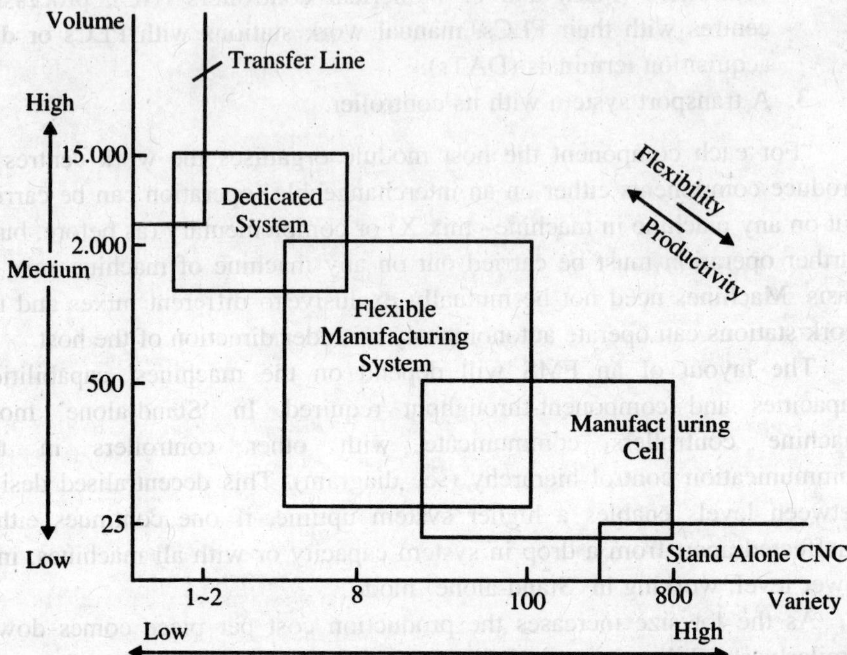

Source : "Innovation of Machine Tool Technology" by Dr. K. Iwata, Kobe University

Figure 22.1 Flexible manufacturing concept.

Manufacturers found it increasingly necessary to build flexibility into their production systems so that they could supply more segmented markets with small lots of components and finished products. Even more important, the growth of competition focused attention on the fact that many manufactured products wait as much as 95% of the total time in the factory just to be processed. New production systems were needed to obtain radical economies between stages. Flexible Manufacturing Systems (FMS) provided the answer to this need. FMSs in operation, at present, are primarily metal cutting systems that execute turning, milling, drilling, grinding and other operations. Advances in Laser Technology have also led to the inclusion of laser processing techniques in these FMSs for performing such operations as cutting, hardening and welding.

FMS is meant only for large scale production, is a wrong notion. FMS is suited for small batch, large variety situations. The spectrum of manufacturing system is shown in Figure 22.1.

In the last few years FMS manufacturers have developed a variety of machines tailored to meet the needs of small to medium sized companies which are willing to automate their production and reduce their cost of production. Many examples have been seen recently of systems designed for producing large variety of components in medium sized batches. These FMSs have been made possible by advances in CNC machine tool technology, technological improvements in the industrial robots that handle tools and workpieces, new developments in unmanned carriers that transport tools and materials between machines and also progress in the computer control technology for managing the systems.

22.3 Classification of FMS

The classification of FMS is done, based on the orientation or layout of the units so as to have three distinct categories, viz., FMS with complementing machines, FMS with substituting machine and hybrid FMS. While Hybrid FMS is the combination of the other two systems, in the first category viz., FMS with complementing machines is the one where units are placed in a complementary fashion in regard to flow path–subsequent operations being done as the work moves from one work station to another. This is similar to line layout in a conventional manufacturing system. In a substituting FMS the work may be transferred from one station to another with substitute tool & equipment.

22.4 "BUILDING BLOCK" Concept

While designing machines with modular concept, it is absolutely necessary to go through the systems approach. Unit construction machines in the nature of 'building block' concept is the basic requirement of a special purpose machine for wider application in a system where flexible manufacturing pattern exists.

Transfer machines, particularly 'rotary' or 'in-line' transfer machine are vital to production units engaged in mass manufacture of similar items. The automatic cycle in such a machine is controlled along the line of travel of executive units. Three different electrical control systems are used, namely (i) 'in-travel', (ii) 'time-sequence' and (iii) 'centralised in-travel systems'. The first one, viz., in-travel control is usually a decentralised one. Control of units in all such cases were by cams, geneva mechanism etc., actuated by limit switches, commanding systems and or hydroservo mechanisms.

Normally transfer machines consist of fixed base with a reciprocating table or rotary table and a number of unit heads placed on the base with proper arrangement for the positional location and travel of the tool-holders or 'unit heads'. A large number of unit heads could be placed, at angular or radial directions to the flow path. Since each unit head has separate drive and feed system, they are sometimes called aggregate machines.

Standardised 'unit heads' may serve the purpose of a special purpose machines, by combination, in tune with the manufacturing requirement. Thus it can be called special-purpose machine, 'tailor made' using 'building block' concept, so that assembling unit heads on a permutation-combination basis, we can rnake a large number of special purpose machine depending upon the requirement and the actual availability of the unit heads. But these types of machines are limited in their functions due to non-flexibility resulting from cams and mechanical linkages. To bring the flexibility, it is necessary therefore to have programmable units connected to computer interface. This necessitated the creation of CNC machining centre.

22.5 CNC Machining Centre

In transfer machines, normaly the work moves from workstation to workstation for processing. But in a machining centre having a large number of tools doing different jobs, the tools are fed to the workpiece stationed at a particular location in sequence, related to the manufacturing programme or operational layout. Even random selection of tools, automatic changes of tool when worn out are some of the many advantages of a machining centre. Bridgeport machining centre having even upto eighty tool holders with random selection operated through CNC are now readily available in the market. Since the workpiece is fixed in a particular location, while various operations are done on it, we achieve excellence in accuracy of the product. Any chances of error creeping in due to thermal instability or dynamic instability due to tool wear or breakage are being effectively compensated now-a-days in all precision CNC controlled machines. Major improvement in accuracy, shorter delays in loading, unloading and tool changes, decrease in storage space and machine layout area, etc., and less time and cost in manufacture (being in flow-line production) heightens the desirability of having CNC-machining centres. In some cases a 5-axis CNC machining centre may eliminate the uses of additional toolings in the form or jigs and fixtures.

Figure 22.2 shows a typical CNC machining centre produced by HMT. Details of such machines are given in Chapter 20.

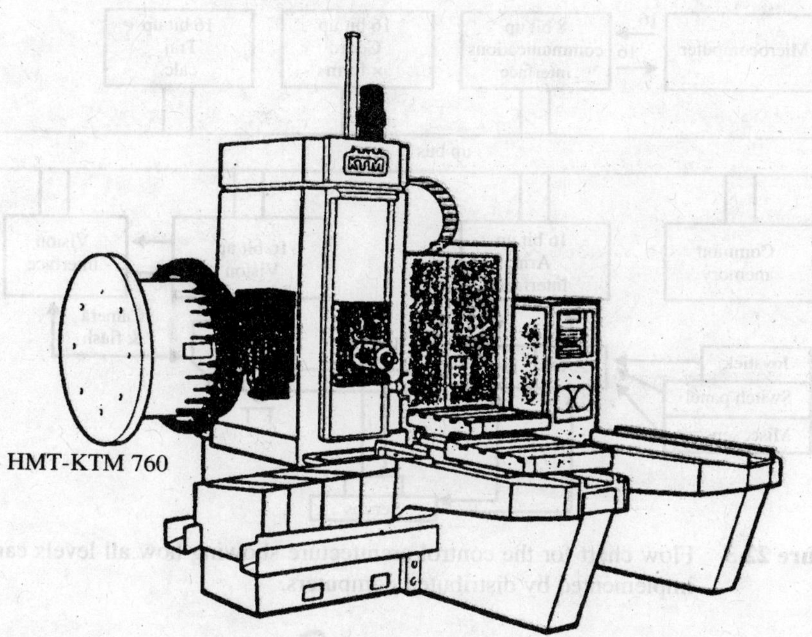

Figure 22.2 CNC machine centre.

22.6 CAD CAM-FMS

Computer Aided Manufacture (CAM) requires a large effort towards modularity and interface standardisation, and though it is not yet recognised much today, its lack will inhibit us to go for full scale Flexible Manufacturing System (FMS). It will also restrict the uses of CAD CAM. In the CAM heirarchial system we normally divide the task into various modules at various levels. Like Major task module, Simple/minor task module, Action module, Servo module etc. "Each one of this module is a 'finite state machine'. Functions at each level are in ;lemented by distributed computers, as shown in Figure 22.3, operating in a language best suited to that level, since the only requirement for the interface is to be able to read and write a given format. Moreover, addition of a new state variable needs solely a definition of where the newcomer is located in the common memory". The control architecture as shown in Figure 22.3 allows scope for addition of new function, elements or other units with little or no impact on the original system. This, therefore, offers sufficient liberty to the manufacturing system to become flexible. It can adjust itself to the operation parameter changes in an effective manner. A typical vision system of the machine has been shown in Figure 22.4. The method of identifying the object can be easily understood from the Figure.

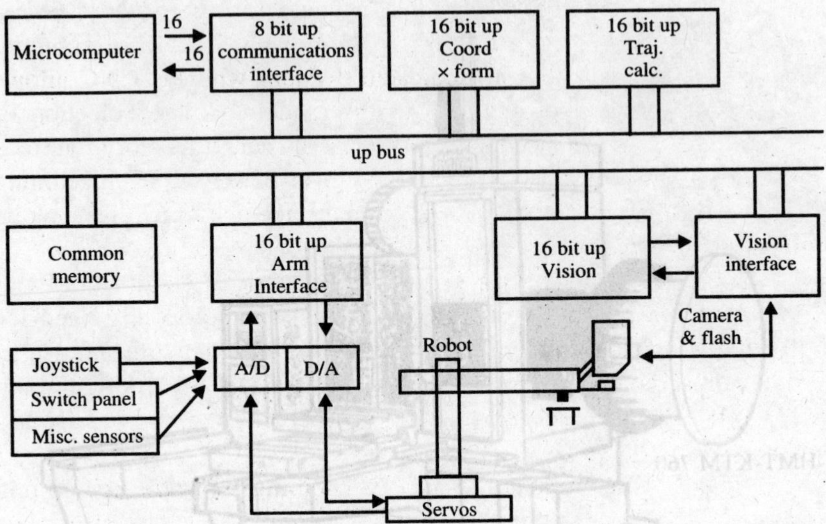

Figure 22.3 Flow chart for the control arcnttecture showing how all levels can be implemented by distributed computers.

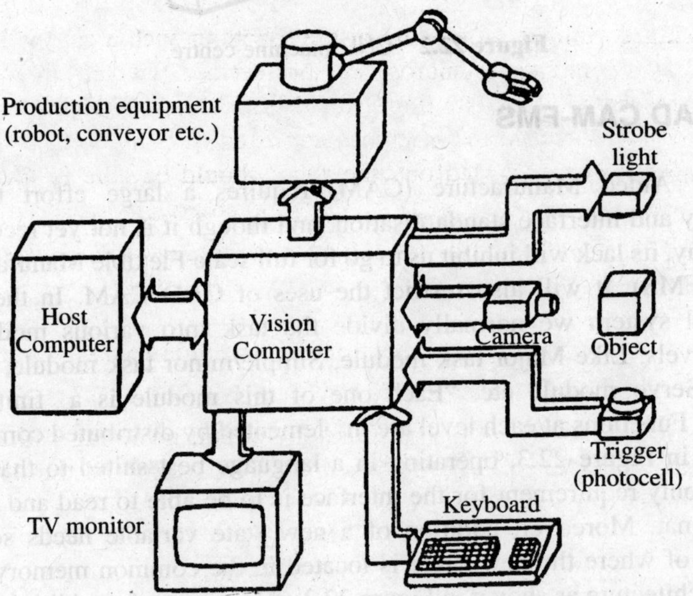

Figure 22.4 A typical machine vision system. The photocell alerts the vision computer when an object enters the camera's field of view. The strobe light enables the system to 'freeze' rapidly moving objects for analysis. The keyboard and video monitor are used primarily for system programming and debugging.[118]

This is the beginning of a Flexible Manufacturing System (FMS). Though the NC machine constitute the first step of FMS, CNC offers more flexibility. NC is more or less a 'dedicated' system whereas CNC allows even interference into the programme at the machine level. Selection of working parameters like speed, feed, depth of cut, trajectory, surface morphology etc., are automatically done by 5-or-6 axis CNC machining centre. Any additional programme in the form of tape or floppy disc may be incorporated at an appropriate stage.

In a Factory Automation System the robots work in conjunction with CNC, NC or DNC machines or carriers. There is an increasing need for robots to readily communicate or respond to local area network, known as LAN with a host computer and other elements as in the configuration. Robotics, CAD CAM link and FMS will soon become the basis of factory auto, mation.

In the major FMS system, which is nothing but an ergono-mic combination CNC/DNC machines and robots or manipulating carriers, the flow of job to and from the work stations are normally done by using conveyors. A robot can easily load or unload a job at the work station and transport the same to the required station with ease, provided it is programmed accordingly. Command of the robots in such a case will come from distributed computer or microprocessor interface. Two arm robots with sensitive gripper and vision are more suited for FMS. In FMS all functions programmed for robot and the machines are listed. It is intended that while the machining is being carried out, the robot should be able to locate the parts, tools etc., for the purpose of transfer and subsequent loading.

22.7 Precision Movement

The accuracy in the precise movement of any unit in a machine tool having NC or CNC is normally dependent on

(i) manufacturing errors of transmission system; involving ball screws, where pitch error, ball diameter error and error of the effective diameter are dominant;
(ii) frictional behaviour and relaxation;
(iii) super-imposed vibration or self-excited one;
(iv) contact compliance and error in topography of the surface due to contact compliance.

Normally the slides in recent machines are coated with PTFE powder, so that behaviour of the friction can be tamed and the system can retain steady positional error (usually very small), within a certain regime of temperature

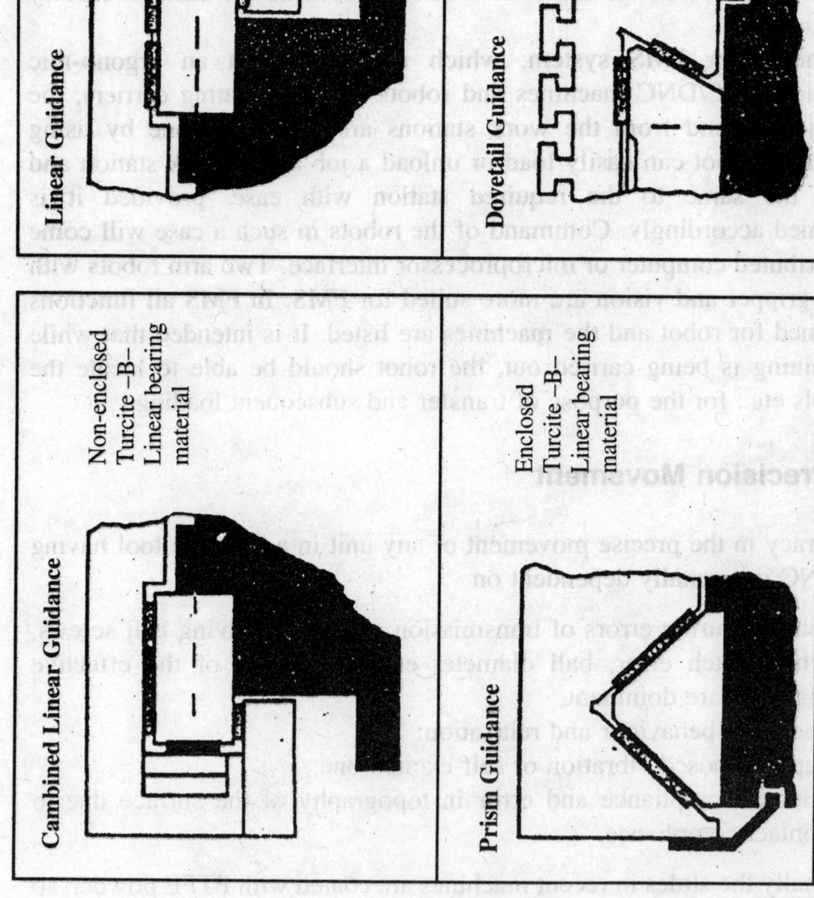

Figure 22.5 Mounting of Turcite B for slide ways. *Courtesy:* Rollon Bearings Pvt. Ltd.[124]

at the interface. Now-a-days bronze reinforced PTFE, known commercially as Turcite B and produced or manufactured by Rollon Bearings Ltd., Bangalore1" are being used most effectively as the material for guideways. This is available in thickness of 1.5 mm to 3 mm to be posted on the surface with adhesives, like Araldite used with hardener inappropriate proportion. Such slide materials exhibit good frictional behaviour, contact compliance etc., under varying conditions of environment, such as temperature, humidity, heavy specific pressure. They also show good wear resistant property and hence indicate a trend towards higher span of life. A few typical installations are shown in Figure 22.5.

The positional error in displacement of an executive element of a machine tool is dependent on the parameter (Δ F/k)

where ΔF = Difference between static and kinetic force of friction
and K = Overall stiffness of the transmitting linkage,

and is directly proportional to it. Since in Turcite B application, ΔF is found to be very little, the positional error due to stick-slip displacement becomes hardly appreciable This enhances the advantages of using the Turcite B as slide material in a NC/CNC/DNC machine.

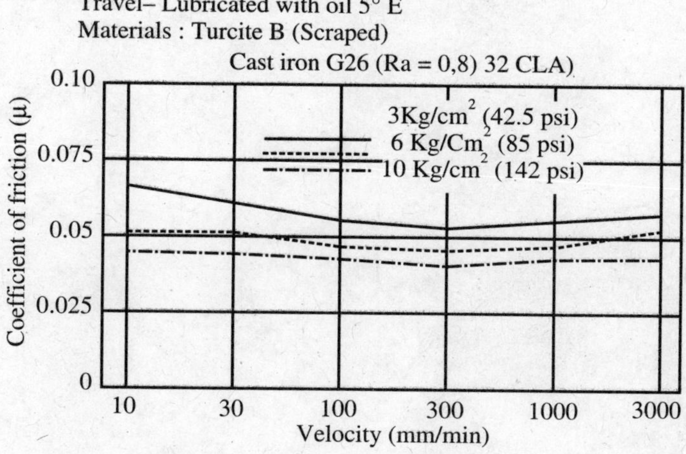

Figure 22.6 The figure shows very low value of Δf and hence avoidance of stick-slip positional displacement is effected.
Courtesy: Rollon Bearings Pvt. Ltd.[124]

Various errors, due to the four points enumerated above may be systematic or random or both. When each contributing parameter may not have interdependence, it is necessary to combine such errors effectively by simulation to know the actual effect on the positional error.

In regard to robot, the repeatability and capability coefficient are considered as performance coefficient. Researches are being done to evolve suitable criteria of dependence for such indices.

CHAPTER 23

Dynamic Analysis of A Few Sub-systems in Machine Tools

This Chapter deals with some of the recent work done in the various systems and sub-system designs of a machine tool, based on dynamic analysis. Some of the recent investigational practices and their modalities have been discussed herewith for the benefit of researchers in modern machine tcols and their applications.

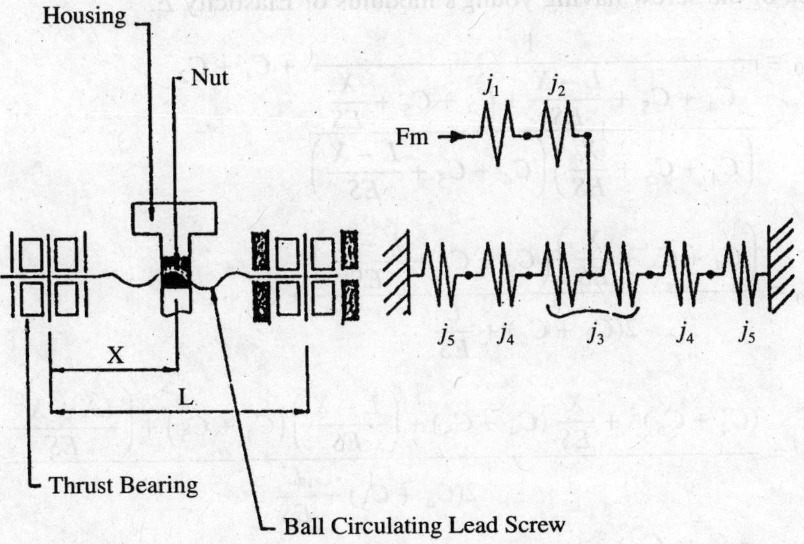

Figure 23.1 Representative diagram for evaluating system compliance.(125)(126)

23.1 System Analysis to Study Dynamic Compliance[125]

The system analysis of the lead screw drive for each axis movement of the CNC machine was done by finding out the equivalent compliance of the system, consisting of various sub-systems, such as, nut and housing, nut and ballscrew; ball lead screw; bearing supports at both ends and bearing and housing with or without preload and for different positions of the traversing table for axis movement. The theoretical analysis was done to obtain generalised equation for the total system compliance and optimise the same for minimum total error in positional displacement.

In the Figure (23.1) J_1 and J_2–Rigidity between nut and housing and rigidity of the thread contact between screw and nut; J_3–rigidity of the Ball leadscrew alone; J_4 and J_5–Rigidities of the thrust bearings, and bearings and their housing on each respectively.

Denoting the coefficients of compliance (reverse of rigidity = $\delta/\sigma = C$, where δ – deformation and σ is the stress causing the deformation) as C_1, C_2 .... C_5 and the combined coefficient of compliance of the system as C_0 we get.[125][126]

$$C_0 = \frac{1}{\dfrac{1}{C_4 + C_5 + \left(\dfrac{x}{ES}\right)} + \dfrac{1}{C_4 + C_5 + \left(\dfrac{L-X}{ES}\right)}} + C_1 + C_2 \qquad (23.1)$$

Where L is the length of the leadscrew and S is the transverse Cross-sectional area of the screw having young's modulus of Elasticity E.

$$C_0 = \frac{1}{\dfrac{C_4 + C_5 + \dfrac{L-X}{ES} + C_4 + C_5 + \dfrac{X}{ES}}{\left(C_4 + C_5 + \dfrac{X}{ES}\right)\left(C_4 + C_5 + \dfrac{L-X}{ES}\right)}} + C_1 + C_2$$

$$C_0 = \frac{\left(C_4 + C_5 + \dfrac{X}{ES}\right)\left(C_4 + C_5 + \dfrac{L-X}{ES}\right)}{2(C_4 + C_5) + \dfrac{L}{ES}} + C_1 + C_2$$

$$= \frac{(C_4 + C_5)^2 + \dfrac{X}{ES}(C_4 + C_5) + \left(\dfrac{L-X}{ES}\right)(C_4 + C_5) + \left(\dfrac{LX - X^2}{ES}\right)}{2(C_4 + C_5) + \dfrac{L}{ES}}$$

$$+ C_1 + C_2 \qquad (23.2)$$

$$\left(\frac{dC_0}{dx}\right) = 0; \left(\frac{d^2C_0}{dx^2}\right) = +ve \text{ For minimum } C_0,$$

We get $x = L/2$; with this value of x

$$C_0 = \frac{(C_4 + C_5)^2 + \frac{L}{ES}(C_4 + C_5) + \frac{L2}{4E^2\ S^2}}{2\,(C_4 + C_5) + \frac{L}{ES}} \tag{23.3}$$

$\ll 1$, Hence neglecting $\left(\frac{L}{ES}\right)$ and $\left(\frac{L^2}{E^2S^2}\right)$

$$C_{0\ \min} = \left[\frac{(C_4 + C_5)}{2} + (C_1 + C_2)\right] \tag{23.4}$$

$$J_{\max} = \frac{2}{2\,(C_1 + C_2) + (C_4 + C_5)} \tag{23.5}$$

Case II: If, further, the amount of magnitude of $\frac{L^2}{4\ ES} \ll 1$, ignoring its effect we can write.

$$|Js|_{\min} = \frac{2}{(C_4 + C_5) + 2(C_1 + C_2)} \tag{23.6}$$

It has been shown by Push, and Tolstoi[39] and push[38] that in a machine tool table moving with a velocity in and around the critical velocity, at which stick-slip vibration, exhibiting discontinuous motion, becomes prominent, the error that may occur in the positional displacement of the table can be shown as:

$$\varepsilon \simeq \frac{\Delta F}{K} \cdot f\left(\theta, \frac{V}{V_c}\right) \tag{23.7}$$

Where of (θ, V/V_c) denotes mathematical function of θ, the coefficient of viscous damping, and V – velocity of table feed and V_c–critical velocity for 'stick-slip'. ΔF denotes the difference between static force of friction and dynamic force of friction at the 'table-guide' interface at velocity V and K is the elastic stiffness of the system. It has been further shown by numerical solution and plotting ($\varepsilon_k/\Delta F$) against $\left(\frac{V}{V_c}\right)$, for various values of parameter θ that

$$\varepsilon = \frac{\Delta F}{K} \cdot \sqrt{\frac{\Pi}{\theta}}, \text{ when } V = V_c \tag{23.8}$$

$$\varepsilon = \frac{\Delta F}{K}(2 - \Pi\theta), \text{ when } V \ll V_c \tag{23.9}$$

From Eqn. (23.6)

$J \simeq K$, Substituting this in Eqn (23.9)

$$\varepsilon \simeq \frac{\Delta F \cdot [(C_4 + C_5) + 2\,(C_1 + C_2)]}{2} \times (2 - \Pi\theta) \tag{23.10}$$

For small viscous damping, we can rewrite this as–

$$\varepsilon \simeq DF\,[(C_4 + C_5) + 2(C_1 + C_2)] \tag{23.11}$$

Thus, the positional displacement error, due to stickslip, in a CNC machine tool, using recirculating ball lead screw the value could be evaluated either from this or from the original equation of static rigidity.

Normally $C_4 = C_5 = C_4'$ and $C_1 = C_2 = C_1'$

$$\varepsilon \simeq \Delta F \cdot [2C_4' + 4C_1'] \simeq 2 \cdot \Delta F \cdot (C_4' + 2C_1') \tag{23.12}$$

Neglecting C_4' in comparison to C_1

$$\varepsilon \simeq 4 \cdot \Delta F \cdot C_1' \simeq \frac{4\Delta F}{J_1'} \tag{23.13}$$

Normally C_4, C_5 cannot be reduced substantially, but C_1 and C_2 can be reduced by adjusting the contact deformation while preloading by adjusting the thickness of the spacer etc. and the combined error of the ball screw under the nut, by improving upon the manufacturing processes involved and quality thread grinding operation.

Positional displacement error of machine lool carriage resulting from 'stick slip' has been shown in equation (23.13) and (23.14).

$$\varepsilon = \Delta F \cdot \left[\frac{(C_4 + C_5) + 2(C_1 + C_2)}{2}\right] \cdot (2 - \Pi\theta) \tag{23.14}$$

$$= \Delta F \cdot (C^*)\,(2 - \Pi\theta), \text{ where,}$$

$$C^* = \frac{(C_4 + C_5) + 2(C_1 + C_2)}{2} \tag{23.15}$$

and $\theta = \dfrac{\beta}{2\sqrt{k.m}}$ – Vibration damping proportional to velocity, K– stiffness of the system and m – mass of the carriage.

Let us put $\Delta F = \hat{f}$ and $(2 - \Pi\theta) = \theta'$ and assume combined compliance $= C^*$

$$\varepsilon = \hat{f} \cdot C^* \cdot \theta'$$

$$\partial\varepsilon = \left(\frac{\partial\varepsilon}{\partial\hat{f}}\right) \cdot \Delta\hat{f} + \left(\frac{\partial\varepsilon}{\partial C^*}\right) = \Delta C^* + \left(\frac{\partial\varepsilon}{\partial\theta'}\right)\Delta\theta'$$

Dividing both sides by ε we get –

$$\left(\frac{\partial\varepsilon}{\varepsilon}\right) = \left(\frac{\Delta\hat{f}}{\hat{f}}\right) + \left(\frac{\Delta C^*}{C^*}\right) + \left(\frac{\Delta\theta'}{\theta'}\right) \tag{23.16}$$

This shows the significant and identical influence of C^* and $\hat{f}$ on the general positional error of the system. Hence both these two have to be tackled simultaneously. For controlling the overall coefficient of contact compliance, namely C^*, it is necessary to do the systematic analysis of the machine tool subsystems and know the magnitude of C for each of the elemental subsystems and its range of variation. And this is why the study of contact deformation of joints is regarded as so much significant.

23.2 Spindle Supports

For the purpose of obtaining drive for the high speed spindle, as used in grinding machines, we often use high frequency asynchronous motor. For high precision surface, of the order of even less than 0.95 μ*m*, it is necessary to have special inertia type drive, which will work and continue to work, even after the application of the braking system.

Selection of the supports, including the bearing is based on the radial gap between the journal and the bearing and the parameter (*dn*) mm/min., where *d* is the diameter of the journal and *n* – rotation of the spindle.

The following table may provide useful data for calculation:

Table 23.1 Basic parameters for selection of supports

Type of Support	*Radial axial deflection in μm*	*Ovality μm*	*(dn)* 10^{-5} *mm/min. maximum*
Journal	1.00	1.0	0-10
Hydrodynamic	0.50	0.5	1.10
Hydrostatic	0.05	0.2	0.15
Aerostatic	0.10	0.5	5.40

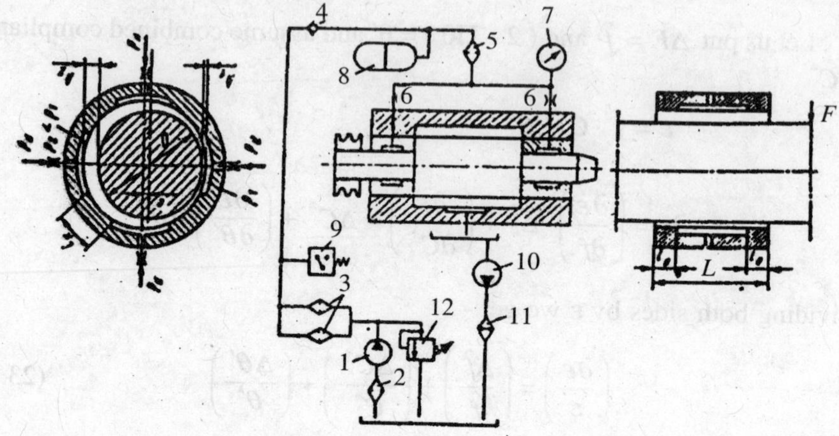

Figure 23.2 Hydrostatic spindle support.

Figure 23.2 shows the sketch of a simple radial hydrostatic bearing. Normally the geometrical size of this type of bearing are depending on the journal diameter d, where –

$$d \geq \sqrt[4]{(0.05 - 0.10)\, l^3} \tag{23.17}$$

l, being the centre distance between the two supports. Figure 23.3 shows the details of a closed loop hydrostatic support to take radial load and thrust. $L = d$, length of pocket $l_K = l_o = O.D.$ and the diametral gap $\Delta = (0.008 - 0$ $001)$. d mm. The working fluid selected in most of the cases is mineral oil with a viscosity $\mu = (1 - 10)\ 10^{-3}$ Pa. sec, for high speed spindle supports, e is the minimum eccentricity selected, as such, a suitable value that the pressure of oil in the pocket $p = 0.5\ p_s$, where is the pressure of the oil raised by the pump.

1. Pump, 2. Filter for large particle, 3. Filter for smaller sized particles, 4. Non-return valve, 5. Filter for very small microscopic particles, 6. Throtle, 7. Manometer, 8. Accumulator, 9. Pressure relay, 10. Pump, 11. Heat exchanger, 12. Relief valve, with reference to Figure 23.2.

The size of the contaminant particles in the oil should be as low as possible and, as such, should not exceed half the size of the radial gap. i.e. it should be between 0.5 – 1.5 µm. Load characteristic of the hydrostatic bearing, depends on radial deformation of the journal and as such[(127)]

$$F_C = p_s \cdot S_{eff} \cdot C_f(\varepsilon, K) \tag{23.18}$$

Where F_C load

p_s – pressure of the pump, MPa;

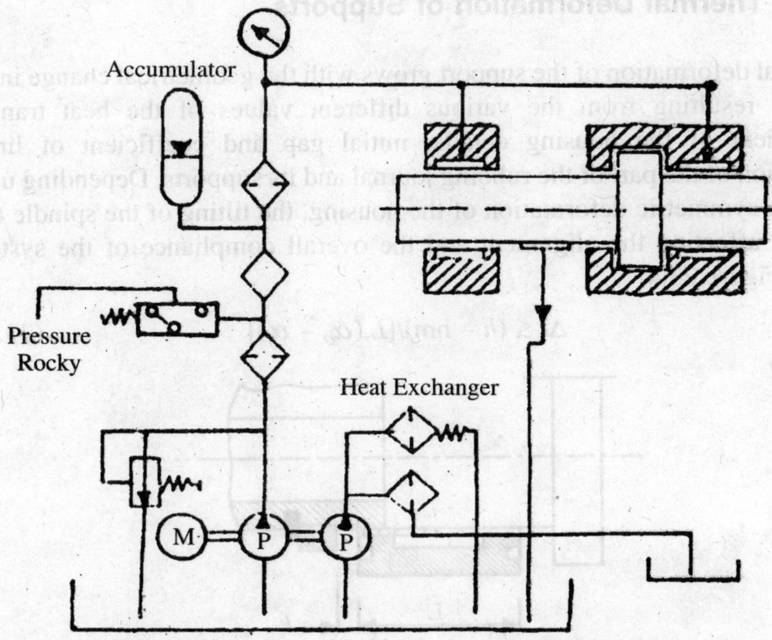

Figure 23.3 Hydrostatic bearing: A closed loop circuit.

– S_{eff} – Effective area of the bearing, mm^2 ($\approx 0.5\ d^2$)

C_f (ε, K) – a parameter depending upon the ratio of deformation of the Journal and geometrical parameter of the support.

$$C_f\ (\varepsilon, K) = \left(\frac{3\varepsilon}{2}\right) \text{ But } \varepsilon = \frac{2e}{\Delta} \text{ where } \Delta\text{–diametral gap}$$

$$F_C = 1.5\left(\frac{\varepsilon}{\Delta}\right) d^2 \cdot p_s \qquad (23.19)$$

Rigidity of the oil film in N/mm is given by

$$J_h = \frac{1.5\ d^2 \cdot p_s}{\Delta} \qquad (23.20)$$

Oil consumption mm^3/sec. is normally calculated from equation below:

$$Q = \frac{10^8 \cdot \prod d\Delta^3 \cdot p_s}{\mu l_a} \qquad (23.21)$$

Here μ is coefficient of viscosity in Pa. Sec.

23.3 Thermal Deformation of Supports

Thermal deformation of the support grows with the geometrical change in the orifice, resulting from the various different values of the heat transfer coefficient of the housing details, initial gap and coefficient of linear expansion of the part of the rubbing journal and its supports. Depending upon the non-symmetric deformation of the housing, the tilting of the spindle axis occurs, affecting the alignment and the overall compliance of the system. Refer Figure 23.4.

$$\Delta t \leq (h - \text{hm})/[L.(\alpha_b - \alpha_j)] \quad (23.22)$$

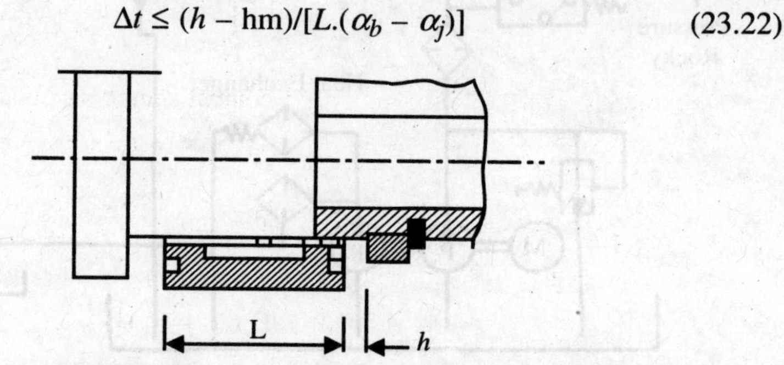

Figure 23.4 Spindle end Calculations.

Where h_m – minimum permissible thickness of the sealing ring, $\Delta . t$ – rise in temperature; and α_b and α_j are the coefficient of linear expansion of bearing and the journal, respectively.

The total power consumed

$$N \text{ total} = 0.072\ 10^{-16}\ \frac{D^4\ \mu m^2}{\Delta} + 314\,(p_{H^2} \cdot \Delta^3 / 3\mu)\ K.W. \quad (23.23)$$

P_s – MPa; μ – Pa. Sec.; – mm d – mm n – Min^{-1}

Considering the Figure 23.5 we see the front end of the main spindle with an overhang of a and an end load, resulting, from the dynamic condition of work = N. Assuming the diameter of the spindle end as, D, we get total deformation at the end as

$$Y_c = \delta + \theta \cdot \alpha \quad (23.13)$$

Where δ is the deflection due to cantilever beam and θ angle of twist in radians.

Using the methodology given by Lecvina and Reshetov[127] and also advocated by Push[125].

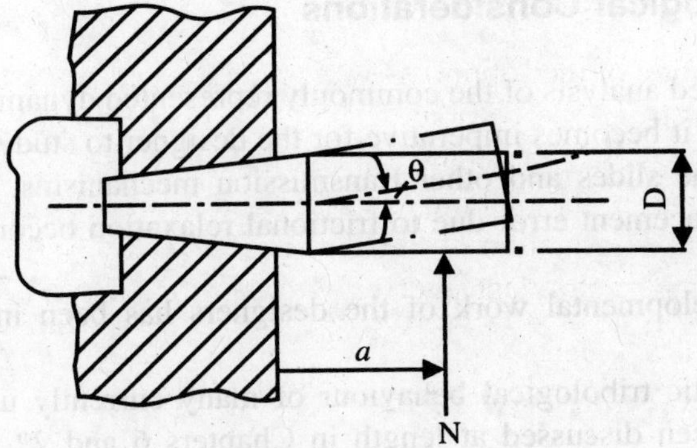

Figure 23.5 Calculation diagram for conical spindle end connection.

$$\left.\begin{aligned}\delta &= \frac{4N\beta.C}{\Pi.D}(a\cdot\beta\cdot C_1 + C_2)\\ \theta &= \frac{4N\beta^2 C}{\Pi.D}(2a\cdot\beta\cdot C_3 + C_1)\end{aligned}\right\} \qquad (23.24)$$

Where $\beta = \sqrt[4]{1/(13CD^3)}$; $C = 0.2$ = Coefficient of contact compliance in $\mu m^2/N$, C_1, C_2, C_3 – Coefficients whose values may be taken as under, based on experimental data $C_1 = C_2 = 1.35$ and $C_3 = I$.

Therefore, Rigidity $\quad J = \dfrac{N}{Y_C} = \dfrac{20D^4}{a^2}$, N/$\mu$m (23.25)

Where D and a are in cm or $C^* = a^2/(20D^4)$ µm/N (23.26)

Approximate calculation consisting of the frequency of spindle in Sec^{-1}, when the mass of the spindle is not appreciable in magnitude, may be done by using the relationship.

$$\omega_c = \gamma\sqrt{\frac{EJ_1}{m(1+\lambda)^3\cdot a^2}} \qquad (23.27)$$

Where m – mass of the spindle; $\lambda = l/a$ ratio of the distance between supports and λ is given by –

$$\gamma = f(\lambda) = \text{coefficient,}$$

which may have a value, for $\lambda = 2.5 – 3.5$, lying between the range 2.3 and 2.4.

23.5 Tribological Considerations

From the detailed analysis of the commonly represented dynamical system in a machine tool, it becomes imperative for the designer to study the frictional behaviour of the slides and other transmission mechanisms, such that the positional displacement error due to frictional relaxation becomes as low as possible.

Major developmental work of the designers has been in this area of significance.

Characteristic tribological behaviour of many currently used materials have already been discussed at length in Chapters 6 and 22. Some recent researchers done by the authors in the area tribological behaviour of some polymeric and reinforced polymeric materials as well as glass ceramic coatings are being presented here – being of much current attention-particularly in view of development of flexible manufacturing systems.

23.5.1 *Friction characteristics and wear of slideway materials using composites*

In the area of slideway wear using polymeric composites, a large quantum of work has been done by Tanaka, Yamada[129] Lancaster[130] and others. Tanaka and Yamada's[129] work deals with wear rates of PTTE incorporating MoS_2, graphite or bronze. It has been shown that the wear rate increases rapidly as the surface roughness value increases beyond a critical value. Such critical value of h_{CLA} varies with the type of the composites. But as far the

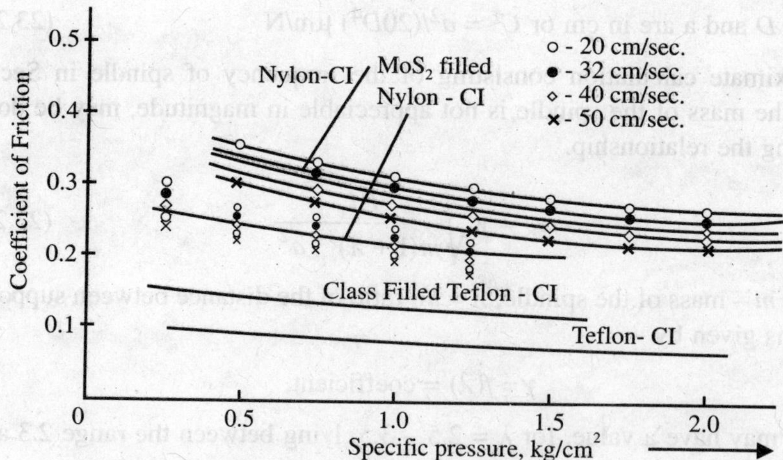

Figure 23.6 Coefficient of friction Vs specific pressure.

friction values are concerned, their characteristic variance with roughness is more or less independent of the roughness value of the mating pairs. Experimental curves have been presented by, Tanaka and Yamada for 'unfilled polyoxymethylene' and 'PTFE-polyoxymethylene'

In case of specific wear rate, as roughness value increases, the specific wear rate also increases. But MoS_2 filled or bronze filled PTFE have much lower rate of specific wear, against roughness value, compared with unfilled PTFE.

Figure shows the variation of the coefficient of friction of various materials used against velocity of sliding in cm/sec. Intensity of pressure on the mating interface Varies between 0.5 kg/cm^2 to 2.5 kg/cm^2.

Figure shows the histogram based on ranges of surface preparation. The study was conducted with all the possible pairs with a starting surface roughness value h_{CLA} 1–4 microns and minimum value of the coefficient of friction occurring corresponding to a particular value of h_{CLA} was considered for plotting. Though Teflon in this shows a minimum coefficient of friction amongst other materials tested, its surface requires to be prepared with utmost accuracy (1-2 micron), preferably between 1 and 1.25 micron.

A large quantum of work has been done by Basu and Shah[131], and Basu, Shah and Prasad[132] in the area of frictional coefficients in case of filled and unfilled plastics. Materials used are mostly Nylon, Teflon, MoS_2 filled Nylon, Glass-filled Teflon etc. to show their dependence on specific pressure of the mating surface, while paired with CI in all cases as shown in Figure 23.6. Similar curves are obtained while plotted against velocity of sliding. The characteristics curves for 'Teflon CI' and 'GF Teflon-CI' show that these are better materials compared to Nylon-filled or unfilled, since their insignificant variation with velocity of sliding or with specific pressure on the interface will prevent occurence of 'stick-slip' motion and hence result in uniform displacement with minimum positional error, as is essential in a N.C., C.N.C. machine. Friction Vs surface roughness for various specific pressure of unfilled and filled plastics has been shown in Figure 23.7.

Figure 23.8 shows the characteristic regime of working for various composites on a p-v diagram where a typical combination of p-v treatment could be chosen. These curves have been plotted after subjecting the data through geometric programming to get optimum values of p and v, corresponding to which there will be minimum frictional energy loss and minimum wear over time. Such type of programming method has been shown in earlier work by Basu and pai[133][134]. Properties of some of the Polymeric materials used are listed in Table 23.2

The properties of the polymers, used for the purpose of machine tools either directly or in filled conditions are given in table below:

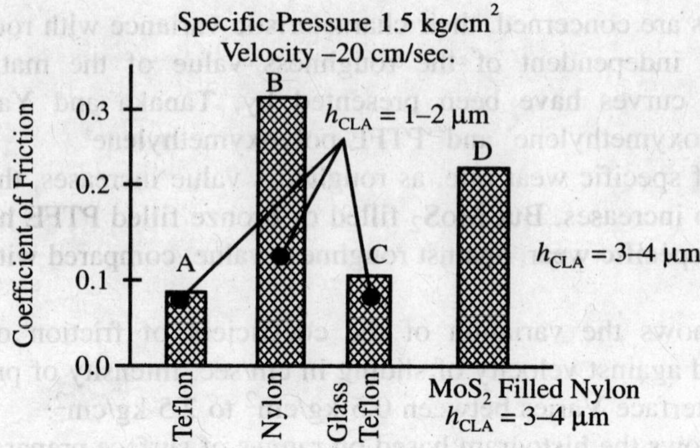

Figure 23.7 Coefficient of friction Vs roughness.

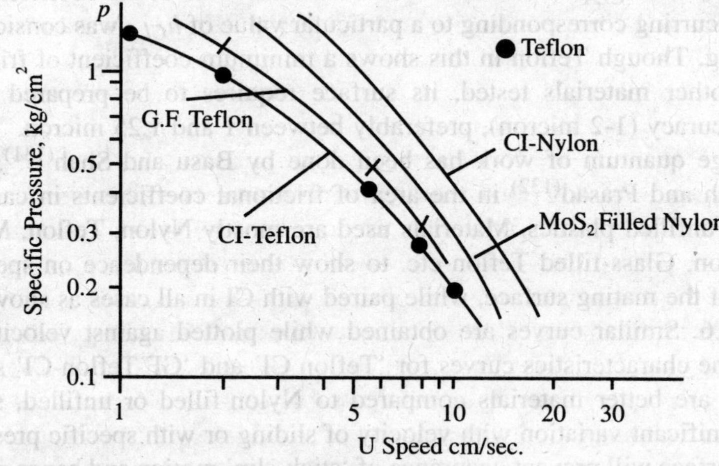

Figure 23.8 p–v Regime.

	Sp. Wt Kgf /Lit	*Hardness*	*Modulus of Elasticity* (10 *kgf/mm*^{2})	*Tensile Strength Kgf/mm*^{2}	*% Elongation*
Nylon	1.12–2.20	108–120	0.126	6.5–8.5	60–300
Teflon	2.13–2.20	85–109	0.004	15–35	200–400
Glass Filled Teflon	1.5	–	56–33	12.1	300
MoS$_2$ Filled Moduli	1.5	–	59.20	11	250–280

Commonly used material, like Turcite *B*, which is nothing but PTFE reinforced with Phosper-bronze, works satisfactorily and they have the distinctive advantage, above the other members of the reinforced polymeric family, due to the fact that contact deformation of slides made by pasting turcite B with araldite and hardener is considerably low and wear and frictional characteristics do not change significantly with temperature or velocity or surface roughness at the interface. Frictional characteristic of Turcite *B* depending on velocity of slide and specific pressure is shown in chapter 22, Figure 22.6.

Coated steel with glass ceramic compared to uncoated steel is much better in minimising the positional error due to displacement. Work done by Sen, Basu, Das and Datta(135) may be referred to. The distinctive advantages of such materials arise due to low wear rate per year, see Figure 23.9-23.11, and temperature insensitiveness. Hence such materials are also being used in robots for handling hot specimens, particularly in the areas of grippcr working on slip sensors. A typical friction characteristic curve has been shown in Figure 23.8

Fig 23.8 shows the variation of specific pressure (p) with the velocity of sliding (v) for various filled and unfilled polymers used. The space surrounded by the curve and the axes, is required to be used for selecting a specific value of V, if p is known, or a specific value of p, if V is known. The peculiarity of this characteristic is that it is tenable for a considerably small range of velocity of sliding. This is precisely the case of heavy duty precision machine tool, where positional error due to stick slip is largely eliminated.

The wear rate for mild steel rubbing against cast iron without any additive was found to be 31 mg/h. While the wear rate for glass-ceramic on cast iron was found to be only 6 mg/h, which is around five times less than the previous one (Figure 23.9, 23.10).

Considering, the co-efficient of friction, it remained more or less constant with time.

Impact strength of Ceramic .Coatings on A, B & C are shown in Figure 23.11.

The Generalised equation for wear obtained empirically by using the experimental results can be given as:

$$W = 6.22\ V^{0.1163}\ P^{0.2007}\ T^{0.2703} \tag{23.28}$$

Where V – velocity cm/sec, P – specific pressure kg/cm^2, T – hours and W – wear in mg.

The typical composition of glass ceramic coating is RO – $R_2'O$ – Al_2O_3 –SiO_2 – (R = Ca, Mg; R = Na, K, Li) With TiO_2 and P_2O_5 as nucleating

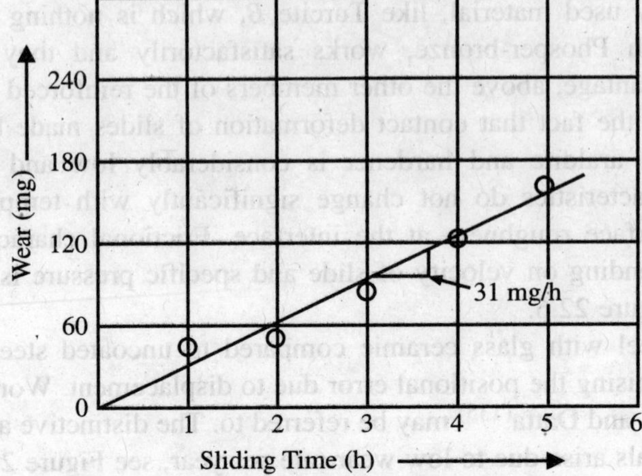

Figure 23.9 Wear-Vs Time for Steel—C.I.

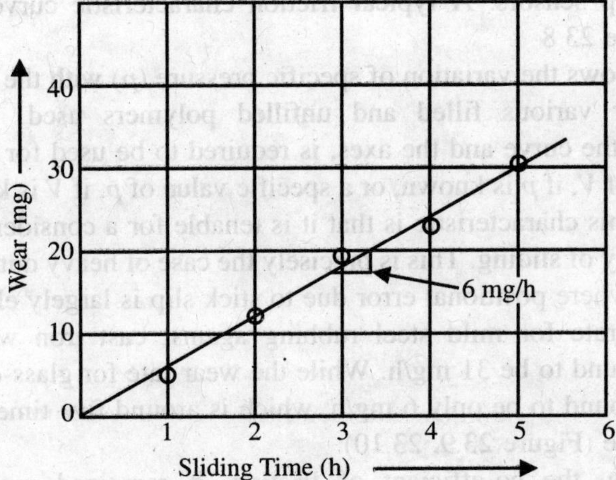

Figure 23.10 Wear-Vs Time for 'Glass-Ceramic'–CI.

agent. The coating shows good adhesion on low carbon steel in comparison to medium or high carbon steel.

Composition in MOL% and impact strength of such material coating is shown in Table 23.3 and Figure 23.11.

23.6 Calculation of Rigidity Considering Finite Element[127]

We shall illustrate a simple example, shown in Figure 23.12 wherein it is necessary to know the displacement of point 2, assuming that each boom has

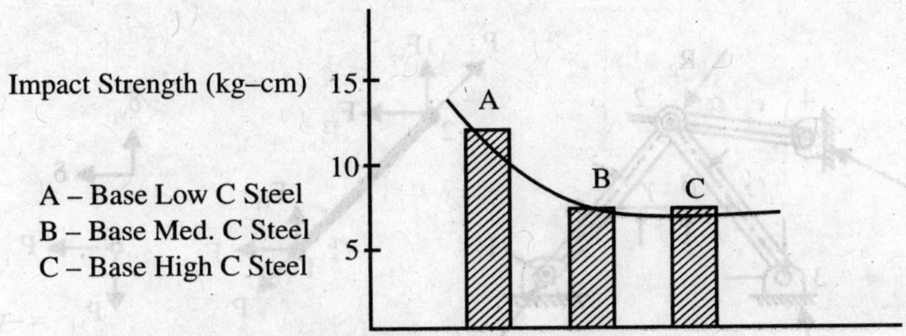

Figure 23.11 Strength of Ceramic Coating on Steel.

Table 23.3 Glass-ceramic

Composition	MOL (%)
SiO_2	30–75
TiO_2	5–20
B_2O_3	0–12
$A1_2O_3$	0–10
Na_2O	0–14
K_2O	0–10
LiO_2	0–5
CaO	0–4
P_2O_5	0–4
Melted at 1200° – 1450°C	

a length L and transverse sectional area S and Modulus E. An axial lead P acts on the element e_1 and its reactions are F_1 and F_2. Let δ_1, δ_2 be the displacement to be found out and their projection on coordinate axes δx_1; δy_1 : δy_2. In the matrix form, we may write:

$$F^{e1} = \begin{vmatrix} F_1 \\ F_2 \end{vmatrix}^{e1} = \begin{vmatrix} Fx_1 \\ Fy_1 \\ Fx_2 \\ Fy_2 \end{vmatrix}^{e1} \tag{23.29a}$$

And

$$\delta^{e1} = \begin{bmatrix} \delta_1 \\ \delta_2 \end{bmatrix}^{e1} = \begin{bmatrix} \delta x_1 \\ \delta y_1 \\ \delta_{x2} \\ \delta y_2 \end{bmatrix} \tag{23.29b}$$

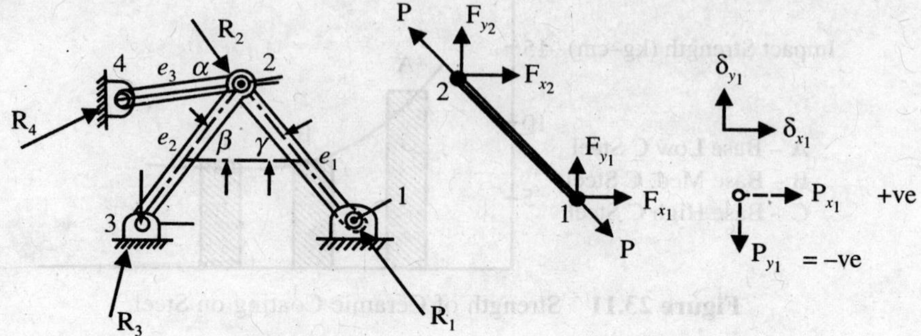

Figure 23.12 Rigidity Calculation.

But

$$\begin{bmatrix} F_{x1} \\ F_{y1} \\ F_{x2} \\ F_{y2} \end{bmatrix} = \begin{bmatrix} P\cos\gamma \\ -P\sin\gamma \\ -P\cos\gamma \\ p\sin\gamma \end{bmatrix} \tag{23.29}$$

Axial force P may be found out from the extention of the bar.

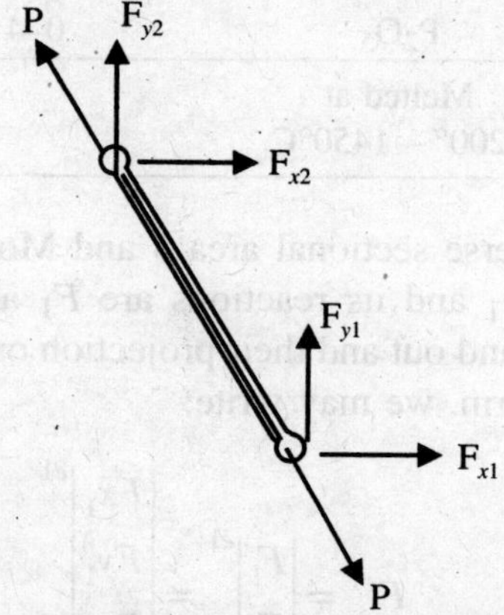

Figure 23.12a Element Freebody.

$$P = \left(\frac{ES}{L}\right)\left[(\delta_{x2} - \delta_{x1})\cos\gamma + (\delta_{y2} - \delta_{y1})\sin\gamma\right] \tag{23.30}$$

Substituting this expression of P as in equation (23.30) in the expression of F in equation 23.29) we get,

$$F_{x1} = \left(\frac{ES}{L}\right)\left[(\delta_{x2} - \delta_{x1})\cos^2\gamma + (\delta_{y2} - \delta_{y1})\sin\gamma\cos\gamma\right]$$

$$= \left(\frac{ES}{L}\right)\left[-\delta_{x1}\cos^2\gamma - \delta_{x1}\cos\gamma\sin\gamma + \delta_{x2}\cos^2\gamma + \delta_{y2}\sin\gamma\cos\gamma\right] \tag{23.31}$$

In the matrix form this may be written as:

$$F_{x1} = \left(\frac{ES}{L}\right)[-\cos^2\gamma. -\sin\gamma\cos\gamma. \cos^2\gamma. \sin\gamma\cos\gamma]\begin{bmatrix}\delta_{x1}\\ \delta_{y1}\\ \delta_{x2}\\ \delta_{y2}\end{bmatrix} \tag{23.32}$$

The four equations of the type of equation (23.32) for FX_1, FY_1, FX_2, FY_2 in the matrix form may be written as under:

$$F^{e1} = \left(\frac{ES}{L}\right)\cdot\begin{bmatrix} -\cos^2\gamma & -\sin\gamma\cdot\cos\gamma & \cos^2\gamma & \sin\gamma\cos\gamma \\ \sin\gamma\cos\gamma & \sin^2\gamma & -\sin\gamma\cos\gamma & -\sin^2\gamma \\ \cos^2\gamma & \sin\gamma\cos\gamma & -\cos^2\gamma & -\sin\gamma\cos\gamma \\ -\sin\gamma\cdot\cos\gamma & -\sin^2\gamma & \sin\gamma\cos\gamma & \sin^2\gamma \end{bmatrix}\begin{bmatrix}\delta_{x1}\\ \delta_{y1}\\ \delta_{x2}\\ \delta_{y2}\end{bmatrix} \tag{23.33}$$

Dividing by $\left(\frac{ES}{L}\right)$ we get

$$F^{e1} = \begin{vmatrix}F_{x1}\\ F_{y1}\\ F_{x2}\\ F_{y2}\end{vmatrix} = \begin{vmatrix} K_{x1,x1} & K_{x1,y1} & K_{x1,x2} & K_{x1,y2} \\ K_{y1,x1} & K_{y1,yl} & K_{y1,x2} & K_{y1,y2} \\ \hline K_{x2,x1} & K_{x2,y1} & K_{x2,x2} & K_{x2,y2} \\ K_{y2,x1} & K_{y2,y1} & K_{y2,x2} & K_{y2,y2} \end{vmatrix}\begin{vmatrix}\delta_{x1}\\ \delta_{y1}\\ \delta_{x2}\\ \delta_{y2}\end{vmatrix} \tag{23.24}$$

$$F^{el} = \begin{vmatrix} K_{11}^{el} & K_{12}^{el} \\ K_{21}^{el} & K_{22}^{el} \end{vmatrix} \times \begin{vmatrix} \delta_1 \\ \delta_2 \end{vmatrix} \tag{23.35}$$

For the other elements also, it is possible to get similar equations:

$$F^{el} = \begin{vmatrix} F_1^{el} \\ F_2^{el} \\ 0 \\ 0 \end{vmatrix} = \begin{vmatrix} K_{11}^{el} & K_{12}^{el} & 0 & 0 \\ K_{21}^{el} & K_{22}^{el} & 0 & 0 \\ 0 & 0 & 0 & 0 \\ 0 & 0 & 0 & 0 \end{vmatrix} = \begin{vmatrix} \delta_1 \\ \delta_2 \\ \delta_3 \\ \delta_4 \end{vmatrix} \tag{23.36}$$

Where K^{e1} – Expanded matrix of rigidity of element e_1
δ – Vector deformation of the units of the system.

The magnitude of force R_2 may be obtained through the projections RX_2, RY_2.

$$RX_2 = FX_2^{e1} + FX_2^{e2} + FX_2^{e3} \tag{23.38}$$

or in matrix form

$$R_2 = \begin{vmatrix} R_{x2} \\ R_{y2} \end{vmatrix} = \sum_{e=1}^{3} \begin{vmatrix} F_{x2}^{e} \\ F_{y2}^{e} \end{vmatrix} = \sum \cdot F_2^e \tag{23.39}$$

Putting equation (10) in equation (12) we get:

$$R = \begin{vmatrix} R_1 \\ R_2 \\ R_3 \\ R_4 \end{vmatrix} = \sum_{e=1}^{3} \begin{vmatrix} F_1^{e} \\ F_2^{e} \\ F_3^{e} \\ F_4^{e} \end{vmatrix} = \sum_{e=1}^{3} F^e = \sum_{e=1}^{3} K^e \cdot \delta \tag{23.40}$$

OR

$$\mathrm{K} \cdot \delta = \mathrm{R} \tag{23.41}$$

This is called Matrix equation of the system and:

$$K = \sum_{e=1}^{3} k \tag{23.42}$$

$$K_{iJ} = \sum_{e=1}^{3} K_{iJ} \tag{23.43}$$

The value of Transformation Matrix $\sum_{e=1}^{3} K_{iJ}$ can be calculated from the family of equations, given in (23.33) to (23.36).

K is normally known as "System Rigidity Matrix". The process of obtaining the matrix equations of the elements and the matrix equations of the system may be formulated with the help of the equation (23.42).

If the force R_2 is known, then the system of the linear algebraic equations (23.41) may be solved by various methods and the values of the deformations may be determined.

The above method of calculation is known as displacement method.

CHAPTER 24

Non-uniform Microdisplacement

In this chapter further analysis of microdisplacement in machine tools sliding mechanism has been done to consolidate the results, so far obtained through investigations, in appropriate cases, by researchers from time to time. Major portions of this in details have been compiled by V.E. Push(126) in his book. This chapter records some of these informations for the better understanding of the phenomenon of microdisplacement and the errors involved therein.

24.1 Calculations for Non-Uniform microdisplacement

Given, the mass of the sliding unit m_1 (kg sec^2/cm); stiffness of the system of transmission k_1 (kg/mm); velocity of driven V (mm/sec) coefficient of vibrational damping θ_1; and difference in the force of friction (between static & dynamic at velocity v) = ΔT in Kg it is possible to summarize the major factors contributing to the errors resulting from microdisplacement, as shown in Table 24.1.

Symbols and Notations used in Table 24.1 are as follows:

Coefficient of friction at a velocity $V = f_v = \phi\left(\dfrac{\lambda}{\lambda_{cr}}\right)$, where

$\lambda = \dfrac{\mu V}{\sigma}$ = characteristic regime, μ is the coefficient of viscosity (kg sec/m^2 or Pa sec) under working temperature T; V-velocity in m/sec, and σ-pressure, kg/cm^2. Regime of work, λ becomes λ_{cr}, when $V = V_c$, V_c being critical velocity.

Table 24.1 Evaluation of significant parameters of positional displacement[127]

Magnitude	*Formula*	*Approximate formula at the least velocity $V < V_c$*	*Approximate formula when $V = Vc$*
Critical Velocity V_c	$V_c = \dfrac{\Delta T}{A_c\sqrt{K_1 m_1}}$ Where $A_c = \sqrt{4\pi\theta_1}$	—	—
Duration of the slip t_1	$e^{\theta_1 \omega t_1} = \cos \omega t_1 + (\theta - A) \sin \omega t_1$ where $\omega = \sqrt{\dfrac{K_1}{m_1}}$, $A = \dfrac{\Delta T}{V\sqrt{K_1 m_1}}$	$t_1 = \dfrac{2\pi}{\omega}$	$t_1 = \dfrac{2\pi - \sqrt{4\pi\theta_1}}{\omega}$
Duration of the stick period t_2	$wt_2 = \dfrac{1 + A^2 - 2A\theta_1}{A - \theta_1 - ct_g\omega t_1}$	$t_2 = \dfrac{\Delta T}{\omega}(2 - \pi\theta_1)$	$t_2 = \dfrac{\sqrt{4\pi\theta_1}}{w}$
Magnitude of stick-slip A_K	$A_K = V(t_1 + t_2) = \dfrac{\Delta T}{K_1} f_s$	$A_K = \dfrac{\Delta T}{K_1}(2 - \pi\theta_1)$	$A_K = \dfrac{\Delta T}{K_1}\sqrt{\dfrac{\pi}{\theta_1}}$

24.2 Non-dimensional Parameters

Following are the non-dimensional parameters, $\Pi_1 = \dfrac{\lambda}{\lambda_{er}}$; $\Pi_2 = \dfrac{N}{\sqrt[v]{k_1 m_1}}$,

Where N-Normal Force; $\Pi_3 = \dfrac{h_c}{\sqrt{k_1 m_1}}$, where h_c - coefficient of resistance force or viscous drag per unit velocity.

When $\Pi_1 \geq 1$, then it approaches to liquid friction in guides. When Π_1, lies between 0.7 and 0.95, then it is necessary to manipulate the value to approximately 1, by increasing the condition for the creation of liquid friction by increasing the number of transverse oil canals on the guide face or by increasing the viscosity of the oil and so on.

Characteristic functional relationship between Π_2 and Π_3 based on various values of Π_2 can be found out from the experimental -cum-theoretical curves plotted in Figs. 24.1 and 24.2.

Figure 24.2 shows the variation of the coefficient of friction against velocity of sliding (V mm/min) and against $\left(\frac{\lambda}{\lambda_{Cr}}\right) \cdot \frac{\lambda}{\lambda_{Cr}} \approx \frac{V}{V_{Cr}}$ as per symbols used in Table 24.1 the dependence of f *Vs* V or f *Vs* $\left(\frac{\lambda}{\lambda_{Cr}}\right)$ is based on materials CI-CI under various specific pressure oh the guide. Lubricant

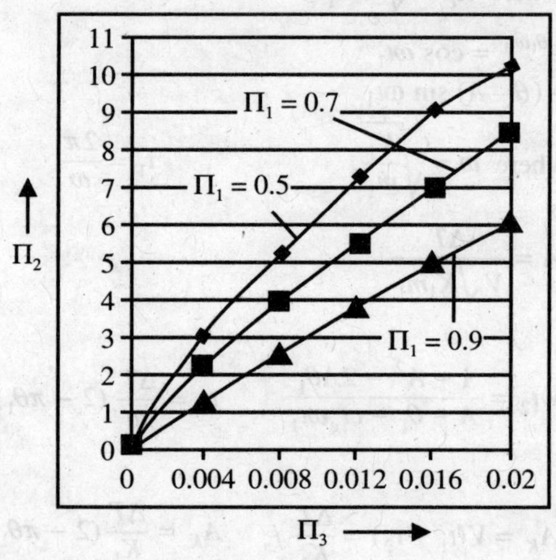

Figure 24.1 P_2 against P_3 for various value of P_1.

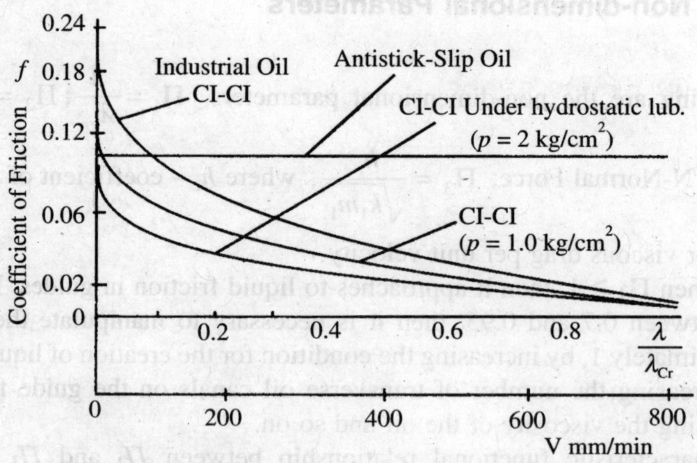

Figure 24.2 f *Vs* $\left(\frac{\lambda}{\lambda_{Cr}}\right)$ Equivalent scale for V has also been shown.

used in all these cases are base oil SAE-20. For antistick slip oil viscosity (N. Sec/m^2) is very low in comparison to base oil and the behavior shows negligible difference between static and dynamic coefficient of friction i.e. ($f_v = 0 - f_v = f =$ almost zero). This therefore reduces the chances of positional displacement error, arising out of "Stick-slip" sliding.

24.3 Minimisation of Positional Displacement Error

Positional displacement error arising out of that may be reduced considerably:

(i) By using oil with polar additives viz., ammonium compound or sulphur compound or by using oelic acid in requisite proportion.
(ii) By reducing the value ΔT, with use of some filled or unfilled plastics. (PTFE phosphor bronze reinforced PTFE commonly known as Turcite-B, glass-filled PTFE, or material coated with PVC powder)
(iii) By using hydraulic relieving.
(iv) By using hydrostatic lubrication and increasing the pressure of the externally pressurized fluid with a very high value of external pressure, so that the dynamic coefficient of friction reduces to practically nil, thereby allowing the slide to be lifted.
(v) By impressing Ultrasonic Vibratory force in the normal direction of the slide (as per mathematical algorithm given by Lomakin[138] with major assumptions and Basu[8] subsequently without any assumption.)

24.4 Distribution of Hydrodynamic Force in Specific Lubrication Pocket

Various types of hydrodynamic pockets have been shown in Figure 24.3(a), (b) and (c).

It has been shown[88] that $$\lambda = \frac{\mu v}{\sigma} = \frac{1 + \left(\frac{L}{B}\right)^2}{LC_P} h_1^2 \qquad (24.1)$$

Where L and B are length and breadth of oil canal respectively. Normally L/B is very large compared to 1 and h_2, h_1-maximum and minimum oil film thickness respectively. C_p is a characteristic coefficient depends on h_2/h_1. The dependence of C_p on h_1/h_2 (=a) has been shown earlier in chapter 6 of this book.

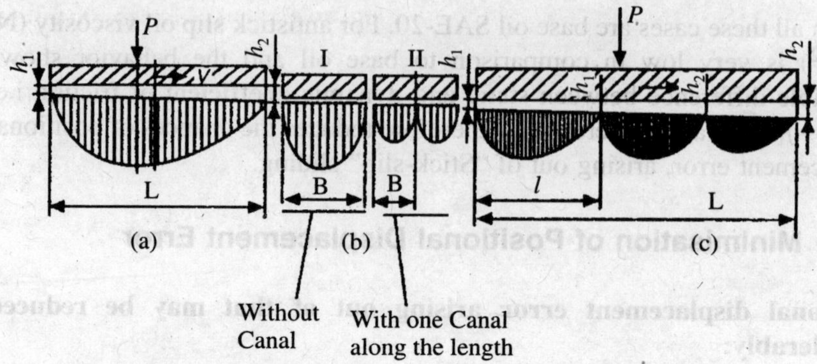

Figure 24.3 (a) (b) and (c): Pressure distribution.

$\therefore \dfrac{\lambda_{cr}}{\lambda} = \left(\dfrac{h_2}{h_1}\right)^2$. Maximum value of Cp and consequently the value of the uplifting force, result in optimum value of $h_2/h_1 = 2.2$ and $C_p = 0.16$.

$$\lambda^*_{cr} = 1175.10^2 \frac{i^2\left[1+\left(\dfrac{L}{iB}\right)^2\right]}{LC_K} h^2_{cr} \qquad (24.2)$$

λ^*_{cr} corresponds to λ_{cr} is the value at which liquid lubrication starts and coefficient of friction has the smallest value and C_k is a coefficient value of which is given in the table below. In this equation (24.2), i is given by $k = i - 1$ or $i = k + 1$ and the coefficient C_k is dependent on k.

k = Number of Transverse canal on the top surface of guide

i = Number of oil pockets

Table 24.1 Value of C_k corresponding to k

k ($k = i - 1$)	0	1	2	3	4	5	7	9	11	13	15
C_k	0.165	0.198	0.210	0.216	0.220	0.225	0.234	0.243	0.249	0.257	0.263

Hence it is possible to calculate f_v from the formula,

$$f_v = f_0\left[1 - k_v \sqrt[3]{\frac{\lambda}{\lambda_{cr}}}\right] \qquad (24.3)$$

Table 24.2 Characteristic regimes of work and coefficient of friction in guides for feed motion.

Length L mm, width of guide B mm, average pressure σ kg/cm^2, velocity of sliding (V) in m/sec.	
Characteristic regime of work, for respective parameters μ, v, and σ.	$\lambda = \frac{\mu v}{\sigma}\left[\frac{\text{Centipoise} \times \text{m/sec}}{\text{kg/cm}^2}\right]$
Critical Minimum Oil film Thickness when f_v is lowest friction $= f_L$ and $\lambda = \lambda_{cr}$	h_{cr} (*mm*)
Number of oil pocket in transverse direction.	K
Critical characteristic of the regime of work for $f_v = f_i$ i.e. boundary condition	$\lambda^*_{cr} = 1175.10^2 \frac{L^2\left[1+\left(\frac{L}{LB}\right)^2\right]}{LC_K} h^2_{cr}$
Critical velocity of sliding	$V_{cr} = \frac{\lambda_{cr}\sigma}{\mu}$ (m/sec)
Relative characteristic of the regime of working	$\frac{\lambda}{\lambda_{cr}}$
Coefficient of Friction $f_v \cong \phi\left(\frac{\lambda}{\lambda_{cr}}\right)$	$f_v = f_0\left[1 - k_v\sqrt[3]{\frac{\lambda}{\lambda_{cr}}}\right]$
ϕ Denotes mathematical function.	where $K_v = 1 - (f_L / f_0)$, and f_0-static coefficient of friction, at $V = 0$

The values of Cp in equation (24.1) depends upon $\left(\frac{h_2}{h_1}\right)$ and have been shown in the Figure 6.10 in chapter 6.

24.5 Dynamic Load Rating of Spindle Supports

The following equation may be used to find out the durability of the spindle supports.

$$C = (RK_1 + mA) \cdot (nh)^{0.3}\ k_b \cdot k_k \quad (24.4)$$

C – Characteristics Coefficient

$$*RK_1 + mA = Q \quad (24.5)$$

R, A being radial and axial components of the load respectively and Q – load in Kg. According to Reshetov[88] K_1 = 0.5–0.6 for radial ball, bearing and for deep groove ball bearing K_1 = 0.65–8; m is approximately equal to 0.5; Life rating h-working hours (normally h = 5000 hours).

$k_b \cdot k_k$ – coefficient considering characteristic of the load and rotation of the inner or outer ring and k_k – is temperature correction factor)

Equivalent Q_e acting on the spindle.

$$Q_e = \left(\sum_J \frac{h_j}{h} \cdot \frac{n_J}{n_r} \cdot Q_J^{10/3} \right)^{0.3} \tag{24.6}$$

h_j–hours of work of bearing under the rotational speed n_J and under load Q_J and n_r is rated r.p.m.

$$\text{Normally } Q_J \approx 0.8\, Q_{max} \tag{24.7}$$

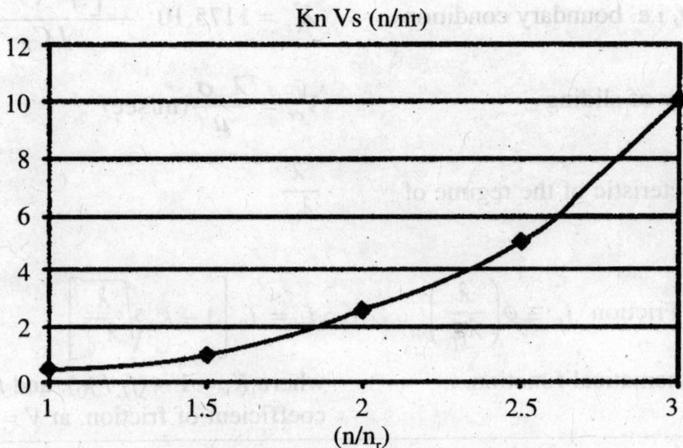

Figure 24.4 K_n *Vs* (n/n_r).

For high speed spindle there is a correction coefficient to be used in equation (24.6) dependent on (n/n_r), n_r being the rated rpm. This correction coefficient K_n is shown against (n/n_r) in Figure 24.4.

24.6 Rigidity of Spindle unit and Optimum overhang of the Spindle end:[126, 127]

Consider a spindle unit as shown in Figure 24.5 (a), with 'a' as cantilever front end and 'b' as bearing centers distance. N being the load at the front end of the spindle. Figure 24.5 (a) and 24.5 (b) show the bending and deformation of the spindle respectively.

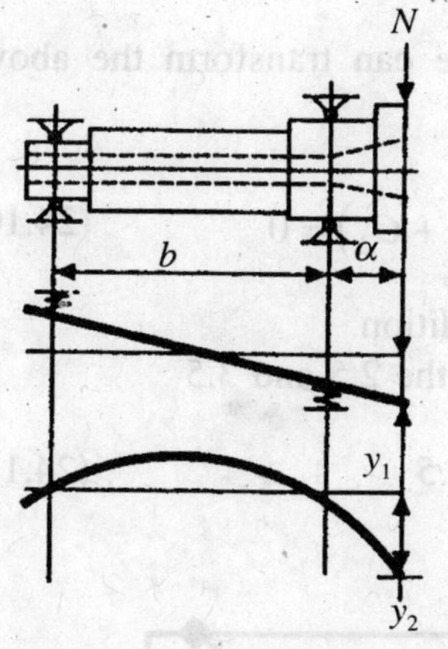

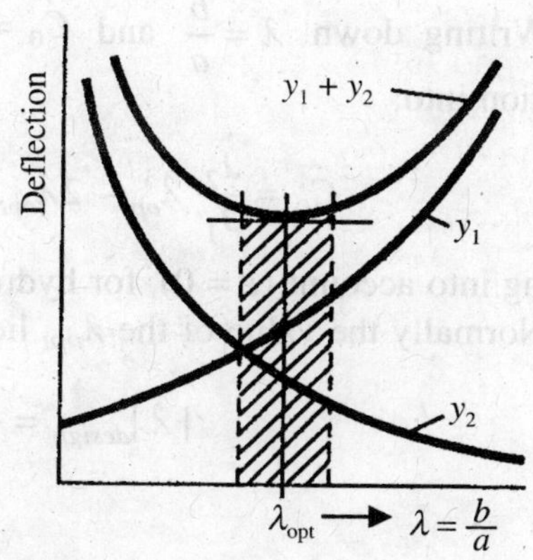

Figure 24.5(a) Spindle and its bending diagram

Figure 24.5(b) Deformation diagram.

Rigidity of the spindle unit is the most essential characteristic and apart from that, it includes the rigidity of all essential sub-units. For spindle having two supports, as shown in the Figure 24.5 above, the sum total of the transverse deflections $y = y_1 + y_2$ and C_1 and C_2 are the coefficient of compliance $\left(\frac{\delta}{P}\right)$ of the two supports (front and rear end respectively).

$$y = \frac{Na^2}{3E}\left[\frac{a}{J_2} + \frac{b(1-\varepsilon)}{J_1}\right] + N\left\{C_1\left[\frac{a(1-\varepsilon)+b}{b}\right]^2 + C_2\,(1-\varepsilon)\frac{a^2}{b^2}\right\} \quad (24.8)$$

N–Radial Force; b-distance between the bearings; a-overhang; E-modulus of Elasticity. J_1 J_2 – axial moment inertia of the cross section of the spindle in the cantilever portion and between bearings. C_1 C_2 – Coefficient due to contact compliance at the front and rear end of the Journal respectively, ε-coefficient due to error in joint compliance at the front end of the support. For the type of the spindle shown in this Figure 24.5(a) ε=0.4, while for hydrostatic support $\varepsilon = 0$.

We may now write down the overall compliance C as under:

$$C = \frac{a^2}{3E}\left[\frac{a}{J_2} + \frac{b(1-\varepsilon)}{J_1}\right] + \left\{C_1\left[\frac{a(1-\varepsilon)+b}{b}\right]^2 + C_2\,(1-\varepsilon)\frac{a^2}{b^2}\right\} \quad (24.9)$$

C_1, C_2, J_1, J_2 depend on only the diametrical gap of the spindle, but for actual construction (E, a, b, ε – being known) C may be taken as a function of d $\therefore\ C = f(d)$, d-being the diameter of the Journal.

Writing down $\lambda = \frac{b}{a}$ and $C_0 = \frac{a^3}{3EJ_2}$, we can transform the above equation into;

$$C_0 = \frac{J_2}{J_1}\lambda_{op}^3 - 2\lambda_{opt}\, C_1 - 2\left(C_1 + C_2\right) = 0 \tag{24.10}$$

Taking into account ($\varepsilon = 0$), for hydrostatic condition.

Normally the value of the λ_{opt} lies between the 2.5 and 3.5

$$|\lambda|_{design} = \frac{b}{a} = 2.5 - 3.5 \tag{24.11}$$

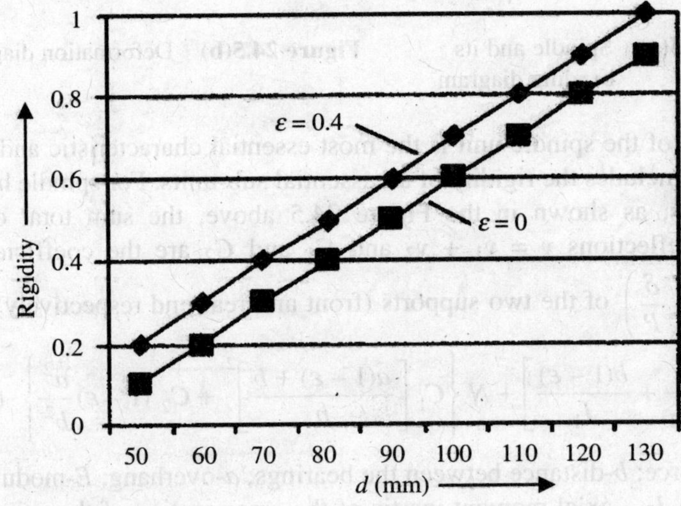

Figure 24.6 Rigidity *j* of the spindle unit depending on diameter of the spindle and coefficient due to error in joint compliance.

24.7 Accuracy of the Spindle Units

Deflection at the free end of the spindle is due to addition of two major deflections: (i) due to bending and (ii) due to shear. Apart from this there is some contact deformations at the support itself. Table 24.3 shows the error occurring on spindle supports in the most of the machine tools.

These values above are normally for the type of spindle supports, commonly used in industries.

Figure 24.7 (a) shows taper roller bearing, preloaded through spring, and (b) shows taper roller bearing, preloaded through oil pressure while, the

Figure 24.7 (c) shows double row roller bearings with temperature compensation – the various parts being marked on the figures are as follows:

1-External ring, 2-Auxiliary ring for locating, 3-Seals, 4-Conical rollers, 5-Springs, 6-lnner ring, 7-Stopping unit, 8-Compensating ring, 9- Outer ring.

Table 24.3 Accuracy of spindle units on various supports

Spindle Support	*Error due to Eccentricity μm*	*Maximum run out μm*	*Eccentricity of workpiece μm*	*Workpiece Surface roughness R_a (Centerline Average) μm*
Rolling Friction Bearings	0.8–1.3	2.0–4.0	3-5	1-2
Hydrostatic Supports	2.0	0.2	0.3	0.06

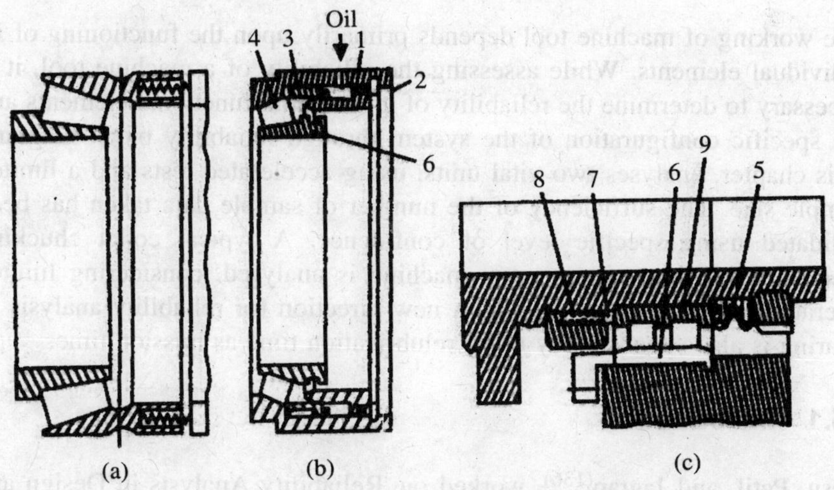

Figure 24.7 (a) (b) and (c) Various types of bearings.

CHAPTER 25

Reliability Analysis of Some Machine Tool Elements[144]

The working of machine tool depends primarily upon the functioning of its individual elements. While assessing the reliability of a machine tool, it is necessary to determine the reliability of its essential functional elements and the specific configuration of the system through reliability block diagram. This chapter, analyses two vital units, using accelerated tests and a limited sample size. The sufficiency of the number of sample data taken has been validated using specific level of confidence. A typical collet chucking mechanism of a semi-automatic machine is analysed, considering limited operating cycles without failure. A new direction for reliability analysis of bearing is also suggested by using relubrication time as mission time.

25.1 Introduction

Basu, Patil, and Jagtap[136], worked on 'Reliability Analysis in Design and Manufacturing through Accelerated method, in some systems, there may be a few failures or even no failure, and as such the failure data is difficult to get. Hence 'accelerated test' method is used.

Basu, Sonawane and Sarawade[137] applied accelerated test for reliability based on 'malfunction', for 'Reliability Assessment of Some Hydraulic Equipment'. Parametric evaluation is time dependant, while non-parametric evaluation does not depend upon time. Application of accelerated test includes assessing component reliability, demonstrating component reliability, detecting failure modes and establishing safe warranty times. The researchers[140] used non-parametric 'variable' method (i.e. K-statistic), though attributes, median and mean ranking methods are also in use.

This Chapter considers situations where conventional method of reliability assessment requires to be modified. The two case studies are presented for chucking mechanism of the semi automatic machine for loading and feeding of the job in the spindle and for analysing reliability of the spindle supports.

25.2 Collet Chucking Analysis

The first case deals with the mechanism for collet chucking as shown in Figure 25.1. The collet opening and closing is controlled by the movement of chucking cam drum which is mounted on the rear cam shaft via a link of elements namely chucking lever, sliding sleeve, spacer, fingers, pusher tube, spring and collet sleeve. Only drawn bars are chucked as they are within controlled tolerance and have even surface texture.

Considering the entire system as a vital unit, the 'time-independent' reliability can be found out by using short sample tests, without failure, following variable method of statistics. This specific short sample test can also be used by truncating the data after one failure. The collet chucking mechanism mainly depends upon finger actuation of the system. This provides a unique method of determining non-parametric reliability.

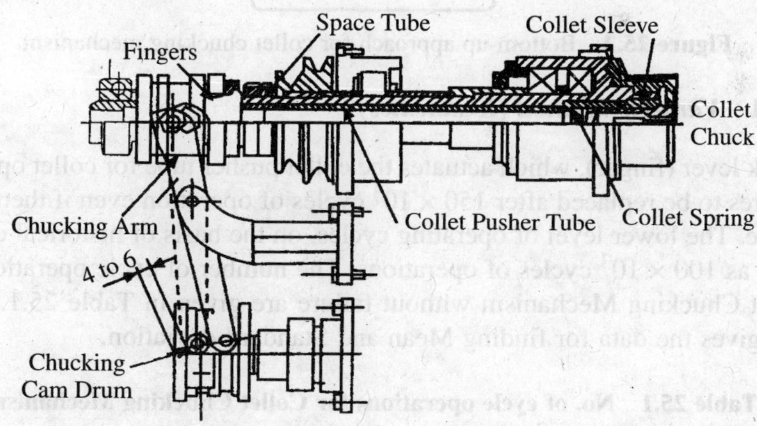

Figure 25.1 Collet Chucking Mechanism[138].

The elements of chucking mechanism (i.e. opening and closing of collet chuck) are represented in a block diagram (Figure 25.2).

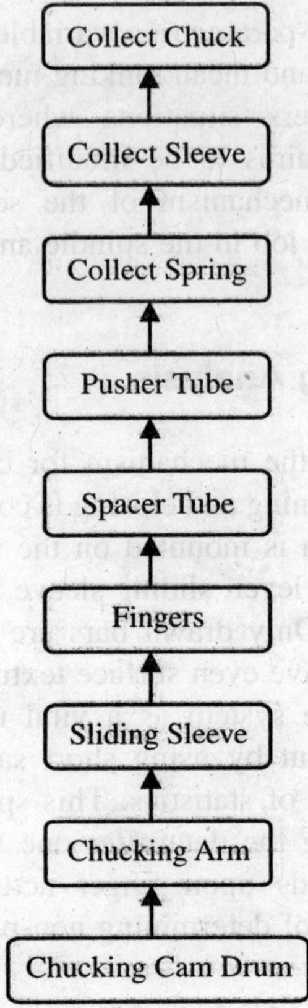

Figure 25.2 Bottom-up approach for collet chucking mechanism.

25.2.1 *Variable Method (K-statistics)*

Chuck lever (finger), which actuates the collet pusher tube for collet opening, requires to be replaced after 150×10^3 cycles of operation even if there is no failure. The lower level of operating cycles, on the basis of historical data, is taken as 100×10^3 cycles of operations. The number of cycle operations for Collet Chucking Mechanism without failure are given in Table 25.1. Table 25.2 gives the data for finding Mean and Standard deviation.

Table 25.1 No. of cycle operations for Collet Chucking Mechanism

Sl. No.	*No. of Operation*
1	144000
2	140000
3	130000

4	136000
5	145000
6	130000
7	128000
8	110000
9	125000
10	105000
11	138000
12	112000
13	150000
14	142000
15	134000

Maximum Cycles (L_u) = 150000
Minimum Cycles (L_i) = 100000

$$\text{Mean } (\bar{X}) = \frac{\Sigma X}{n} = \frac{196000}{15} = 131266.67$$

$$\text{Standard Deviation } (\sigma_x) = \sqrt{\frac{\sum (X - \bar{X})^2}{n - 1}}$$, where n is the number of observations.

$$\sigma_x = 13456.1015$$

Table 25.2 Short Sample Data for Finger

SI. No.	*No. of Operation* (X_1)	$(X_1 - \bar{X})$	$(X_1 - \bar{X})^2$	X^2
1	144000	12733.33	162137776.93	20736000000
2	140000	8733.33	76271110.53	19600000000
3	130000	–1266.67	1604444.53	16900000000
4	136000	4733.33	22404444.13	18496000000
5	145000	13733.33	188604443.53	21025000000
6	130000	–1266.67	1604444.53	16900000000
7	128000	–3266.67	10671111.33	16384000000
8	110000	–21266.67	452271112.53	12100000000
9	125000	–6266.67	39271111.53	15625000000
10	105000	–26266.67	689937779.53	11025000000
11	138000	6733.33	45337777.33	19044000000
12	112000	–19266.67	371204445.73	12544000000
13	150000	18733.33	350937776.53	22500000000
14	142000	10733.33	115204443.73	20164000000
15	134000	2733.33	7471110.93	17956000000
	Σ1969000		**Σ2534933333.33**	**Σ2.60999E+11**

Using K statistics, the values of K_u and K_l are found out by equation (25.1) and (25.2) respectively.

$$K_u = \frac{L_u - \bar{X}}{\sigma_x} = \frac{150000 - 131266.67}{13456.1015} = 1.392 \approx 1.4 \quad (25.1)$$

$$K_l = \frac{\bar{X} - L_i}{2} = \frac{131266.67 - 100000}{13456.1015} = 2.3236 \quad (25.2)$$

The minimum value of k i.e. $K_u = 1.40$, for 15 samples, is considered for finding the reliability. Referring the graph (Figure 25.3)(139) of K statistics for the 15 sample readings, for 70% level of confidence, the reliability of finger is 0.89.

25.2.2 *Sufficiency of Readings*(140)

For the validity of short sample, it is necessary to take readings a number of times repetitively. The number depends upon the level of confidence needed and the value of standard deviation of the universe for the given element.

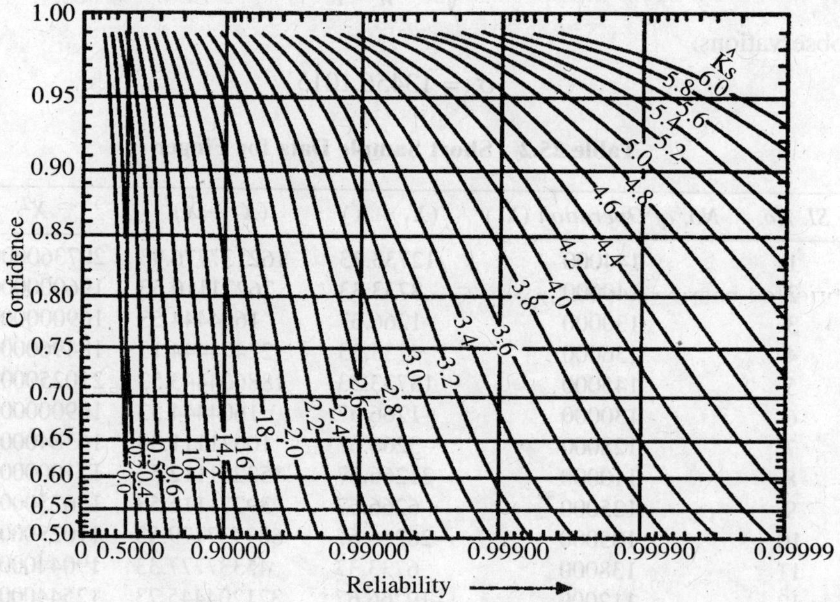

Figure 25.3 Reliability determination from safety margin computed from 15 sample units(139)

Considering the observations (N) on a particular time element, it is presumed that the study should be reliable with 95% level of confidence. The observations needed for the analysis can be found out from the statistical formula given by equation (25.3).

$$N_{\text{exp}} = \left(\frac{40\sqrt{N \times \Sigma X^2 - (\Sigma X)^2}}{\Sigma X} \right) \tag{25.3}$$

Where $N_{\text{exp.}}$ – Expected no. of observations to give the 95% level of confidence

N – Observations already taken.

X – the value of each observation.

Keeping all values, the expected number of observations obtained is 15. This gives the check for sufficiency of the number readings.

25.2.3 *Reliability Analysis for Bearing*

For Bearing No. 6211

The maximum operating speed of the automat is 4000 rpm. For accelerated testing, the standard speed is increased by 1.33 times of maximum speed.

Std Testing Speed (n) = 1.33 × Maximum operating speed.

$$n = 1.33 \times 4000 = 5320\ rpm$$

For reliability estimation, we can take 5320 rpm as the operating speed.

n_g is permissible speed for bearing[141]

For the bearing no. 6211 the maximum permissible speed for grease-lubricated bearing is 6700 rpm. (Figure 25.4), n_g = 6700 rpm.

$$\frac{n}{n_g} = \frac{5320}{6700} = 0.794$$

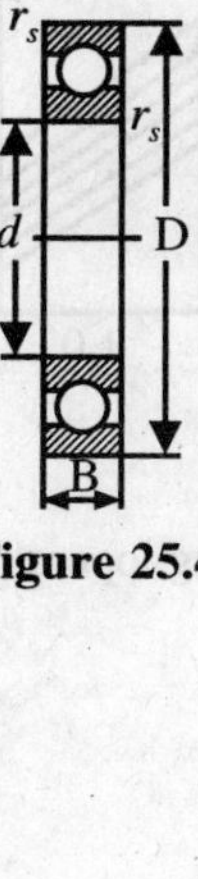

Shaft Dimension d D B r_3 D_n a_n b_n r_n (mm, min)	Load rating dyn. stat. C C kN	Limiting speed Grease Oil min^{-1}	Number Bearing FAG	Snapring FAG	Weight kg
55 10021 1.5	43 25.5	6700 8000	6211		0.607
55 10021 1.5 96,83,282,70,6	43 25.5	6700 8000	6211.C3	SP100	0.607
55 10021 1.5	43 25.5	6700 8000	6211N		0.607
55 10021 1.5	43 25.5	4300	6211RSN		0.607

Figure 25.4 Deep groove ball bearing, shaft dimensions and limiting speed for grease lubrication[141].

Referring to the manufacturers catalogue[142], the following points are considered for filling grease in the housing.

If $\dfrac{n}{n_g} \le 0.2$ Full housing space filled with grease

$\dfrac{n}{n_g} \approx 0.2 - 0.8$ 1/3 housing space filled with grease

$\dfrac{n}{n_g} > 0.8$ No grease filling in housing space.

In the present case the ratio $\dfrac{n}{n_g} = 0.794$, which is in between 0.2 – 0.8.

Therefore 1/3 housing space should be filled with grease.

Relubrication interval '*T*' in hours can be found easily from following graph (Figure 25.5).

Relubrication interval = T ≈ 2000 Hr.

Safety Margin using similarly to Load-Strength interaction for the bearing = *Z*

$$Z = \frac{(\bar{S} - \bar{L})}{\sqrt{\sigma_S^2 + \sigma_L^2}} \tag{25.4}$$

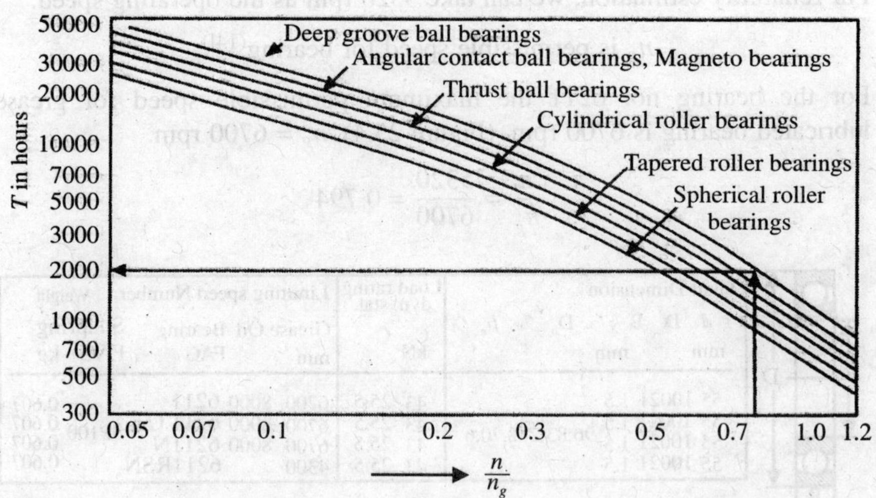

Figure 25.5 Relubrication interval '*T*' for grease lubricated bearing operating under good ambient conditions ($q = 1$)[142].

Where $\bar{S}$ – Maximum Permissible Speed = n_g = 6700 rpm

$\bar{L}$ = Maximum Operating Speed = Std Test Speed n = 5320 rpm

In the absence repeated data under same condition. Standard deviation 'σ' may be taken as 10%[(146)].

$$\therefore Z = \frac{(6700 - 5320)}{\sqrt{(670)^2 + (532)^2}} = 1.613$$

From standard statistical data table, for Z = 1.613, Reliability of component is 0.946

The failure rate of the component can be found out by the formula

$$R_s = e^{-\lambda t}$$

Where λ is failure rate of bearing No. 6211,

t = Mission time in Hr.

In this case the Relubrication interval is taken as Mission time.

$$\therefore \quad t = 2000 \text{ Hr.}$$

$$R_s = e^{\lambda \times t}$$

$$0.946 = e^{-\lambda \times 2000}$$

$$\lambda = 0.0000277 \text{ per hr.}$$

Hence the Failure for the Bearing No. 6211 is 0.0000277 per hr.

25.2.4 *Discussion*

- For evaluating the reliability of the collet chucking mechanism, it is necessary and easier to apply non-parametric method of analysis through K-statistic.
- The number of observations taken is limited to 15, being a short sample test without failure,
- The number of observations has been validated and found to be sufficient for this set of observations,
- Bearing reliability can be estimated by using characteristic curve of dependence between (operating rpm / standard rated rpm) and life interval between relubrication.

25.3 Analysis of Spindle with Linear Elastic Supports and Viscous Damping

Let us consider the details on the spindles containing pulley, gears, faceplate, in the form concentrated point loads, distributed on the boundary of the parts, having mass m_i and moment of inertia J_{ix}.

Calculation of the dynamic characteristics of the spindle unit is based on determining the amplitude of vibration of the spindle from the plane 0 to the plane 3, i.e. from end to the rear, the forces being considered are due to the cutting, (magnitude and direction), as well as due to drive etc.

Linear formulation of the problem requires the uses of principle of superimposition and consequently determining the characteristics of the unit in all predetermined direction. The methodology and the systematic analysis as adopted by Xomiakov B.C.[145] and reported by Push (126) are represented here.

25.3.1 *Analytical Methodology*[127]

Calculation for dynamic characteristics of a spindle may be done by considering the spindle as an elastic shaft on elastically deforming supports with some quantity of concentrated mass.

However for spindle with multiple supports or for spindles having a few bearings at one support, it is necessary to calculate the various criteria of static loading.

The spindle shown in this sketch Fig 25.6 has three different parts on the end it takes the concentrated load (for example, chuck, pulley of belt transmission etc). On the right hand side of the spindle acts, force on the part responsible for cutting - Force Fo(t), in the third portion the driving force $F_3(t)$.

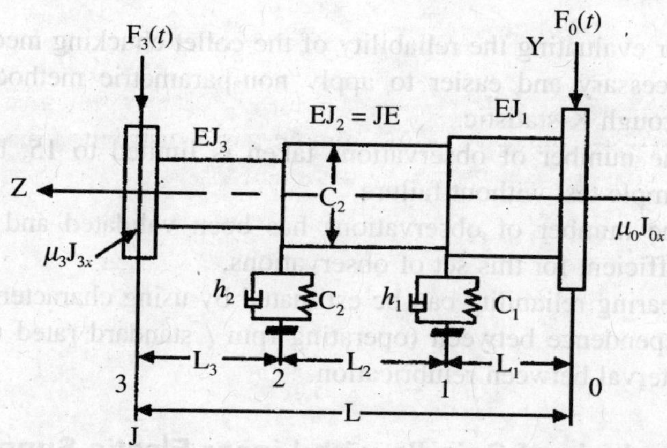

Figure 23.6 Dynamic characteristics of spindle.

Let us now indicate the amplitude in the ith section as y_i. and the angle of rotation ϕ_0, bending moment M_i. and the transverse loading Q_i. ssIt is possible now to right down the differential equation of the vibration of the spindle through parameters, y_i, $\phi_0 l$, $M_i l^2/EJ$, $Q_i l^3/EJ$ and (Since $EJ = EJ_2$ the rigidity of span between the support is uniform and L-length of the spindle.

Parametric vectors on the free end of the spindle (in the initial and the 3rd section) without considering the outside loading can be written in the form:

$$\overline{Y}_0 = \begin{vmatrix} y_0 \\ \phi_0 l \\ M_0 l^2/EJ \\ Q_0 l^2/EJ \end{vmatrix} = \begin{vmatrix} y_0 \\ \phi_0 l \\ 0 \\ 0 \end{vmatrix}, \quad \overline{Y}_3 = \begin{vmatrix} y_3 \\ \phi_3 l \\ M_3 l^2/EJ \\ Q_3 l^2/EJ \end{vmatrix} = \begin{vmatrix} y_0 \\ \phi_3 l \\ 0 \\ 0 \end{vmatrix}, \tag{25.5}$$

The method helps to relation the parameters of the initial cross section of the spindle, through transformation matrix, with the parameters of 3rd cross section and 3rd part of the spindle.

Under the influence of vibration in the initial section act the load and moment of inertia forces of the faceplate, the magnitude of the amplitude of which equal to $\mu_0 w^2 y_0$ and $J_{0x}\, \phi_0 w^2$. Where μ_i is the mass of the part i and J_{ix} when i has values 0, 1, 2, 3 depending upon the section 0, 1, 2, 3 respectively.

Matrix of the subsequent loading:

$$L_0 = \begin{vmatrix} 1 & 0 & 0 & 0 \\ 0 & 0 & 0 & 0 \\ 0 & -\delta_0 & 1 & 0 \\ \gamma_0 & 0 & 0 & 1 \end{vmatrix}$$

where $$\delta_0 = \frac{J_{0x} w^2 l}{EJ}, \gamma_0 = \frac{\mu_0 w^2 l^3}{EJ} \tag{25.6}$$

Parameter $\overline{Y}_0$ may be worked out by transformation through initial cross section. Further, the first part of the spindle has a distributed mass m_1 and rigidity EJ_1 Dependence between the parameters on its

$$T_1 = \begin{vmatrix} A_1 & \beta_1 B & \dfrac{\beta_1^2 C_1}{\alpha_1} & \dfrac{\beta_1^3 D_1}{\alpha_1} \\ \dfrac{\lambda_1^4 D_1}{\beta_1} & A_1 & \dfrac{\beta_1 B_1}{\alpha_1} & \dfrac{\beta_1^2 C_1}{\alpha_1} \\ \dfrac{\alpha_1 \lambda_1^4 C_1}{\beta_1} & \dfrac{\alpha_1 \lambda_1^4 D_1}{\beta_1} & A_1 & \beta_1 A_1 \\ \dfrac{\alpha_1 \lambda_1^4 B_1}{\beta_1^3} & \dfrac{\alpha_1 \lambda_1^4 C_1}{\beta_1^2} & \dfrac{\lambda_1^4 D_1}{\beta_1} & A_1 \end{vmatrix}. \tag{25.7}$$

where $\alpha_1 = EJ_1/EJ$; $\beta_1 = l_1/L$; $\lambda_1^4 = m_1 L_1^4 w^2 / EJ_1 = k_1 \lambda w^2$ and functions A_1, B_1, C_1, D_1 may be expressed in the form of a stepped series.

Under the support l, the variable force changes the reaction force due to 'slip' (considering the elasticity and dissipating component of the reaction). This slip while occurring at the support can be studied by reducing parametric vectors on the matrix of the support.

$$Q_1 = \begin{vmatrix} 1 & 0 & 0 & 0 \\ 0 & 1 & 0 & 0 \\ 0 & 0 & 1 & 0 \\ (-\varepsilon_1 - f_1) & 0 & 0 & 1 \end{vmatrix} \quad (25.8)$$

Where $\varepsilon_1 = C_1 l^3/EJ$; $f_1 = J\, h_1 w l^3/EJ = J\, k_1 f w$ (e_1 and h are coefficients of rigidity and damping of the support, respectively $j = \sqrt{-1}$).

Transforming by this method, from one part to another, we move to the left end of the Spindle, where parameters can be expressed by Vector $\overline{Y}_3$.

$$\overline{Y}_3 = \Pi \overline{Y}_0 \quad (25.9)$$

Consequently we get the matrix equations (without considering internal loading).

Transformation matrix for the scheme Π is given by

$$\Pi = L_3\, T_3\, Q_3\, T_2\, Q_1\, T_1\, L_0 \quad (25.10)$$

where L_1 and L_3-matrices of the concentrated load on initial and third section of the spindle. T_1, T_2, T_3 are the matrices of the 1st, 2nd, 3rd part of the spindle respectively. Q_1, Q_2 are the matrices of the linear elastic support of the spindles with viscous damping.

$$\overline{Y}_3 = \Pi(\overline{Y}_0 - \overline{\psi}_0) - \overline{\psi}_0 \quad (25.11)$$

Where $\overline{\psi}_0 = \begin{vmatrix} 0 \\ 0 \\ 0 \\ F_0 l^3/EJ \end{vmatrix} \qquad \overline{\psi}_3 = \begin{vmatrix} 0 \\ 0 \\ 0 \\ F l^3/EJ \end{vmatrix}$

For obtaining the Transformation Matrix Π is done by using the dimensional matrix 4×4. Normally at the ends of the spindles two out of the 4 parameters in matrix table $\overline{Y}_0$ and $\overline{Y}_3$ equal to zero, which appears to rationalize the calculation work. Let us explain it by an example. While calculating the transfer function W_0 (*jw*) of the system acting in the direction of the quantity process (when $F_3 = 0$) the resulting equation could be written as:

$$\begin{vmatrix} y_3 \\ \phi_3 l \\ 0 \\ 0 \end{vmatrix} = \begin{vmatrix} * & * & * & * \\ * & * & * & * \\ a_{31} & a_{32} & * & a_{34} \\ a_{41} & a_{42} & * & a_{44} \end{vmatrix} = \begin{vmatrix} y_0 \\ \phi_0 l \\ 0 \\ -F_0 l^3 / EJ \end{vmatrix} \qquad (25.12)$$

As seen, in Matrix Π we get six elements out of sixteen, which are distributed in the sections along the zero values of the vector $\bar{Y}$ and in the table the data not equal to zero are seen along the direction of the vector $(\bar{Y} - \bar{\psi}_0)$ (The remaining elements which are not used in calculation are shown 'stars') Subsequently, we can write two linear equations, for finding out the parameters Y_0 and $\phi_0 l$ respectively:

$$a_{31} y_0 + a_{32} \phi_0 l = a_{34} \frac{F_0 l^3}{EJ} \qquad (25.13)$$

$$a_{41} y_0 + a_{42} \phi_0 l = a_{44} \frac{F_0 l^3}{EJ} \qquad (25.14)$$

Where coefficients 'a' happens to be function of frequency, inertia, w dissipation, elastic and geometric parameters of the spindle units. From the two equations above we can get the expression for frequency transformation function of the system:

$$W_0(jw) = \frac{y_0}{F_0} = \frac{l^3}{EJ} \frac{a_{34} \cdot a_{42} - a_{32} \cdot a_{44}}{a_{31} \cdot a_{42} - a_{41} \cdot a_{32}} \qquad (25.15)$$

Analogically it is possible to find the frequency transformation function $W_3(jw)$ of the system acting in the direction of the drive (under $F_0 = 0$)

$$W_3(jw) = \frac{y_0}{F_0} = \frac{l^3}{EJ} \frac{a_{32}}{a_{32} \cdot a_{41} - a_{31} \cdot a_{42}} \qquad (25.16)$$

Since some of the element of the matrix Q_1 and Q, of 'elasto-damping' support happen to be complex numbers, transfer matrix $\Pi = S_1 + j\,S_2$, which consists of two matrices S_1 and S_2. Equation for frequency transfer function $W_0\ (jw)$ or $W_3\ (jw)$ can be shown in the form:-

$$W(jw) = \frac{P_1 + jP_2}{P_3 + jp_4} + \frac{P_1P_3 + P_2P_4}{P_3^2 + P_4^2} + J\frac{P_2P_3 - P_1P_4}{P_3^2 + P_4^2} \qquad (25.17)$$

$$= Re\ [W\ (jw)] + Jm[W\ (jw)],$$

The above equation is used for determining P_1, P_2, P_3 and P_4 elements of the matrices S_1 and S_2.

After calculating for the operational part of the equation it is possible to find out the components $Re[W(jw)]$ and $Jm[W(jw))$, 'amplitude-phase' frequency characteristics of the dynamic system of the spindle unit.

CHAPTER 26

Questions

In this chapter, authors make a modest endeavour to provide a few possible questions that the students may expect from various chapters. An attempt has also been made to give model answers to many a question, specifically those having some sort of numerical problems.

(A) QUESTIONS

Chapter 2

Q. 1. Starting from the first principle and using non-dimensional analysis, derive an equation for cutting force while machining steel with HSS single point cutting tool. Experiments performed shows that the major vertical component of the cutting force is independent of the relative cutting speed. State the assumptions made if any.

Q. 2. Develop a generalized empirical relationship for the thrust and torque in drilling and compare the same with turning to validate Optiz's hypothesis that the principles of cutting mechanism fall into the same category.

Q. 3. Find out the method of differentiating a special purpose machine from a general-purpose machine based on the kinematic structure.

Chapter 3 and 4

Q. 1. Discuss briefly the silent features to be considered for selecting and designing a suitable drive system in a machine tool.

Q. 2. Graphically represent the speed ray diagram of a machine tool having 6 spindle speeds. Draw at least 4 alternative speed ray diagram and use 'minimum nodal sum' method to select the best ray diagram out of the alternative ray diagrams available.

Q. 3. Show that in' twin gear block' or 'triple gear block' the difference between the consecutive numbering teeth should be greater than or equal to 4.

Q. 4. (a) Make a compact speed ray diagram for compromise gear box having 15 nos. of spindle speeds between 90 rpm and 2100 rpm.

(b) Make a layout of a gearbox and draw the structure diagram.

(c) Discuss the method of determining the dimensional size of the main spindle.

Q. 5. What is meant by 'preferred number' and why it is so very important in designing speed ray diagram ? If $\phi = \sqrt[e1]{2} = \sqrt[e2]{10}$, write down the standard values of e_1 and e_2, noting e_2 as standard random number varying between 20 and $\frac{20}{6}$.

Q. 6. It is necessary to design a variable stepped speed gear box for machine tool having 6 speeds ranging from 120 rpm to 1100 rpm, drive having a.c. synchronous motor of 8Kwatts.

(i) Draw the best speed ray diagram

(ii) Calculate the gear ratios and draw the structure diagram.

(iii) Find out the size of the pulley on the input shaft if the motor rpm is 1480.

(iv) Make a layout of the gearbox

Q. 7. By drawing a few possible ray diagrams for a 6 speed machine tool gearbox, explain the method of selecting the best ray diagram using minimum nodal sum.

Q. 8. Draw the speed ray diagram (compromise gear box) having 6 speeds in the upper range and 8 no. speeds in the lower range. The spindle speed ranges from 2600 to 180 rpm. What are the limitations of such a compromise gearbox and what are its advantages?

Q. 9. Discuss with neat sketch the feed gearbox, operated by Tumbler gears, showing three positions forward, neutral and reverse.

Q. 10. Explain the functions of feed change gears operated through sliding key.

Q. 11. It is necessary to design the shaft of a gearbox, considering :

(i) Bending moments in horizontal as well as vertical planes

(ii) Torsional moments

(iii) Direct stresses etc. on the principle of permissible shear stresses based on principal stresses.

Q. 12. Discuss the method of designing a Bevel gear transmission system. What are the essential considerations for designing such a system?

Chapter 5

Q. 1. (i) Discuss the method of obtaining stepless speed variation of a machine tool having regulation upto 20, using epicyclic mechanism.

(ii) Can it be called stepless altogether or just a near approach to stepless?

Q. 2. Explain the principle of formation of

(i) Conical pressure variator

(ii) Torso variator

(iii) Ball variators.

Q. 3. Discuss the essential requirements while designing a pressure variator.

Q. 4. How can you make a variator using internal cone and conical roller, a positive self locking unit?

Q. 5. With a neat sketch explain the function of Kopp type P.I.V. drive and a compact ball variator for machine tool having low range of regulation.

Q. 6. Find out an expression for the power loss in friction in a friction variator with flat surface contact due to loss in sliding velocity. How does it change the transmission ratio of speeds?

Q. 7. Figure shows the calculation diagram of a stepless self tightening pressure variator. Analyse the forces to show that for all relative positions of the friction cone $R \geq \frac{mZ_1}{2}\left[\frac{\sin\gamma\cos\alpha}{f} - \cos\gamma\right]$

Where *f*-coefficient of friction, m-module of gear with Z_1 teeth and R is the radius of the contact point of the variator from the center line of the driving cone.

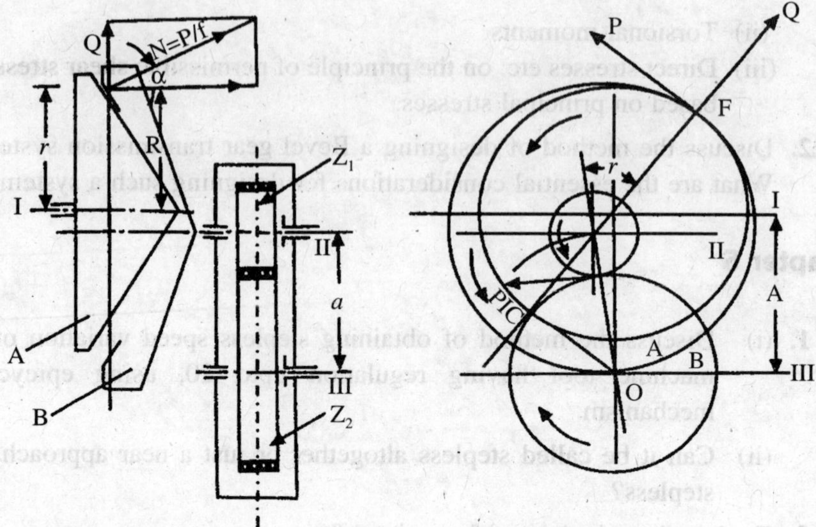

Figure 26.1 Stepless self locking pressure variator.

Chapter 6

Q. 1. Classify various types of guide shapes commonly used in machine tools with their characteristics and uses.

Q. 2. Explain the specific merits and demerits of plastic guides commonly used in machine tools. Name some of the filled and unfilled plastic guides.

Q. 3. What are the advantages of Turcite *B* as a suitable guide material for *NC*, *CNC* machines ? State the frictional characteristic behavour of such a material against speed in comparison to other traditional guide materials. Explain its suitability for use in regard to micro-positional displacement error.

Q. 4. Show that a straight flat guide, with hydrodynamic lubrication can uplift a load given by $P = \dfrac{1.333\mu\, v\, B^3}{h_0^2}$, where μ-viscosity in Nsec/m^2, v-speed in m/sec, B-width of guide in meters, h_0-CLA values of surface roughness in meters and P is in Newton.

Q. 5. Find out an expression for the load bearing capacity of a hydrostatically lubricated guideways. Explain how does the coefficient of friction due to oil film in a hydrostatic slide come down to zero, dependent on the externally pressurized oil film.

Q. 6. What is meant by a rigidity of a lubricated slideways? Show that the rigidity of a hydrostatic slideway is 50% more than that of a hydrodynamic slideways.

Q. 7. Classify the various types of configuration of the guides used in machine tools, based on material, lubrication system, drives control etc.

Q. 8. Discuss the variation in coefficient of friction of a table against relative sliding speed based on coefficient of viscosity and specific pressure on guides. Write down a generalized empirical equation for coefficient of friction dependent on velocity, viscosity and intensity of pressure.

Q. 9. Explain the static coefficient of friction f_s, when velocity of the slide $V = 0$ and when $V \neq 0$ but starting from zero. Show the variation of coefficient of friction under boundary lubrication and under partial as well as full hydrodynamic lubrication.

Q. 10. In a surface grinding machine it is necessary to design table guides having straight flat configuration so as to obtain hydrodynamic lubrication. If the viscosity of the film is 0.07Ns/m^2 and minimum oil film thickness is 1.5 micron, find the width of the guide. Table moves with a feed velocity of 0.4 mm/sec and the uplifting hydrodynamic force balances half of the vertical load acting on the table. Weight of the table and vertical component of cutting force together comes to be 25KN and the specific pressure on guide on each side is 1.5 N/cm^2. Design a guide.

Q. 11. Discuss merits and demerits of using filled and unfilled plastics for machine tool guides. What is meant by Turcite *B* and why it is universally used as a suitable material for machine tool slideways?

Q. 12. Show the variation of coefficient of friction of a hydrostatic slideway against external pressure to which the oil is raised. Is it possible to get a condition when the coefficient of friction becomes zero? If so, determine the value of oil film thickness.

Q. 13. Find out the suitable expression of coefficient of friction in a machine tool table with combination of rolling and sliding friction.

Q. 14. Discuss the method of designing a table using ball guides or roller guides. Give expression for estimating the total load that can be sustained by this type of guide. Discuss the merits and demerits as well as field of applicability of such system.

Q. 15. Based on minimum factional energy, show that hydrostatic guide should be pocketless and that the pure hydrostatic lubrication may not be possible and as such the system may work on hydrostatic-cum-hydrodynamic lubrication.

Q. 16. With a neat sketch show the circuit diagram of aerostatic lubrication system, explaining the importance of each element in the circuit diagram. Write down an equation for flow of air through restrictor orifice in terms of supply pressure, as well as pressure in the pocket, and absolute temperature of air supply.

Q. 17. Discuss the silent features and uses of antifriction guides, employing balls and /or rollers as rolling elements. State their principles of construction and assembly and specific fields of applications.

Q. 18. Calculate the maximum specific pressure bourne by a ball type antifriction guide, having following particulars: Load on each ball is 250 N, diameter of ball 7mm, Modulus of elasticity of balls and guides respectively 21×10^6 N/cm^2. Poisson's coefficient of the material of ball and guides are 0.35 and 0.32 respectively.

Q. 19. While designing guides of the table, designer also likes to have the resultant of all forces acting on the guide faces, pass through a distance of X_A from the centroid where $X_A < L/6$, where L is the length of the table.

Chapter 7

Q. 1. In designing the bed of a machine tool, it is often found that the hollow rectangular cross-section is the most suitable one. Make a comprehensive evaluation of the various types of cross sections commonly used in machine tool on the basis of stress and deflection in both bending and torsion.

Q. 2. Discuss the method of designing a horizontal circular table of a vertical boring machine, where the cylindrical workpiece is clamped. Average diameter of the guide is d and the width of the guideways is b.

Q. 3. State the various systematic steps involved in designing a milling machine column having hollow rectangular cross-section.

Q. 4. Explain the method of evaluating the stiffness and natural frequency of vibration of a machine tool bed having two-tier cross-section with stiffeners, using Krylov's function. State the final expression for frequency of vibration.

Q. 5. Make a comparative evaluation of advantages and disadvantages of the following materials used in Machine Tool Beds:

(i) Reinforced concrete.

(ii) Epoxy concrete.

(iii) Steel fabricated.

Chapter 8

Q. 1. Show that the total error in the pitch of a sliding friction lead screw can be denoted by $\Delta = \Delta_1 \left[1 + \frac{1}{2\eta} \left(\frac{p}{D} \right)^2 \right]$, where p is the pitch, η-efficiency, D-effective diameter and Δ_1-pitch error due to direct axial stress in the lead screw.

Q. 2. Find the total axial load to which a ball screw can be subjected, if 4 mm balls are used with a pitch circle radius of 20 mm. Allowable contact stresses is limited to 25000kg/cm^2. Assuming semicircular thread profile and circuit consists of two threads only and the combined young's modulus E = 2.1 × 10^6 kg/cm^2.

Q. 3. A BLS with polycircular groves of a CNC machine is working with a range of speed varying between 100 to 150 rpm. The screw has ball circle dia. 30 mm and pitch of the screw is 6 mm. Diameter of each ball is 3.5 mm. The screw has a calculated life or durability of 500 hrs. The loading cycles per unit revolution of the screw = C_L having a value given by

$$C_L = 0.5\, Z_1 \left[1 + \frac{r_1}{r_0} \cos \alpha \right]$$

where Z_1 is number of balls in one thread groove of ball screw, r_1, r_0–ball radius and ball circle radius resp. and α is the semi contact angle. The screw has a ratio of min. axial load to max. axial load as 0.6 and there is probability of 0.70 for working with maximum axial load. Determine the dynamic load that the BLS can take, circulation of balls being over two full circuits.

Q. 4 Show, with neat sketches, atleast two methods of preloading a ball lead screw. Also deduce an expression that the magnitude of preload is normally equal to 1/3 of the total load.

Q. 5 Show that in a sliding friction lead screw the distribution of load per tooth is non-uniform. Write down an expression for efficiency of a sliding friction lead screw, assuming included angle of the thread as 2β. How will this expression be changed, in the case of a Recirculating Ball Screw? State clearly the reasons thereof.

Q. 6 Find out the dynamic load of a ball lead screw having balls 4 mm in dia, at a ball circle radius of 18 mm and pitch 6 mm. It has 4 circuits of recirculation with a permissible contact stress 25×10^2 N/mm^2 and $E_1 = E_2 = E = 21 \times 10^6$ N/cm^2. The screw has gothic arc profile of thread and runs at a max. and min. speed of 300 and 100 rpm resp. Probability of occurrence of max. load is 0.60. Any other data, not given may be assumed.

Q. 7 In a recirculating ball screw show that the optimum value of the design parameter R_2/R_1 is normally equal to 1.115, where R_2, R_1 are the radius of the raceway and the ball respectively. A few data of (R_1/R_0), where R_0 is the radius of the ball circle are given in the Table below.

IAR Model Lead Screw	R_1/R_0
IAR 1202	0.1 Assumed
IAR 2005	0.1587
IAR 2505	0.1270
IAR 3205	0.0990
IAR 3210	0.1984
IAR 4005	0.0294
IAR 4010	0.1587
IAR 5010	0.1270

$$m_\sigma = \left[1.32 - 3.49\left(\frac{A}{B}\right)\right]^2$$

$$\frac{A}{B} = \left(1 - \frac{R_1}{R_2}\right)\left(1 - \frac{R_1}{R_2}\cos\alpha\right)$$

Chapter 9

Q. 1 Make a sketch of atleast two different types of spindle ends and make a comparative evaluation of their characteristics. State the method of analyzing the heat generated in the sleeve bearing in k calories per second and the method of finding out the consumption of oil per unit time.

Q. 2 Analyse the load taken by the balls in a ball bearing used as a spindle support and show that due to contact deformation not more than 80% of the balls take the entire thrust.

Chapter 10

Q. 1 What is meant by static rigidity of a machine tool and how can it be measured? State any method of analyzing the dynamic rigidity of a machine tool based on self excited forced and damped vibrations.

Q. 2 Discuss the systematic steps for selecting the lubrication oil to be used in machine tool operations and what are their influences on overall design of the machine tools. Explain the frictional behavior of an anti-stickslip oil comparing it with the frictional behavior of traditional lubricating oil. Name some additives which normally tame the behavior of friction against velocity of sliding.

Q. 3 State the characteristic specification of lubrication oil. If V_{cr} denotes the magnitude of the critical velocity wherefrom starts the regime of liquid friction. State the method of obtaining this value.

Chapter 11

Q. 1 Classify the essential control systems, with particular reference to shifting of gears in a gear box. Explain the difference between:
(i) centralized control,
(ii) selective control and
(iii) preselective control system.

Q. 2 State the method of realizing preselective control system using:
(i) rack and pinion (i.e. mechanical means)
(ii) hydraulic shifting.

Chapter 12

Q. 1 From the first principle show that time for braking, time for starting the motor in a machine tool is given by

$$t_{st} = \frac{J.2.\pi.n}{60(M - Mc)},\ t_{br} = \frac{J.2.\mu.n}{60Mc},$$

where torque at any instant is M and Mc resisting torque, n-rpm of the motor and J is the mass moment of inertia of the rotor.

Q. 2 Show the circuit diagram for effecting 'push button' control system in a machine tool.

Q. 3 With neat sketches of circuit diagrams show the functioning of a thermal relay and an electrical braking system.

Q. 4 Write short notes on :
(i) Electromagnetic chuck for forward and reverse motion of a planning machine table.
(ii) Ferromagnetic powder clutch
(iii) Motor type time relay.

Q. 5 A machine tool works with a following 'power duration'. Find out the total power of the motor to be selected, assuming overall efficiency as 0.80.

Power (KW)	15	10	15	Rest	20	10	Idle	Time for Braking	Time for Starting
Time (Sec)	15	20	30	15	30	20	20	5	6

Q. 6 Make an electric circuit diagram for realizing the forward and return motion of a slide using a reversible speed motor and limit switches.

Q. 7 With neat sketches explain the functioning of:
(i) Decentralised In-travel control system
(ii) Centralised time sequence control system
(iii) Centralised In-travel control system

Q. 8 Plot the curve and select the proper motor for the following intermittent duty of a machine tool.

$P_1 = 10$ KW for $t_1 = 2$ sec

$P_2 = 20$ KW for $t_2 = 5$ sec

$P_3 = 15$ KW for $t_3 = 10$ sec

$P_4 = 20$ KW for $t_1 = 4$ sec.

Between P_3 and P_4, there is a pause of duration 8 sec and for starting and stopping motor has a response time of 3 seconds each. Find out the duty factor and minimum power required of the motor, if the overall efficiency-is 0.88.

Chapter 13

Q. 1 State the main advantages and disadvantages of the hydraulic drives and control system in machine tools. Also state characteristic properties that a hydraulic fluid should possess for being used in machine tool drives and control mechanisms.

Q. 2 Draw the entire hydraulic circuit for obtaining forward as well as quick return motion of a shaping machine. Indicate all the parts and elements and describe briefly their characteristic functions.

Chapter 14 and 15

Q. 1. Explain the principles of numerical control of machine tool using stepper motor and elementary pace, as done earlier in MIT system. Also explain the terms character, block, parity check.

Q. 2. Describe a built-in auto inspection system for measuring the outside diameter of a cylindrical grinding machine. With a neat sketch explain the functioning of various elements of the system.

Q. 3. Discuss the methods available for checking, on-line the size of the internal hole being ground for its size, in an internal grinding machine.

Chapter 16

Q. 1. (a) Discuss briefly the self-excited vibration in a machine tool. Why such a vibration is called 'self — excited' and what are the causes for their occurrence?

(b) 'It is necessary to have high damping coefficient and large stiffness of the tool to reduce such vibration'—Discuss the statement giving specific example of a turning operation.

Q. 2. Discuss briefly the theories put forward by Pappenhuysen, Push Deriaguin and Tolstoi, to explain the stick-slip type positional displacement error of a machine tool.

Q. 3. Explain 'critical velocity' and its significance.

Discuss, based on Lomakin's principles, the method of reducing stick slip error by applying normal sinusoidal load vibrating with ultrasonic frequency.

Q. 4. 'Stick-slip' error can be reduced by using hydraulic relieving of the table of a vertical boring machine. Prove that in such a case the oil film thickness is given 1.47 h_o, where h_o is the minimum oil film thickness, mainly due to surface asperities.

Q. 5. Enumerate the various methods, used in practice to reduce the positional displacement error due to 'Stick Slip'.

Chapter 17

Q. 1. Explain the methods of obtaining microdisplacement in a machine tool with the help of magnetostrictive materials. Plot the curves showing the relative magnetostrictive strain against ampere turns for various materials, such as permalloy, gold, cobalt and nickel. How can the error in the lead screw be compensated by using magnetostrictive principle?

Q. 2. Show that the resultant positional displacement error of a slide is given by 'b', where $b = C_b \dfrac{\Delta F}{K}$, where C_b is a constant whose value depends on θ, damping characteristic coefficient, usually varying between 1 and 2. ΔF is the difference between static and dynamic force of friction and K is the stiffness of the slide. State the various assumptions made.

Q. 3. Discuss a suitable experimental way of finding out the contact deformation at the contacting surface of two bodies in contact with a normal load acting on it.

Chapter 19

Q. 1. What is meant by 'Teach Robot' or 'Master–slave' configuration of robot? With a neat sketch the entire method of programming an industrial robot using teach method.

Q. 2. Explain the various functions performed by a sensor, in its application in robotic manipulation. Classify the various types of sensors commonly used in robotic control.

Chapter 20

Q. 1. Explain the methods classifying a *CNC* or *NC* machine through *P*, *L* and *C*. what do they signify? Explain the degrees of freedom for specifying such a machine.

Q. 2. With a neat sketch of the block diagram explain the principles of functioning of *CPU* (Central Processing Unit) in a *CNC* machine. Also show the servo-mechanism involved in a closed loop control system.

Q. 3. Explain the methods of programming a *CNC* machine with *APT* language.

Q. 4. What are the various types of transducers commonly used in a *CNC* machine? Explain the functioning of Linear Transducer and Angular Transducer.

Q. 5. Explain the functioning of a shaft encoder and Why gray code rather than binary used in a shaft encoder in some specific cases of encoder design.

Q. 6. A finished part of a product is shown in sketch below. Make a programme for the manufacture of the part using *APT* language.

Q. 7. What are the essential requirements in retrofitting an existing machine tool into a *CNC* system?

Q. 8. Compare stepper motors with DC servo motors, in regard to their uses in NC, CNC machines.

Q. 9. Figure 26.3 below shows a particular M.S. part to be machined in a lathe. Determine the original position of the

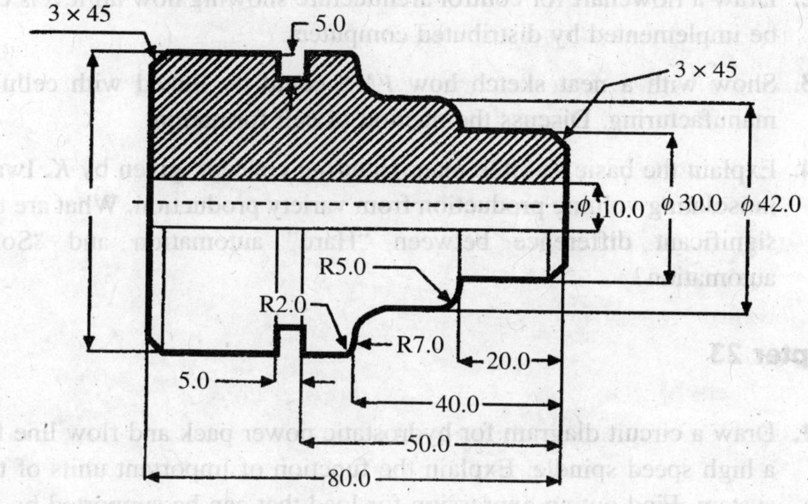

Figure 26.2

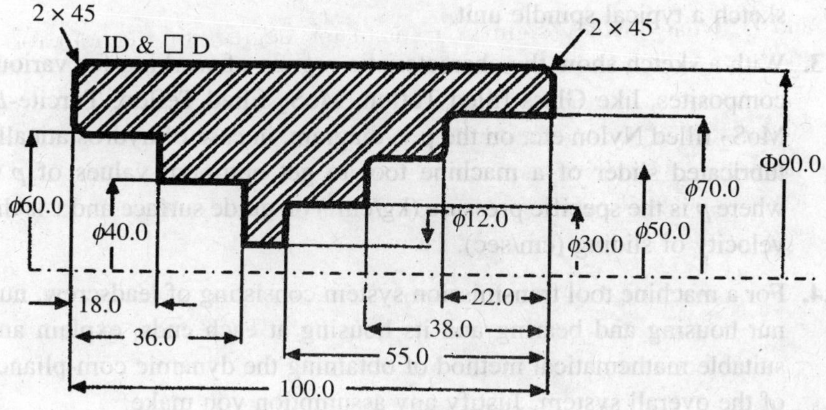

Figure 26.3

tools and the datum for the workpiece. Write down the programme to be used in computerized manufacture, for machining only the surface marked (✓). Show the dummy running of the tool.

Chapter 22

Q. 1. What is meant by *FMS*? Classify *FMS* into various categories and explain the advantage of *FMS* in modern manufacturing industries.

Q. 2. Draw a flowchart for control architecture showing how all levels can be implemented by distributed computers.

Q. 3. Show with a neat sketch how *FMS* could be linked with cellular manufacturing. Discuss the method of cell formation.

Q. 4. Explain the basic flexible manufacturing concept, given by *K*. Iwata in isolating volume production from variety production. What are the significant difference between "Hard" automation and "Soft" automation?

Chapter 23

Q. 1. Draw a circuit diagram for hydrostatic power pack and flow line for a high speed spindle. Explain the function of important units of the system. Find out an expression for load that can be supported by it.

Q. 2. Describe the procedure of designing a machine tool spindle and sketch a typical spindle unit.

Q. 3. With a sketch show the characteristic regime of working for various composites, like Glass filled Teflon, MoS_2 filled Teflon, Turcite-*B*, MoS_2 filled Nylon etc. on the p.v. diagram, in case of hydrostatically lubricated slider of a machine tool to get optimum values of *p.v.* where *p* is the specific pressure (kg/cm^2) on guide surface and *v* is the velocity of sliding (cm/sec).

Q.4. For a machine tool transmission system consisting of leadscrew, nut, nut housing and bearing and its housing at each ends, explain any suitable mathematical method of obtaining the dynamic com-pliance of the overall system. Justify any assumption you make.

Chapter 24

Q. 1. Coefficient of friction at any velocity V is a function of $\left(\frac{\lambda}{\lambda_{cr}}\right)$ and as such empirical equation can be written in the form $f_v = f_0\left[1 - K_v \sqrt[3]{\left(\frac{\lambda}{\lambda_{cr}}\right)}\right]$. Explain each term in this equation and state the method of estimating the value of Kv.

Q. 2. State the various methods of reducing positional displacement error. A *CNC* machine is using *PTFE* reinforced with Phosphor Bronze commercially known as Turcite-*B*. Show the various characteristic features of this material, such that the displacement error is reduced to the maximum.

Q. 3. A spindle unit uses supports as roller bearing at the front end and ball bearing at the rear end. Write down the equations for overall compliance assuming the bearing centers distance as and front end overhang as '*a*'. How can we find out the optimum value of $\left(\frac{b}{a}\right)$.

Chapter 25

Q. 1. A Roller bearing fitted on a Journal dia. 55 is meant for carrying a radial load as estimated from company's catalogue is 50 KN. But actual study shows it varies between 47 an 44 KN. Experiments were done to find the radial load and test reading are noted below. Find out the reliability of the bearing.

Sr. No.	1	2	3	4	5	6	7	8	9	10	11	12	13	14	15
Radial Load	47	44	46	47	44	47	46	46	45	45	46	44	45	44	46

Q. 2. The maximum speed of an automat is 4500 rpm, but for some sudden heavy demand the speed is increased by 20%. But the maximum speed permissible with grease lubricated spindle support is 7000 rpm. Find out upto what level of grease should be filled for the bearing. Find out also the reliability, assuming the type of the bearing as deep groove ball bearing.

(B) ANSWERS

Chapter 3

Ans. Q. 2

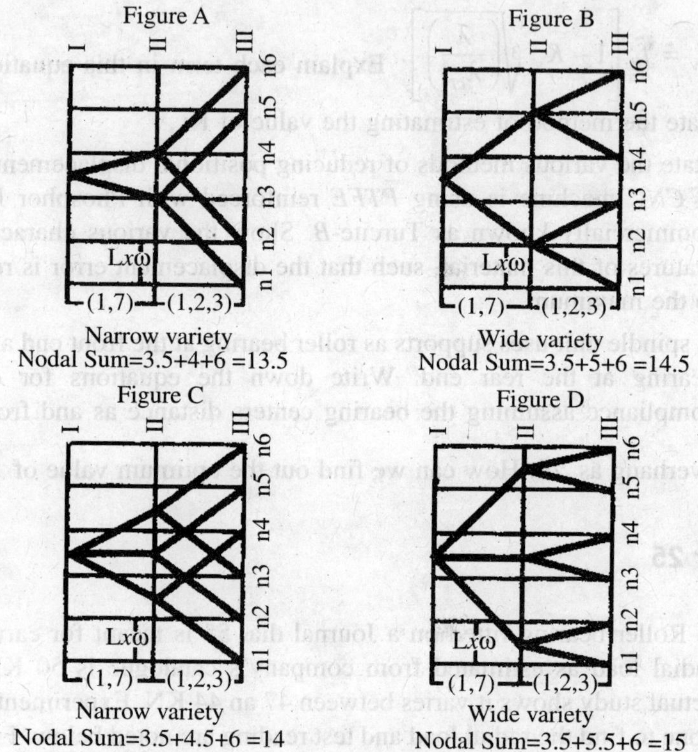

Figure 26.4 Speed Ray Diagrams.

Type A is having minimum nodal sum and this is the best Speed Ray Diagram out of the 4 alternative Ray diagrams drawn.

Ans. Q. 4

Max. Speed = 2100 rpm, Min. Speed = 90 rpm.,

Z = No. of steps = 15

Step 1 : Calculation ϕ

$$\frac{N_{\max}}{N_{\min}} = \frac{2100}{90} = \phi Z^{-1} = \phi^{14} \text{ Where Z = No. of steps}$$

$$\phi = 23.3^{(1/14)}$$

$$\phi = 1.246$$

Step 2 : Calculation of speeds

$n_1 =$ 90 rpm
$n_2 =$ 112.5 rpm
$n_3 =$ 140.6 rpm
$n_4 =$ 176 rpm
$n_5 =$ 220 rpm
$n_6 =$ 275 rpm
$n_7 =$ 344 rpm
$n_8 =$ 430 rpm
$n_9 =$ 537.5 rpm
$n_{10} =$ 672 rpm
$n_{11} =$ 840 rpm
$n_{12} =$ 1050 rpm
$n_{13} =$ 1313 rpm
$n_{14} =$ 1641 rpm
$n_{15} =$ 2051 rpm

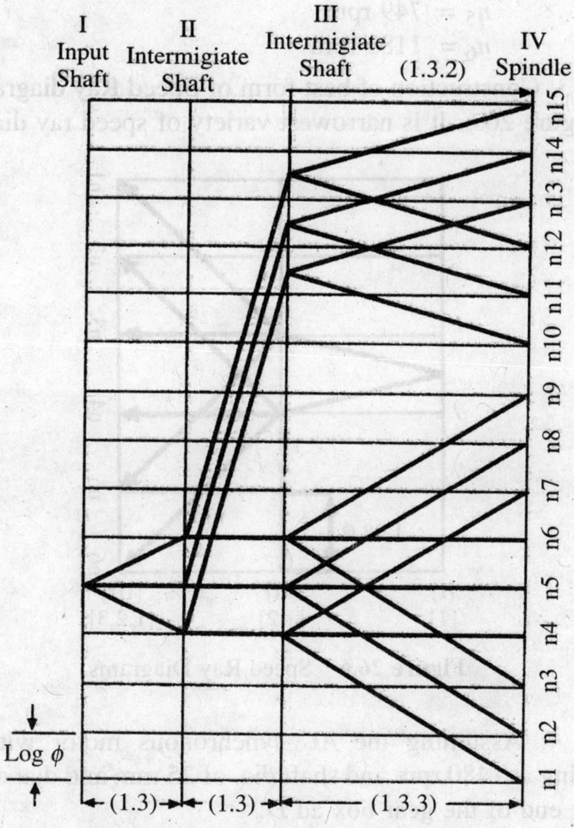

Figure 26.5 Speed Ray Diagram.

Step 3: Construction of best form of Speed Ray diagram. (Figure 26.5)

Ans. Q. 6

Max. Speed = 1100 rpm, Min. Speed= 120rpm.,
Z = No.ofsteps = 6
Step 1: Calculation of ϕ

$$\frac{N_{max}}{N_{min}} = \frac{1100}{120} = \phi^{Z-1} = \phi^5 \text{, Where Z = No. of steps}$$

$$\phi = 9.17^{(1/5)}$$

$$\phi = 1.557 \approx 1.58$$

Step 2 : Calculation of sppeds

n_1 = 120 rpm
n_2 = 190 rpm
n_3 = 300 rpm
n_4 = 474 rpm
n_5 = 749 rpm
n_6 = 1183 rpm

Step 3: Construction of best form of Speed Ray diagram is as shown in Figure 26.6. It is narrowest variety of speed ray diagram.

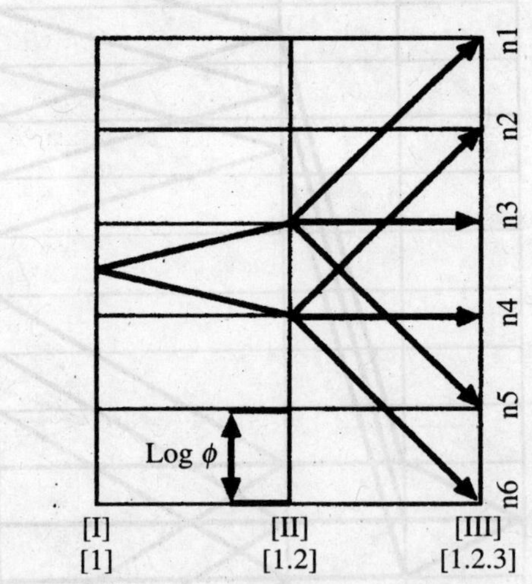

Figure 26.6 Speed Ray Diagrams.

Step 4: Assuming the *AC* synchronous motor with normal slip running at 1480 rpm and shaft dia. of 25 mm and dia. of pulley on the input end of the gear box ad *D*.

$$\frac{431}{1480} = \frac{25}{D}$$

$$D = 85.84 \approx 90$$

Selecting dia of pulley as $D = 90$ mm.

Step 5: Layout of the gear box has been shown in Figure 26.4.

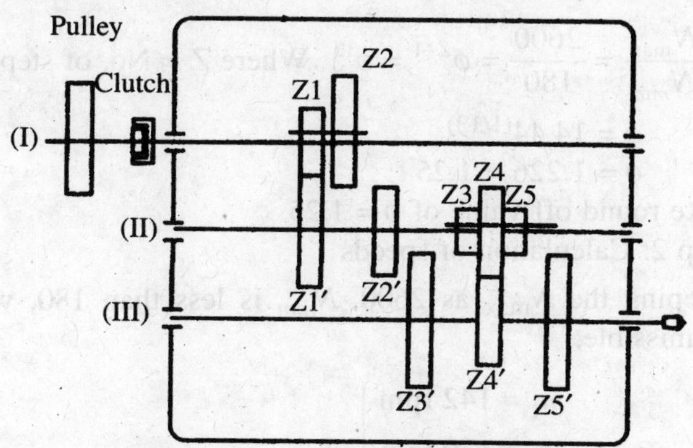

Figure 26.7 Layout of the gearbox.

Step 6: Teeth calculation and structure diagram (Figure 26.8)

$$\frac{Z_1}{Z'_1} = \frac{431}{300} = 1.44 \qquad \text{If } Z_1 = 30,\ Z'_1 = 20$$

$$\frac{Z_2}{Z'_2} = \frac{474}{431} = 1.1 \qquad \text{If } Z_2 = 26,\ Z'_2 = 24$$

$$\frac{Z_4}{Z'_4} = \frac{467}{1130} = 0.41 \qquad \text{If } Z_2 = 20,\ Z'_4 = 20$$

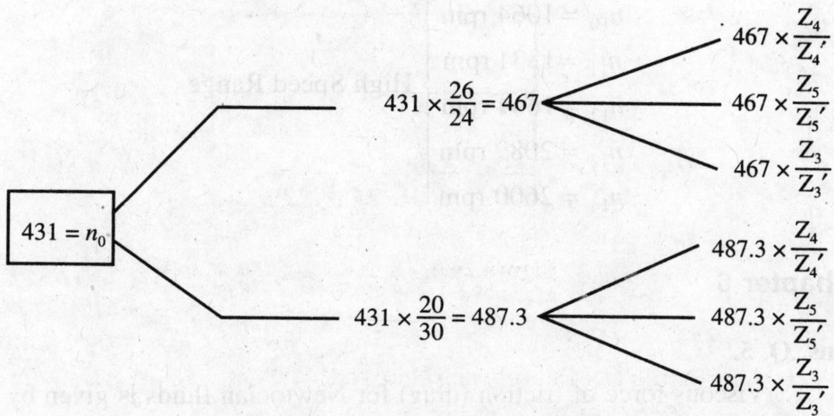

Figure 26.8 Structure of the gear box.

Q. 7 See the solution in Q. 2

Ans. Q.8

See the compound speed Ray diagram as given in 3.10 and 3.13

Speed to be calculated as under :

Step 1: Calculation of ϕ

$$\frac{N_{\max}}{N_{min}} = \frac{2600}{180} = \phi^{Z-1} = \phi^{13}, \text{ Where Z = No. of steps = 14}$$

$$\phi = 14.44^{(1/13)}$$

$$\phi = 1.226 \approx 1.25$$

Take round off value of $\phi = 1.25$

Step 2: Calculation of speeds

Keeping the $N_{\max}$ as 2600, N_{mm} is less than 180, which may be permissible.

$$\left.\begin{array}{l} n_1 = 142 \text{ rpm} \\ n_2 = 179 \text{ rpm} \\ n_3 = 233 \text{ rpm} \\ n_4 = 279 \text{ rpm} \\ n_5 = 349 \text{ rpm} \\ n_6 = 436 \text{ rpm} \\ n_7 = 545 \text{ rpm} \\ n_8 = 682 \text{ rpm} \end{array}\right\} \text{Low Speed Range}$$

$$\left.\begin{array}{l} n_9 = 852 \text{ rpm} \\ n_{10} = 1064 \text{ rpm} \\ n_{11} = 1331 \text{ rpm} \\ n_{12} = 1664 \text{ rpm} \\ n_{13} = 2089 \text{ rpm} \\ n_{14} = 2600 \text{ rpm} \end{array}\right\} \text{High Speed Range}$$

Chapter 6

Ans. Q. 5.

Viscous force of friction (drag) for Newtonian fluids is given by

$$F = \frac{\mu VA}{h_0}$$

where F – coefficient of viscosity in N. Sec/m^2
V – Velocity of table in m/sec
A – area of the layer of oil in m^2
h_0 – Minimum oil film thickness, in m

Total vertical load that could be carried = Oil pressure in the pocket (p) × Area (A)

$$f = \frac{F}{N}$$

$$f = \frac{\mu VA}{h_0 p \cdot A} = \frac{\mu V}{h_0 p \cdot}$$

If the oil is externally pressurized, and we increase the pressure, the coefficient of friction approaches zero value.

Ans. Q. 6

In hydrodynamic lubrication the uplifting force is given by the equation

$$P = \frac{1.333 \mu V B^2}{h_0^2} \tag{1}$$

With reference to Figure 6.24 in chapter 6, in hydrostatic lubrication, the overall uplifting force is given by

$$P_{ov} = \frac{p \cdot L\,(a+b)}{2}$$

$$p = \frac{12\mu \cdot Ql}{h_0^3 \cdot L}$$

where Q-consumption of oil over a length L and unit breadth.

$$P_{ov} = \frac{12\mu \cdot Q(b-a)}{2 \cdot h_0^3 \cdot L} \frac{L\,(a+b)}{2}$$

$$P_{ov} = \frac{3\mu \cdot Q(b^2 - a^2)}{h_0^3} \tag{2}$$

Rigidity is $\frac{dP}{dh_0}$ in such cases.

$$\text{From equation (1) } \frac{dP}{dh_0} = Rigidity = -\frac{2P}{h_0} \quad (3)$$

$$\text{From equation (2) } \frac{dP_{ov}}{dh_0} = Rigidity = -\frac{3P_{ov}}{h_0} \quad (4)$$

This hydrostatic slideways have rigidity 50% more than the slideways working on hydrodynamic lubrication. In equation (3) and (4), the negative sign is notional. If only show with the increase in h0, the value of P and P_{ov} decreases.

Ans. Q. 10

Given data:
Viscosity of the film (μ) = 0.07 Ns/m^2
Minimum oil film thickness (h_0) = 15 micron = 15×10^{-6} m
Feed velocity of table (V) = 0.4 mm/sec
Uplifting hydrodynamic force balances half of the vertical load acting on the table.
Weight of table and vertical component of cutting force = 25KN
Specific pressure on guide on each side is 1.5 N/cm^2

Hydrodynamic force needed = $\frac{1}{2} \times 25 = 12.5$ KN

$$P = \frac{1.333\mu VB^3}{h_0^2}$$

$$12500 = \frac{1.333 \times 0.07 \times 0.4 \times 10^{-3}\ B^3}{15 \times 10^{-6}}$$

$$B = 0.79 \text{ m} = 79 \text{ cm}$$

$$1.5 = \frac{12500}{79 \times L}, \quad L = \frac{12500}{79 \times 1.5} = 105.4 \text{ cm}$$

Ans. Q. 18

Given data:
Load on each ball, P = 250 N.
Diameter of ball, d = 7 mm
Youngs modulus, $E = E_1 = E_2 = 21 \times 10^6$ N/cm^2
$\mu_4 = 0.35$
$\mu_2 = 0.32$

$$P_{mas} = 91.8 \sqrt[3]{\frac{P}{K^2\ d^2}} \text{ Kg/cm}^2$$

(where P is in Kg, d is in mm and K is in mm^2/Kg)

$$K = \left(\frac{1-\mu_1^2}{E_4} + \frac{1-\mu_2^2}{E_2}\right) \text{mm}^2/Kg$$

$$K = \left(\frac{1-0.35^2}{2.1\times 10^4} + \frac{1-0.32^2}{2.1\times 10^4}\right)$$

$$K = 0.8453 \times 10^{-4} \text{ mm}^2/\text{Kg}$$

$$p_{mas} = 9.18 \sqrt[3]{\frac{25}{(0.8453\times 10^{-4})^2 \times (7)^2}}$$

$$p_{\text{mas}} = 91.8 \times 414.86$$

$$p_{\text{mas}} = 38085.2 \text{ Kg/cm}^2 \text{ (Ans.)}$$

Chapter 8

Ans. Q. 2

Given data:

Diameter of ball, $D_1 = 4$ mm. $R_1 = 2$ mm

For Semicircular thread profile, Assuming $R_2/R_1 = 1.1$

Hence, $R_2 = 2.2$ mm

Pitch circle radius $R_0 = 20$ mm

Young's modulus, $E = 2.1 \times 10^6$ Kg/cm^2

Allowable contact stresses = $\sigma_c = 25000$ Kg/cm^2

$$\sigma_c = m_\sigma \sqrt[3]{\frac{PE^2\,(R_2 - R_1)^2}{R_1^2 \times R_2^2}}\; Kg/cm^2$$

$$\frac{A}{B} = \left(1 - \frac{R_1}{R_2}\right)\left(1 - \frac{R_1}{R_0}\cos\alpha\right) \quad (\alpha = 45°)$$

$$\frac{A}{B} = \left(1 - \frac{2}{2.2}\right)\left(1 - \frac{2}{20}\cos 45\right)$$

$$\frac{A}{B} = 0.0836$$

$$m_\sigma = \left[1.32 - 3.49\left(\frac{A}{B}\right)\right]^2$$

$$m_\sigma = (1.32 - 3.49)\ (0.0836))^2$$

$$m_\sigma = (1.028)^2$$

$$m_\sigma = 1.057$$

$$\sigma_c = m_\sigma \sqrt[3]{\frac{PE^2\ (R_2 - R_1)^2}{R_1^2 \times R_2^2}}$$

$$25000 = 1.057 \sqrt[3]{\frac{P(2.1 \times 10^6)^2\ (0.22 - 0.2)^2}{(0.2)^2 \times (0.22)^2}}$$

$$\frac{25000}{1.057} = \sqrt[3]{\frac{P(2.1 \times 10^6)^2\ (0.22 - 0.2)^2}{(0.2)^2 \times (0.22)^2}}$$

$$P = 14.5211 \text{ Kg}$$

$$Q = i\ P\ Z \text{ Sin } \alpha_k \text{ Cos } \lambda$$

From given data, i = no. of grooves = 2

$$Z = \frac{2\pi\ R_0}{D_1} = \frac{2\pi.20}{4} = 10\pi$$

$Q = 2 \times 14.5211 \times 10\pi \times \text{Sin } 45° \text{ Cos } 0$ (Assume Helix angle $\lambda = 0$)

$Q = 645.106$ Kg (Ans.)

Ans. Q. 3

Given data:

Min. Speed (n_1) = 100 rpm

Max. Speed (n_2) = 150 rpm

n = Avg. Speed = 125 rpm

Ball Circle dia. D_0 = 30 mm

Pitch of Screw = 6 mm

Diameter of each ball, D_1 = 3.5 mm.

Hours of working = T = 500 Hrs

Loading cycles per unit revolution of the screw (C_L)

$C_L = 0.5\ Z_i \left[1 + \frac{r_1}{r_0} \cos\alpha\right]$ where r_1 = ball radius, r_0 ball circle radius, Z_i = no. of balls in one thread groove of ball screw. Semi contact angle $\alpha = 45°$

Ratio of min. axial load to max. axial load = $0.6 \left(\because \frac{Q_{min}}{Q_{max}} = 0.6\right)$

Probability for working with Max. axial load = 0.7

No. of full circuits of ball = $i = 2$

$$C_L = 0.5Z_i\left[1+\frac{r_1}{r_0}\cos\alpha\right]$$

$$Z_i = \frac{2\pi\, R_0}{D_1} = \frac{2\pi \times 15}{3.5} = 26.928$$

$$C_L = 0.5 \times 26.928\left[1+\frac{1.75}{15}\cos 45\right]$$

$$C_L = 14.574$$

Coefficient of dynamic load (K_σ)

$$K_Q = 0.7 + 0.3\frac{Q_{\min}}{Q_{\max}} = 0.88$$

$$K = K_Q\sqrt[3]{\frac{60 \times T\; n\; C_L}{10^7}}$$

$$K = 0.88\sqrt[3]{\frac{60 \times 500 \times 125 \times 14.574}{10^7}}$$

$$K = 0.88 \times 1.761$$

$$K = 1.55$$

If $K \le 1$, $Q_{dynamic} = Q_{Static}$

If $K \le 1$, $Q_{dynamic} = \dfrac{Q_{Static}}{K}$

If present case, $K > 1$, hence $Q_{dynamic} = \dfrac{Q_{Static}}{K}$,

$$Q_{Static} = i\; P\; Z\; \text{Sin}\; \alpha_k\; \text{Cos}\; \lambda \tag{1}$$

From given data, i = of grooves = 2

$$Z = \frac{2\mu\; R_0}{D_1} = \frac{2\mu.15}{3.5} = 26.928 \approx 27$$

The value of P is calculated as per the steps followed in question 2. The Q_{static} is calculated from equation 1 and then $Q_{dynamic}$ is calculated.

Ans. Q. 7

Considering Hertz- Beliaev's equation for contact stress, we have

$$\sigma_c = m_\sigma \sqrt{\frac{PE^2\ (R_2 - R_1)^2}{R_1^2 \times R_2^2}} \tag{1}$$

where P is the load in each ball,

E-Equivalent Young's modulus, given by $E = 2E_1E_2/E_1 + E2$, E_1 and E_2 are Young's modulii referring to material of screw and nut respectively, R_2 and R_1 – radius of the raceway and the ball respectively and m_σ – is the characteristics constant.

$$P = \frac{\sigma^3}{E^2} \frac{R_1^2 \cdot R_2^2}{m_\sigma^3\ (R_2 - R_1)^2} \tag{2}$$

$$P = \frac{\sigma^3}{E^2} \frac{R_1^2}{\left[1 - \left(\frac{R_1}{R_2}\right)\right]^2} \frac{1}{m_\sigma^3} \tag{3}$$

But from empirical relationship we know

$$m_\sigma = \left(1.32 - 3.49\left(\frac{A}{B}\right)\right)^2$$

where

$$\frac{A}{B} = \left(1 - \frac{R_1}{R_2}\right)\left(1 - \frac{R_1}{R_0}\cos\alpha\right) \quad \text{(Refer to para 8.11)}$$

Here R_0 – Ball circle radius

Putting $\left(1 - \frac{R_1}{R_2}\right) \lambda$ and $\left(1 - \frac{R_1}{R_0}\cos\alpha\right) = K$

we can write,

$$P = \frac{\sigma_c^3}{E^2} \frac{R_1^2}{\lambda^2} \frac{1}{[(1.32 - 3.49\ \lambda K)^2]^3} \tag{5}$$

$$P = \frac{\sigma_c^3}{E^2} \frac{R_1^2}{\lambda^2} \frac{1}{[(1.32 - 3.49\ \lambda K)^3]^2}$$

$$P = \frac{\sigma_c^3}{E^2} \frac{R_1^2}{[\lambda(1.32 - 3.49\,\lambda K)^3]^2}$$

In order to get the advantage, namely maximum P, the denominator should be minimum, i.e. function $|\phi|$ given by :

$$|\phi(z)|_{min} = \lambda(1.32 - 3.49\,\lambda K)^3 \text{ should be minimum} \quad (6)$$

$$\frac{d\phi(z)}{d\lambda} = 0, \text{ and } \frac{d^2\phi}{d\lambda^2} = +ve$$

$$|\phi|_{min} = \lambda^2 (1.32 - 3.49\,\lambda K)^6$$

Optimum value $\left(\frac{R_2}{R_1}\right)$ for semicircular / Gothic arc profile could be obtained as under,

$$\lambda.3(1.32 - 3.49\,\lambda K)^2 (-3.49\,K) + (1.32 - 3.49\,\lambda K)^3 = 0$$

$$-10.47\,\lambda K(1.32 - 3.49\,\lambda K)^2 \text{ A } (1.32 - 3.49\,\lambda K)^3 = 0$$

$$(1.32 - 3.49\,\lambda K)^2 [-10.47\,\lambda K + 1.32 - 3.49\,\lambda K] = 0$$

$$(1.32 - 3.49\,\lambda K)^2 [-13.96\,\lambda K + 1.32] = 0$$

either $(1.32 - 3.49\lambda K) = 0$ or $[-13.96\,\lambda K + 1.32] = 0$

$$\lambda = \frac{1.32}{3.49\,K} = \frac{0.3782}{K}$$

$$\lambda_{avg} = \frac{0.3782}{0.9117} = 0.4148$$

From the data given in the problem.

Model of Lead screw of IAR	$\frac{R_1}{R_0}$	*K*	K_{av}	λ_{av} *from equation*	*Average optimal* $\left(\frac{R_2}{R_1}\right)$
IAR 1202	0.1	0.9292			
IAR 2005	0.1587	0.8877			
IAR 2505	0.1270	0.9102			
IAR 3205	0.0990	0.9298	0.9117	0.4148	1.115
IAR 3210	0.1984	0.8597			
IAR 1202	0.0294	0.9792			
IAR 1202	0.1587	0.8877			
IAR 1202	0.1270	0.9102			

Chapter 12

Ans Q.5

$$\text{Equivalent Power} = \sqrt{\frac{P_1^2 t_1 + P_2^2 t_2 + ... + P_n^2 t_n}{t_1 + t_2 + ... + t_n}} \quad (1)$$

$$\sum_{i=1}^{n} t_i = 15 + 20 + 30 + 30 + 20 = 115 \text{ sec}$$

$$\textit{Idle Time} = t_{rest} + t_{Idle} + t_{st} + t_{br} = 15 + 20 + 6 + 5 = 46 \text{ sec}$$

$$\text{Duty factor } (\varepsilon) = \frac{\sum_{i=1}^{n} t_i}{\sum_{i=1}^{n} t_i + \textit{Idle time}}$$

$$\text{Duty factor } (\varepsilon) = \frac{115}{115+46} = 0.7142$$

Using equation (1) we can write

$$P_{Equivalent} = \sqrt{\frac{15^2 \times 15 + 10^2 \times 20 + 15^2 \times 30 + 20^2 \times 30 + 10^2 \times 20}{115}}$$

$$P_{Equivalent} = \sqrt{\frac{26125}{115}}$$

$$P_{Equivalent} = 15.07 \text{ KW}$$

$$|P_{Equivalent}|_{Intermittent\ duty} = \frac{15.07}{\sqrt{0.7142}} = 17.83\ KW$$

Assuming overall efficiency as 0.80, the total power of motor is given by

$$P_{Motor} = \frac{17.83}{0.8} = 22.28 \approx 23\ KW$$

Ans. Q. 8

$$\text{Duty factor } (\varepsilon) = \frac{\sum_{i=1}^{n} ti}{\sum_{i=1}^{n} ti + \textit{Idle time}}$$

$$\text{Duty factor } (\varepsilon) = \frac{2 + 5 + 10 + 4}{2 + 5 + 10 + 4 + 8 + 3 + 3} = 0.6$$

$$\text{Equivalent Power} = \sqrt{\frac{P_1^2\ t_1 + P_2^2\ t_2 + \ldots.. + P_n^2\ t_n}{t_1 + t_2 + \ldots\ldots. + t_n}}$$

$$P_{Equivalent} = \sqrt{\frac{10^2 + 2 + 20^2 \times 5 + 15^2 \times 10 + 20^2 \times 4}{21}}$$

$$P_{Equivalent} = 16.973\ KW$$

$$|\ P_{Equivalent}\ |_{Intermittent\ duty} = \frac{16.973}{\sqrt{0.6}} = 21.91\ KW$$

Assuming overall efficiency as 0.88, the Minimum power of motor is given by

$$P_{Motor} = \frac{21.91}{0.88} = 24.89 \approx 25\ KW$$

Chapter 20

Ans. Q. 6

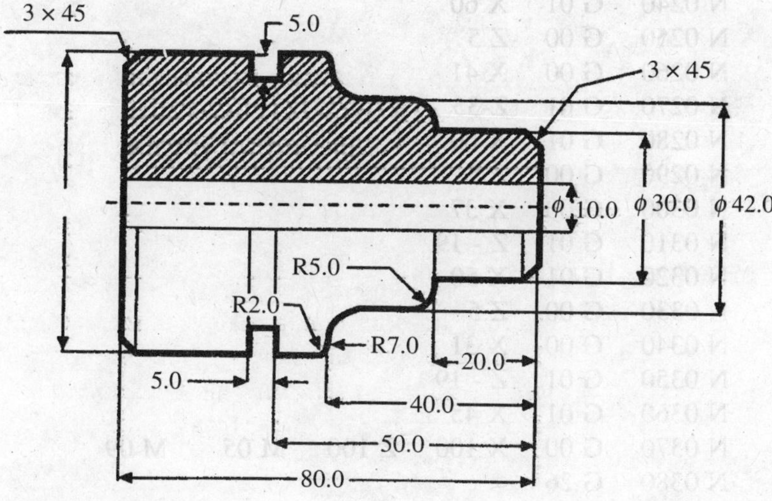

NC Part Programme for the above part
(Without using cycles) (First side)

```
N 0010   G 54   X O     Z 100
N 0020   G 90   G 94    S 4000   F 150   T 0101   M 4   M 8
N 0030   G 00   X 65    Z 0
N 0040   G 01   X-1.5
N 0050   G 01   Z 5
N 0060   G 00   X 61
N 0070   G 01   Z-60
N 0080   G 01   X 65
N 0090   G 00   Z 5
N 0100   G 00   X 57
N 0110   G 01   Z-39
N 0120   G 01   X 65
N 0130   G 00   Z 5
N 0140   G 00   X 53
N 0150   G 01   Z-39
N 0160   G 01   X 65
N 0170   G 00   Z 5
N 0180   G 00   X 49
N 0190   G 01   Z-39
N 0200   G 01   X 65
N 0210   G 00   Z 5
N 0220   G 00   X 45
N 0230   G 01   Z - 37
N 0240   G 01   X 60
N 0250   G 00   Z 5
N 0260   G 00   X 41
N 0270   G 01   Z-35
N 0280   G 01   X 60
N 0290   G 00   Z 5
N 0300   G 00   X 37
N 0310   G 01   Z - 19
N 0320   G 01   X 50
N 0330   G 00   Z 5
N 0340   G 00   X 31
N 0350   G 01   Z - 19
N 0360   G 01   X 45
N 0370   G 00   X 100   Z 100    M 05     M 09
N 0380   G 26
N 0390   T 0103         M 04     M 08
N 0400   G 00   X 24    Z 5
N 0410   G 01   Z 0
```

```
N 0420   G 42
N 0430   G 01   X 30    Z-3
N 0440   G 01   Z-20
N 0450   G 01   X 32
N 0460   G 03   X 42    Z-25
N 0470   G 01   Z-33
N 0480   G 02   X 49    Z-40
N 0490   G 01   X 56
N 0500   G 03   X 60    Z-42
N 0510   G 01   Z-60
N 0520   G 01   X 65
N 0530   G 01   X 100   Z 100   M 05   M 09
N 0540   G 26
N 0550   T 0208
N 0560   G 00   X 00    Z 5     M 04   M 08
N 0570   G 01   Z-20
N 0580   G 01   Z 5
N 0590   G 01   Z-40
N 0600   G 01   Z 5
N 0610   G 01   Z-55
N 0620   G 01   Z 5
N 0630   G 01   Z-70
N 0640   G 01   Z 5
N 0650   G 01   Z-85
N 0660   G 01   Z 5
N 0670   G 01   X 100   Z 100   M 05   M 09
N 0680   M 30
```

NC Part Programme for the above part
(Without using cycles) (Second side)

```
N 0010   G 54   X 0     Z 100
N 0020   G 90   G 94    S 4000  F 150  T 0101  M 04  M 08
N 0030   G 00   X 65    Z 0
N 0040   G 01   X-1.5
N 0050   G 01   Z 5
N 0060   G 00   X 61
N 0070   G 01   Z-20
N 0080   G 01   X 65
N 0090   G 00   X 100   Z 100   M 05   M 09
N 0100   G 26
N 0110   T 0103
```

N 0120	G 42	G 00	X 60	Z 5	M 04	M 08
N 0130	G 01	X 65				
N 0140	G 00	X 100	Z 100	M 05	M 09	
N 0150	G 26					
N 0160	T 0215					
N 0170	G 41	G 00	X 65	Z-27.5	M 04	M 08
N 0180	G 01	X 56				
N 0190	G 01	X 62				
N 0200	G 01	X 52				
N 0210	G 01	X 62				
N 0220	G 01	X 50				
N 0230	G 01	X 65				
N 0240	G 01	X 100	Z 100	M 05	M 09	
N 0250	M 30					

Chapter 21

Ans. Q. 7

The following points need due consideration which retrofitting an existing machine into a CNC machines:

(i) In transmission it is necessary to have a separate motor to each axis movement.

(ii) Driving motor should be either a stepper motor or D.C servo system motor.

(iii) But stepper motor is not to be used if closed loop servo control is needed.

(iv) All lead screws should be replaced by preload recirculating ball screw commonly known as BLS.

(v) Guiding surface should give minimum frictional coefficient and the difference between static and dynamic coefficient of friction at the speed of the table or executive organ should be reduced to a minimum value, to avoid positional displace-ment error to a great extent.

(vi) Guide surface of the table and the bed need to be filled with Turcite *B* (Polytertra fluoroethylene) reinforced with phosphor bronze for reasons quoted in (*v*).

(vii) Turcite B is commercially available in sheet form with thickness 1mm, 1.5 mm or 0.5 mm needed to be pasted on the machined surface using adhesive (Araldite and hardener in equal proportion) and allowed to rest forcuring over 24 or 48 hours.

(viii) It is necessary to have transducers or encoders fitted for measuring in each axis or rotational movement.

(ix) The retrofitted machine should be provided with a machine controller.

Ans. Q. 8

Comparison between Stepper motors and DC servo system motors.

Stepper Motor	*DC Servo Motor*
1. Torque changes not possible.	1. Torque proportional to armature current
2. Small size compared to DC motor for same maximum torque.	2. Large size for same maximum torque.
3. Position is quantized and precise.	3. Precise arbitary positioning possible
4. Response time is poor	4. Rapid response
5. Can operate in both open and close loop.	5. Operates in closed loop
6. Synchronous movement.	6. Nonsynchronous movement
7. No maintenance needed	7. Commuter brushes need maintenance.
8. Operates from single battery	8. Requires bi-polar power supply.
9. Accelerates, decelerates during each step, hence motion jeiky.	9. Drives smooth.

Chapter 25

Ans. Q. 1

Company's estimated Radial load = 50 KN = $\overline{S}$

Actual load (average) = $\overline{L}$ = 45.46

$$S.D.\ \sigma_L = 1.07$$

$$S.D.\ of\ Strength = \sigma_s = 10\%(50) = 5$$

$$\Phi(Z) = \frac{\overline{S} - \overline{L}}{\sqrt{\sigma_s^2 + \sigma_L^2}} = \frac{50 - 45.46}{\sqrt{25 + 1.145}} = 0.888$$

With this value and using statistical normal CDF value table, we get reliability as 0.814.

Ans. Q. 2

Max. speed = 4500 rpm

Speed for testing (n) = accelerated speed= 1.2 × 4500 = 5400 rpm

Permissible speed (n_g) = (Rated speed) = 7000 rpm

$$\frac{n}{n_g} = \frac{5400}{7000} = 0.771$$

Since $\frac{n}{n_g} = 0.771 \cong 0.2 - 0.8$ as per bearing manufacturers catalogue therefore 1/3rd of the space in the bearing will be filled with grease.

For reliability estimation, we use the 'load – strength' interaction method.

$$\Phi(Z) = \frac{n_g - n}{\sqrt{\sigma_{ng}^2 + \sigma_n^2}}$$

Assuming a in each case is 10% of the values of n_g arid n respectively.

$$\Phi(Z) = \frac{7000 - 5400}{\sqrt{700^2 - 540^2}} = 3.592$$

For $\Phi(Z)$, from Std. normal *CDF* chart, the reliability comes to 0.9998.

Using this value of reliability we may find out the failure rate λ from the equation $R(t) = e^{-\lambda t}$

Mission time, can be determined from the graph (Figure 25.5), as 2000 hours.

$0.9998 = e^{-\lambda\, 2000}$

$\lambda = 1 \times 10^{-7}$ failures/hour

Glossary

Symbol	*Description*
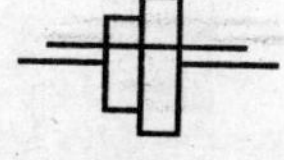	Sliding Gear Block
	Fixed Gear (fixed on the shaft)
	Lead Sctew and nut
	D.C. Motor
	Three phase A.C. Motor
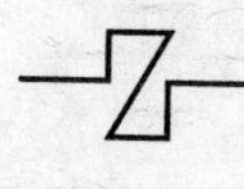	Coil or Solenoid
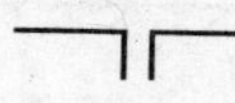	Block Contact
	Motor fitted with thermal relay
	Normally closed Contact

Normally closed Contact

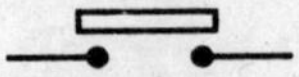

Normally open Cantact

Electro magnet

Solenoid with Core

Coupling between motor and Generator, etc,

Rheostat

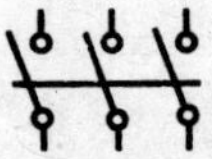

Working contact

Resistance

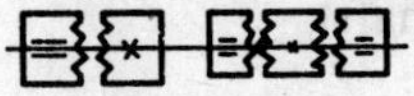

Clutch

References

1. Acherkan, N.S.: *Design and Calculation for Metal Cutting Machine*, Mashgiz, 1952 (in Russian).
2. —: *Metal Cutting Machines*, Mashgiz, 1938 (Russian Translation of Schelsinger's book).
3. —: (Chief Editor): *Metal Cutting Machines*, Mashgiz, 1958 (in Russsian).
4. —: (Chief Editor): *Handbook of Metal Worker*, vol. 1 -5, Mashgiz, 1960 (Russian).
5. Arnold, R.N.: *Mechanism of Tool Vibrations in Cutting of Steel. Proceedings of the Institution of Mechanical Engineers* (Lond.), vol. 154, 1946.
6. Basu, S.K.: *Ball Recirculating Nut: Machine Tool and Tool Industry*, No. 3, 1960 (in Russian).
7. —: "Trend in the Design of Machine Tool Guides", *Journal of Science and Engineering Researc`*, vol. IV, Part 2, 1960.
8. —: "Minimisation of Stick Slip in Machine Tools," *Journal Institution of Engineers* (India), vol. XLI, No. 2, Part 2, Oct. 1960.
9. —: "Frictional Behaviour of Machine Tool Guides", *Journal Institution of Engineers* (India), vol. XLII, No. 5, ME 3, Jan. 1962.
10. —: "Rigidity of Recirculating Ball Screw," *Journal Institution of Engineers* (India), vol. XLII, ME 2, Nov. 1961.
11. —: "Behaviour of the Power Screw Assembly of a Machine Tool under Rolling as well as Sliding Friction," *Journal Institution of Engineers* (India), Vol. XLII, No. 9, ME 5, May, 1962.
12. —: "Some Aspects of Programming a Machine Tool," *Journal Institution oj Engineers* (India), No. 3, ME 2, Nov. 1962.

13. Basu, S.K. and Mishra, R : "Trend in the Manufacturing Techniques and Design of Machine Tools-a Challenge to Production," Journal Institution of Engineers (India), ME 2, No. 4, Jan. 1963.
14. Basu, S.K. : "Progress in Machine Tool Engineering," *Indian and Eastern Engineer*, Anniversary Number, 1963.
15. Bowden, F.P. and Leven, L : "Friction of Lubricated Metal," *Phil. Trans-action of the Royal Society*, Lond. vol. A239, Nl, 1940.
16. Bhattacharya, S.G.: "Analysis of Self-induced Vibration During the Process of Metal Cutting," *Journal Institution of Engineers* (India), N 3, ME 2, Nov., 1962.
17. Doi, S: "On the Chatter Vibration of Laths Tools," *Momoirs of Engineering*, Nagoya University, vol. 5, N 2, Sept. 1953.
18. Dobrowolsky, V.A. and Others: *Machine Elements* (English translation by International Book House, Moscow), 1961.
19. Ermakov, V.V.: *Fundamental Calculations for Hydraulic Drives*, Mashgiz, Moscow, 1951 (in Russian).
20. Fridlender. I.G.: *Problems of Accuracy of Production on Machines*, Harkov, 1959 (in Russian).
21. Harizamenov, I.V.: *Electrical Equipments in Machine Tools*, Mashgiz, Moscow, 1958 in Russian).
22. Hanovitch, M.G.: *Supports of Liquidand Combination Friction*, Mashgiz, Moscow, 1960 in Russian
23. Haimovitch, E.M. : *Hydraulic Drives and Hydrauln Automation in Machine Tools*, Mashgiz 1959 (in Russian).
24. Isaev, P.P. and Bogdanov, A.A.: *Metal Cutting*, Oborongiz, Moscow., 1959 (in Russian).
25. Ilnitzkii, A.B.: *Vibration in Machine Tools and Means of Their Limination*, Mashgiz. 1958 (in Russian).
26. Iaxin, A.B.: *Antomation of'Machine Operations*, Moscow, 1957 Publication Trudrezcrvizdat.
27. — and Chernishea, A.N. : *Automation of Machine Operations by the Application of Programme Control Systems*, Mashgiz, 1957, Moscow (in Russian).
28. Kovan, Y.M. : *Technology of Machine Building*, Mashgiz, Moscow, 1959 (in Russian).
29. Kamenskaya, V.V. and Others : *Beds and Box Details of Machine Tools*, Mashgiz, 1960 (in Russian).
30. Kucher, I.M. and Kucher, A.M.: *Modernisation and Automation of Machine Tools*, Mashgiz, 1958 (in Russian).
31. Levit, G.A. and Tsirlin, M.N.; "Working Characterstics of Circular Guides of a Vertical Lathe, *Stanki-i-Instrument*, N6, 1956.

32. Levit, G.A.: "Hydraulic Calculations for Straight and Circular Guides," *Stanki-i-Instrument*, N9, 1958.
33. Lapiduce, A.S. and Others: "Materials, Construction and Systems of Lubrication in the Table of Heavy Vertical Boring Machine," *Stanki-i-Instrument*, N5, 1958.
34. Lapiduce, A.S.: "Fabricated Plastic Guides for Machine Tools." *Stanki-i-Instrument*, N 11 and N 12, 1955.
35. Merchant, ME.: "Characteristics ot Typical Polar and Non-polar Lubricant Additives under Stickslip Conditions," *Lubrications Engineer*, vol. 2. N3, 1946.
36. Papenhuyscn. P.J.: "Wrijvingsproven in Verband met het slippca van autovanden," *De Ingcriear*, N 53, 1938.
37. Pinkus, O. and Sternlicht B.: *Theory of Hydrodynamic lubrication*, McGraw Hill, 1961.
38. Push, V.E.: *Small Displacement in Machine Tools*, Mashgiz, 1961 (in Russian).
39. Push, V.E., Tolstoi, D.M.: "Theory of Stick Slip Sliding of Solids," *Conference on Lubrication and Wear, I. Mech. E. Proceedings* (Lond.), 1957.
40. Reshetov, D.N.: Machine Elements, Mashgiz, I960 (in Russian).
41. Rozenoerg. U.A. and Vinogradova, I.E.: *Lubrication of the Mechanisms of Machines*, Mashgiz. Moscow, 1960 (in Russian).
42. *Report No.* 1 *on High Temperature Machining Methods*, Physical Research Department, the Cincinnati Milling Machine Co. Ohio, U.S.A.
43. Singh B.R. and Push V.E.: "Stick Slip Sliding" *Journal Inst. of Engineers* (India), vol. 38, N 7, Part 2, March, 1958.
44. Shibel : *Sliding Theory of Guides and Bearings*, 1934. (Rusian translation from original book in German language).
45. Shyvalov, U.A. and Vedcnskii V.A.: *Metal Cutting Machines*, Mashgiz, Moscow, 1951 (in Russian).
46. Sisoev, V.I.: *Metal Cutting*, Mashgiz, Moscow, 1960 (in Russian).
47. Sokolov, U.N.: *Calculation of Temperature Fields anil Temperature Deformation of Machine Tools*, ENIMS, 1958 (in Russian).
48. Treior, V.N.: *Machine Elements Design for Life and Durability*, Mashgiz, Mascow, 1956.
49. Tobias, S.A. and Fishwick W: "Eine Thcorie des Regcnerative Raterns, An Werkzengmachinen", *Maschinenmarkt*, No. 17, 1956.
50. Tlusty, J.: "Die Bercchnung des Rahmens der Workzeng-machine" *Schwerindustric der C.S.R.* Heft 1, 1955.

51. Vladzievskij, A.P.: *Automatic Lines in Machine. Building*, vol 1, Mashgiz, 1958 (in Russian).
52. Volosov, S.S.: *Automatic Attainment of Accuracy of Dimensions while Grinding*, Mashgiz, Mascow, 1958 (in Russian).
53. Wright Baker, H.: *Modern Workshop Technology*, Part II, I960. Cleaver Hume Press (London).
54. Zakarov. B.: *Heat Treatment of Metals*, Moscow, 1962.
55. *Encyclopedia of Machine Building* : vol. 2, vol. 9, Mashgiz, Moscow, 1949.
56. *Metal Castings Hand*: American Foundrymen's Society Publication, 1944.
57. *Design Calculation for Machine Tools Guides on the Basis of Frictional Properties*, Publication EN1MS (G.A. Lcvit), Moscow, 1960.
58. Levtt, G.A. and Lure, B.G.: Hydrostatic slideway calculations, *Machines and Toolings* No. 10, 1963.
59. Levit, G.A. and Lure, B.G.: Calculation of hydrostatic slideways, Machines and Toolings, Vol. XXXV No. 6, 1964.
60. Levit, G.A. and Lure, B.G.: Hydraulic load relief for slideways. *Machines and Toolings*, No. 5, 1965.
61. Pio, J. (Czechoslovakia): Pressure in machines slideways, *Machines and Toolings*.
62. Pal, D.K.: Analysis of design parameters and performance characteristic of plastic slideways in Machine Tools. Ph.D (thesis), Burdwan Univ. 1970.
63. Basu, S.K.: Load distribution on the threads of the nut of recirculating ball screw assembly, *Journal of Science and Engineering Research*, Vol. 7, Pt. 1 Jan. 1963.
64. Birger, I.A. and Arutunyan, E.P.: Load distribution on the turns of a ball thread. *Russian Engineering Journal*. No. 3, 1971.
65. Levit, G.A.. ct al Design and Construction of Rolling Friction Lcadscrew (Ball recirculating). *Machines and Toolings*, 1963, No. 5.
66. Levina, Z. M., and Reshetov, D.N.: Contact stiffness of machine tools. Machine Building Publication, Moscow, 1971.
67. Pyasik, I.B.: Ball Icadscrew mechanism, *Mashgiz* M, Kiev 1962.
68. Pavlov. B.I.: Ball Icadscrew mechanisms in instruments manufacture. *Machine Building*. M-L, 1968.
69. Rodinov, I.V. et al: Load distribution between turns of thread in a rolling screw nut transmission. *Machines and Toolings*. No. 6, Vol. 36, 1965.

70. Shulga, Y.U. I.: Load distribution pattern on the turns of a ball load screw pair No. II, Vol. 56, 1976.
71. Zukovskii, N.E.: Pressure distribution on threads of screw and nut. Coll. of Works, Vol. 8 ONTI. 1937.
72. Yakobson. M.O. and Funberg, A.L.: Modern methods of manufacturing recirculating ball screw pairs. *Machines and Toolings*. No. 3. 1965. Vol. 36.
73. Push, V.E.: Micro displacements in Machine Tools, Mashgiz, Moscow. 1961.
74. Connolly, R. and Thornley. R.H.: Significance of joints on the overall deflections of the machine tools structures. Proc. MTDR Conference 1965 (Pergamon Press).
75. Chikate, P.P. and Basu, S.K.: Contact stiffness of machine tools joints, *Tribology International*, February, 1975.
76. Agarwal, R.A., Patki. G.S. and Basu. S.K.: Effects of surface topography on contact deformation in joints. *Jour of Engineering Production*, Vol.1, No. 4, 1977.
77. Agarwal. R.A., Patki, G.S. and Basu, S.K.: Deformations of surface irregularities under static loads. *Wear*, Vol. 58, 1980.
78. Palei, L. YA.: The Stiffness and Natural Frequency of Machine Tool Beds. *Machines and Toolings*, Vol. XXXVI No. 7.
79. Mitra, G. and Basu, S.K : Unconventional approach in design and performance analysis of Modern Machine Tools, Mcch. Engg. Bulletin, No. 4. December 1971, pp. 127-131.
80. Vragov, Y.D., Lapin, T.U. and Nefedeev, V.S.: Probability Method of determining speed characteristic of Milling Machines, *Machines and Toolings*, Vol. 34, No. 6, June, 1963.
81. Thierauf R.J., and Groose, R.A.: Decision making through operations research, John Wiley and Sons, 1970.
82. Czichos, H. and Solomon, G.: The Application of Systems Thinking and Systems Analysis to Tribology-Bam-Berichte, No. 30 Sept. 1974.
83. Burney, F.A., Pandit, S.M.. Wu, S.M.: A New Approach to the Analysis of Machine Tool System Stability under Working Condition. A.S.M.E Paper No.76-WA/Prod-11, 1976.
84. Wu, S.M.: Dynamic Data System-a new modelling approach ASME, *Jour. Engg.* for *Industry*, Vol. 99, No. 3 1977.
85. Gracia-Gareda, E., Burney, F.A. and Wu, S.M.: Determination of True Cutting signal by separation of Instrumentation Dynamics from measured response. ASME. No. 78-WA/Prod-16,1978.
86. Burney, F.A., Pandit, S.M. and Wu, S.M.: A stochastic approach to characterisation of Machine Tools System Dynamics under Actual

Working Conditions. ASME, *Jour of Engineering for Industry*, Vol. 98, No. 2.

87. Sinha, B.K. Chenchenna, P. and Sengupta, S.N.: Rigidity of grinding wheel-workpiece system in an internal grinder. *Journ. Mech. Engg. Bulletin*, No. 3, Sept 1973, pp 87-94.
88. Reshetov, D.N. (ED.): Elements and Mechanisms of Metal Cutting Machine Tools Vol. I and 2. Moscow, 1972. Published by 'Mashino-stroienne'.
89. Acherkan, N.S. fEd.): Machine Tool Design (English Publication), Moscow, Vol. 1, 2, 3, 4.
90. NCST: Sectoral Science and Technology Report on Machine Tools, Dec. 1975.
91. Machine Tools Development Council: Sectoral Technology Reports, May 1978 on (a) Metal forming machines, (b) Heavy Machine Tools, (c) Development of Machine Tools of Modular Construction, (d) Precision Machine Tools and (e) Plastic Processing Machinery.
92. Mansukhani, T.V., Status pf Machine Tools Industry, Machine Building Industry, May 1977.
93. Union Ministry of Industrial Development-Guidelines for Industry 1975-76-Machine Tools, *Machine Building Industry*, Oct 1975.
94. Chinoy, E.J., Machine Tools-Hurdles in the way of Increasing Exports, *Machine Building Industry*, Oct 1975.
95. DAS, D.R.: Marketing of Machine Tools, *Machine Building Industry*, Jan. 1970
96. Engelberger, Joseph F.: Robotics in Practice, Kogan Page & Avebury Publishing Co.. U.K., 1980.
97. Cugy, Andre & Page Kogan: *Industrial Robot Specifications*, Kogan Page Publication, U.K., 1980.
98. Pugh, A.: *Robotic Technology*, Peter Peregrinus Ltd., UK., 1983.
99. Rooks, Brian: *International Conference on Robot Vision & Sensory Control*, 3rd. Cambridge (MAS). 1983, North Holland Publishing Co.. Amsterdam.
100. Warnecke, H.J. and Schraft, R.D.: *Industrial Robots*, IFS Publishing Co., U.K., 1982.
101. Kochhar. A.K. & Burns N.D.: *Microprocessors and Their Manufacturing Applications*, Edward Arnold (Publishers) Ltd., U.K., 1983.
102. Sattnaufeldt.: *Industrial Robots Mauual*, Sweden, 1984.
103. Parent, M and Iaurgeau, C: *Logic and Programming*, Kogan Page Ltd., London, 1984.
104. 'Hitachi Review', Vol. 34. No..1, Feb. 1985.

105. An Overview of Artificial Intelligence & Robotics, Robotics Vol. II, National Bureau of Standards, U.S. Department of Commerce.

106. Datta, U., Dassarma, A , & Basu, S.K.: *Robot Language–State of the art.*, I.E. (I), Jour. P.E.. Vol. 67, July, 1986.

107. Jones, Borag Leatham. : *Introduction to Computer Numerical Control*, Pitman, London, (John Wiley & Sons), 1986.

108. Kundra, T.K., Rao, P.N. and Tewari, N.K.: *Numerical Control And Computer Aided Manufacturing*, Tata McGraw-Hill, 1985.

109. Martin, S.J.: *Numerical Control Machine Tools*, The English Language Book Society and Hodder and Stoughton, 1970.

110. Ruthmill, K., et at.: *The Implementation of N.C. Plant In Functional And Cellular Grouping*. Proc. J. Mech. E. (Lond.), Vol. 195, 1981.

111. Simon, Wilhelm: *Numerical Control of Machine Tools*, Edward Arnold Publishers, London.

112. Parrish, D.J.: *Commissioning Flexible Manufacturing Systems*, CME, I. Mech. E. Publications (Lond.), Nov. 1987.

113. Kearny & Trecker Corpn: *KT'S World of Manufacturing System*, Milwakee, Wisconsin, 1980.

114. Basu, S.K. & Pal, D.K.: *Design of Machine Tool* (Revised Edn.), Oxford & IBH Publishing Co. Pvt. Ltd.. 1983.

115. Rembold Ulrick, Blume, Christian & Dillmann Ruedicer: Computer Aided Manufacturing Technology and System, Marcel Dekker Inc., N.Y., 1985.

116. Simpson, John, A.: *Automation for the Small Job Shop*, C.M.E., June 1984, Pub. I. Mech. E. (London).

117. Sento, H. & Fukuchi. F.: *Recent Trends and Future of Hitachi Robots*, Hitachi Review, Oct. 1983. Vol. 32, No. 5.

118. Balasubramanium, T.A. (edited): *Vision System at Work*, Plus– *the total Computer Magazine*, Oct. 1984.

119. Basu, S.K. & Pramanik, D.K.: *Design of Recirculating Ballscrew Assemblies*–Pub. C.M.E.R.I., Durgapur, India, 1982.

120. Appa RAO, G.V.: *Presidential Address Delivered At the Twelfth All India Machine Tool Design & Research Conference*, 10th Dec. 1986 I.I.T., New Delhi.

121. Plummer, J.C.S.: *Making Full Use of Solid Model Database* C.M.E., July-Aug. 1985.

122. DE Barr., A E.: *The Development of Numerical Control*–Past, Present and Future,. The Production Engineer, Sept. 1971.

123. H.M.T. Limited. Catalogues: SB-CNC 35; STC 25; NH-CNC; Electronic Die Sinking Machine; VTC, HTC-600; HMT KTM

Machining Centres; CNC-Flexible Manufacturing line: CNC Wirecut EDM; GNC-18; T-70; HMT CNC Booklet 1986.

124. Rollon Bearings Pvt. Ltd., Bangalore: Pamphlet of Turcite B Slideway materials
125. Phatak H.V., Tillu S.G, and Basu S.K.: Error Analysis in Machine Tool Based on Compliance and Tribological Parameters. Jour. Inst of Engrs., Pt. Prod. Engg. Nov. 1994.
126. Push V.E., Pigert P. and Sosonkin B.L.: Automatic Machining Systems, Mashinostroenie, 1982, Moscow.
127. Push V.E. Ed. Metal Cutting Machine Tools, Machinostrenie, Moscow 1986.
128. Levnia Z.M. and Reshetov D.N.: Contact Rigidity in Machine Tools, Mashinostroenie, Moscow 1971.
129. Tanaka K. and Yamada Y.: Influence of Counter Face Roughness on Friction and Wear of PTFE etc.: C 138/87, Proc. I. Mech. E/(Lond), 1984.
130. Lancaster J.P. and Biltrow J.P.: Role of Counterface in the Friction of Wear of Carbon - Fribre Reinforced Thermoosetting Resin.
131. Shah N.C.F. and Basu S.K.-. ConYacX Compliance of Filled and Unfilled Plastics: Proc. Int. Trib. Conf. Melbourne, 1987.
132. Shah N.C.F., S. Prasad and BASU S.K.: Tribo Characteristics of Filled and Unfilled Plastics. 13th AIMTDR Conf. 1988.
133. Basu S.K. and PAL D.K.: Friction and Wear Characteristics of Plastic Guides. Jour. Inst, of Engineers, vol. 50, July 1960.
134. Basu S.K. and Pal D.K.: Wear Analysis of Plastic Guides. Wear 21,1972.
135. Sen R. Das S.K. and S. Datta and Basu S.K..- Evaluation of glass ceramic coatings for machine tool slides. Wear 1989.
136. Patil GS., Jagtap M.S., Basu S.K.: Reliability Analysis in Design and Manufacturing through Accelerated Method – 'A case study on meter-mix dispensing machine', National Conference on Advances in Manufacturing Systems, Jadavpur Univ., Kolkata, March-2003.
137. Basu S.K., Sonawane D.R., Sarwade R.N., "Reliability Assessment of Some Hydraulic Equipment", IE (1) Journal– PR Vol. 79, May 1998 (pp. 1-4).
138. "Instruction & Spare Part Manual", Sumangal Engg. Company, Bhosari Industrial Estate, Pune - 411026.
139. Amstadter Bertram L., "Reliability Mathematics", McGraw Hill Book Company, 1971.
140. Basu S.K., Sahu K.C., Datta N.K., "Works Organisation and Management", Third edition, Oxford & IBH Publishing Co. 1984.

141. FAG Standard Programme, Catalogue 41510 EA.

142. "The lubrication of rolloing bearings", FAG Kugelfischer Georg Schafer & Co. Schweinfurt, Germany, Pub. No. 81, 103 EA.

143. Rao S.S., "Reliability-Based Design", McGraw Hill Inc. 1992.

144. Jaybhaye M.D. and Sonawane B.U., "Reliability Analysis of Some Machine Tool Elements", Journal Inst, of Engrs. (I), PR Vol. 87, Sept. 2006, pp. 7-10.

145. Xomiakov VS., Sabirov F.S. and Dosko S.I., "Experiment and Investigation on Machine Tools", ENIMS, Moscow, 1988.

146. Zukovsky, N.E (1987), Distribution of Pressures in Screw Threads. Complete collection of works. Vol 8.

147. Jacquet, E. (1981), Uber Eine neuratge Schraubenverbindung, Schweiz, Beauzeitung, 98page 207.

148. Maduschka, L. (1986), Beanspruschung Von Schranbenverbindung and Zweekmassige Gestaltung der Gewindetrager, Forsch, Geb. ingenierwesseus.

149.Galerkin, B-G, (1929), stresses on trapezoidal profiles ,Bulletin of the Leningrad Institute (Discussions) XCIX.

150. Kuklin, B.V (1957), Accurate Calculations of Screw Thread Connections. VestnikMashinostroenie , N7.

151. Birger, IA. (1959), Calculations of Screw Associates, Moscow.

152. John (Mrs) J, Basu S.K. (1997), Contact Deformation of Filled and Unfilled Plastics under Dry as well as Lubricated Condition, Institution of Engineers (I), Journal PR, Vol. 78 No.5

153. Tamaka H, Machida H, Half toroidal Traction Drive for Continuously variable Power Transmission; Proc. I.Mech.E., Vol. 210, 1996.

Index

A

ALARP; APT; APT **II**; Autospot, 407
AL, 427
AML; 427
ARL, 426
Antistick-slip oil, 133
Automatic inspection of in-process parts, 310
Automatic operation of a horizontal drilling machine, 273
Automatic controls, 283
Automation in machine tools, 297

B

Ball guides, 126, 131
Ball recirculating power screws, 175, 180, 192
    Classification and types, 169
    Dynamic load, 213, 166, 465
    efficiency of, 180, 296
    pre-loading methods, 217
    rigidity of, 24, 62, 202, 224, 226, 237, 244
    standard sizes, 26
    strength calculations, 59, 356
Bearings in Machine Tools, 222
    Ball-load calculations, 221–222
    Methods of adjustment for backlash, 222–224
    Selection of, 50, 259
    Sleeve-adjustment, design calculations, 171
Beds, 153, 164
    classification and types, 169
    Design considerations, 156, 157
    Materials, 110, 133, 450
    Methods of heat treatment,
    Natural frequency, 63, 65, 164
    Sections (typical),
Braking, 36, 259
    electrical, 7, 265, 266
    Self, 320
Building block concept, 5, 15, 433
Built-in inspection units, 305, 310, 314

C

CAD-CAM, 431
Centralised system of controls,
Change gear drive, 76, 77
Character, 193, 232

Chatter in machine tools, 319, 320
Classification of NC, CNC, 4
Closed loop, 386, 397
Clutch, 79, 267
  Electromagnetic, 79, 267
  ferromagnetic, 70, 300
Clutches, 57, 79, 400
  gear box with, 53, 77
CNC, 356, 382, 412, 434
CNC control, 431, 434
CNC machining centre, 5, 11, 434
CNC shideways, 417
Collet chucking - analysis, 471
  mechanism, 98, 271
Columns, 71, 73, 153, 380
  design for milling machine, 160, 161
Compensation of errors in power screws, 183, 184
Cone pulley drive, 76
Compliance - dynamic analysis, 442
  -joints, 74, 351, 372
Composites, 5, 450, 451
Controls, 5, 6, 248, 283
  electrical, 7, 265, 266
  hydraulic, 139, 217, 253
Control of shifts in gear box, 248–257
  individual system, 5, 37
  pre-selective system, 5
  selective system,
  simple system, 248, 250
  mechanical, 38, 90, 150
  programme, 5, 37, 382
Coordinate setting in numerical control, 302–307
Copying machine, 6, 290, 301,
  Hydrocopying, 289, 290
Photoelectric copying, 300, 301
Critical velocity evaluation, 461
  duration of slip, stick period, 461

D

Deviation diagram, 47, 85
Deformation, thermal, 111, 112, 146
Disc floppy, 437
DNC, 404–406, 437
Drives, basic classification, 262, 268
  Change gear, 51, 58, 77
  Cone pully and line shaft, 77–78
  Design considerations, 77–78
  electrical, 7, 265, 266
  electromechanical, 37
  hydraulic, 139, 217, 253
  Stepless, 5, 37, 91
  variable speed, 37, 38, 99, 264,
  with epicyclic gear, 98
  with variators, 97–105
Dynamic data system, 9, 364, 523
Dynamic load Rating, Spindle, 465
  power screws, 169
  of spindle supports, 465, 468

E

Efficiency, 180, 296
  of belt drive,
  of driving motor, 34, 259
  of gear train, 25
  of power screw with rolling friction, 176, 179–182
  of power screw with sliding friction, 174
Electrical automation, 273
Electrical equipments in machine tools, 520

Electromagnetic clutch, 79, 267
Electromagnets, 265, 267
Electromechanical regulation, 37
Encoders, 388, 391–393
End effectors, 367, 369, 372
Errors in coordinate settings, 304, 305
  Ordinary lead screw, 169–172
  pitch, 38, 50, 171
  recirculating ball screws, 350, 388, 398
Errors, stick slip, 183, 241, 110, 117
Ex APT, 407

F

Fabrication of plastic guides, 110
Feedback-positional, velocity, 389
Ferromagnetic powder clutch, 270, 492
Finite element analysis of elastic support with damping, 477–478
  rigidity, 1, 62, 67
  transformation matrix, 459, 479, 480
Floppy disk, 416
FMS, 378, 413, 430, 433
Forced and damped vibration, 245, 322, 491
Forced vibration, 316, 331
Forces, determination in broaching, 18, 24, 31
  drilling, 162, 273,
  grinding, 27, 62, 63
  milling, 2, 24, 26
  shaping, 2, 32, 37
  turning, 2, 6, 47
Forces, hydrodynamic on guides, 114–117
  hydrostatic on guides, 134–143
Forces on single point cutting tool, 17–34
Factional behaviour of guides und lubrication, 128–129, 231

G

Gain, 417
Gear box design, 56
Gear design, bevel, 83, 84, 183
  spur, 7, 59
Graphical representation of speeds, 96
Gratings, 388
Grippers, 373, 377, 422
GT, 431
Guides, 106, 111, 112
  aerostatic, 5, 142
  ball, 21, 104, 126
  boundary lubrication in, 232
  calculations in, 32
  circular, 110, 112, 131
  classification of, 2, 4, 106, 283
  combination, 12, 96, 97, 232
  fabrication of, 110
  frictional behaviour of, 450, 519
  fundamental types of circular guides, 131
  hydrodynamic lubrication of, 234, 236
  hydrostatic lubrication of, 134, 137
  lubrication of, 126, 234, 236
  materials, 110, 133, 450
  merits and demerits of plastic guides, 486
  preloading of, 184, 339
  pressure distribution in, 121, 124

regime of working, 234–235, 277, 278
roller guides, 110, 126, 331
specifications, 26, 110, 125, 524
straight flat, 113, 133, 234
temperature deformation, 110, 111

H

Hardware, 402, 416, 427
Hierarchial Configuration, 377
Horizontal table, 157
design calculation for strength, 156, 157
Hunting, 417
Hydrodynamic lubrication, 7, 113, 117, 236

I

Introduction, 1, 230, 286, 297, 310, 316, 343
Individual drive, 5, 37
Individual system of control, 248
Inductosyn, 386, 388, 395, 396
In-process inspection of parts, 310
IRL, 427
Isolators, 318, 319

J

Journal bearings, 26, 135, 220
Calculations, 24, 27, 32, 59
design consideration, 219
methods of adjustment, 217
preloading, 184, 208, 332

K

Kinematics of machine tools, 36
basic, 37, 173, 258
classification, 2, 36, 89, 169
considerations, 100, 153, 230
deviation diagram, 47, 85
ray diagram, 41, 44, 48, 50
structure diagram, 43, 47, 48

L

Language-Robot, 419, 426, 427
Loop-Closed and open, 386
Lubrication, 5, 7, 110, 119
boundary, 128, 145, 232
classification of oils, 232, 235, 236
hydrodynamic, 5, 79, 113, 117
hydrostatic, 5, 9, 134, 138
of machine tools, 1, 2, 5, 7, 14, 34
quantity consumption of,
specifications, 26, 110, 125, 524
systems of, 5, 76, 289, 311
temperature, 1, 110, 111, 160
Lubrication pocket, 120, 463
Linear recess,

M

Machine tools, 1, 2, 5, 7, 14 34,
classification, 2, 36, 89, 169
developments, 1, 6, 14, 137, 433
Machining, 434, 412, 308, 215, 36
power consumption in, 17, 27, 31
vibrations, 9, 74, 166, 314
Magnetic box, 387
Magnetic tape, 297, 299, 417
Magnetostrictive drive, 343
Magnet, electro, 79, 117, 267
Magnetic clutch, 74, 267, 269

tape control, 299, 300
Material, 110, 133, 450
for ball screws, 350, 388
for beds, 154
for columns 164
for tables, 164
MCL, 428
MCU, 382
Mechanical friction drives, 90
regulation of speed, 41, 89, 225, 260
Methods of increasing range of regulation, 95
of eliminating back lash, 183, 184
of compensation of errors, 184
of eliminating stick slip motion,
Micro-displacement, 7
Micro-displacement - non-uniform, 123, 146
Positional displacement evaluation of error, 5, 347, 386
Microprocessor, minicomputer control, 8, 373–375, 402
Minimization of stick slip sliding,
Moire fringe pattern, 394, 395
Motor-Stepper, 372
Motor tape relay, 274–275

N

NC Machine, 4, 11, 347, 382
NC Retrofitting, 398
NEAPT, 407
Numerical control in machine tools, 301–309

P

Part programming, 381, 405–407, 409
Photoelectric tracing, 300
Positional feedback, 386, 389
Power in machining, 21, 24, 25, 26, 27
of driving motor, 21, 24, 25, 26, 27
Preloading of circular guides,
Programme control, 6, 7, 37, 175, 241, 300
Probability in design, 357

Q

Quantity of, lubrication, 236
oil flow in circular guides, 335, 336
oil flow in hydrodynamic lubrication wedge pockets, 120

R

Range of speed, 37, 38, 96, 286, 489
Regulation obtained in different machines, 42, 43
Regulation of speed electrical, 261
electromechanical, 37
mechanical, 5, 10, 38, 90, 263
Relays in machine tools, 271, 274, 275
Reliability analysis, 470, 475
collet chucking, 470, 471, 472
bearings, 471, 472
failure rate of bearing, 477
by variable method (K-statistics), 472, 473
by load-strength Interaction, 476
Relubrication of bearing, 475–477
Resolver transducer, 395–397
Rigidity, 24, 62, 202, 224, 226, 244
concept, 5, 9, 123, 325, 357

dynamic, 2, 5, 7, 73, 213
static, 62, 67, 110
Rigidity of recirculating ball screw assembly, 519
Robot, 8, 367, 369, 370
Robot-teach, 376, 377
Robot-Reliability, 372

S

Screws, 150, 169, 350, 388
Self-tightening, 94
Self-excited vibrations, 314, 316, 320
Sensor, 367, 369, 378, 418
Servo system, 386, 387, 514
Shock absorbers, 342
Sigla, 428
Signal, 305, 306, 308, 386
Specification of lubrication oil, 235, 491
of motor, 34, 259, 277, 278
of power screw, 169
Speeds, 1, 5, 23, 37, 44
control, 5, 6, 248, 283
graphical representation, 43, 44, 86, 96, 401
selection, 50, 57, 71, 131
variable, stepped, 484
stepless, 5, 37, 91
calculation, 24, 27, 32, 47
Spindle – bearing distance, overhang, 21, 23, 448, 466
rigidity, 24, 62, 202, 224, 226, 244
dynamic characteristics, 478
Spindle units, 222, 224
accuracy, 222, 224, 468
deformations, 123, 147, 148, 193
equivalent load, 466
Spindle units, 22, 224
adjustment of bearings,
Stick slip vibration, 110, 117, 119, 126, 325, 331
Surface topography and contact stiffness, 351
types, 47, 131, 153, 154
typical units, 288
Software, 382, 400, 405, 416, 419
Stepper, 372, 375, 493, 514
Synchro position transducer, 395
Synchro resolver position transducer, 395, 396

T

Tables, design, 156, 157
Specifications, 26, 110, 125, 524
Temperature, 1, 110, 111, 160
of lubrication oil, 117, 128, 134, 235, 236
deformation of circular tables, 112
Thermal relay, 272, 273, 491
Thermodynamic drive, 347
Time for starting and stopping the motor, 259
Time relays, 117, 300
Time series analysis, 364
Tool, lifting arrangement, 276, 277
Transducer, 386, 387, 388, 389
Turcite B, 400, 439, 453, 463
Typical bed section, 155, 156
Typical hydraulic circuits, 288

U

Ultrasonic vibratory force, 463
Unified system approach, 259

V

VAL, 427
Variators, 96, 97, 101, 104
  design considerations, 153
  types, 47, 131, 153, 154
Velocity, critical for guides, 316
  critical for stick slip accurance,
Velocity Feedback, 389
Vibration of machine tools,
  forced, 245, 316, 322
  forced and damped, 245, 322, 491
  isolation, 319
  minimisation, 50, 349, 463
  self induced or chatter, 320
  stick-slip, 117–119, 236, 305
Vision System, 425, 435, 436, 525

W

Wearing of guides, 106
  amount wear rate on conventional guides, 131
  consideration in designing, 316